William Henry Smyth

THE MEDITERRANEAN

A MEMOIR

PHYSICAL, HISTORICAL AND NAUTICAL

\Elibron Classics
www.elibron.com

Elibron Classics series.

ISBN 1-4021-9703-9 (paperback)
ISBN 1-4021-5380-5 (hardcover)

This Elibron Classics Replica Edition is an unabridged facsimile of the edition published in 1854 by John W. Parker and Son, London.

THE

MEDITERRANEAN.

THE

MEDITERRANEAN

A MEMOIR

PHYSICAL HISTORICAL AND NAUTICAL

BY

REAR-ADMIRAL WILLIAM HENRY SMYTH, K.S.F., D.C.L.,

ONE OF THE BOARD OF VISITORS OF THE ROYAL OBSERVATORY;

SOME TIME

FOREIGN SECRETARY AND VICE-PRESIDENT OF THE ROYAL SOCIETY,

VICE-PRESIDENT AND DIRECTOR OF THE SOCIETY OF ANTIQUARIES, PRESIDENT OF THE ROYAL ASTRONOMICAL SOCIETY,

PRESIDENT OF THE ROYAL GEOGRAPHICAL SOCIETY,

AND VICE-PRESIDENT AND CHAIRMAN OF THE UNITED SERVICE INSTITUTION OF LONDON:

CORRESPONDING MEMBER OF THE INSTITUTE OF FRANCE;

HONORARY MEMBER OF THE ROYAL IRISH ACADEMY; AND OF THE SCIENTIFIC ACADEMIES OF NAPLES, PALERMO, FLORENCE, BOSTON, WASHINGTON, AND NEW YORK.

LONDON

JOHN W. PARKER AND SON

MDCCCLIV

TO

REAR-ADMIRAL SIR FRANCIS BEAUFORT, K.C.B.

F.R.S., F.G.S., F.R.A.S., ETC.

HYDROGRAPHER TO THE ADMIRALTY.

St. John's Lodge, near Aylesbury,
Jan. 24th, 1854.

MY DEAR BEAUFORT,

I KNOW not to whom I could in any case have addressed my hydrographical treatise so appropriately as to you, who have so long and ably presided over the surveying department of the navy. But nearly half a century of professional acquaintance, including thirty years of intimate friendship, with the knowledge of you which they have given as a man, a seaman, and an officer, leave me wholly unable to say whether I ought rather to inscribe this work to you as a public compliment, or as a mark of private regard. To you, then, I submit the present exposition of the state of our knowledge of the Mediterranean Sea at the time of my return to England in the close of the year 1824; only regretting, on various counts connected therewith, that you were not holding office at that period.

This work, as you are aware, has long been meditated, but has 'hung fire' for the completion of the surveys in the Archipelago, so that it might only just precede a complete Sailing-Directory for the whole Inner Sea. The unexpected breaking off of Captain Graves, however, towards the very close of those operations, and their consequent suspension, determined me to proceed with my part at once. For this and two or three other reasons, together with the loss occasioned by the destructive conflagration of the printing-office of Messrs. Savill and Edwards, the book is later in its publication than was intended. Still the delay has been of no actual detriment to the Service, since my charts—which, in fact, are working diagrams of all the labour—have long been engraved and circulated by the Admiralty; while, as you can testify, my observations and memoranda have always been accessible to inquirers.

The undertaking, though heavy, is nevertheless not wanting either in interest or importance: the Mediterranean Sea, so secondary in extent compared with others, being, *per se*, of vast surface, with many of its characteristics on the grandest scale. Besides, viewing it as the actual site where the intellectual culture to which we are most directly indebted was first developed, it cannot but be regarded for its portentous historical occurrences; nor will a sailor forget that it is the sea whereon the fleets of Carthage, Greece, and Rome contended in former days, and those of Spain, France, Italy, and England in later times. 'The grand object of travelling,' said Dr. Johnson to General Paoli, 'is to see the shores of the Mediterranean. On those shores were the four great empires of the world; the Assyrian, the Persian, the Greek, and the Roman. All our religion, almost all our arts, almost all that sets us above savages, has come to us from the shores of the Mediterranean.'

It might appear strange that a coast of such paramount interest should still have required surveying in the present day; but the following pages may, in part, account for the necessity. To what is hereinafter mentioned, it may be prefaced, that useful knowledge has recently spread as largely over the waters as upon land; much of which must be ascribed to sundry of its older trammels being thrown overboard, and much to the progressive improvement which time naturally and surely creates. Since you and I first dabbled in these matters, a vast stride has been made in hydrography; and the results are seen in more efficient instruments, better astronomical tables, sounder nautical directions, and more correct charts. Instead of a *Lunarian* being a 'rara avis' in our ships, as of erst, there is now a very host of them afloat; and chronometers, then so scarce, are at present to be found in every ship of consideration. Meantime, the various branches of available science have been so steadily advancing among seamen of all nations, that besides a higher practice in mechanical navigation, they possess a more accurate information respecting the phenomena of winds and oceanic currents than heretofore. Already the elements are nearly reduced to subjection by the union of science and practical seamanship, so that sea-passages are wonderfully shortened within memory; and these beneficial effects are on the eve of being strengthened in utility by a systematic arrangement and impartial discussion of connected facts, as proposed by the energetic Lieut. Maury, of the United States Navy. In a word, although incompetence may sometimes appear in the van, to the serious detriment of the public interests and character, the true place and substantial advantage of real talent is fast gaining recognition; whence it must follow that inchoate notions, and arbitrary assumptions, will inevitably succumb to experienced skill, and the logical reasonings of induction.

It would be deeply important to our knowledge of the terrestrial attributes of our globe, were the profundity, form, and physical nature of the ocean ascertained; but an enormous amount of labour and money must be consumed, before positive conclusions can possibly be arrived at over such an immensity. Indeed, a complete oceanic survey may be beyond human power; but, to the best of my opinion, a sub-aqueous map of the Mediterranean is within our compass. Towards an object so truly valuable to science, the following pages, it is hoped, may prove a trustworthy pioneer; albeit my inquiries were mainly directed to our maritime requirements in 1810, the date of my commencing operations in that sea. Still, in conducting examinations and gathering every information in my power, neither toil, responsibility, nor personal expense, were ever spared by

Your truly attached friend,

CONTENTS.

Part I.

A Chorographical View of the Shores of the Mediterranean Sea, with especial reference to their Produce and Commerce.

Part II.

Of the Currents, Tides, and Waters of the Mediterranean Sea.

Part III.

Of the Mediterranean Winds, Weather, and Atmospherical Phenomena.

Part IV.

Of the Surveys and Geographical Investigations in the Mediterranean Sea.

Part V.

Of the Orthography and Nomenclature adopted; the Geographical Points—or Co-ordinates of Latitude, Longitude, and Height—of the Mediterranean Shores; with the Variation of the Magnetic Needle, and other Notanda.

Appendix.

ERRATA.

The reader is earnestly requested to correct, with his pen, the following oversights of the press.

Page 46, line 20, *for* '*Chinuera,*' *read* '*Chimera.*'
— 47, last line of the note, *for* 'futile' *read* 'fictile' fragments.
— 67, line 10 *ab imo*, *for* 'Psitoriti,' *read* 'Psiloriti.'
— 92, last line, *for* '1820,' *read* '1821.'
— 149, after line 22, in the heading of the last column of the table, *for* cubic 'inches,' *read* 'miles.'
— 218, line 2 *ab imo*, Brewster's formula, insert × before cos. lat.
— 375, *for* Assistant-Surgeon 'Beg,' *read* 'Begg.'
— 396, No. XIV., *for* Port 'Cross,' *read* Port 'Cros.'

THE MEDITERRANEAN SEA.

PART I.

A CHOROGRAPHICAL VIEW OF THE SHORES OF THE MEDITERRANEAN SEA, WITH ESPECIAL REFERENCE TO THEIR PRODUCE AND COMMERCE.

§ 1. The Mediterranean Shores of Spain.

THE Mediterranean Sea, equally remarkable from its position in the midst of the most civilized nations, and its connexion with many memorable events in ancient and modern history, is that vast central gulf emphatically styled in the Sacred Scriptures the Great Sea; justly receiving that appellation, as being the largest assemblage of waters known to the earliest writers of those records: and indeed its importance was truly paramount among the ancients, as it was the grand key to both portions of the then known world. Mediterranean Sea.

By the word Mediterranean, or midland, we understand water enclosed either wholly or nearly by land; but the term was not applied to this sea by any classical writer. The ancient Greeks seem to have had no general name for it,—Herodotus merely calls it 'this sea,' and Strabo the 'sea within the columns,' that is, within Calpe and Abyla. By their present descendants it is called *Aspri Thalassa* (Ἄσπρι θάλασσα) the White Sea, to distinguish it from the Euxine, which they call *Mavri Thalassa* (Μάυρη θάλασσα) the Black Sea. It was gradually designated the Grecian Sea, and then the *Mare internum;* while Mela terms it *mare nostrum.* Though some of the Arabians described it as the Green Sea, it was *Bahr-Rúm*, the

Greek or Roman Sea, with most of them; among our own seamen it has long been specialized as the Straits. Geographically speaking, however, the term Mediterranean is now more strictly applied to the whole expanse between the South of Europe, Asia Minor, and the north coast of Africa; extending from the Straits of Gibraltar to the shores of Syria, including the Sea of Marmora, the Euxine, and the Palus Mæotis. It is separated from the Red Sea by the Isthmus of Suez; from the Atlantic Ocean, though their waters unite, by the Strait of Gibraltar, and communicates with the Black Sea by the Dardanelles and the Canal of Constantinople.

The political and social events which have occurred on the shores of this remarkable part of the ocean, are closely connected with the history of almost every country in the world; but independently of its classical and historical associations, the Mediterranean still confers invaluable advantages upon the numerous occupiers of its coasts, and through them on the interior of the surrounding continents. It is moreover the great bond of intercourse between the nations of Europe, Asia, and Africa, although it appears as if it had been destined to keep them asunder. Beautifully diversified with islands, and bounded by almost every variety of soil, its products are proportionally various; and from its communication with the Atlantic, it facilitates commerce with every part of the globe. Here navigation made its earliest efforts; and the comparative shortness of the distances between port and port, by rendering the transit easy, even to imperfect vessels, tended to promote and diffuse civilization, it being an unquestionable axiom, that whatever is calculated to make men better acquainted with each other, whether the inhabitants of distant lands or neighbours, must inevitably produce beneficial results for the whole. But though commerce is and has been both vast and various in this sea, its energies cannot be said to have attained their full development, clogged as they have been by impolitic curbs and impositions; nevertheless under numerous vexatious restrictions, direct employment and subsistence are still afforded to many hundred thousands of those people who have access to its shores.

Mediæval Commerce.

Our ancestors had acquired some acquaintance with Mediterranean traffic as early as the time of the Crusades; but like all other nations of Europe at that period, they were ignorant of the principles of commerce, as well as destitute of the capital by which it is made steady and lucrative. In the 12th year of Henry VII. (1497), as shown by an act of Parliament, our goods were exported to Genoa and Venice, yet they seem to have been carried entirely by foreign ships and traders; the argosies of Shakspeare being Ragusan vessels. According to Hakluyt, our first trade of moment with this sea, in English bottoms, began in 1511, just before the Turks obtained possession of Chios, to which port our vessels traded, and where, two years afterwards, a consul* was appointed to superintend our interests. The year 1550 found the tall ships of our merchant adventurers carrying on a commerce with Sicily, Candia, Cyprus, and Syria; by which the germs of the Levant Company were matured.

Spain.

Sailing from England, the first Mediterranean feature the voyager falls in with is Spain (*Iberia*), a country of proud recollections and interesting story. But the political events of the last half century have severed her colonies, destroyed her fleets, and irretrievably damaged her commerce. Still her national character for persevering effort in warfare is remarkable, whether shown in her struggles, as with the Saracens and in the succession campaigns, or in the arduous contentions which have taken place on her own soil, as those between Rome and Carthage, or those in which she took part when overrun by the armies of France, and occupied by her English allies. Yet her formidable barrier on the north,

---

* It is therefore erroneous to call John Tipton, who was appointed to Algiers in 1581, the first English Mediterranean consul.

her being posited as it were between Europe and Africa, and her station between two seas, would show that nature intended that Spain should remain integral, extend commerce, and facilitate intercourse with other nations. Still the enormous chains of mountains which traverse the peninsula, and a want of navigable rivers, added to a host of moral causes, have been serious obstacles to her prosperity. Spain indeed, as Wilson shows in his *History of Mountains*, may be considered to be composed of a series of elevation-terraces, which, projecting successively their rugged edges towards the south, present a flight of gigantic steps from the Pyrenees to the Strait of Gibraltar, and from the Rock of Lisbon on the Atlantic shore to Cape Creux on the Mediterranean.

Gibraltar. The northern shores of the Strait of Gibraltar, after passing Cadiz (*Gadir*) and Cape Trafalgar,* are marked by the isle of Tarifa—the southernmost point of Europe—and the fortified rock of Gibraltar (*Calpe*), which is a vast body of limestone of the oolitic period, elevated 1430 feet above the level of the sea, with a handsome free-port town at its base. This peninsular mass—about two miles and a quarter long by nearly three quarters of a mile broad, with a circuit of between four and five miles—is joined to the continent by a low sandy neck, which towards the Mediterranean has an elevation of several feet above its opposite side abutting on the bay, an effect occasioned by the strong Levant winds and waves along what is termed Back-strap Bay. Here a fair system of government, full toleration in worship, and the prosperous results of an open free port, together with the energy and taste of its English residents, have made an otherwise barren and burning rock a scene of commercial activity and luxurious abode; but as such minute circumstances are hardly admissible in so cursory a glance as this, we must hurry onwards, especially because the

* Between Cape Trafalgar and Tarifa, and near Bolonia tower, are the ruins of *Bælon;* this was the ancient place of embarkation for Tangier (*Tingis*), in Africa, apparently to avoid being set through the Strait by the current, which might happen more to the east.

statistical details of Gibraltar, as a British colony, will be found at the close of the present chapter. To the south, this important ocean-strait is bounded by that part of the coast of Africa which extends from Cape Spartel to the fortress of Ceuta; and between those two points lie the port of Tangier and the lofty cliffs of the Sierra Bullones, or Ape's Hill. Here it is fabled that Hercules set up the pillars on *Calpe* and *Abyla,* to commemorate the extent and termination of his territorial conquests; hence the geographical terms *Fretum Herculeum, Fretum Gaditaneum,* and *Columnarium,* formerly applied to a spot which was long deemed to be the *ne plus ultra* of navigation, if not the extremity of the earth. It was called *Bab-ez-zakák,* the Gate of the narrow Passage, by the Arabs, then the *Gut* by our seamen and pilots. Through this strait a stream from the Atlantic is continually flowing; but before we speak of the waters of this sea, it will be as well to take a rapid view of the lands forming its boundaries.

The Mediterranean shores of Spain extend about 780 geographical miles in a north-east direction from Gibraltar to Cape Creux, where the French domain commences, and present a great variety of plains and mountains bounded by a highly fertile though not well-wooded coast, indented by numerous harbours and bays. Some of the mountains in Catalonia consist of granite, but the prevailing formation of the coast is limestone. In Granada, among other mineral riches, are many valuable varieties of marble; at Tortosa are the celebrated quarries of jasper; and the hills of Becares and Filabres, near Alicant, are reported to be entirely composed of a pure white statuary marble. Most of the streams descending from the heights are rather torrents than rivers, swollen to a great width in the winter and spring, yet very low in the summer. Their mouths generally form petty trading anchorages for the small coasters which carry for the larger and more frequented ports. The principal products are corn, maize, rice, wine,

Spanish Coast.

Produce.

oil, brandy, olives, wool, salt, alum, kermes, barilla, potash, esparto mats and cordage, turpentine, soap, dates, raisins and other dried fruit, aniseed, flax, saffron, honey, wool, cottonades, cotton, linen, silk, iron, lead, zinc, antimony, manganese, copperas, and grana—a species of cochineal. Though the materials for commerce are thus numerous and valuable, from the mistaken system of restriction so long followed by the Spanish government, the prosperity of its subjects is not only crippled, but almost annihilated. Entering into details, there will be found between Gibraltar and Malaga a chain of lofty mountains, parts of which, as near Ronda (*Acinipo* vel *Arunda*), for example, offer the scenery termed by the French *les belles horreurs;* the country is fine, but with various waste and barren spots, nor are there many littoral points of note. Estepona and Marbella (*Salduba*), are two of the loading places for the coasters alluded to above, and Frangerola, a very ancient fortress, is noted for its anchovy fishery. Malaga (*Malacha*) itself is a fine large city, with striking public buildings; but its harbour is rapidly filling up by the detritus poured in during the freshes of the Guad-al-Medina. Between Sacratif, or Carchuna Point, and Cape de Gata, and even to Moxacar, the coast is in general very high, the only interruptions being at the mouths of the small rivers, where are formed *playas*, or triangular plains, from one to several miles in extent, open to the sea, and extending to the foot of the mountains, the washings of which have formed the rich alluvial soil generally found in these plains.

Malaga.

Ports and Trading Places.

Passing the loading places of Almunécar (*Menoba*), Salabreña (*Salembina*), Motril (*Hexi*), Castel de Ferro, and Adra (*Abdera*) we arrive at Almeria (*Murgis*), once renowned for commercial enterprise, but now merely exporting some barilla and lead. To the north-north-east of Cape de Gata (*Prom. Charidemum*) is Carthagena (*Carthago nova*), which, though one of the three royal ports of the kingdom, is miserably neglected, and its marine arsenal, constructed at

Carthagena.

a vast expense, is nearly ruined. This is a consequence of the inertness of the government, the apathetic dulness of the heavy Murcians, and the deleterious miasma of its marshes, by which it is subjected to the severe visitations of an endemic bilious fever. The whole district was called *Campus Spartarius*, from the abundance of esparto, or Spanish-broom, which grew there. Between Carthagena and Valencia the coast is generally low and sandy, but broken by various loading places and ports, of which the most important is Alicant (*Lucentum*), a commercial town situated in the northern extremity of the bay, and at the foot of a castled hill. Its vale of Huerta is fertile but unwholesome; few of its cultivators escaping fever or ague, not a little assisted by the waters of the vast tank, or rather artificial lake, *el Pantano*. It is, however, the *entrepôt* of the productions of Valencia and Murcia, and its custom-house was long the most valuable that the Spanish monarchy could boast. Passing Altea, the coves of Cape S. Martin, Denia (*Dianium*), and the River Xucar (*Sucro*), we arrive at the great plain on which the city of Valencia stands, the finest of the whole coast, and so fertile that it is called *La Huerta*, the garden;* but there is no port in the whole Sucronian Gulf where any shelter is afforded in onshore winds, except within the moles of the Grao, at the mouth of the Guadalaviar (*Turias*), nearly three leagues off. Just below these moles is the great lake of Albufera, four leagues long and two broad, full of fish, and separated from the sea by a narrow sandbank. Pestilential exhalations arise from many parts of this otherwise beautiful and productive plain, but especially in the extensive rice-grounds. Valencia, which has borne its present appellation since the time of the Romans, is one of the smallest provinces of Spain,

Alicant.

Valencia.

* It is necessary to keep in mind that a slight difference in this phrase makes a great modification of its meaning. *La Huerta* is a kitchen or market-garden, for the growth of vegetables and pot-herbs; *El Huerta* is a walled fruit-garden, or orchard. Quantities of dates, and etiolated palm-branches for festivals, are brought from Elche (*Ilicis*).

though one of the richest and most populous, from the fertility of its soil; but even here—from the languid system of husbandry, the extent of waste lands, the want of easy communication with the interior, and the irregularities of a feeble government—there is no immediate danger of population pressing upon subsistence. In comparison with the other Spanish provinces of the Mediterranean shore, its numbers are very striking; yet in submitting them it should be remembered that by Andalusia is meant what is now restricted to the kingdom of Seville; for in the Middle Age statistics that rich and powerful province included Seville, Cordova, Granada, Jaen, and the districts of Sierra Morena, an extent of 27,550 square miles, comprehending a space so beautiful and delicious, that the Moors fancifully imagined heaven to be suspended over it. These are the figures for 1810:—

	Total inhabitants.	No. to a square league.	Leagues of Sea Coast.
Andalusia	755,000	1009	20
Granada	700,000	860	57
Murcia	383,226	582	22
Valencia	830,000	1285	69
Catalonia	859,000	856	67

Murviedro. Murviedro (*Muros viejos*) is more remarkable for the beauty of its prospects, its Moorish walls, and its being the site of the ancient *Saguntum*, than for the trade which is now carried on; its exports being, as well as those of its neighbours Benicarlo and Vinaroz, a small quantity of wine and brandy. Much benefit was expected from these places when the attention of government was drawn to the capacity of the mouth of the Ebro, in 1792; but, as with most of the works and projects of that day, the attempt was abortive, owing both to deficient means, and the prevalent dogma that had Providence intended such things, they would have been so ordered. River Ebro. The Ebro (*Iberus*), the largest river in Spain, rising in the heights of the Asturias, and pursuing its course in an easterly direction between the Pyrenees and one of their secondary branches, passes Tortosa (*Dertosa*), and enters the Mediterranean from the shore which we have now reached, after flowing nearly 400 miles. As this river had accumulated two marshy peninsulas and several sandy islets at its mouth, an attempt was made to improve the port of Alfaques, formed by a peninsula, by building the town of San Carlos, and cutting a canal to Tortosa, to carry large vessels to a point which the velocity of the stream precludes their reaching by the river during great part of the year. This spot, and the adjacent low grounds, are severely scourged after the summer months by *el fiebre periodico*, arising from the bad air exhalations. **Alfaques.**

Between the Ebro and Barcelona are numerous small towns, with coves where vessels take in cargoes; but the principal place is Tarragona (*Tarraco*), still surrounded by ancient Roman walls: this place gained a melancholy notoriety in the late war with France, from the slaughter of its inhabitants by Marshal Suchet. Barcelona (*Barcino*), lying in rather too moist a situation between the rivers Llobregat (*Rubricatus*), and Besos, is the capital of Catalonia, and boasting 160,000 inhabitants, is the second city of Spain in population and commerce. It is well built, and possesses manufactories of silk, gauze, lace, cotton, canvas, leather, woollens, cutlery, paper, fire-arms, soap, and glass, which together with wine, spirits, cork, and fruit, form the great articles of export. The port is made by art, having little depth, and that depth is daily diminishing by the sand thrown up during easterly gales, the mole preventing any offset; while the anchorage in the roads is exposed to all sea-winds. The coasts of Valencia and Catalonia have gained considerably on the sea, from the incessant agency of the Ebro, the Llobregat, the smaller streams, and the numerous torrents, which deposit vast quantities of silt, and form lines of shoals parallel to the shores, lessening the general depth of water for some distance out. **Tarragona.** **Barcelona.** **Produce.** **Changes of Coast.**

From Barcelona to the north-eastwards, the shore presents a quick succession of small towns and villages, possessing little trading coves, and remarkable for pleasant situations and cleanliness. Among these, the place of most consideration is Mataro (*Illuro*), which has 25,000 inhabitants, various manufactories, and a thriving trade. But from Mataro to the frontiers of France, we scarcely meet any town deserving of notice, though the country continues fine, and there are many neatly built places of traffic. The roads of Pálamos and Rosas are good and extensive; and as the worst wind, the *tramontana*, comes from the land, it never occasions serious mischief. The comparatively little fortress of Santa Trinità, which governs the town and anchorages of Rosas, has more than once made a resolute defence against assailants. Its capture was a difficult exploit to the French, when Figueras, the boasted Bulwark of Catalonia, though full of ammunition and provisions of all kinds, with a garrison of 9000 men, disgracefully fell before General Perignon, in 1794, and with it the important district of Ampurdan, of which Port Ampurias (*Emporiæ*) is the capital. Beyond Rosas, there is a rocky peninsula or head-land, of which the chief projecting points are the Capes Norfeo and Creux. From this, the most eastern part of Spain, the Pyrenees, stretching westward in a straight line of 270 miles, form a strong natural barrier between the French and Spanish territories: while from the sea their lofty summits produce a continually varying series of striking objects. Alpine limestone, old red sand-stone, and transition rocks, reposing on mica, slate, or granite, are the principal constituents of these mountains.

Mataro. Pálamos. Pyrenees.

§ 2. The Spanish Islands.

Baleares.

THE Balearic Islands, comprehending Majorca, and Minorca, with Cabrera, Ayre, and several smaller islets, lie off the coast of Valencia, and hence that part of the Mediterranean was formerly called the Iberian Sea. Some recent legislative enactments have included Iviza and its dependencies among the Baleares; but the geographical distinction is warranted by Strabo, who distinguishes the two groups under the names *Gymnesiæ* and *Pityusæ*. Pityusæ. Yet it is possible, that the Cretan geographer meant to include both Pityusæ and Gymnesiæ under the generic name of Balearides, his words being 'of the isles lying in front of Iberia (we find or have) two Pityusæ and two Gymnesiæ, but they call them Balearides.' Whether he means all four or only the two last, is not clear. However, from what Diodorus Siculus and Pliny say, it is probable that he meant to give the name of Balearides only to the Gymnesiæ Considering that this last group possesses but about 275 miles of coast, and the Pityusæ only 73, they are densely populated, the numbers standing thus in 1810:—

	Total inhabitants.		No. to a square league.
Majorca	140,700	...	1256
Minorca	31,000	...	1257
Iviza and Formentara .	15,290	...	1019

Majorca, the larger island, is nearly square in shape, Majorca. with a mountainous surface, and generally rocky coast, with deep water around: hence its aspect is varied, with a delicious climate, insomuch that an almost vernal temperature is to be found on the higher grounds in the greatest heat of the summer, and except on the mountainous ridges, the winters are mild and pleasant. There are no rivers, properly speaking; but the *arroyos*—streams or rather torrents—are often impetuous during rains, and the Rierra, the largest of them, has frequently occasioned great damage and loss of life in the vicinity of Palma, the capital of the island, and still preserving its ancient name. This is a fine town, with a haven and good road-stead; but though in the thirteenth century one of the chief markets of Europe, it has now comparatively but little commerce. The other principal resorts for shipping are Alcudia and Pollenza (*Pollentia*), both of which are extremely unhealthy in autumn; and the chief exports are wine, oil, salt, canvas, silk, coarse linens, and woollens, dried fruits, honey, mill-stones, limestone, and marble.

Minorca. Minorca is smaller than Majorca, as implied by the names (*major et minor insula*); and it is more level, its only remarkable elevation being a central hill, Mount Toro, with a convent on its summit. Besides the Creek of Ciudadella—the capital, Port Fornelles, and several resorts of minor consideration, this island possesses the capacious harbour of Port Mahon (*Portus Magonis*), one of the finest and most commodious places for shipping in this sea, and, save occasional summer visitations of marsh fever, one of the most healthy. Here, during the late war, have I seen a potent and magnificent fleet of English men-of-war wintering from before Toulon, each ship in a roomy berth; although there were no fewer than six large three-deckers, and twenty-five two-deckers, besides numerous frigates, sloops, and brigs, in the highest state of efficiency. It was a glorious sight!

Pityusæ. In mid-distance from Majorca to Cape San Martino of Valencia, lie the Pityusæ islands, a classical name, supposed to have been derived from the pine trees with which the larger one was covered. This denomination comprehends Iviza, Formentera, Conejera, Bledas, and various smaller islets and rocks. Iviza, the *Ebusus* of the ancients, is hilly and stony in many parts, but in others very fertile, producing corn, oil, wine, and fruits of many kinds; and the mountains are well wooded with pines, firs, and junipers. There are several ports affording good anchorage to moderate-sized vessels; but the best is before the town of Iviza, the capital, where much salt and timber are embarked.

Columbretes. Between the Balearic Islands and the coast of Valencia, at about the distance of ten leagues from the latter, lie the Columbretes (*Colubraria*), a group of volcanic rocks, which, for the reasons I have given in the first volume of the Journal of the Royal Geographical Society, I cannot but deem to have been the *Ophiusa* of the earlier ancients.

The harbour on the east side of the largest islet, is evidently the mere broken mouth of an ancient crater. Finding that they were unknown in detail, in order to distinguish the several rocks, I named them after the most scientific officers in the Spanish navy, as they now appear on the plan published by the Admiralty.

---

§ 3. The Mediterranean Coast of France.

FROM the Mediterranean shores of France, as well as of Spain, the sea has retired in several places, and gained upon none. Indeed, in no other *closed* sea, are there so many well ascertained accessions of land at the mouth of large rivers. Thus Aigues Mortes, which in the thirteenth century was a sea-port, is now five miles inland; Miquelon and Psalmody were islands in the year 815, and in 1820 they were two leagues from the sea; and even some of the present vineyards of Agde, were covered by the sea only a century ago. By the advance of the land upon the sea, owing to the alluvium of the Argeus, the ancient port of Fréjus was converted into a pestilential marsh, and at a later period into terra firma, half a mile from the sea. But the greatest accession of new land is that which forms the delta between the two mouths of the River Rhone, where old lines of guard-towers and sea-marks occur at different distances from the present beach, and indicate the successive retreats of the sea in comparatively recent times. Changes of Coast.

Between Cape Creux and the River Var, a space of about 300 miles, the Mediterranean is bounded by the southern shores of France—a mighty state, whether empire, kingdom, or republic—shores which I have visited both in war and in peace. On passing the Cape, the coast of the Department of the Eastern Pyrenees continues mountainous and rocky; then succeed the low and marshy flats of the France.

rivers Aude and Hérault, in which are many extensive étangs or salt lagoons communicating with the sea, the principal ones being those of Leucate (*Leucata*), Sijean, Gruisson, and Thau, which are intersected by a navigable canal. Coast-ways, after leaving Spain, the first town of

Port Vendre.

France is that which gives a name to the insignificant Port Vendre (*Portus Veneris*), and the second is its congener Callioure. A low coast and bad lee-shore, as well as an insalubrious one, extend along the front of the ponds, to which the principal inlet is the port of Narbonne (*Narbo Martius*), once a flourishing maritime place, now ten miles inland. East-north-east of this, the fort of Brescou, on an insulated rock, announces the approach to the ancient town

Cette.

of Agde (*Agatha*), formerly the capital of a county. Cette (*Setius*), was built on a narrow tongue of land which separates the pool of Thau from the sea, its haven being the principal outlet of the great Canal of Languedoc in 1666: it has

Montpellier.

also become the port of the adjacent city of Montpellier (*Mons Pessulanus* vel *Puellarum*), so celebrated for its excellent climate, and its school of medicine. Here they embark the wines of Lunel and Frontignan; the perfumery, preserves, liqueurs, wines, calicos, woollens, snuff, soap, cream of tartar, vitriol, and verdigris of Montpellier; and the salt, as also tunnies, sardines, anchovies, and other fish of the neighbour-

Rhône.

hood. Beyond this, the mouths of the Rhône (*Rhodanus*) form a number of islands, a great part of which are nearly level with the sea. The largest of them, named Camargue, has nearly the shape of a Delta, with the city of Arles (*Arelate*), at its apex: but it is properly a mere assemblage of little marshy islets and sand-banks accumulated along the former seashore, with the great brackish lagoon of Valcares occupying the centre. Between this and Port de Bouc, is

La Crau.

the singular stony desert called La Crau (*Campus Lapideus*), a space of more than 150,000 acres, entirely covered with boulders and rolled shingle. This name is not to be confounded with Les Graus du Rhone (*Gradus Rhodoni*), a designation of the mouths of the river; this word, 'Graus,' seems to have meant a landing place at such locality, as with the Grao of Spain, and the Grado of Italy.

The principal mouth of the Rhône is its eastern branch, or that which flows into the Gulf of Foz, where, during freshes, it disembogues with such force and rapidity, as to carry its own waters far into the sea. This was practically proved by the watering of Sir Edward Pellew's fleet in the late war, when his ships anchored in the offing south-east of the Tour de St. Louis, and skimmed as much potable water as there was occasion for: at three quarters of a mile from the shore, the fresh run was about three feet deep, and it was taken during the morning lulls. Gulf of Foz.

Between the Gulf of Foz and the River Var, the *Comté de Provence* of former days, the coast is undulatory, and indented with numerous ports, coves, and bays, where vessels of all descriptions may find a berth. Of these, Martigues (*Maritima*) at the entrance of an extensive lagoon communicating with the sea, often called *l'etang de Berre*, is of importance from furnishing both fish and salt for commerce. Marseilles (*Massilia*) is a maritime city of the first class, having upwards of 120,000 inhabitants, and maintaining a commercial activity second to none in France; the approach to its secure haven is marked by the rocky isles of Planier, Ratoneau, and Pomegue. Eastwards of Marseilles, the towns of Cassis and Ciotat are resorted to by coasting traders, and the muscat wines of this department are largely exported. Passing these, and rounding that well-known headland Cape Sicie,* we enter the harbour of Toulon (*Telo Martius*) the second naval port of France, so celebrated for its arsenal and fortifications; but the business of the place, independently of the government establishments, is not very great. Flour, wine, brandy, oil, Martigues. Marseilles. Toulon.

* I say well-known, because it was in the sight of our officers and seamen of the blockading fleet for months together. I had a tolerable spell of it myself in 1811 and 1812.

olives, dried fruits, tobacco, and other productions of the neighbourhood, besides soap, coarse woollens, cordage, morocco-leather, chocolate, vermicelli, and other manufactures, are the principal articles of commerce; and some merchant vessels are built at this place.

From Toulon eastward, we enter the great bay or road of Hyères. Hyères, all the projecting points of which are strongly fortified; lying within the Porquerolles (*Stœchades*) islands, and having the beautiful town of Hyères (*Olbia* vel *Areæ*) and its ever-fruitful gardens on an adjacent eminence. It should be remarked, *en passant*, that this is said to be the only place in Europe north of Italy, where the orange flourishes without artificial shelter in winter. The whole of this expanse of water forms a superb anchorage for any number of ships; and in August, 1811, the English blockading fleet, though fired at in passing, took up a berth there out of gun-shot. A French division consisting of three first-rates and ten two-deckers, came out of Toulon on the following day; but on the speedy appearance of some of our fleet, one of which was the Rodney, of seventy-four guns, (the ship in which I was then serving,) they retreated to their moorings, and we returned to the Bay of Hyères, where we remained for three weeks at single anchor. During this time the topsail yards were swayed to the mastheads every evening, strong divisions of guard-boats properly stationed, with every possible preparation for any sudden emergency. Such was the contrast afforded to this scene when I last sailed among those islands in the Adventure, that scarcely a vessel was to be seen; and I passed with perfect impunity close to the point of Porquerolles, where I had formerly seen those noble first-rates the Caledonia, Ville de Paris, Téméraire, and Hibernia, receive and return a heavy cannonade, while they drifted past under a light morning air!

The coast from Hyères Bay to the Italian boundary is more or less elevated; and has many bays, coves, and indentations between the rocky headlands, as those of St. Tropèz or Grimaud (*Athenapolis*), Frèjus (*Forum Julii*), Napoule or Cannes, Gourjean, and Antibes (*Antipolis*); which are improperly termed gulfs by the native pilots and mariners. From Antibes a low sandy beach turns to the north-eastward, where, and at the distance of about two leagues, is the mouth of the river Var (*Varus*), which separates France from Savoy, the boundary being marked by a toll-house in the middle of the floating bridge across the stream. This turbid river, which rises in the ramifications of the Alps, between Barcelonette and Colmars, has a course of about twenty-five leagues to the sea, where its bed is above a mile in breadth. It runs so rapidly during freshes from the mountains or the melting of snows as to prevent the erection of a permanent bridge; at such times, the floods of the Var, freighted with silt, discolour the sea-waters to a considerable distance from the shore.

Coast towns.

The Var.

§ 4. The Coast of Western Italy.

EXCEPT illustrious Greece, the classical region of Italy (*Hesperia*) is more calculated to awaken enthusiasm and contemplation than any other on earth; for to that most interesting and beautiful country, the rest of Europe is largely indebted for the practice of various branches of polity, science, art, and rural economy. Italy, like the expiring eagle, has been sorely wounded by arrows feathered from her own wing, and has long since descended from her high estate; but though morally degenerate, she still boasts all those physical advantages which raised her to a world-wide distinction. What a potent arm of power and civilization she might still exert, were not all the several states at utter enmity with each other, so that she could not become an integral kingdom, nor scarcely a federative republic! Although

Italy.

inhabitants of a soil where climate, religion, language, manners, and customs are all in substance nearly uniform, however varied in community and government, there is no bond of common national feeling; all being alike in one quality only, that capacity of indulging hatred of one another which feeds implacable factions, and engenders distrust and disunion. I have lately been assured that the population has rapidly increased, and now amounts to upwards of twenty-five millions, an assertion for which proof is not given. The numbers furnished by my friend and colleague, General Visconti, for the year 1820, are founded on a census taken by order of Napoleon about the year 1810.

	Square Miles.	Population.	To the square mile.
Naples and Sicily	43,600	6,750,000	162
Kingdom of Sardinia	27,400	3,976,000	146
Lombardy and Venice*	18,920	4,054,600	212
Ecclesiastical State	14,500	2,350,000	168
Grand Duchy of Tuscany	8,500	1,182,900	140
Duchy of Parma	2,280	377,000	121
States of Modena	2,060	370,000	190
Duchy of Lucca	420	138,000	328
Republic of San Marino	40	7,000	175

Gulf of Genoa.

Arriving at the coasts of what was anciently termed Liguria, the space between the Var and Genoa is known as the Riviera di Ponente (*Western beach*); and that from Genoa to Spezzia as the Riviera di Levante (*Eastern beach*); in both cases with relation to Genoa as a centre. The whole consists of rocky precipices flanking lofty mountains, intersected by fruitful valleys, and varied with a succession of picturesque towns and villages through a length of about 175 miles. There is a considerable coasting-trade along the shores of the Gulf of Genoa, the produce being oil, rice, fruit, hemp, silk, velvet, anchovies, and palm-branches, with which last articles, San Remo has the exclusive privilege of supplying Rome for its religious festivals.

* By a census taken in 1825, the population of Austrian Italy had increased to 4,237,000; of whom 500 were Armenians, 700 Greeks, 5600 Jews, 66,500 Germans, and 4,163,700 Italians.

Nizza, or Nice (*Nicæa*), the first maritime town on the Riviera di Ponente, with its commodious little artificial port, is agreeably placed on the banks of the Paglion, a mountain torrent, at the foot of Mont Albano, the citadel of which protects both Nice and the capacious harbour of Villa Franca on the east. The whole of this neighbourhood indicates that great geological changes have occurred, not only in the position of rocks and strata, but even in the relative height of land and water. The hollows and fissures in the rock now above water are frequently found to contain shells similar to those which now exist in the Mediterranean; and strong marks of upheaving were pointed out to me by that obliging local naturalist, Dr. Risso. Between Nizza and Ventimiglia (*Albium Intemelium*), lies the little principality of Monaco (*Portus Herculis Monæci*), and further eastward are the considerable trading towns of San Remo, Port Maurizio (*Portum Maurici*), Oneglia, Alassio, Albenga (*Albium Ingaunum*), Finale, Noli, Vado (*Vada Sabata*), and Savona (*Savo*); which last, beautifully situated, strongly fortified, and commanding as well the waters as the Cornice, or great road of the Riviera, has been absurdly and unwisely sacrificed to the commercial jealousy of Genoa, to please whose prejudices the harbour, which was roomy and good, was partly choked up in 1525. Notwithstanding this and other injuries it carries on a considerable trade, the port being still very secure, though fit only for vessels of about two hundred tons. In the beautiful ravines and valleys of this space both intermittent and remittent fevers are occasioned by the summer exhalations; and it will be recollected that Napoleon, who must have had early knowledge of malaria at Ajaccio, was charged with having sent Pius VII. to Savona, that he might fall a victim to the marsh miasmata.

Nice. Villa Franca. Geological changes. Coast towns. Savona.

Genoa (*Genua*), one of the handsomest cities in Europe, is built on the declivity of a hill flanking the Ligurian Apennines, in the deepest recess of the wide gulf which

Genoa.

forms a crescent from the frontiers of France to those of Tuscany. Its port, *fatto per forza* as the Italians term it, is formed by two moles, sixteen or eighteen feet above the level of the water, the heads of which are about three hundred fathoms asunder. This people have been well known for ages, as essentially commercial, and although its fleets are no longer the terror of the Levant, its tendency to maritime superiority is still manifest, its ships, though few, being in excellent order, and its sailors among the best of Italy. Even in its palmy days, although the government was highly aristocratic, the nobility of Genoa were allowed to be engaged in manufactories of velvet, silk, and cloth; to farm the duties; to speculate in foreign commerce, and to hold shares in merchant vessels; but all other business and handicrafts were strictly forbidden.

To the east of Genoa, along the populous shores of the Riviera di Levante, the succession of towns and villages is almost as quick as to the west. Running past Bisagno and Nervi, the first place of consideration in a nautical point of view is Porto Fino (*Portus Delphini*), a cove with a small pier-haven between two lofty promontories, which form conspicuous guides for the coasting voyager. Rapallo, Lavagna (*Entella*)—where the only quarry of slate in Italy is worked—Chiavari, Sestri di Levante (*Segeste*), Moneglia (*Monilium*), and various sea-side villages, lead to Porto Venere and the capacious Gulf of Spezzia (*Portus Lunensis*), which latter has good soundings throughout, and is one of the finest and safest bays in the Mediterranean. The importance of this place being represented to Napoleon, he was led to contemplate making it one of his great naval stations; and the plans consequently drawn up by his engineers being obtained by Lord William Bentinck, at the surrender of Genoa in the year 1814, his lordship, to whom I acted as naval aide-de-camp, kindly submitted them to my examination.

Porto Fino.

Gulf of Spezzia.

Although the independent states of Massa and Lucca

are considered as forming the eastern boundary of the duchy of Genoa, it must geographically be said that Spezzia is succeeded by the coast of Tuscany; and indeed it was formerly held to be the limit of Liguria to the south. Passing, therefore, the Magra (*Macra*), a mountain-stream which enters the sea at the Marinella of Luni (*Lunæ Portus*), and running by Pietra Santa (*Lucus Feroniæ*), Via-Reggio (*Fossæ Papirianæ*), and the mouth of the Arno (*Portus Pisanus*), we arrive at Leghorn—the mouth of Tuscany, and one of the busiest ports of Italy. The prosperity of the place is owing to the sound judgment which prompted the establishment of a free port with full religious toleration and liberal immunities. Boasting no splendour of antiquity, although the site of *Portus Herculis Liburni*, it became a place of importance on the fall of the neighbouring Porto Pisano, which succumbed to the joint operations of the Arno, the sea, and the Genoese, in filling and choking it. By careful drainage the malaria around it has been greatly diminished, but still the climate is so damp as to justify the proverb—*Pisa pesa a chi posa.* Leghorn suffered greatly under the rule of Napoleon; but it is now a prosperous Porto-Franco again, supplying a large part of the interior with foreign goods, and exporting coarse woollens, cotton, silk, maize, oil, iron, paper, potash, marble, alabaster, coral, anchovies, platted straw for hats and bonnets, and artificial flowers. The fishery of anchovies is very productive. This fish enters the Mediterranean in large shoals by the Strait of Gibraltar in the spring, for the purpose of breeding, after which it retires again to the depths of the Atlantic.

Tuscany.

Leghorn.

From Leghorn the coast of Tuscany extends south-eastwards as far as the bold promontory formed by Mount Argentero; and there are various loading-places between, as Vada (*Vada Volaterrana*), Cecina, Porto Baratto (*Populonium*), Piombino, Fullonica—where there are the principal smelting bloomeries or works of the iron from Elba—Castiglione, Telamone (*Portus Telamo*), Port San Stefano,

Small ports.

and Orbitello, in the middle of the lake of that name, the eels of which are equally celebrated and profitable. These places are on the margin of the sea-side marshes, known as the Maremme, unwholesome lowlands, diffusing with more or less virulence their pestilential exhalations along the whole west coast of Italy. Those of Tuscany, through drainage and tillage, have been of late greatly improved, but continue to be very deadly in summer. Of the portion now before us, the river Ombrone (*Umbro*) and its affluents form a principal drain, if such a term can properly be used in speaking of the Tuscan *maremme* and *paludi*, the pestiferous exhalations of which furnished Dante with his disgusting parallel of the tenth gulf in hell.* In the heights around, most of the women have been married two or three times, because only the men leave their houses to labour in the marshes, where they are tempted by fatigue sometimes to sleep, though at the risk of illness and death. The sea has retreated from Telamone, and left a mere morass; while Domitian's Port, under the northern cliffs of Monte Argentero (*Mons Argentarius*), is submerged. The coast of Tuscany, from Carrara to this port, may be estimated at about 140 geographical miles.

Maremme.

Changes of coast.

The Tuscan islands lying opposite to this coast are, Gorgona (*Urgo*), Capraja (*Ægilon*), Elba, Giglio (*Igilium*), Pianosa (*Planasia*), Monte Christo (*Oglasa*), and some smaller islets. Of these, Elba (*Æthalia* sive *Ilva*) merits a distinct mention on account of its excellent harbours and bays, abundance of iron and other mineral productions, its picturesque beauty of scenery, and its having been allotted by the Congress of Vienna as a Barataria for the

Elba.

* Dante, in Canto xxix. of the *Inferno*, visits the *decima bolgia*—

Qual dolor fora, se degli spedali
Di Val-di-chiana, tra 'l luglio e 'l settembre,
E di Maremma, e di Sardegna i mali
Fossero in una fossa tutti insembre;
Tal era quivi; e tal puzzo n' usciva,
Qual suol venir dalle marcite membre.

fallen Napoleon. Elba itself is somewhat less than fifty miles in circuit, and the islets immediately adjacent are its dependencies.

Rounding Argentaro to the southward, the first Roman town is at Port Ercole (*Portus Cossanus* sive *Herculis*); between which and Terracina, a space of 150 miles, lies the west coast of the Papal States, for the most part an open and exposed beach on which the sea-winds drive a heavy surf. The shore consists principally of low unhealthy flats, interrupted at certain points by a bolder and better country. Nearly in the mid-distance of this space, the celebrated river Tiber discharges its waters by two principal branches through its marshy delta still called Isola Sagra (*Insula Sacra*); off which there is good anchorage within sight of St. Peter's lofty cross. The scantiness of the Mediterranean tides renders the estuaries of its rivers nearly useless for the purposes of navigation and commerce; a great and lasting disadvantage to these countries, especially as most of the streams have a rapid descent: and that aid which causes the current to turn, and bear the laden ship to the busy mart, is here wanting. My schooner-rigged barge, riding in the middle of the river opposite the Dogana—with its ensign and pendant, its morning and evening gun, and its well-disciplined crew—was an object of great admiration to the Romans of all ranks and conditions. Though trading vessels resort to Fiumicino, the southern mouth of the Tiber, and to various other little ports in this territory, the main harbour of the States of the Church on this side is Civita Vecchia (*Centum-cellæ* sive *Trajani Portus*), a town fortified by that universal genius, Michael Angelo; and possessing fine specimens of Trajan's marine works.*

Roman coast.

Tiber.

Civita Vecchia.

* Besides the havens of Centum-cellæ, Portus Trajanus in Etruria, and Ancona, I am inclined from examination to think that Trajan greatly enlarged and improved the work of Claudius, at the Port of Ostia: wherein I agree with the Scholiast of Juvenal, in his commentary on the passage where that poet describes the narrow escape of his friend Catullus from shipwreck.

But between the Tiber and Terracina (*Anxur*), especially near Porto d'Anzo (*Ceno Portus*) and Nettuno (*Antium*) are ruins of piscinæ, baths, and villas of the old Romans, among which is Astura, so long the residence of Cicero, and still retaining its name, now submerged in the sea, which thus proves its encroachment on this part of the coast. It will be recollected that this same space is the margin of the Pontine Marshes (*Pontinæ Paludes*), a district so notorious for its *aria cattiva*, and consequent deserted state, although once the principal element of the power of the Volsci, and the prosperous site of thirteen cities on a territory now without even a village, despite of its being rich beyond conception in cattle, timber, and vegetable wealth. These *paludi*, or fens, are occasioned by the quantity of water carried into the plain by innumerable streams that rise at the foot of the mountains to the east of Rome, which, for want of sufficient declivity, creep sluggishly over the level space, and sometimes stagnate in pools, or lose themselves in the sands. Here, fermenting with decayed vegetable matter, and acted on by a fervid climate, malaria is produced;—that invisible enemy which poisons the fairest portions of Italy, otherwise so salubrious, and renders man a sufferer from his cradle to his early grave. There is every appearance that the basin of these marshes was once a gulf of the sea, which has been gradually filled up by the alluvium from the mountains; and that Monte Circello (*Circeii prom.*) was an island when Homer wrote, whether he ever meant the place or not.

Pontine Marshes.

The kingdom of Naples, extending over the southern part of Italy, abounds in beautifully varied scenery, and is remarkably fertile, insomuch that its commerce, though pretty good, might easily be made much better, but for the arbitrary system of duties and impolitic restrictions on trade. The western coast of this kingdom, which we have now reached, is generally bold, and indented by many deep bays—usually termed gulfs by the native pilots,—

Coast of Naples.

such as those of Gaeta, Naples, Salerno, Policastro, Sant Eufemia, and Gioja: and its extent, from Terracina to Cape Spartivento is about 370 miles. In this space there are numerous small ports and caricatori, or loading places, and various tolerable anchorages for large ships in all winds from the eastward. The chief exports are wine, fruit, oil, olives, cheese, maccaroni, silk, aloes, wool, argols, lichens for dyeing, pozzolana, potash, hemp, and leather.

Gaeta (*Caieta*) is a strong fortress on the rocky promontory of La Santa Trinità, which is joined to the mainland by a narrow isthmus, whence it has obtained the appellation of little Gibraltar: there is a haven for small vessels, and its road offers excellent anchorage, especially in front of Mola (*Formiæ*). Before the bay is a group of small volcanic islands,—namely, Palmarola (*Palmaria*), Ponza (*Pontia*), Gianuti, Zannone (*Sinonia*), Vandotina, and some smaller rocks: they are bold, and the channels among them very deep; but all have evidently lost much of their original extent by the destructive force and degradation of surf and atmosphere. The Bay of Naples (*Parthenope* sive *Neapolis*) is of a semicircular form, surrounded by mountains, among which the still-active volcano, Vesuvius, rises to the height of 3880 feet; and by its smoke acts as a pharos to vessels in the offing. The limits of this bay, still called *Crater* by the native hydrographers, are Cape Miseno and the isles of Ischia (*Ænaria*) and Procida (*Prochyta*) on the north, and Cape Campanella (*Minervæ prom.*) with the isle of Capri (*Capreæ*) on the south. The combination of such scenery with objects so memorable and celebrated, is as remarkable as it is striking: the beautiful city winding along the shores of this bay, its villas and villages, its picturesque heights and islands, its intensely interesting remains of classical antiquity, and even its earthquakes and volcanic eruptions, all combine to fill the spectator's mind with delight. Although this is not the place for archæological recollections, yet

Gaeta.

Naples.

who can forget Baiæ and Herculaneum, Pompeia and Stabia! Nor can we but give a glance to the southward, and ask who can believe in the idle story of Pæstum's stately temples having only been discovered in 1755? These interesting relics may have been unnoticed by antiquarians—perhaps from the moral and physical impediments of lawless men and pestiferous air—but were assuredly well known to seamen, being conspicuously visible from every offing of the Bay of Salerno (*Salernum*); and 'i pilieri di Pesto' were given as a sea-mark, long before Sebastian Gorgoglione, the pilot, wrote his very popular Portolano.

Pæstum.

Continuing along the Lucanian coast, from the Galli rocks (*Sirenusæ insulæ*) and picturesque shores of Amalfi to the southward, a rich line of coast is presented, with all varieties of beach and cliff, hill and valley, and towns and villages. Of the latter the principal are Policastro (*Pyxus*), Amantea (*Lampetiæ*), Tropea (*Prostropræa*), Scylla (*Scyllæum*), and Reggio (*Rhegium*). The territory affords a substantial trade in corn, wine, oil, honey, wax, silk, fruits, and legumes of all kinds; while the shores yield tunnies, sword-fish, pilchards, sardines, and various other fishes; but all the lower grounds are exceedingly unhealthy. The great sea-mark of the coast is the elevated Monte Cocozzo, which is the highest of the Calabrian Mountains, for the greater part of the year covered with snow.

Coast towns.

M. Cocozzo.

Geological changes.

It is decided, upon what appears to be sound geological evidence, that a great part of the Italian coast has been raised and lowered several times within the historical era, while the sea must have ever maintained the same level; although it may once have washed the foot of the nearest Apennine. The mouth of the Tiber has advanced greatly, and lowered its level, even in the last eighteen centuries; for Ostia, and the port Claudius, are now far inland; while nearly the whole delta, called Isola Sagra, has been formed since that period. Between the Tiber and Terracina, the old port of which has long been filled up, the coast is strewed with remains of ancient villas and works, which in some places—as at Nettuno and Anzo—are far in the sea. Homer appears to have heard Monte Circello described as an island, since his account of the place tallies very fairly with what the Pontine marshes would now suggest; and in the age of Theophrastus, this hill was a mile from the shore. By the joint actions of the marl and peat of the marshes, the vast alluvial deposits, and perhaps the detritus of the decreasing Ponza islands thrown up by the sea, this mass of transition limestone has become a promontory; and its cliffs, as well as some of those on the coast of Calabria, bear unequivocal testimony of a former submersion, from being thickly perforated by the borings of recent mollusks at the height of more than one hundred feet above the present level of the sea. In the Abbate Romanelli's description of Capri, published in 1816, he says—'Near the eastern summit there is also a singular calcareous mass, closely pierced by *mitoli* and *vermi litofagi*, which indication of the pholades proves the sea to have formerly reached that height:' and this was confirmed to me by Professor Scipio Breislak, at Milan, on his showing specimens which he had taken from a summit of that island, *colla suo propria mano.* But the master-key in evidence of geological cycles of great extent on this coast, is offered in the interesting ruins of the temple of Jupiter Serapis, near Pozzuoli, in the bay of Baiæ, and about one hundred feet from the sea. Of this temple there are three columns still standing, which are profusely drilled, to the height of from twelve to eighteen or nineteen feet, by the perforations of the lithodomus, a bivalve still existing in the adjoining sea. When I first visited these ruins, in the spring of 1814, the pavement seemed to be rather above the level of the Mediterranean; yet it would appear to be slowly sinking again, since in 1850 there were upwards of two feet of salt water over it. The whole line of the adjacent coast between the downs and the beach, is of modern formation, consisting of beds of pumice and sand, with recent marine shells, lateritious fragments, and water-worn pieces of pottery. From every evidence, direct as well as inferential, it may be safely concluded that the land has risen and fallen twice since the Christian era; and that each movement of elevation and subsidence has exceeded twenty feet.

---

§ 5. Of the Italian Islands.

THIS term, in chorographical parlance, does not include the isles which lie near the coasts of Tuscany or Naples; but is especially applied to Corsica, Sardinia, and Sicily, with their dependencies. The two first lie north and south of each other off the west coast of Italy, in nearly a straight line, stretching across the sea between Genoa and Tunis for 80 leagues; and though separated only by the Strait of Bonifaccio, exhibit many remarkable differences, both moral and physical. Corsica may be considered as one vast mass of granite rising to a height of 8700 feet above the level of the sea, with a *banda* or plateau on each side, of about 310 miles in circuit; it possesses a great variety of minerals, and is clothed with forests of oak, beech, fir, cedar, cork, ash, and chestnut trees. It holds the third rank among Mediterranean islands, and its produce is corn, wine, oil, olives, legumes, carobs, fruits, silk, honey, wax, marbles, coral, tunny, botarghe, and salt.

Corsica.

Sardinia.

Sardinia, although from the small sinuosity of its coasts it has a circuit of little more than 500 miles, is upwards of 140 miles in length from north to south, with an average breadth of 60 miles; and, as I have elsewhere said, until I had myself established the admeasurement, I considered Sicily, from a very prevailing error, as the largest of the

Mediterranean islands; and though the difference is trifling, I now subscribe to the assertion of that very early hydrographer Scylax, who is somewhat technically called, by my venerable friend Major Rennel, 'the Pilot,' and who, according to Cluverius, says, 'Maxima est Sardinia, secunda Sicilia, tertia Creta, quarta Cyprus, quinta Eubœa, sexta Corsica, septima Lesbus.' It is a much lower island than Corsica, few of its mountain summits exceeding 3000 feet in height; and Gen-Argĕntu, its culminating peak near the centre, being only 5276 feet. A chain of primitive rock runs from north to south down the east side of the island, but there is a large volcanic district extending through its centre, and jutting out in many places to the west coast. This is its great physical diversity from Corsica, and the principal moral peculiarity may be said to be, that Sardinia still retains the feudal system; but there are also various points of difference in other respects which struck me during my occasional visits, some of which are recorded in my published account of that island. Their produce however is similar, as well as their fisheries, and both have their coasts indented with excellent bays, harbours, and roadsteads; unhappily there is another point of resemblance, since all the low grounds are fatally infected in the summer months with intemperie, or malaria, which in some spots is truly deadly.* The trading places of Corsica are Bastia (*Mantinum*), Porto Vecchio, Bonifaccio, Ajaccio, Calvi, and San Firenzo: of Sardinia the principal ones are Cagliari (*Caralis*), Sassari, Alghero, Oristano, San Pietro, Ogliastro, Terra-nova, La Maddalena, Longo-Sardo, and Castel-Sardo: the eastern shores of both islands are less indented with bays than the western, and it is especially so with regard to Corsica; where the sea is receding

---

* For instance, in my Sketch of Sardinia (page 295) I mention the pestilential atmosphere on the western coast, as authorizing the oft-repeated proverb:—

A Oristano che ghe va,
In Oristano ghe restà!

from that side, so much that Alleria, once a Roman seaport, is now upwards of a mile inland.

Sicily.

The island of Sicily is separated from the continent by the celebrated Faro, or Strait of Messina (*Stretto Mamertino*), where every appearance justifies the popular belief that a violent disruption, or subsidence of strata, has taken place at some one remote period. Although this island is actually in surface rather smaller than Sardinia, it has a circuit of 550 miles of winding coast, and is commonly deemed as of greater importance; and may be so justly, whether its geographical position, historical celebrity, climate, or produce, be the circumstances considered—in each and in all of these, it assuredly is one of the most interesting and important islands in the world. As is implied by its ancient name, Trinacria, it is terminated by three remarkable promontories, and intersected throughout by ranges of hills, between which are valleys and plains of the most exuberant fertility; but unhappily these are the usual seats of malaria, some of them being notoriously pestiferous. None of the hills are of any very considerable height except Mount Etna, the most remarkable volcano of Europe, which rises to the elevation of 10,874 feet.* It appears from the sea, on every side of the island, like a vast dome towering over all the other mountains.

Ports of Sicily.

The ancient ports of Palermo (*Panormus*) have been filled up in comparatively recent ages, but it possesses a very capacious mole harbour, by which a great traffic is

* I must here show a gratifying coincidence in the determinations of the several stations on this mountain, made by two observers without any knowledge of each other's operations:—

	Smyth, 1814.	Herschel, 1824.
Grotta delle Capre	5,362	5,423·6 *English feet.*
Bishop's Snow-store	7,410	7,103·8
The English House	9,592	9,592·7
The summit of Etna	10,874	10,872·5

Several other places were evidently as corroborative, but that the observations were taken at different spots; as, for instance, my height of Nicolosi is 2449 above the sea at the Convent, and Herschel's 2,232·8 at Gemellaro's House. (See CAPTAIN BASIL HALL'S *Patchwork*, vol. iii. ch. 3.)

maintained. Besides its numerous caricatori, or authorized loading-places, and its artificial ports, Sicily boasts the fine harbours of Messina (*Mesana* vel *Zancle*), Augusta, Syracuse, Trapani (*Drepanum*), and Milazzo (*Mylæ*); and there are also various other excellent road-anchorages for the largest ships. Hence there is a busy traffic; and, notwithstanding the moral causes which deaden the physical energies of its resources, the exports are still both valuable and various. An enumeration of its principal branches of trade sufficiently proves this, for they consist of corn, wine, oil, fruit, manna, honey, wax, saffron, carobs, liquorice, sumach, marbles, sulphur, nitre, barilla, salt, linseed, amber, cantharides, coral, cork, flax, rice, silk, hides, soap, cheese, squills, rags, cotton, wool, madder, orchil, timber, fish, botarghe, tobacco, and all kinds of leguminous vegetables. Produce.

Hard beaches.

The arenaceous and shingly conglomerate to which I have alluded in my Memoir on Sicily, occupies a great part of the beaches around the island; and is very observable between Cape Granitola and Sciacca. Indeed a compound of this kind, replete with shells, fills up the hollows in most of the older rocks in Sicily. On facts connected with this, I have also stated that it is unlikely that the Faro of Messina has increased in width for many ages; from thence to Scaletta, the beach is generally hardened into a compact conglomerate of which small mill-stones are made on the spot. This may probably proceed from the water which percolates through the fiumare, holding carbonate of lime in solution, and precipitating travertine.

Lipari islands.

On the north coast of Sicily are situated the Lipari Islands, so long famed as the *Æolian*, or *Vulcanian* group; consisting of Alicudi (*Ericodes*), Felicudi (*Phœnicodes*), Salina (*Didyme*), Lipari (*Lipara*), Vulcano (*Hiera*), Panaria (*Hicesia*), Stromboli (*Strongyle*), and some smaller rocks. The islands are generally precipitous and bold-to, except in the vicinity of Panaria; but as all the dangers are marked on the chart, they are easily avoided. The group is entirely volcanic, and yields sulphur, nitre, alum, arsenic, pumice, various salts, and specular iron. Lipari is the largest, most fertile, and best inhabited; but Stromboli is the most remarkable on account of its unceasing eruptions, which have gained it the name of the lighthouse of the Mediterranean. A feature in the hydro-geology of these islands may be noticed. Most of them present steep cliffy fronts on the west, and descend in moderate slopes on the eastern side; the former plunging at once into deep water, and the latter offering a gradual suite of regular soundings. A similar peculiarity is also found in many other parts of the world.

Ustica. To the west of the Liparis, and north of Palermo, lies Ustica, a small but well-cultivated volcanic island, where the best barilla of these markets is prepared. Off Trapani are the isles of Maretimo (*Hiera*), Levanso (*Bucinna*), and Faviguana (*Ægusa*); and to the south of Sicily are Pantellaria (*Cossyra*)—a place of exile for state delinquents,—the uninhabited Linosa, Lampedusa—Prospero's enchanted island—and Lampion Rock.

Every one is aware of the vastness of phlægrean eruptions in these districts, and the numerous extinct as well as active cones exhibit the cause of a long series of changes. And I think the term 'volcanic agency' may be also applied to those emissions of mud, petroleum, and sulphuretted hydrogen, of which an example is given in my account of Sicily, at the Maccaluba near Girgenti (page 213).

Malta. Malta (*Melita*) also was anciently an appanage of the crown of Sicily, but having been granted by Charles V. to the Knights of the Order of St. John of Jerusalem on their being driven out of Rhodes in 1530, its superb harbour was impregnably fortified as a barrier against the Turks, and it became one of the most celebrated spots of modern times until within a century ago, when unequivocal symptoms of moral degeneracy began to be manifested. From its position between Sicily and Africa, it commands the channel which connects the two great basins of this sea, and was therefore, in the recent struggles, too important a station to be left in the paralyzed hands of the Knights. In 1798, the French made an almost unresisted conquest of it; but after a blockade of almost two years—vigorously maintained by the inhabitants themselves on land, and by the British fleet at sea—they were compelled to relinquish it to the English, to whom it was finally ceded at the general peace of 1815. By Scylax, the Maltese isles were reckoned among the appendages to Carthage, and *Melita Africana* was distinguished from Melita Illyrica; but by a British act of parliament, Malta is now included in Europe, notwithstanding that the customs, language, and simple mode of life of the natives are a very decided evidence of their affinity to the Arabs in Barbary.

The island of Malta—as well as its dependencies, Gozo (*Gaulus*), Comino, and Filfla—is composed of calcareous rock, abounding in petrifactions and fossil remains, and generally of an undulating surface, but with some hills, as the Benjemma range of above 500 feet, and the Guardia of Gozo, which is 570 feet high. There is sufficient local evidence that these isles, with a present circuit of about sixty miles, have lost much by disintegration. The utmost industry has been exerted to fertilize every interstice among these otherwise sterile rocks, the soil being—except in a few favoured spots—not more than eight or ten inches deep; and their *campi artificiali* afford proofs of laborious industry. Yet moral energy here overcomes physical defects, for the isles are exceedingly productive, although their corn is barely sufficient for a five months' subsistence of the numerous population; the fruits are finer than those of the adjacent countries, and luxuriant crops of sulla (*Hedissarum coronaria*) form the most substantial and nourishing fodder for horses and cattle. But the principal branch of their industry is the cultivation and manufacture of cotton, of which the two best kinds—the Gallipoli and Nankin—

the one white, and the other yellowish brown, have a staple combining length and silkiness in a superior degree. We refer the reader to the close of this chapter, for further statistical particulars respecting this colony.

---

§ 6. The Adriatic Sea.

ALTHOUGH archæology is not our object in this sketch of the borders of the Mediterranean, a geographical definition of the term Adriatic must be given, lest it might be supposed that the views of Veryard, Giorgi, and Bryant, can be assented to by one who has critically examined both Malta and Meleda.

Adria. The Adriatic Sea is considered to have gained its name either from the very ancient city of Adria, or Hadria, now some fifteen miles inland, at the farther end of the Gulf of Venice, or from Atri in the Abruzzo; but the latter seems untenable. The name is, perhaps, first mentioned by Herodotus, who, however, seems to apply it rather to the country around the coasts than to the sea itself; although he asserts (*Clio*, 163) that it was first of all explored by the Phocæans. Thucydides tells us (lib. i.) that Epidamnus, now Durazzo, is a city on the right hand as you sail into the Ionian Gulf; it is the Hadriacus Undas of Virgil, while Horace makes the *Arbiter* Adriæ wash the Calabrian coast; and Pliny, who calls the Adriatic the second gulf of Europe, expressly places Cape Lavinium and the town of Croton—both of Calabria Ultra—on its shores. Strabo describes the Iapygian and Ceraunian shores as the line of separation in these divisions. He admits that the mouth or strait belongs to both, yet it is obvious that the lower part was colonized from Ionia, the upper from Adria; the name therefore of the first part of this sea is termed Ionian, and the

Adriatic gulf.

inner part up to its recess, Adriatic; 'but now,' he says, (*circa* A.D. 18,) 'the latter is the name even of the *whole* sea:' and this statement is strengthened by the fact of the Gulf of Venice being called the Upper Sea (*mare superum*) by the Latin writers. In a splendid copy of Ptolemy, lent to me by his late R.H. the Duke of Sussex, which was printed at Rome in the year 1478, Mare Adriaticum appears in uncial characters on *tabula secunda*, in the space between Sicilia and Corcyra; on *tabula sexta* it is below Bruttium and Messene; on *tabula septima* it is marked in the offing of Leontium, in Sicily; and on the tenth plate it is opposite to the space between Zacynthos and the Strophades.

Hence it is evident that the Adriatic Sea was held to be that vast expanse of waters contained in the Upper, the Ionian, and the Sicilian Seas—in fact, that it extended both to the north and south, from the narrows which some have chosen to assume as its mouth. But these were convertible terms; for, as we have just seen, Thucydides cites the position of Epidamnus as in the Ionian waters, and St. Paul's ship was driven up and down in Hadria: the Adriatic Sea, says Heschius, 'is the same with the Ionian Sea,' a definition that might have suppressed arguments which have been conducted with more vehemence than judgment. The upper portion of this space, so appropriately designated the Gulf of Venice, is of moderate depths—from twelve to twenty fathoms between Istria and Venice, and about 100 near its centre; between this and the entrance there is a basin which has upwards of 500 fathoms; and at the narrows between Otranto and Valona are 350 fathoms, deepening suddenly towards the Ionian Sea. The southern division of the Adriatic is as yet unfathomed; at least, I have had occasion, in searching for reported dangers, to try for soundings with from 400 to 700 fathoms of line, without at any time striking the bottom. The flat lands around this sea are subject to malaria in summer. Depths.

Cape Spartivento, or wind-splitter (*Herculeum prom.*), Eastern Calabria.

is the south-eastern extremity of Calabria, and between it and Cape Santa Maria di Leuca, the coast is indented by the bays or gulfs of Squillace (*Scylleticus Sinus*) and Taranto, with the petty ports and coves of Gerace (near *Locri*), Catanzaro, Cotrone (*Croto*), Strongoli (*Petilia*), Roseta, Cesareo (*Sasina*), and Gallipoli (*Callipolis*). Taranto (*Tarentum*), seated in the north-west angle of the gulf named after it, was once the rival of Rome, and had an excellent port at the mouth of a fine river, which becoming choked up from neglect, commerce deserted it: still, however, it boasts nearly 20,000 inhabitants, and derives some consideration from its fisheries. Indeed, the bays of the whole coast, from the Faro of Messina, abound with excellent fish.

Hard beaches.

The Calabrian beaches offer many specimens of arenaceous conglomerate with the calcareous cement, so largely occurring in Sicily. Off Cape Rizzuto, a two-fathoms shoal may possibly be the remains of the *Ogygia* vel *Calypsus Insula* of Pliny, which, with four others on this coast, are considered as having been swallowed up. On the west side of the Gulf of Taranto there are symptoms of the sea's having receded from the coast, owing to the alluvia carried down by the rivers, and the marine deposits thrown up. This is well shown at the margin of the once fertile plains of *Metapontum*, between the rivers Bradano (*Bradanus*) and Basiento (*Casuentus*); where a square tower called Torre di Mare, built by the Angevine kings as a station for coast-guards, is now above a mile distant from the shore.

Changes of coast.

Gulf of Venice.

Doubling Cape Santa Maria di Lucca (*Iapygium prom.*), under which a black rock called Maleso marks the boundary of the Bay of Taranto, we enter the Gulf of Venice by the narrow mouth denominated the Strait of Otranto. On examination, the navigator will perceive the wide difference which exists between the two sides of this sea, the eastern shore being generally rocky, replete with islands and ports of bold approach, but deficient in inhabitants, provisions, and, in many parts, also in potable water; the western coast, on the contrary, is comparatively shallow, and almost without any ports of capacity, yet—excepting some parts of Puglia—populous, and abounding in provisions, water, and articles of trade.

The west side of the Adriatic is bounded by Italy, in the beaches of which are frequent specimens of the calcareous concretion above-mentioned; and here the soundings are more regular than on the opposite coast, with an approach of considerably less boldness—a consequence of the main current's setting along the shores of Albania, Dalmatia, and Istria, and returning by those of Friuli, Venice, Romagna, the Abruzzi, and the Capitanata. Besides numerous road-anchorages between Otranto (*Hydruntum*) and the mouth of the Po, there are the ports of Brindisi (*Brundusium*), Monopoli (*Egnatia*), Bari (*Barium*), Barletta, Manfredonia (*Sipontum*), Viesti (*Apenestæ*), Ortona and Ancona —which retain their ancient names and sites—Sinigaglia (*Sena Gallica*), Fano (*Fanum Fortunæ*), Pesaro (*Pisaurum*), Rimini (*Ariminium*), Comacchio, Chioggia (*Fossa Claudia*), and some places of less note, to which busy coasters resort. The exports are corn, rice, legumes, fruits, vegetables, oil, wine, cotton, wool, silk, manna, salt, hemp, cheese, soap, timber, glass, and liquorice. The great lakes between Peschichi and Termoli (*Interamna*), named Lesina (*Lacus Pantanus*) and Varano (*Portus Garnæ*), have immemorially been celebrated for the abundance, variety, and excellence of their fish; but their borders are unhealthy.

East coast of Italy.

Trading ports.

Produce.

The uniformity of this western line of coast is broken in three principal places,—namely, first, at Testa di Gargano, or Mount Sant' Angelo (*Promontorium Garganum*), near which lie the four Tremiti Isles (*Insulæ Diomedeæ*); secondly, at Mount Conero (*Cumerium prom.*), between Loretto and Ancona; and thirdly, at the Delta formed by depositions at the mouths of the Po (*Eridanus* and *Padus*).

River Po.

This river, which has been celebrated even as 'rex fluviorum,' has its source in the Grison Alps, and after flowing from west to east for more than 280 miles, discharges itself into the Adriatic by seven different channels, sometimes, during freshes, with such violence, that Tasso says it carries war, not tribute, to the sea. At these times the Po renders the water of the sea brackish, to a considerable distance out, by diluting its saltness: at which opportunities our frigates cruising off Goro used to replenish their water by skimming the surface just out of gun-shot. The ravages of this river have made a great exertion of hydraulic engineering requisite for preventive purposes; and the embankments, rendered absolutely necessary by its repeated deposits, have raised its bed many feet above the plain through which it flows, keeping the whole country of Ferrara and the Polesino in constant fear of a flood; and therefore it is a perpetual source of anxiety and expense.

The Apennines. It should be observed that, between the heights of Ancona and Mont Sant' Angelo, the celebrated chain of the Apennines—the true mountain-system of Italy—runs nearly parallel to the sea-line of the Abruzzi, and comparatively near; thereby influencing the seasons and agriculture of the intervening space. The population of the plains may be said to have nearly reached the utmost verge of subsistence; but the slopes of the mountains are extremely populous, and the immense forests of sweet chestnuts maintain a great proportion of the inhabitants of the district—and this at an elevation where no food for man could be procured in our climate. Among the summits seen from this part of the Adriatic, the peaks of Monte Corno (*Precuti*), or the Gran Sasso d'Italia, 9500 feet high, and Monte Majella (*Palenus*), nearly as high, are very striking.

Venice. The territory of Venice extends from the northern mouth of the Po, across the head of the gulf as far as the bay of Trieste; the greater part of this extent is composed of low marshy islets and lagoons, formed by the many neighbouring rivers; as the Brenta (*Medoacus major*), the Adige (*Athesis flumen*), the Piave, the Tagliamento (*Tilaventum*), &c. In the midst of these rises the once potent city of Venice, between which and the Adriatic —however frequently the gorgeous ceremony of marriage has taken place—there may now be said to have been a lasting, if not final divorce, although a semblance of union is still maintained. This singular capital is built upon half a hundred little islets or banks, consolidated by piles, and intersected in every direction by canals and smaller watercourses (*canaletti*), serving the purposes of streets and lanes, but navigated by the gondola; which, notwithstanding its sable, hearse-like appearance, is a commodious, well-furnished boat, swiftly sculled by a single waterman.

Beyond point Sdobba, the land becomes steeper and the water deepens, as we approach the territory of Istria; which is bounded on the north-west by the river Isonzo, and on the south by the Gulf of Quarnero; and unlike the parts we have just left, has a coast which is generally bold, broken, and irregular, with a mountainous interior formed by an offset of the Julian Alps. Such, at least, are the hydrographical boundaries; but the portion usually designated the Peninsula, is only about forty-five miles long. The chief occupation of the inhabitants, who are mostly of Sclavonian origin, is agriculture; they also attend to some minor manufactures, and to fishing. The country produces in small quantities, oil, wine, fruits, corn, honey, wax, silk, leather, tallow, timber, and salt; and has abundant quarries of freestone and marble, whence excellent lime is abundantly obtained by calcination. **Istria.** **Produce.**

The chief loading-places of Istria are Castel Duino (*Pucinum Castellum*), Trieste (*Tergeste*), Capo d' Istria (*Ægida*), Pirano, Parenzo (*Parentium*), Orsera, Rovigno, and Pola; there are also many smaller resorts. Of all these, Trieste, a flourishing seaport, claims the pre-eminence; having lately become the commercial victor of Venice, and **Istrian ports.**

the most important of the Austrian port-marts. It has a very secure artificial harbour, and an outer road with a moderate depth of water; but this anchorage is exposed to the wind from the west and south-west quarters, and specially subject to the violent gust called Bora. Now

Pola.

Pola is a roomy haven, with many advantages, yet, from the malaria on its shores, it is all but deserted; and it is not a little singular, considering its position, that it should be exempt from the visitations of the Bora. The tonnara, or tunny-fishery, claims its chief marine attention, and is a source of considerable profit. Most of the tunnies are sent fresh to Venice, but the surplus is eviscerated on the spot, cured with the highly-prized Istrian salt, and packed off for general markets. It is singular that Pola still preserves its most ancient name, for that which the Romans gave it —*Pietas Julia*—has long disappeared.

Croatia.

The east coast of the Adriatic, from Cape Promontore to Ragusa, is more bold and picturesque than that of Istria, being flanked by numerous islands, some of considerable size, and others mere rocks; here exhibiting productive cultivation, there neglect and barrenness. The water between them is deep, and the shores mostly bold and precipitous, insomuch, that a fleet may generally work to within half a cable's length on either side; and sailing among the sinuous channels of the Quarnero is easy and pleasing, except that the Bora often renders it an unsafe navigation—nor are the gusts off Monte Maggiore, or Caldero (4530 feet high), to be disregarded. Between Istria and Dalmatia lies the State of Croatia (*Liburnia*)—*Horv'áth Ország* of the natives—of which the principal coast-towns along the Morlachian shores are Fiume—the great seaport proper of Hungary—Porto Re, and Karlopago. Between the two latter is Segna, the whole plain of which was evidently once a harbour.

Monte Maggiore.

Gulf of Quarnero.

The Gulf of Quarnero (*Sinus Flanaticus* or *Liburnicus*) takes its name from the four principal islands, Cherso

(*Cripsa* or *Crexa*), Veglia (*Curicta* or *Cyractica*), Arbe (*Scardona*), and Pago (*Cissa*, afterwards *Paganorum insula*), the two latter of which are close to the mainland. Cherso is joined by a causeway bridge to the island and mount called Osero (*isle*), whence they are generally considered as one; their soil is uneven and stony, but they abound in cattle, vines, olives, and honey; and possess, among other ports, the fine harbour of Lossin Piccolo, which is at once spacious and land-locked. Veglia is the largest, as well as the most fertile and populous of the Croatian group, though some of the arable grounds of Arbe are in greater esteem. Pago is not more noted for the tortuosities of its form than for the extraordinary variability of its climate, and the wildness of the inhabitants. It is singular that those neighbours—Cherso, Osero, Lossin, Canidole, and Sansego—should abound with fossil bones;* and there are symptoms of the whole of the islands having once joined the continent. It should also be recollected Osero was formerly *Absorus* and *Auxerum*; whence the immediate isles of the vicinity were called Absyrtides.

Veglia.

Pago.

Leaving Croatia, we enter the province of Dalmatia, the government of which, including Ragusa, extends from Obrovazza—south of Karlopago—to Lastua, beyond Budua, the last being properly in Albania, but that there is about the Bocche di Cattaro a sort of variable frontier. The whole district is mountainous and generally barren; though towards the interior there are extensive forests of timber. Its chief ports are Novigradi, on a sea lake; Zara (*Jadara*), the fortified capital, with a spacious harbour, well-furnished arsenal, and 8000 inhabitants; Scardona on the Kerka, which flows into the Adriatic near Sebenico, after forming

Dalmatia.

Zara.

Scardona.

* The bone-breccia of these islands appears to be identically the same conglomerate with those of Gibraltar, Cerigo, and other places in the Mediterranean. A large collection had recently been sent from Lossin Piccolo to Vienna, just before my arrival in that port, in which Dr. Capone, my informant, found relics of oxen, deer, and other rodentia.

several cascades and five magnificent falls in its course of about fifty miles: Sebenico, on the declivity of a rocky hill, near the lake of Kerka (*Titium*), is a well-built town of 4000 inhabitants, with a castle; Ragosniza, a good port, with a poverty-struck village; Trau (*Tragurium*), a town on the main, with a suburb on the Isle of Bua, having 3000 inhabitants, the two connected by a mole with a draw-bridge for the passage of vessels; Salona, still enjoying its ancient name; Spalatro (*Palatium*), one of the most commercial ports of Dalmatia, is a fortified city, with a population of 8000 people, bearing many traces of former prosperity and Diocletian's munificence; Almissa (*Oncœnum*), at the mouth of the river Cettina (*Nestus* or *Tilurus*); Macarska (*Rataneum* or *Rhœtinum*), a little, open town, with a small port; Fort Opus and Sabioncello, on the shores of the gulf into which the Narenta empties itself; the once powerful maritime city of Ragusa,* and its splendid canal of Calamota; Ragusa Vecchia (*Epidaurus*); Cattaro, with its unique and noble broad of waters, called *le bocche*—formerly *Sinus Rhizonicus*—meandering amid precipitous mountains; and the small ortified town and harbour of Budua (*Buthoe*).

Towns and ports.

Aspect and produce.

The mountainous tract at the back of these towns—for the most part wild, rugged, and barren—is industriously cultivated towards the shore. A general want of water, with an arid soil, however, render Dalmatia unsuitable for agriculture, and therefore it was of old better known for piracy than for commercial enterprise; yet it has long exported considerable quantities of corn, wine, oil, figs, almonds, cheese, salt, wool, brandy, maraschino and other liqueurs, honey, fruits, sardines, and tunny. There is much timber in the interior, but the forests near the coast have

* Ragusa politically ceded certain portions of her territory to the Turks, in order to avoid a more dangerous intimacy with Venice. From the qualities and disposition of the inhabitants, the city has been termed the Paris of the Adriatic.

been exhausted. A principal feature of the whole is the range called Montenegro (*Czernagora*), consisting chiefly of the cretaceous or Mediterranean limestone, so extensively developed from the Alps to the Archipelago; and which is commonly remarkable for its bare and craggy character. The general height is about 3000 feet, with a few higher summits; and the slopes are gentle in the direction of the inclination of the strata, with precipices at the outcroppings, which give a fine variety to the scenery. It is inhabited by a race of hardy and warlike mountaineers, who have managed to maintain their independence between the Turks, whom they abhor, and the Austrians, for whom they care not; and no man moves without his gun and poniard. They are under the rule of a Vladika, or Prince-Bishop, by whom I was received, in 1818, with marked kindness and hospitality, in the fortified convent of Stagnevitch, on an elevated slope on the south side of Mount Giurgvitch. This was the celebrated Peter, of the clan Petrovitz, who succeeded to his dignity so far back as the year 1777; and who so heroically defeated Mahmoud Pasha in 1795. His dominion is perhaps the only independent country in Europe which does not contain either a town, or any village, or cluster of habitations, large enough to be compared to one, although with a surface of more than 400 square miles. Conformity of religion, decorations of knighthood and presents from the Emperor Paul, and the distant position of the power, combined to induce the Montenegrini to prefer the friendship of Russia to that of Austria; and the politic predilection was increased by the favours and courtesy of the Emperor Alexander. They are, however, surrounded by extremely jealous, and even inimical neighbours.*

Montenegro.

* It was here that the existence and views of the secret society, called the Hetæria, for the emancipation of the Greeks, were revealed to me, and which, as the trial-outbreak was to occur in the Ionian islands, I was in duty bound to disclose to the British government, through Sir Thomas Maitland. This certainly placed the projects of the Hetærists in peril, but

Dalmatian islands.

The numerous islands appertaining to this line of coast appear to have originated in the breaking up of the lower grounds, by some violent action, leaving their limestone-summits above water: from the salient position of the promontory terminating in Point Planca, they are divided into two distinct groups, which the Greek geographers designated the *Absyrtides* (above-mentioned) and *Liburnides* (Strab. vii.). They trend north-west and south-east, greatly longer than broad, and form various fine channels, here called *canale*, and named from the nearest adjacent island; which being bold, with scarcely a hidden danger, give a variety of secure passages for ships between them. These islands are generally poorly supplied with potable water, and some of them suffer greatly from the want of it; they are, therefore, not fertile, although scantily affording oil, wine, honey, wax, olives and other fruit. Some of them are miserably off, so that I found many families unable to afford themselves the use of bread, except on festivals. The principal islands are Scardo, Grossa (*Lissa*), Incoronata, Zuri (*Crateæ*), Solta (*Olynta*), Brazza (*Brattia*), Lesina (*Pharos*), Lissa (*Issa*), Curzola (*Corcyra Nigra* vel *Meloena*), l'Agosta (*Tauris*), Melada (*Melita*), and many smaller ones; replete with ports and harbours, some of which are upon an extensive scale. South of Lissa, and nearly in the centre of the Adriatic Sea, stands the rocky

Pelagosa.

isle of Pelagosa, and west of the latter is Pomo, a pyramidal rock 100 feet high, with a dangerous shoal off its north end: as Pelagosa is a very important sunset point of departure for all passing Adriatic traders to take the bearing of, I requested of Baron Prochasca, in 1818, on stated grounds, that a lighthouse might be erected there by the Austrian Government, which has been done.

---

no choice was left me; and within sixteen months afterwards I was called upon to co-operate with General Sir Frederic Adam, in suppressing the dangerous insurrection at Santa Maura. This was precisely what I had been informed, although the plea was about a new tax.

Between Dalmatia and the Gulf of Lepanto, the eastern shores of the Adriatic are formed by the coast of Albania, which, in the greater part of its northern portion, is of moderate height, and in some places even low and unwholesome, as far as Valona, or Avlona (*Aulon*), where it suddenly becomes rugged and mountainous, with precipitous cliffs descending rapidly to the sea. This is the Khimára range, upwards of 4000 feet high, once much dreaded by ancient navigators as the Acro-Ceraunian promontory. Some of the inland hills are so clearly seen over the intervening land between Durazzo and Avlona, that many vessels coming down this sea have been deceived, and consequently wrecked on Samana, the shelvy point formed by the river Toberathi, or Krevasti, the *Apsus flumen* of yore. Among those heights none is more remarkable than Monte Pegola, which has an altitude of 7764 feet; and is perhaps thus named from being near the beds of asphaltum, or mineral pitch, of Selenitsa. Albania. Mount Khimara. Point Samana.

The coast of Albania, though its limits are not strictly defined, is generally held to extend from Antivari on the north, to the Gulf of Lepanto on the south: the space between the former and Avlona answers to the ancient Illyricum, and Lower Albania to Epirus; both are still inhabited by a desperate race, who usually are at once soldiers and robbers. The principal Adriatic ports of this district are—Antivari, which is thought to have been named from being nearly opposite to Bari, in Italy; Dulcigno (*Olcinium*), long a nest of pirates, who by means of the river Boïana, frequently ravaged the shores of Lake Scutari; Alessio (*Lissus*), a town of fishermen on the banks of the Drino (*Drilo*), the largest of the Illyrian rivers, which communicates with the Ocrida Lake (*Lychnidus Palus*); Durazzo (*Epidamnus*, postea *Dyrrhachium*), a fortified town at the head of an excellent bay for anchorage; Valona, a very tolerable town on the east side of a spacious and beautiful gulf, which is rendered Ports.

additionally secure by the isle of Sasseno; and Port Palermo (*Panormus*), a fortified cove at the foot of the Khimára range. Of these places, Valona is the first in maritime consideration, since the bay will accommodate fleets with anchorage, water, wood, provisions, fish, and refreshments in abundance; and its exports are timber, gall-nuts, corn, oil, wool, mineral pitch, and salt. The site of *Oricum*, which stood in the southern part of this bay, was (1818) occupied only by two or three huts, among vestiges of an aqueduct.

Valona.

Palæste.

Between Valona and Port Palermo, the coast is indented with numerous little coves, which were heretofore the resort of piratical vessels lying in wait for their passing prey: but none of them recalled the 'quiet station for ships amidst the rocks and dangers of the Ceraunian coast,' which Cæsar describes (*Bel. Civ.* iii. 6). On the hills above Aspri Rouga—Strada Bianca of the Italian pilots—is Paleassa, which may be the site of *Palæste*, from which Cæsar marched in one day to Oricum, and took it. Near it is Chimara (*Chinuera*), which gives name (*Chimariots*) to the inhabitants of the whole mountain-range.

The basin of the Adriatic Sea seems to be a continuation of the original trough-shaped longitudinal valley of the Po; separating the parallel ranges of elevated secondary strata of the Apennines, and of the Illyrian mountains. The head of this gulf consequently receives all the waters that flow from the southern descent of the Alps and the mountains of Carniola, between the Po and the Isonzo; a space in which the sea also receives the Adige, Brenta, Piave, Livenza, the Tagliamento, and numerous minor streams, each carrying down, in freshes, enormous quantities of mud and gravel into the lagoons, or vast extent of shallows which border the intervening shore. By these means Aquileia, which once may have stood near the sea, has long been an inland town; Adria, which was a station for the Roman fleet, is now more than fifteen miles inland; and Ravenna, formerly

Geological changes.

on piles surrounded by lakes and saltpans, and only bearable from being purified by the tide, as Strabo says, is now in the midst of gardens and meadows; while *Portus Classis*, its ancient harbour, has become a marsh four miles from the sea, from which it is separated by the celebrated Pineto, or forest of pines. Spina, with its adjacent *Ostium*, a Pelasgic town at the most southern branch of the Po, was, in the time of Scylax, about two miles and a half from the sea; but in less than six hundred years afterwards, Strabo describes it as being ninety stadia, or more than eleven miles inland: nor could Strabo or Pliny find any vestiges of the two islands called *Electrides*, which the more ancient historians placed at the mouth of the Po,— or of the amber from which they derived their name.

Still, although the draining of so large a portion of the Alps and Apennines may, with the successive depositions of the sea, have formed the greater part of old Lombardy,* and though there are many circumstances favourable to the encroachment of the land on the sea and rivers, I am not inclined to think the increase has been so great or so rapid as some of my Italian friends have inferred. In mentioning ancient ports, it is not always meant that they were close to the sea; swamps, ditches, and stagnant pools formed, in fact, the principal feature of all the tract in question; and there is nothing to prove that these marshes were ever covered by the Adriatic within the period of history. The lagoons may have been contracted, but Padua, as in the time of Livy, is seventeen miles from the

---

* In the recent operation of boring for an Artesian well at Venice, four different beds of peat were passed through, at the respective depths of 18, 29, 48, and 126 metres; proving that at four different epochs, the surface, which appears to have been slowly subsiding, was covered with fresh-water lakes, of small depth. Again, at Adria, where it has been shown by excavation that it stands on the ruins of two former towns, the progress of alluvial deposits which may have occurred within a space of 3000 years, is demonstrated: the first town arrived at being on a level several feet below the present surface, exhibited Roman vestigia; the second, at a greater depth, appeared to be earlier, the futile fragments being wholly Etruscan.

sea, while Brondolo and Chioggia remain the same as described by Pliny; and even at the delta of the Po, which has so vast a power in transporting mud and silt, a comparison of the best old map of Ferrara, shows an increase of about twenty-five yards per annum between the years 1200 and 1600, latterly accelerating; certainly a very considerable rate of increment, but greatly inferior to what is now occurring, for instance, at the head of the Gulf of Persia.

Assumed river.

One more point. Scylax, who we may also remark is followed by Scymnus Chius, assumed a very disproportionate extension of the Adriatic, placing its innermost angle near the Ister, an arm of which falls into it; and Pomponius Mela assumes that Istria thus derived its name. Apollonius Rhodius—the Alexandrian poet, and no great authority in such a case—makes Jason's fleet fly before that of Æta, across the Euxine, up the Ister, and thence into the Adriatic; and the Abbate Fortis thought he perceived in the fluviatile sands of Sansego and Ossero, evidence of the arm which Jason descended. Having inspected the locality, I am as much surprised at Fortis as Pliny was at Cornelius Nepos, for believing in the existence of such a river. Aristotle seems to have believed that the fish called *Trichias* passed from the Danube into the Adriatic.

§ 7. The Shores and Islands of Western Greece.

Western Greece.

THIS designation is applied generally, rather than in a strictly geographical sense, to the space between Avlona and Cape Malea, including the Septinsular and the Greek Islands—all of which have had, in our times, a resurrection from that political death with which, for many ages, they had been struck, and which, Greece being the land—κατ' ἐξοχήν—of enthusiastic aspiration, gave joy to every civilized country.

The principal Albanian ports on the Ionian sea, after quitting Port Palermo, and passing the loading-place under Agioi Saranta (*Onchesmus*), are first—Butrinto (*Buthrotum*), opening into the Channel of Corfu; Gominitse, near the mouth of the Calamis (*Thyamis*); Mourtzo, the outer isle of which still bears the name of *Sybota;* and then the once piratical Parga (*Torone*), which, with other Venetian possessions, was ceded to the Porte in March, 1800, by a treaty, in which England had no part; but, from circumstances, this cession became obligatory on the English. It forms no portion of my present plan to revive particulars, but as I was in Sir Thomas Maitland's confidence, and was actually at Parga after it was left by its inhabitants, as well as before Ali Pasha's troops were admitted, I feel it proper to state, that the accounts given of that unfortunate event in the *Edinburgh Review** are completely erroneous. **Albanian ports.** **Parga.**

Between Parga and Previsa there are the little ports of San Giovanni and Phanari, the last being at the mouth of the Glyki (*Glykis limen* and *Acheron*), and on the margin of the mountainous district of Suli, emphatically named Kakosouli by the Turks, from the calamities and evils they encountered in its subjection. Previsa (near *Actium* and *Nicopolis*), the chief commercial place of Lower Albania, stands at the mouth of the Gulf of Arta (*Ambracius sinus*), a sheet of water navigable for vessels of the largest size when the bar is passed. Near the south-eastern extremity of the gulf, and on a hill in command of the Port of Kervasara, are the Cyclopian walls and other remains of *Argus Amphilochicum;* and from thence round by Ruga and Vonitsa to the western point (*Anactorium*), the whole shore exhibits traces of former importance. The sanitary condition must at that period have been better, and the country morally more safe; for now, in addition to a veer-and-haul government, the gulf is unhealthy during the **Phanari.** **Gulf of Arta.**

* No. LXIV. Vol. 32.

summer months, at which time remittent fevers of a dangerous type are common, especially in the lower grounds.

The state of the pavement of a Roman road on the northern shore of the gulf, with indications marked by the clay-levels, and other signs of submergence, give an idea of local depression. Politically, the centre of this gulf is now the boundary between Turkish Albania and the new kingdom of Greece; but agreeably to local hydrography, we shall continue the name of Albania to Lepanto, although the portion between Previsa and the River Aspropotamo (*Achelous*) is named Karnia (*Acarnaniæ*). On this coast we meet Port Kandela (*Alyzia*); the excellent Bay of Dragomestre (*Astacus*), once crowned by a large town and fortress, but now nearly deserted; and then the embouchure

Achelous.

of the Aspropotamo, the most considerable river of Greece. Herodotus described it as gradually connecting the Echinades with Acarnania, 2300 years ago; and Thucydides predicted (l. ii. § 102) that as the river was rapid, and brought down great quantities of sand, those islands must in time form part of the mainland. The distance between them and the main has become considerably contracted since that prediction was made; and the present designation of the Achelous—'White River'—is from the turbid tint of its muddy waters, which so whiten the sea around the Kurzolari (*Echinades*) group, that I was somewhat startled on first sailing through them. Oxia, the largest of these, and the nearest to the mainland, was probably the *Dulichium* of Homer; but I can find no confirmation for the conjecture. To the south of this stream lies one of its

Missolunghi.

accretions, Port Scropho; and to the east of it is Missolunghi (*Melitepalus*), at the entrance of an extensive salt lagoon, having Natolica (*Cyniapalus*) at its head. With sufficient littoral advantages for active commerce, the Albanians have hitherto confined their energies to piracy and a petty trade in timber, oil, wool, valoni or dye acorns, fish, botarga, and general provisions.

We now enter the Gulf of Lepanto, or Corinth (*Sinus Corinthiacus*), a sheet of water above seventy miles long from east to west, and about twelve or thirteen miles broad towards the middle, exclusive of the gulfs of its northern inlets under Parnassus and Delphi; having bold shores, carrying from seven to ten fathoms close in, and a central depth of more than 250 fathoms—no bottom having been struck with that quantity of line out. The entrance of this gulf is defended by two castles on projecting points, which are not much more than a mile and a half distant from each other, and are known as the Dardanelles of Lepanto (*Rhium* and *Antirrhium*); the town of which name (or rather towns), once *Naupactus*, stands on the side of a hill a little within the northern castle called Roumili—properly Rúm-ili-Kisár, i.e., castle of the Roman (Greek) land. To the east of this place, and on the same shore of the gulf, are several bays, affording good anchorage for large vessels, as Salona (*Crissa*), Galaxidi (*Tolophon*), Aspra-Spitia (*Anticyra*), Port Sarandi (*Mychos* vel *Tiphæ*) under Mount Zagora (*Helicon*), Dobrena (*Thisbe*), Ghermano (*Ægosthena*), and Livadostro (*Creusa*). The eastern extremity of the gulf terminates in two bays; that of Livadostro on the north, that of Corinth (*Corinthus*) to the south, where the Morea is joined to Greece by an isthmus, over which, and in the city of Corinth, the air is so bad, that all those inhabitants who can, abandon the place during the summer months. From the wretched dogana,* which is the sole representative of the once busy *Lechæum*, to the Morea castle, at the mouth of the gulf, is a comparatively depopulated district; yet at *Sicyon*—the ruins of which are on a hill between two streams, the ancient *Asopus* and *Helisson*—luxury was in the ascendant, and there the fine arts took their birth, as

Gulf of Lepanto.

Depth.

Ports.

Corinth.

* Before the insurrection, my friend, Kyamil Bey, ruled in Corinth, where his family had governed for above a century, during which the district was as prosperous as any in Greece. I received much attention from him, and regretted his fate.

was well testified by its illustrious citizens, Zeuxis, Lysippus, Apelles, and Timarchus. The present chief place of consideration is Vostitsa (*Ægium*), from whence the produce of the adjacent country is conveyed in boats to Patras. When I was there in 1820, it was a prosperous and busy town, the landing-place of which was marked by a plane tree of forty feet girth, and 100 feet in height, around which were fourteen brass cocks for supplying water from the purest of springs; a few months afterwards, the whole was desolated by the Turks, even to the destruction of this noble tree. Between Vostitsa and the Dardanelles of Lepanto, the extensive sandy point of Drepano induced a distinguished antiquary to think that it marks the site where Bura and Helice were swallowed up, as mentioned by Pausanias (*Archaics*, ch. xxiv.), and Ovid (*Metam.* lib. xv.); but I see no reason to doubt that catastrophe's having occurred at the base of Mount Meliala, to the east of Vostitsa, according to the usual supposition. In all the lower grounds of this district, malaria is to be expected; and the fine vale of Kalavryta is singularly unhealthy in the fall of the year.

Vostitsa.

Bura and Helice.

Ionian Islands.

Directly off the coasts of Lower Albania and the Morea lie the Ionian Islands, or Septinsular Republic; a group formerly subject to Venice, but occupied during the late wars by different belligerents in succession, and finally assigned by the Congress of Vienna to the protection of Great Britain. The United Ionian Islands—in order of constitutional precedence—are Corfu (*Corcyra*), Cephalonia, Zante, Leucadia, Ithaca, Cerigo, and Paxo, together with their numerous dependent islets. Their population amounts to about 200,000, all of whom have toleration in religion, and equal rights in the eye of the law: and these assuredly form the basis of true liberty. The Ionian flag bears the lion of St. Mark, but with that proof of especial protection—the British union—in the upper angle; the appearance of which in those seas, was the signal of dissolution to various hordes of pirates.

Corfu.

Corfu, called Korphí (in the plural) by the present

Greeks, is the seat of government, and though of rugged surface, abounds with olive-trees, and has some very fertile, but unhealthy plains,* producing corn, oil—which is its chief export, wine, fruit, and flax; and salt is obtained in considerable quantities by desiccation in some extensive and shallow lagoons, which communicate with the sea. The anchorage is at once roomy, convenient, and secure; but Port Govino, the former arsenal, has become so unhealthy, from the increase of malaria through defective drainage, that its use is discontinued. The island is about thirty-five miles long by twelve miles at its greatest breadth; it is extremely picturesque, the west shore being an abrupt precipice, with exuberant foliage overhanging the sea. Off its north end there are some rocky islets, the most important being Fano (*Othronus*), which has sometimes been called the key of the Adriatic.

Port Govino.

Fano.

Eight miles south of Corfu, and about ten miles west of Epirus, is Paxo (*Paxos*), the smallest of the septinsular group; steep and rocky, but well covered with olive-trees, producing the best oil in the Ionian Islands. Quitting its excellent little port, Gaio, and passing the almost desert isle of Anti-Paxo, we arrive at *Leucadia*, or Levkádhia, an island about sixty miles in circumference, which has long been called Santa Maura by the Italians; it is very mountainous, yet, being cultivated in every possible part, is tolerably productive, and has a considerable export of wine and oil. The north-east extremity of the island is separated from Acarnania only by a narrow channel, which is supposed to have been cut by the Corinthians when Leucadia was a peninsula. The slip of land thus severed is called the Placca, and resembles a work of art, but it is a body of gravel and sand cemented by calcareous matter into so compact a mass, that excellent mill-stones are made from it. The strong castle of Santa Maura stands close to this, and is divided from

Paxo.

Santa Maura.

The Placca.

* The large and fertile plain called Val di Roppa, is delineated as a spacious harbour in the early maps.

Amaxiki, the head town of the island, by extensive lagoons, which are crossed by light canoes, appropriately called monoxyla. Among the dependent islets, Meganisi (*Aspalathia*) holds the first place, as its name imports; but since, in the insurrection of 1819, it became a station for spies, I was under the necessity of disarming the inhabitants, and, for a time, restricting intercourse with their neighbours.

Cephalonia.

Cephalonia (*Cephalonia*) is the most considerable in extent of all the Ionian Islands, being 180 miles in circuit, and its coasts are indented with deep bays and ports, of which the harbour of Argostoli is the most important, being spacious enough for the largest fleets, and secure in all winds; on it stood the very ancient cities of *Palle* and *Kranii.* The highest elevation of the island, and indeed of the Ionian group, is that anciently named Mount *Ænos*, the Montenero of modern geography, which is 5300 feet above the sea; it was formerly clothed with a fine forest, of which vestiges still remain, but the greater part was wilfully burnt by the natives. The destruction of timber during the conflagration was enormous; and though the fire occurred before the occupation by the English, the mountain still presented a singular appearance of desolation when I visited it in 1820; nor should it be forgotten, that by this wanton ruin an injurious effect is considered to have been made on the climate. This lofty mountain crosses the island, the ramifications of it spread over the whole space, and jut out into the sea in various parts, forming bold headlands; while, among the lower projections, the valleys are tolerably well cultivated, producing currants, oil, cotton, fruit, wine, brandy, and liqueurs: their corn suffices only for half the annual consumption, the deficiency being supplied from the Morea; but as there is a pretty sure market for their currants—which not unfrequently amount to upwards of 4,000,000 lbs. in one year—they are not dissatisfied.

Ithaca and its islets.

The small island of Ithaca stretches along the north-east side of Cephalonia, divided by a narrow channel clear of dangers. It is a rugged, broken, calcareous mountain, yet carefully cultivated in all places of promise, and producing excellent currants, wine, and oil, which are embarked at the secure and—as implied by the name—deep port of Vathi. Between Ithaca and the mainland there are numerous little uninhabited islets which afford pasturage for the sheep and goats of the Ithacan peasants; the principal of these are Atoko, Provati, Pondico, Modi, Mokrì, and Oxoi. The northernmost of these may have been the *Taphii* or *Teleboæ* of Homer, and the southern are the Kurzolari group already mentioned. Ithaca has always been called Itháki, or Theaki, by the natives, thus unequivocally retaining its ancient name; but the Italian geographers have dubbed it Val di Compare, and Cefalonià-piccola.

South of Cephalonia, and opposite to Castel Tornese in the Morea, lies the fine and fertile island of Zante (*Zacynthus*), about seventy miles in circumference, with a population of upwards of 40,000 souls. The aspect of Zante is highly picturesque: two chains of mountains, and the sea to the south, enclose an extensive plain of about ten or eleven miles in length by nearly eight broad, and beautifully interspersed with villages and country-seats. It is entirely covered with gardens and vineyards, producing corn, wine, oil, fruits, vegetables, and the currants so renowned for their excellence, as well as the quantity annually exported—abundant years yielding above 6,000,000 lbs. Moreover, the springs of mineral pitch at the end of this plain—opposite the small isle of Marathonisi (*Marathe*)—visited and described by Herodotus so long ago as the fifth century before our era, are still skimmed for economic purposes, so that about 100 tons of bitumen are procured from them every year. The epithet, however—ὑλήεις, *nemorosa*—of Homer and Virgil, is no longer applicable to Zante, the only wood on the island being the olive-groves on the great plain. The town is on the north side of Monte Scopo (*Elatos mons*), and has a capacious mole. **Zante.**

Stamphane. About fourteen miles to the south of Zante, lie the two small islets vulgarly called Stamphané or Strivali; on the largest of which there is a strongly-fortified convent, with a capital garden, and an abundant supply of excellent water. They were anciently assigned to Elis, under the name *Strophades;* but they are now the property of Zante.

Cerigo. Quitting Zante for the remaining island included in the Septinsular Republic, we have to sail upwards of forty leagues to reach Cerigo,* the ancient *Cythera*, a mountainous island with well-cultivated valleys, and a fair produce in corn, wine, oil, cotton, fruit, cattle, sheep, and goats. Midway between Cerigo and Candia are some lesser dependent islets, of which Cerigotto (*Ægilia*) is the only one of consideration. We must now return to the mainland; but further statistical details of the Seven Islands will be found at the end of this chapter.

Cerigotto.

Though the arts, sciences, virtues, and glories of Greece have waned, and many of her river-gods have nearly exhausted their urns, still her soil, her climate, mountains, and valleys, remain as of old; and though the sometimes mawkish ecstasy of the classic enthusiast may be eschewed, he is not to be envied who can traverse such a country without emotion. Thus, from vivid recollections, the whole shores of the Gulf of Corinth which we have so recently passed, teem with interest to scholars, patriots, and artists; and we return to the southern castle of the Dardanelles of Lepanto, to resume the clue with invigorated incitement.

The Morea. The Morea (*Peloponnesus*) has been compared in shape to a mulberry-leaf; but more probably it derived its name from the Slavonian word for maritime, though others insist that it is so called from this country having been the first to which silkworms, with the *morus*-tree, were imported

* This Italianized orthography is here used, because it is adopted by the Ionian government, though Tzerígo (pronounced Cherigo) is used by the modern Greeks. It is a Slavonian name, introduced by the settlers called Tzacones, in the eighth and ninth centuries.

from Persia. Its coasts are deeply indented with bays and islets: while the interior forms an elevated table-land, traversed by numerous ridges of hills, which enclose spacious basin-plains; and there are also other extensive and fertile grounds, producing, even under imperfect culture, corn, cotton, silk, oil, flax, tobacco, gums, galls, currants, and most other fruits. Timber is obtainable, notwithstanding the lamentable devastation which the forests have undergone, in great part by the wanton rapacity of the inhabitants themselves; and fine pines, planes, chestnuts, and oaks, still clothe the inland mountains, especially in Arkadhia (*Arcadia*). The acorns of the *Quercus ægilops*, sold as a mordaunt in dyeing black, and known as the valania of commerce, are exported in considerable quantities.

From the Morea Castle, a sandy beach turns south-westward to the landing-place and mole of Patrás (*Patræ*), which beautifully-situated city stands on an elevated ridge projecting from a declivity of Mount Voidhiá (*Panakaicum*); but the grounds around are unhealthy in the season of malaria. The semicircular bay here presented is bold and clean, with gradual soundings and good anchorage all along, where the largest ships can ride in perfect safety. Rounding Cape Papas, the south-west extremity of the Gulf of Patras, a deep bay is found between it and Cape Tornese, in the bottom bight of which are the Venetian ruins of Klarenza, near the ancient *Cyllene*,—a place said to give their title to the English dukes of Clarence. From Cape Tornese (*Chelonatas prom.*) to the next projecting headland, Cape Katacolo, the shores are low and wooded, and on the summit, near the middle, stands Castel Tornese, an old Venetian fortress, with an inconsiderable village; below it there is a creek that affords occasional shelter to small vessels, into which the Iliaco (*Peneius*) discharges its waters, after flowing through the vestiges of Palæopoli (*Elis*), and past the unhealthy town of Gastúni (*Œnoe?*).

Patras.

Castel Tornese.

Passing Cape Katacolo (*Ichthys*), and standing to the south, we enter a large open bay called the Gulf of Arkadhia, on the southern part of which the shores rise in a succession of woody hills, while the northern merges into the low maritime plains of the Eleia, through which flow the river Ruféia and its tributaries. This stream, the principal river of the Morea, is the ancient *Alpheius*, the name of which is but slightly traceable in Ruféia; it has its source a little beyond Mount Pholöe, and flowing in a westerly direction through the vales of *Elis* and *Olympia*,* disembogues into the sea through a marshy beach, below the thriving town of Pyrgo (*Pyrgi*). South-eastward of this mouth, and below the Skala Rufeia, are the extensive lagoon fisheries of Giagiapha and Kaiapha: where, by good engineering, this river might be made to facilitate the trade of the interior. But the unwholesomeness of the air must not be forgotten.

Alpheius.

Prodano.

Passing Cape Konello (*Cyparissium prom.*), the southern point of the Gulf of Arkadhia, and the coast-isle of Prodano (*Prote*), we arrive at the excellent and spacious harbour of Navarino (*Pylus*), formed by the main on the east, by the peninsula of Paléo Avarino on the north, and in front by the long narrow isle called Sphagia (*Sphacteria*), which defends it from the sea-winds. The entrance is at the south end of this isle, nearly opposite to the new town and fortress of Navarino (*Neo-Castro*), which stand on a promontory running out from the foot of Mount Lykódamo (*Temathia*).

Navarino.

Four miles and a half to the south of Navarino is the fortified town of Modon (*Methone* prius *Pedasus*), with the islands of Sapienza and Skhiza, or Cabrera (*Œnussæ*), near it. The space between this and Cape Gallo (*Acritas*

Modon.

---

* From the probability of votive offerings to the Alpheius, at the Olympic games, and other sacrifices, my late friend, Sir Patrick Ross, then Governor of Zante, was desirous of promoting a party to drag that river in 1820. The Greek outbreak, in that very year, thwarted the project.

prom.) forms a large bay, carrying deep water very nearly to its respective shores. Off Cape Gallo lies the isle of Venetico (*Theganusa*), with the Mourmaki rocks (*Thyrides*) to the south of it; and at about thirty-four miles south-eastward of the latter is Cape Matapan (*Tænarum prom.*), forming the headlands of the Gulf of Korón (*Messeniacus sinus*). Near the head of this gulf the rivers Bias and Pyrnatza (*Pamisus*) empty themselves, not far from the town of Kalamáta (*Calamæ*); between that and Matapan, the most noted places are Kitries (*Gerenia*), Porto Vitylo (*Œtylus*), and Djímova, or Tzimova: the two last forming a united firm of very free traders, who dread nothing but the south-west gales, which beat dead upon their shores, and shut their vessels in,—yet these co-acting 'powers' cordially hate each other. The two sides enclosing this expanse of waters present widely different features: on the west stands the city of Korón (*Æpea ?*), one of the most commercial places in the Morea, in a fertile country covered with olive grounds and gardens; while the eastern side is formed by the mountainous and craggy declivities of Mount St. Elias, or Makrynó (*Taygetus*), inhabited by the Maïnotes, an unsubdued people of the *Braccio*, or district of Maïna, the tract we are treating of. The shores are exceedingly bold-to, there being a depth of 120 fathoms at a short distance from the shore; and at a league out I found 479 fathoms. The coast of Maïna, on both sides, is serrated with coves, and inaccessible retreats, which till very recently were the resort of the most determined and barbarous pirates in the Mediterranean.

Koron.

Vitylo.

Mount Taygetus.

Maïna.

Geology.

Respecting the geological condition of this neighbourhood, M. Bobbaye, in his memoir on the alterations produced by the sea on calcareous rocks along the shores of Greece, by examining the littoral caverns worn in the limestone cliffs, and noticing the lithodomous perforations, came to the conclusion that there are four or five distinct ranges of ancient sea-cliffs, one above the other, at various elevations in the Morea, which attest as many successive elevations of the country. There is a volcanic mass at Modon, which was described by Strabo, resembling the Monte Nuovo near Baiæ, in Italy.

Kolokythia. Rounding Cape Matapán, and between it and the isle of Cervi (*Onugnathus*), lies another spacious gulf, namely, that of Kolokythia (*Laconicus sinus*), with water altogether deep, there being no bottom to be found with 350 fathoms of line at two leagues from the shore, the lead only striking ground at a near approach. The Maïnote shore, by Porto Káio or Quaglio (*Amathus*), and other coves, presents a cheerless assemblage of rugged precipices and rocky mountains. But after passing Marathonisi, or Fennel Isle (*Cranaë*), we arrive at the head of the gulf, where the Eurotas. Vasíli-potamó (*Eurotas*), or Royal River, discharges itself, near the three islets Trinisi (*Trinasi*), after its course through the long valley that slopes between the two ranges of mountains, which detach themselves from the central highland near Megalopolis, the principal town. These projecting into the sea, form the promontories of Matapan (*Tœnarium*) and St. Angelo, or Kavo Malea (*Malea*). The Vasíli-potamó — sometimes called Irí —is navigable for boats at its entrance; and on the banks of a small tributary about eight leagues inland, is the city of Mistrá (*Messe* of Homer), an important post for the defence of the ancient *Sparta* or *Lacedœmon,* from which it lies about four miles, nearly due west. The best anchorages on the east side of the gulf, are Port Rupina (*Asopus*) and the Bay of Vaticà (*Bœaticus sinus*), between Cervi and Cape St. Angelo.

§ 8. The Archipelago, Black Sea, and Levant.

IN consequence of the hydrographical treaty I made at Paris in November, 1820, which is particularized in Part IV., my own acquaintance with the regions above named was restricted to a couple of cursory visits along the east coast of the Morea, the shores of Attica, some of the outermost of the Cyclades, and the west end of Candia. But as I possess exact information from such authorities as Gauttier, Beaufort, Graves, and other experienced friends, the rapid sketch here given may be relied upon; at least, so far as it serves to fill up the proposed outline of the Mediterranean Sea.

East coast of the Morea.

From Cape St. Angelo (*Maléa prom.*) the east coast of the Morea commences; and passing the singularly situated town of Monembasía, commonly called Napoli di Malvasía (*Minoa* near *Epidaurus Limera*), it trends northward to the Gulf of Nauplia (*Argolicus sinus*); and as the city of Argos is a mile and a half inland, the principal town and port of the gulf is Napoli di Romanía (*Nauplia*), with a fortress of some strength at the foot of Mount Palamides, admirably placed both for defence and commerce. The eastern side of this gulf—along the shores of the ancient Hermione—has many bays and islets; and between it and the Gulf of Ægina, are the barren islands of Spetzia and Ydhra or Hydra (*Tiparenus* and *Aperopia*), with their dependent rocks, among which the most remarkable is that of Poro (*Calaurea*). Here the activity and industry of the natives have wrung advantages from sterility, proving the triumph of moral over physical action; and when I visited them, although nearly all Greece was in a state of torpor, they had formed a kind of independent republic, and were

Spetzia.

Hydra. the carriers of a large portion of the Levant trade. Hydra then had upwards of 4000 excellent seamen, and about 150 ships, of which no fewer than 80 were of 300 tons burden and upwards, and most of them well manned and armed. Such was the rock of which it was said, that its layer of soil was so thin, as not to afford the Hydriotes sufficient earth to bury their dead.

Ægean Sea. From the Gulf of Nauplia we will now run round the coasts which form the periphery of the Ægean Sea, and afterwards glance at the numerous islands with which that space is studded. The name of this sea, by the way, has undergone various corruptions. Tradition delights in referring its designation to the death of Ægeus; but Strabo deduces it from an islet called Ægæ (Αἰγαι). Some derive it from Αιγαιον πελαγος, fancifully assumed to mean the *Goat Sea;* but the Venetians of the Levant seem to have first used the term 'Arcipelago,' for it does not appear that the Greeks ever used such a word, nor is it likely they would. From that came the general use of Archipelago; and this led to the Arches of English sailors!

Gulf of Enghia. Between the capes of Skyllo and Colonna (*Schyllæum* and *Sunium*) the Gulf of Enghia (*Saronicus sinus*) separates the Morea from the continent of Greece on the east. It is serrated with bays and good anchorages, the most frequented of which are Kalavria (*Calaureia*), Pidavro (*Epidaurus*), Kenkries (*Cenchreiæ*), Kalamáki (*Schœnus*), Salamis. Koluri (*Salamis*), and the famous Porto Leone (*Peiræus*) of Athens. Nearly in the middle of this interesting gulf, stands the Enghia of Venetian seamen; but Dapper is right in saying that name is not known to the natives (*Archipel.*, p. 138), among whom it seems to have always borne its ancient name, *Ægina.* It is a hilly island with fertile valleys, about six leagues in circumference, and with Peiræus. a population of nearly 4000. Porto Leone, with its dependent coves (*Munychia* and *Phalerum*), though small, and exporting little except oil, is a very convenient and completely sheltered port for a limited number of vessels of size, which can ride in from four and a half to nine fathoms water; at least, such were the soundings early in 1820, and there is not much alteration to be apprehended from the small percolation of the supplies now yielded by the nearly exhausted urns of the Cephyssus and Ilyssus. The air, however, of Athens is singularly dry and elastic, delicious and wholesome.

Rounding Cape Colonna and the Temple of Sunium, we pass the long rocky Macronisi (*Helena*) and its port of Mandri; and at about three leagues more to the northwards, we enter Port Raphti (*Prasiæ*), the finest harbour on the Athenian coast, taking its modern name from a sedent statue, considered to be in a tailor's attitude. Still further on, we reach the beach which margins the famous plain of Marathon, and arrive at the boundary of Bœotia. Lying along these coasts, is the island of Negropont (*Macris post Eubœa*), the south-east point of which is Cape Mandili (*Geræstus prom.*), and the north-west is Cape Lethada (*Cenæum prom.*). They are about ninety miles apart, the intervening land being elevated; insomuch that Mount Elias, above Karystus, is 4750, and Mount Delphi (*Dirphe*) 7300 feet above the sea. Negropont is separated from the main by the Egripo (*Euripus*), a channel so narrow at about half its length, that the two shores are connected by a bridge; whence it has been inferred that Eubœa was torn from the Bœotian coast by an earthquake, or some other convulsion. Running along the outer and precipitous, iron-bound shores of Negropont, we find the narrow channel of Trikhiri separating Thessaly and the northern end of Eubœa; the form of which will be best understood by a reference to the charts. Trikhiri is a busy commercial town on the eastern shore of the entrance to the great gulf of Volo (*Pagaseticus sinus*); but leaving this on our right, and sailing westward, we enter the bay of Zitúni (*Maliacus sinus*), in the south-west angle of which, and on the side

Cape Colonne.

Negropont.

Euripus.

Trikhiri.

Thermopylæ. of Mount Œta, is the celebrated pass of *Thermopylæ*. Passing the Lithada islets, and turning down the channel Talanta. of Atalante or Talanta (*Opuntius sinus*), the depth of the water appears to correspond with the height of the land, for under Mount Telethrius no bottom is found with 220 fathoms of line within half a mile of the shore It then shoals gradually towards Egripo, where the channel is only 100 yards wide; south of this it again opens out, and there are several good anchorages on the coast of Negropont, especially among the Petalio (*Petaliæ*) Islands.

Skyro. Off the middle of the outer coast is the island of Skyro (*Scyros*), which is, as some imagine Σκῦρος implies, both rugged and rocky; but its shores are bold-to. North-Khillidromi islands. west of Skyro, and off the Trikhiri Channel, is the Khillidromi (*Peparethus*) group of islands, trending to the north-east across the entrance of the Gulf of Salonica, to the extent of about forty miles. These are Skiatho (*Sciathus*), the westernmost isle, Skopelo (*Scopelus*), Khelidromi (*Halonesus*), Sarakino, or Peristeri (*Eudemia*), Seanghero (*Skandyle*), Pelago (*Solymnia*), Ioura (*Ios*), Piperi (*Peparethus*), and several other stony and uninhabited islets.

Returning to the main, north of this rocky group, and between Cape San Dimitri (*Sepias prom.*) and Kassandra (*Posidium prom.*), an extensive gulf penetrates into Macedonia; at the head of which is the important harbour Salonica. and *entrepôt* of Salonica (*Thessalonica*), a city of 60,000 inhabitants, and one of the most commercial places in Turkey. The coast which forms the west side of this fine inlet exhibits a magnificent range of mountains, including Mount Pelion. Plessidi (*Pelion*) 5200 feet high, Kissavo (*Ossa*) 6100 feet, and Elymbo (*Olympus*), which last is 9850 feet above the sea. The river Salambria (*Peneus*) runs through the Bogaz, or celebrated vale of *Tempé*, thence between the bases of Elymbo and Kissavo, and into the Gulf of Salonica: from the exposure of the strata in its course through the hills, it is suggested that some great cataclysm broke through the range, and drained the great basin of Thessaly: this being the process by which Nature has often contrived to get rid of lakes, and to substitute dry lands in lieu of them; *factas ex æquore terras*—thus, in time, yielding bread in place of fish.

Eastward of this gulf, and separated from it by the peninsula of Kassandra (*Pallene*), the deep bay anciently called the *Toronaicus sinus* is passed, and that of Monte Santo (*Singitius sinus*) entered. The latter gulf is divided from the following one of Contessa, or Réndinà (*Strymonicus sinus*), by the Agion-oros, or Holy Mountain of the modern Greeks, the famous *Athos* of their forefathers. This has been considered an extraordinary mount in all ages; among the ancients, from the fancies of Xerxes, Dinocrates, and those who told of its extravagant elevation; and among the moderns, from its numerous churches, monasteries, and monks. Its height is now pretty well known to be about 6500 feet, and its summits are seen even from Cape Sigeum and the plain of Troy. Its precipitous slopes descend at once into an almost unfathomable sea, as from 80 to 100 fathoms are carried to within a quarter of a mile of it; and many trading vessels keep under sail while embarking their cargo of nuts and other fruit. **Monte Santo.**

The waters between Monte Santo and the Dardanelles are broken into two large divisions by the mountainous island of Tasso (*Thasos*); and the mainland is again indented by the Gulfs of Ænos (*Stentoris palus*) and Saros (*Melus sinus*). The former receives the River Maritsa (*Hebrus*) the source of which is in the Balkan mountains (*Hæmus*): it is navigable for large boats to Adrianople, the second city in Turkey. Off Saros lie the abrupt islands of Samothraki (*Samothrace*) and Imbros; and in the mid-distance, between Monte Santo and the Dardanelles, is the quadrilateral Stalimini, or Lemno (*Lemnos*), with two ports, of which that on the south has considerable capacity, though it has not yet obtained the importance which its maritime **Thasos.** **Lemnos.**

population would appear to demand. About six leagues to the south-west of this port lies the isle of Ayio-Strati (*Nea*), with a village and roadstead, and the very small adjoining islets, called Roubos and S. Apostoli, whence it had anciently a name in the plural number—*Neæ*. It was sacred to Minerva, but latterly has only been notable for its export of velanídi, or valanía.

Running our coast-directory eastward from Saros, we shall pass the mouth of the Dardanelles without stopping for the present; but going between the straight coast which trends from Cape Janizary (*Sigeum*) to Cape Bábá (*Lectum*), Tenedos. and Taushan-ádássi (*Lagussæ*) or Hare isles, and Tenedos—which still retains its ancient name—we pass the sandy beach skirting the Plain of Troy, extending to the base of Mount Gargaráh (*Ida*), which is 5700 feet high. Here commences the Anatolia. western coast of Anatolia, the country of the East anciently called Asia Minor, which stretches away to the southwards as far as Cape Krio (*Cnidus*), and beyond it. Throughout its whole extent the shore is indented by a rapid succession of bays and coves, sprinkled with islands and islets, and teeming with anchorages and loading-places, though there is little trade except in timber, oil, wool, and valanía. Some Scala. of these stations have *scala* as an addition to their names: a term signifying ladder or stairs, because such aids were common in harbours. This term is so much used in the Levant that the phrase *fare scala*, in the language of the native seamen, means to touch at any port; and in many places on this coast, where the shore is very steep, there were flights of steps cut in the rocks, to facilitate landing.

Among the bays and ports of this interval, the first and most important, in a maritime and commercial view, is Smyrna. Smyrna, the third city in Turkey, at the head of the fine gulf of that name. Here the population is estimated at not less than 70,000; and it is the great emporium of the Levantine trade. The nations of Europe have each a consul resident at Smyrna; and there has long been a distinct quarter of the town allotted to the European inhabitants, who are under the especial protection of their respective consuls, and enjoy great privileges. The beautiful suburb of Bournabad is interspersed with handsome houses, in the midst of gardens and vineyards. Between Smyrna and Cape Krio are several deep bays, the first being that of Scala-nova (*Neapolis*), between which and the ruins of *Claros* on the north, the river Mendere (*Caystrus*), after flowing under the remains of *Ephesus*, enters the sea: then follow the Gulf of Mandeliýah (*Bargyliacus* vel *Jassicus sinus*), into which flows the tortuous Maddro (*Meander*); the Gulf of Kos, or Boudrúm (*Ceramicus sinus*); and the Gulf of Doris (*Doridis sinus*). These, from the abundance of fine havens and anchorages, possess extraordinary capability, although, in a maritime and commercial sense, they are greatly neglected; but nothing can lessen their interest to the antiquary and the scholar, for the whole space teems with vestiges of ancient skill, energy, and power.

Candia

Such are the shores which bound the waters of the Archipelago on the west, the north, and the east. The south is marked by a range of islands extending in a crescent from the Morea to Asia Minor, with its convexity towards the Levant Basin. The western branch of this curve has the large island of Candia, or Kriti (*Crete*), as its bulwark, an island which, though mountainous, is fairly cultivated and very productive. In the centre rises Mount Psitoriti, from 'Υψηλορείτιον (*Ida*), which rises to the elevation of upwards of 6700 feet, with a bare summit, but the ramifications are covered with forests; and it not only serves as an excellent landmark to sailors, but is also a means of ascertaining the state of the atmosphere, and consequent weather at sea. The north coast is serrated with ports and bays, but the south side presents nearly a rugged front to the on-shore winds; passing vessels, therefore, are not wont to go between the Gozze isles (*Claudos* of St. Paul), although there is ample room between them and the main. The country is singularly beautiful; but although it affords wine, oil, fruit, cotton, silk, honey, wax, cheese, soap, liquorice, and timber for export, its trade—under Ottoman rule and restriction—is comparatively small in regard to its fertility and capacity. Candia is forty-six leagues long by about ten broad, at its widest part. The principal maritime resorts are Grabusa (*Coryca*), in the Castro or fort of which I obtained permission to make a station; Canea (*Cydonia*); and from thence rounding Cape Maleca (*Ciamon prom.*)—called Acrotiri by the natives—we enter the most spacious harbour of the island, sailing between the fortified islet of Suda (*Leucæ*) and the paleo-castro of *Aptera*. Keeping to the east we find Armyro (*Amphimalla*), Retimo (*Rithymna*), Candia (*Cytæum*), Megalo-castron (*Matium*), Spinalonga (*Chersonesus*), and Sitia (*Etia*); all of which bear substantial evidence of the skill, wealth, and power of the Venetians.

Rhodes. At the opposite or north-east portion of the crescent in question stands Rhodes, a very considerable island, and the key of the important pass which it commands. Its northern shores are low, rising inland to a high and tabled mountain; the southern declivities of which end in a sandy but tolerably fertile soil. On the north-east extremity of the island are its two well-known harbours, over the entrance to the smaller of which stood the celebrated Colossus of brass: but Rhodes is of higher interest in a nautical point of view, on account of its inhabitants—who for ages ruled the Mediterranean Sea—having promulgated a very early code of Laws, which became the standard for the decision of controversies relative to maritime affairs throughout the whole of Europe.

Scarpanto. Nearly midway from Rhodes towards Candia, the barrier between the Archipelago and the south-eastern part of the Mediterranean is completed by the mountainous and arid island of Scarpanto (*Carpathus*) and its dependants, Caxo (*Casos*), Caxopulo, with some smaller rocks: the word *pulo* is of frequent occurrence in modern Greek, and is merely a diminutive form—perhaps derived from the Latin *ulus*—used to express a subsidiary islet.

The ancients, in order to systematize the wilderness before them, and facilitate reference, divided the islands of the Archipelago into two distinct portions—the Cyclades and the Sporades: the former were thus named from their lying in something like a circular position around Delos; the latter signifying dispersed, from their scattered position along the coast of Anatolia. Those on the left of a navigator sailing through the middle of the Ægean towards the Hellespont, were considered as belonging to Europe; those on the right, to Asia Minor. And this has generally been observed, saving that Dionysius Pariegetes, in his geographical hexameters, expressly claims Delos and its neighbours for Asia. Since the recent recognition of Hellenic independence, these groups may be denominated the Greek islands, and the Turkish. **Division of the islands.**

The Cyclades comprehend about half a hundred isles and islets, besides many smaller rocks, of which few only are worthy of consideration. They are generally hilly and arid, with a bleak aspect; but though few of them have many trees, and there is a general sterile appearance, most of their levels and valleys are productive, especially of fruits. The first in modern importance is Milo (*Melos*), having a very capacious harbour, and a nautical population, from whom the pilots for this sea are generally selected. It is of volcanic origin, rising at Mount St. Elias to the height of 2000 feet, and though without running water, is fertile. But Axia, called by the Italians Naxia (*Naxos*), is the largest component of the group, of which it is styled the Queen; it is without a port for shipping, but its surface is diversified by hills, valleys, and plains, and it is tolerably well wooded and watered. Paros, two leagues west of Naxia, so celebrated for its white marble, although possessing the best harbours in this sea, has but a trifling commerce, its exports being **Cyclades.** **Milo.** **Axia.** **Paros.**

confined to a small quantity of cotton, and a little wax and honey. Its finest port is Naussa (*A'gusa*), on the north, but its shores are said to be so very unhealthy, from the malaria of the neighbouring marshes, that it is compara-

Siphanto.

tively deserted.* Siphanto (*Siphnos*) is supposed to be full of mineral wealth—Serpho (*Seriphus*) yields iron—Thermia (*Cythnus*) is valued for its mineral springs and excellent fruits—Policandro (*Pholegandros*), though rocky,

Santorini.

affords good wine—Santorini or St. Irene (*Thera*), which, as well as its adjacent subsidiaries, was thrown up by volcanic agency within the reach of history, some even in the beginning of the eighteenth century, exports wine and clothing—Nio (*Ios*), though rocky and rugged, has an in-

Syra.

dustrious population—Syra (*Syros*) is a place of trading enterprise—Tzia or Yea (*Ceos*) has one of the finest ports in the Archipelago—Tino (*Tenos*) is a rugged but well-inhabited place, in good repute for honesty and industry—

Myconi.

Myconi (*Myconus*), with its commercial population, is divided by a narrow channel from the small but celebrated,

Delos.

and once sacred *Delos*, now Sdili. There are also Argentiera (*Cimolus*), Amorgo, known by its former name, Nio (*Ios*), Stampalia (*Astypalœa*), Ghioura (*Gyarus*), Sikino (*Sicinos*), Polina (*Polyœgos*), Skino (*Schinussa*), and many smaller islets: whilst the most northerly and one of the largest of

Andros.

the Cyclades is Andros—still retaining its ancient name; fertile, but not possessing the advantage of a safe harbour, it was rarely visited by strangers. It is separated from

* Sonnini, who was here in 1780, a very few years after the Russian fleet, under Alexis Orloff, had made the Port their station, says, that 'such is the rapidity with which everything is destroyed in Turkey, that not a vestige remained of the Muscovite works when he was there.' Now, I am assured by my friend, the Rev. G. C. Renouard, formerly our chaplain at Smyrna, that when he visited the place in 1815, the Russian establishments were so little dilapidated, that immediate occupation might have been taken: and even since their violent destruction in the late war of extermination, Captain Graves informs me that the ruins are still very extensive, and easily to be traced! We shall have to return to M. Sonnini in the next chapter for assertions of still greater freedom.

Negropont by a strait named the Bocca Silota, or Capo d'Oro passage; but how 'Bocca Silota' came upon the maps and charts is a mystery, for it is neither Italian nor Greek: the Turkish Bóghâz may have brought Boccasi, but the lota is unknown. Between Andros and Scio, lie the two dangerous rocks called Kaloyeri, which appear to have been thrown up by a volcano.

The Sporades, the other great Ægean group, are scat- Sporades.
tered to the east of the Cyclades, with which in fact several of them are intermixed, where they form, as already mentioned, a chain along the coast of Anatolia, between Samos and Rhodes inclusive; a space comprehending the ancient Icarian and Carpathian seas. Of these islands, the most important, next to Rhodes, is Samos, the inhabitants of Samos.
which are in higher esteem for industry than for honesty; they export silk, wool, fruits, wine, and oil; and the flanks of the snow-clad Mount Keris are clothed with good timber-trees. Pátino (*Patmos*), interesting as the place of the Patmos.
Apocalypse, is rugged and unproductive; Stanco (*Cos*) is exuberantly fertile, and much frequented by traders; Ni- Nicaria.
caria (*Icaria*) is not in very great repute for industry; Kalólimno (*Calymna*) is very mountainous, and celebrated for its excellent honey; and next to it is Lero (*Leros*), a Leros.
stony spot, producing fine fruit. There are also the Kharkí (*Chalce*) islets, Piscopi (*Telos*), Nísari (*Nisyrus*), and many scattered smaller ones. Among these, several trachytic rocks have risen from the bottom of the sea, and added to the general number.

Of the islands on the coast of Anatolia not included in the Sporades, three are entitled to be here especially named; because, together with Samos, Cos, and Rhodes, political geography assigns them to Asia Minor. These are Mytilini (*Lesbos*), a fertile, well-wooded, and healthy island, furnished Lesbos.
with spacious and safe harbours, whence its produce is largely exported; Ipsara (*Psyra*), a barren rock brought Ipsara.
by the maritime energy of its natives into high importance; and thirdly, Scio (*Chios*), covered with beautiful groves and Scio.

gardens, and esteemed the most fruitful and fertile spot in the Archipelago. This terrestrial paradise enjoyed great immunities from the Porte; but the inhabitants having, though reluctantly, joined their brother Greeks in the recent insurrection, were indiscriminately massacred by their revengeful masters the Turks, and the whole island was reduced to one scene of hideous desolation. Indeed the blind and bigoted fury of their Musulman adversaries caused the destruction of Scio to be among the most tragic events of the late dreadful struggle.

Navigation of the Archipelago.

Such is the Archipelago: the navigation of which is easy and pleasing enough in general, most of the islands being high, as well as precipitous and bold-to, with a delicious climate. But a good look-out must be kept, for there are very sudden and fresh squalls; and at times there is much bad and even dangerous weather in the winter. In such cases, the waves, having little room to extend themselves, make a confused sea, rising to a considerable height, and breaking with fury against opposing coasts and rocks. Moreover, there is a very great depth of water between the isles—usually no bottom with 150 fathoms of line out, at a short distance from the shore. These interesting islands are thinly peopled, and some of them may indeed be considered as scarcely inhabited. There is, however, an animated traffic, the imports being suited to the wants and wishes of the islanders, to most of whom necessity has given a seafaring disposition; while their own moderate but diversified exports consist of corn, wine, oil, raisins, olives and other fruits, honey, wax, wool, silk, cotton, sponges, iron, alum, pitch, turpentine, sulphur, salt, timber, mastic, gall-nuts, kermes, and velanidi. These articles enable them to supply much which is wanted by their more wealthy neighbours.

Produce.

Geological changes.

Besides the volcanic ravages at Santorin (*Thera* prius *Calliste*), other great geological actions and reactions have taken place in this sea. The long valley through which the Meander makes its tortuous course, was clearly once a gulf, reaching through the present brackish lake called Thalassa Bastarda (*Latmus sinus*), which washes the vestiges of Heraclea; besides, the whole of its soil consisting of both sea and river deposits, affords further evidence of the fact. By the action of the waters of this river* it is that the isle of Laide, where the Athenian fleet took up a station A.C. 412, must have become part of the great alluvial plain before Miletus, at the spot where, between the remains of that city and the present beach, a hill rises upwards of 300 feet above the general level: and the inhabitants both of Miletus and Ephesus were repeatedly obliged to change the sites of their towns, and follow the receding sea. Pausanias (*Arcadics*, ch. xxxiii.) says: 'There was an island, Chrysæ, at no great distance by sea from Lemnos; where they say, that in this island the misfortune from the hydra happened to Philoctetes. The waves have overwhelmed this island so that it has entirely disappeared, being lost in the abyss of the sea. But there is another island, called Hiera, which at that time did not exist. So temporary are human things, and far from being durable.' The isle of Minoa, on the coast of Megara, is lost; and the harbour of Kos has been filled up, as conjectured by Sir F. Beaufort, from the action of the two great currents, the one sweeping westward from the Levant, the other descending from the Dardanelles: these two meeting here, deposit the soil and materials with which they are fraught. Besides the silting up of Kos, they have raised an extensive alluvial point. But in earlier days this region is suspected of having experienced changes of a still vaster character; for many geologists have been of opinion, that the islands of the Ægean Sea are really only the summits of a country submerged by the irruption of the

River Meander.

Laide.

Chrysæ.

Kos.

* Strabo informs us that the Meander was indictable for mischief done to the neighbouring lands by its floods: if any damages were granted against the river, they were paid by those who rented its ferries.

Deluge. Black Sea; and this notion is supported by their general aspect, most of them appearing to have been exposed to the ravages of a violent inundation, which, washing away the soil, left only the denuded surfaces of rock. Two of the floods which may have effected this are on record,—namely, the Ogygian deluge by which Bœotia and Attica were overflowed, and that which ravaged Samothrace and the coast of Asia Minor; and both are usually ascribed to an irruption of waters from the Black Sea. The Samothracian deluge is described in a fragment of the lost work of Strato of Lampsacus, which is preserved in Strabo, and which led Eratosthenes to investigate, though without a satisfactory result, the problem of the uniformity of level in all external seas flowing round continents.

Dardanelles. Returning to the Dardanelles (*Hellespontus*), this beautiful strait, which forms the avenue, as it were, to the Sea of Marmora (*Propontis*), separates Europe from Asia at this particular point; and—unlike Homer's *broad* water—it resembles an immense river flowing majestically between two chains of elevated and exuberantly fertile mountains. It is strongly fortified, and without rocks or hidden dangers, having in some parts a depth of sixty fathoms, but generally eight or nine fathoms within a mile of the shore; it is narrowed towards the middle by the opposite points of Sestos and Abydos, where the strait is diminished from six Gallipoli. or seven miles' breadth to 2700 yards. Passing Gallipoli (*Callipolis*), the principal trading-town of the Dardanelles, Sea of Marmora. we enter the Sea of Marmora, so called from the modern name of Proconnesus, an island to the north of the peninsula of Artaki, formerly the well-known *Cyzicus*. Hence, with a fair wind, no further obstruction is presented to a ship's progress towards the ancient *Demonesi*, now Prinkipos or Prince's Islands, a group lying just beyond the southern point of the entrance to the Canal of Constantinople, and about ten miles distant from that city.

Constantinople. Upon a point of land washed by the Sea of Marmora on one side, and by its port—the far-famed Golden Horn (*Chryso-ceras*)—on the other, stands the grand city of Constantinople (*Byzantium*), the Stambúl of the Turks, on an undulating series of gentle declivities, and with a population of about 600,000 souls, including the suburbs of Galata, Pera, and Topkhána. The Golden Horn has an active trade by land and by water, traffic for which it is exceedingly well adapted, from the facilities of its excellent quays, and its easy ingress and egress. Moreover, the port constantly cleanses itself, for the current which issues from the Black Sea, striking against the Seraglio, or west point of entrance, enters the Horn on one side, and, making a circuit round it, sweeps out again along the opposite shore; this rotatory current, combined with that produced by several streams of fresh water emptying themselves into the head of the harbour, carries off all the silt and impurities which would otherwise damage it and cause obstructions.

Between Constantinople and Scútari or Uskiudár, its suburb on the Asiatic shore, is the entrance of the Thracian Bosporus, now called the Canal of Constantinople: it is here rather more than a mile in breadth, with depths varying from sixteen to thirty fathoms, and the western shores mostly bold-to. From this mouth the channel extends, in a serpentine form, to the Black Sea, a distance of sixteen miles, never narrowing to less than half a mile, with a great mid-channel depth of water throughout, and a stream named *Sheïtan Akandi-si*, that is, Satan's Current, setting southwards, at times very strong. It thus winds like a large river between two chains of mountains, the summits of which are clothed with wood, their sides with cultivation, and their bases with towns, villages, and fortified posts. The tower of Leander, Kiz Kal'eh-sí or Lady's Castle of the Turks, is on a rock in the canal, nearly opposite to Seraglio Point, and just off the point of Scutari; and at the entrance of the Black Sea is a lighthouse on each shore,—one, the Roúm-illi fanár, that is, European lantern or lighthouse,

Thracian Bosporus.

Tower of Leander.

being on the ancient *Panium promontorium*,—the other, that on the Asiatic side, standing on the ancient *Prom. Ancyreum*, so named, it is pretended, from the fragment of rock taken from thence by the Argonauts, to be used as the first anchor. The group of volcanic islets once supposed to float, are here yet, still retaining, among western Europeans, their classic name Cyaneæ; and on one are the remains of an altar, dedicated to Augustus: they were also called the *Symplegades*, and were the terror of ancient navigators.*

Cyaneæ.

Black Sea.

The Black Sea (*Pontus Euxinus*) is an inland basin with a margin of coast generally elevated and rocky, having a transverse diameter of about 650 miles from west to east, a conjugate one of more than 300, and an area of 172,000 square miles. Its modern name is supposed to originate from the dense fogs which occasionally cover it, or the danger of its navigation arising from these fogs: at all events, it was much dreaded by the ancients, who placed their Cimmerian land of utter darkness on its northern shores. Besides the fresh water from Asia Minor, it receives some of the largest rivers in Europe, including the Danube (*Ister*), Dnieper (*Borysthenes*), and Dniester (*Tyras*), the Don (*Tanais*), and the Kouban; its waters are in consequence only brackish; and it is singular that, with such a large and constant accession of fresh streams continually pouring into it, any saltness should be retained. Its depth in general is great, no bottom being struck with 150 fathoms of line; but off the mouth of the Danube the water deepens very gradually, and nearly as much so from Serpent's Isle by Odessa to the Crimea. The streams of the great rivers produce strong currents, particularly in the beginning of summer, when they are increased by the melting of the

Rivers.

Depth.

* There were several islands called the Asiatic Cyaneæ, near the *Prom. Ancyreum*, vaguely mentioned by Strabo, Arrian, and Dionysius Périégetes: according to Petr. Gyllius (*De Bosporo Thracico*, ii. cap. 24) they were rather more than seventy Roman paces (*passus*) from those of Europe. Where are they now?

snows; and when strong winds act against these flowings, a chopping sea is produced, which in foggy weather is dangerous to small craft. Independently, however, of such chances, the Black Sea is free from any dangers; having, with a trivial exception or two, neither islands, rocks, nor reefs in the general track of navigation: and almost everywhere there are excellent anchorages, affording good riding for the largest ships. Its trade consists of grain, wine, timber, charcoal, pitch, potash, fish, caviar, isinglass, shagreen, salted provisions, cheese, poultry, butter, wool, hides, hemp, tallow, honey, tobacco, salt, iron, copper, and saltpetre; but especially corn. **Weather.** **Trade.**

The large body of water on the north-east of the Euxine, called the Sea of Azof (*Palus Mœotis*), the Azák-deniz-í of the Turks, has a surface of rather more than 13,000 square miles: and from the action of its rivers, its waters are rather brackish than salt. The navigation of this subdivision of the Black Sea is impeded by the freshes of the Don, its general shallowness, numerous shoals, and occasional ice; nor can it be entered by shipping otherwise than by the narrow strait of Taman or Yenikaleh (*New Castle*), the ancient Cimmerian Bosporus. But notwithstanding these physical impediments, such are the advantages of moral exertions, that Taganrog, its chief port, is a place of considerable and increasing consequence, the value of its import trade in 1850 being upwards of £380,000, and its exports about half a million. **Sea of Azof.** **Taganrog.**

It seems agreed among cosmogonists, that the Black Sea, at a remote period, extended much further to the east and north than it now does, occupying the whole of the vast plains and steppes that surround the Caspian and the Sea of Aral, neither of which had then a separate existence; the difference of their levels having arisen at later periods. Their depth must probably alter materially, since the beds of the rivers above-mentioned are charged with an extraordinary quantity of sand and *slime*, which from the rapidity **Geological changes.**

of their course they hold in suspension till they approach the sea, where, spreading over a wider area, and flowing in a more gentle current, they deposit the substances brought down, so gradually that the elevation of their beds is almost imperceptible. Polybius, who states this as a cause for predicting the filling up of the Euxine in process of time, describes a shoal one thousand stadia in length before the mouth of the Ister, at one day's sail from the land: this having long since disappeared, has no doubt become a part of the delta of the Danube. The Sea of Azof has manifestly contracted its boundaries; but this subject will be resumed in the next chapter.

Karamania. To return to the Levant Basin. Proceeding from the Archipelago eastwards along the shores of Asia Minor, the space between Cape Symi—opposite Rhodes—and the Gulf of Iskanderún is called, by European geographers, the Coast of Karamánia—from Karamán-ílí, the land of Karaman Aghá. It is broken into deep bays and gulfs, backed by high ranges of mountains, the Peak of Takhtahlu (*table-topped*), on the west side of the Gulf of Adalia, being 7800 feet in height; and beyond it, to the eastward, are the still higher and ever-snowy summits of Taurus.

Between the gulfs or bays of Symi and Makri (*Glaucus sinus*), are several small ports; but the north-west portion is occupied by the extensive and land-locked haven of Marmorice. Mermerícheh or Little Marmora (*Physais*), the Marmorice of our charts, which is a beautiful basin capable of affording a safe anchorage to the largest fleets. This was happily proved by the timely refuge it gave to our weather-stricken expedition under Lord Keith and General Abercrombie: yet it was actually unknown to the pilots of our armament, the ships of which ran in before a furious gale, solely on the authority of a note from Sir Sidney Smith. The immediately neighbouring harbour of Kara-aghátch, though less Yedi Burun. commodious, is easier of access; and beyond it the Yedí-Búrun (*Cragus M.*), or rugged peaks of the seven capes, below Makri, bound the bay into which the River Kodja-chai (*Xanthus*) discharges its waters, after flowing through the pashalik of Meis (*Lycia*). Between that and Cape Khelidonia there are also numerous ports and creeks, whither ships of any size may resort and refit with safety and facility, the access being everywhere easy, from the boldness of the shores. The principal of these, Kastelorizo or Castello Rosso, and Port Tristomo (*Three Mouths*), according to Admiral Beaufort, 'may be considered the more valuable, as from hence to Syria there is but one land-locked harbour.' Water and refreshments, however, in the present desolate condition of the country, are scarcely obtainable.

Castello Rosso.

Capes Khelidonia (*Sacrum prom.*) and Anamúr (*Anamurium prom.*), form the headlands of the large Gulf of Adalia, the *Pamphylian* Sea of the old geographers. Off the pitch of the former headland is a cluster of five islands, two of which are large, and from 400 to 500 feet high, containing some creeks, in which small vessels may be sheltered. Passing several coves and islets which fringe the coast under the magnificent mountain of Takhtahlu, we arrive at Adalia (*Attalia* vel *Olbia*), the largest city on this coast. From hence a lower shore, with occasional sandy beaches, extends south-eastward to Cape Anamur, where the land becomes bold and bluff. The produce of these parts, principally timber, gall-nuts, wax, honey, camel's-hair, and liquid storax, is usually carried to Cyprus, and thence re-exported: corn is embarked, though under prohibition.

Gulf of Adalia.

Produce.

From Cape Anamúr, the southernmost point of Asia Minor, a broken shore extends by Cape Cavaliére (*Sarpedon prom.*), Provençal or Manavat isle, and the projecting sands of the ancient *Zephyrium prom.*—the deceitful Lingua di Bagascia of Frank navigators, also stigmatized by the Turks as Lisán-el-Kahpeh (*Harlot's mouth*)—to the Tersús-chái, river of Tarsis; the maritime town (*scala*) of Tersús, the present representative of the once powerful city of *Tarsus*, about twelve miles inland. This river, the

Tarsus.

Cydnus. *Cydnus* of old, which once received the stately galleys of Cleopatra, is now inaccessible to any but the smallest boat. To the east of it is a deserted and marshy tract of country, with a sandy beach, extending to Cape Kara-dutash (*Megarsus*), or black rock, where the Gulf of Iskanderún or Alexandretta (*Issicus sinus*) may be said to commence; most parts of it are unwholesome to a deadly degree. In this bight is the boundary of Asia Minor.

Geological changes. The south coast of Asia Minor exhibits indications of gradual changes in its littoral line, especially towards its eastern extremity, where the river Jaïhun (*Pyramus*), by its volume of deposits, has produced an extensive arid plain. 'The low sandy point,' says Sir Francis Beaufort, 'pushed out by the Jyhoon, has already (1811) advanced six miles beyond what appears to have been the original line of the shore.' In the Gulf of Makry, on the contrary, is the stately mausoleum of a warrior, which assuredly was erected on the shore; but it is now upwards of thirty yards from it, and the sea covers at least two feet of its base. The walls of *Telmessus* also, in the *Glaucus sinus*, were undoubtedly built originally on dry land, but are now likewise surrounded by water: and at Kakara, in some places three or four of the lower steps of house-doors, and the foundations of the walls, are now beneath the surface of the water. Caunus, which was a seaport in the time of Strabo, is now two miles inland, and its harbour has become a fresh-water lake, from whence the waters have a fall towards the sea. Also the alluvial plains of Xanthus, Phineka, Myra, and Makry, have increased considerably in thickness of soil, since the time when the cities on those plains were flourishing. The borings of marine animals show marks of upheaving; and in the Gulf of Iskanderún, are still to be seen the walls of a castle erected by the Saracens, now one mile and a half from the shore, in which there remain the rings to which the ships were formerly made fast. Sir Francis thus describes a geological effect which he examined at the efflux of a lake near Cape Phineka:—'This lake is separated from the sea by a narrow ridge of sand and gravel, the shape and limits of which are evidently prescribed by the opposing efforts of the currents within, and of the sea without; the former sweeps along its interior edge, and, perhaps, supplies it with fresh accession of matter from the mountains; while the external surf rolls back the loose gravel, and piles it up like a wall. It was pleasing to observe in action, the causes which can thus enable a neck of fragile sand to resist the impetuosity of the ocean, while every day furnished instances of the most compact rocks yielding to its violence.' The same intelligent officer also noticed what he designates a 'petrified beach,' at several places on this coast, where the upper slopes, to some distance into the sea, had become a solid crust of pudding-stone; for which he assigns a similar cause to that which is already given for the consolidation of the Sicilian beaches: and he adds, that 'the unwary boat that should mistake it for a common beach of yielding materials, and should run on it before a following surf, might be fatally apprized of its error.'

Coast of Syria. The sea-board of Syria is an extent of about 440 miles, being bounded on the north by the mountains called Al-Lokám (*Mons Amanus*), which fall precipitously into the sea at Cape Khynzyr (*Rhossicus Scopulus*), the crown of which (*M. Pieria*) is 5500 feet above the sea; and it extends from the river Bayás (*Issus*), in the Gulf of Iskanderún, to the torrent Al-'Arísh, which last separates it from Egypt. A chain of lofty hills lines its whole length, receding from one to eight leagues from the shore. Among them Mount Lubnám, or Libanus, the far-famed *Lebanon*, rises conspicuously to the height of 7100 feet. The shores between Tripoli and Tyre are principally hilly; but present in many places a large extent of low and flat coast, the plains of which suffer severely from remittent fever and dysentery in summer and autumn, from want of drainage.

The most frequented ports and trading-places are, the
Iskanderún. unhealthy and dilapidated Iskanderún; Swaïdiyah on the Nahr-el-'A'si (*Orontes*); Latakia (*Laodicea ad Mare*); the fair town of Tarabolus (*Tripolis*), or Tripoli in the East;
Súr. Beïrút (*Berytus*); Saïdá (*Sidon*); Súr (*Tyre*); 'Akká or Acra (*Ptolemaïs*); Kaipha, under Mount Carmel; Kaïsariyah
Jaffa. (*Cæsarea*), a tolerable anchorage near a heap of ruins; Jaffa (*Joppa*), the port of the western pilgrims of the Holy Land; Scalona (*Ascalon*); and Ghazza (*Gaza*), which is backed by very fertile grounds. These places are resorted to by small craft only, in the fine season, for the whole is a dreaded
Exports. lee-shore in westerly gales. The principal exports are wine, olives, tobacco, cotton, silk, wool, fruit, sesamum, galls, and medicinal plants; but, from mismanagement, the trade is not brisk,—or rather, its commerce is far inferior to what, from its resources, might be maintained.

Geological changes. The sea has considerably receded from some parts of the coasts of Syria; while, on the other hand, at Beïrút, there is a tower standing in the water, and remains of the ancient marine works at Jaffa and Kaïsariyah are submerged. The island of Tyre is now united to the continent, and some portions of its peninsula bear evidence of submergence. While this coast, on the whole, affords an instance of elevation, the vast adjacent valley from the Jordan, through El Ghor to the Gulf of Akaba—the Aulona of the Greeks, and Cœlo-Syria of the Romans—offers a most remarkable instance of the depression of land: it being reckoned at the Sea of Galilee to be 628 feet below the waters of the Mediterranean, and at the Dead Sea more than 1200 feet. This must have been the effect of what is termed some violent 'convulsion of nature,' either by means of fire, water, or subsidence of strata.

Cyprus. In the north-east part of the Levantine Sea, at ten or twelve leagues south of the coast of Karamania, and about twenty leagues to the westward of Syria, is the large and once famous island of Cyprus (Κυπρος); once an important kingdom, now a mere appanage of the Sultan's Grand Vizier. Its length is 140 miles, by 50 at its greatest breadth, narrowing gradually to the east; it is traversed from east to west by a range of woody mountains, of which Oros Troados (*Olympus*), the principal summit, is 6590 feet above the sea. It possesses the ports of Famagusta (*Arsinöe*), Limasol, Baffa (*Paphos*), Larnaka, and Ghyrna (*Ceryneia*), of which Famagusta is the chief. But though the range mentioned extends through Cyprus, the greater part of the island consists of fine plains, of which the soil is excellent; and even under imperfect cultivation yields corn, wine, oil, carubbas, and other fruits; and among its exports are also silk, cotton, wool, morocco-leather, soda, salt, coloquintida, gum, laudanum, madder, cochineal, turpentine, tar, and pigments. The resources of the island are, however, sadly depressed by misgovernment, the Grand Vizier acting only by proxy. Ports. Produce.

Having thus finished this section, which is principally a compilation, but from sources upon which implicit reliance may be placed, I now proceed to resume the result of my own observation and experience.

§ 9. The North Coast of Africa.

ALTHOUGH it has been usual to commence Egypt at Tineh (*Pelusium*), some geographers have restored it to the ancient point El Arish (*Rhinocorura*), the southern boundary of Syria, as ordered by Joshua nearly 3400 years ago (ch. xv. ver. 4 & 47); who also well described its position on the 'river of Egypt,' a ravine receiving the pluvial waters of various torrents. Between this and Tineh are the moving sands called by the Hebrews Shúr, and by the Arabs Al Jofár, bordered by the Serbonian Pool,—a district never yet occupied by an enemy, and which, says Abulfedá, Egypt.

is commonly known as the 'Sands of Egypt' (*remel Misr*). From this notable land-mark the shores of Egypt extend—at least did when I was there*—to Rasal Kanáïs, about 115 leagues to the westward (*Hermea extrema*): they are generally low and arid, with occasional vast sandy downs and extensive marshes, thickly sprinkled with the round hillocks called dhahars (*rough hard backs*). The central portion of this apparent waste is the far-famed Delta, formed by the mouths of the Nile, the fertilizing nurse of the whole country of Mizraim. The annual inundation of this beneficent river is occasioned by the periodical rains of Central Africa: it commences about the summer solstice and continues till September, during which period the outpouring is very powerful, insomuch that fresh water may be skimmed off the surface of the sea, at the distance of two or three miles out in the offing.

Damietta. The Egyptian ports are—Damyát, or Damietta (*Tamiathis*), a trading town among the marshes of the eastern Rosetta. or Phatnitic mouth of the Nile; Rosetta, or Rashíd, beautifully surrounded by palm-groves and gardens on the western or main mouth, known as the Bolbatic branch; Al Bekur (*Canopus*), a castle and loading-place on the old Canopic mouth, in the bay to the west of Rosetta; and the Alexandria. harbours of Alexandria, with two or three insignificant coves between the last-mentioned place and Ras al Kanáïs (*Cape Churches*). At these ports the European commerce of the country is carried on, and great quantities of imported goods are conveyed to markets in the interior. Trade. The chief exports are grain, rice, dates, fruit, cotton, flax, silk, fine stuffs, wool, hides, ivory, ostrich feathers, gums, spices, and drugs — the corn being in quantity sufficient still to stamp Egypt a granary; and, unlike the time of

* This remark is made, because the Basha of Egypt has, since then, laid a kind of claim to the sovereignty of the whole coast of Marmarica, an extent of about 320 miles.

Herodotus, beans are sown abundantly, and exported in large cargoes. The great emporium is Alexandria, a city and port which the late Mehemet Ali rescued from a state of torpid decay, and raised to maritime importance; and when I visited him he had just completed that gigantic undertaking, the Mahmúdíyeh Canal, as described in my *Ædes Hartwellianæ*, by which the trading vessels of the Nile avoid the dangers of the Rosetta (*Bóghaz*) Mouth. **The Great Canal.**

It was anciently supposed that a great gulf once penetrated from the Mediterranean into Egypt as far as Thebes; that the isle of Pharos was at a very considerable distance from the main; and that, therefore, the whole Delta is the gift of the Nile.* This must have been the work of many ages, for the general coast of Egypt, except in the secondary changes,—as the silting up of Damietta since the thirteenth century, the draining of Lake Mareotis, and the filling of lagoons,—still answers the description which Herodotus gave of it 2300 years ago. But by observing the coast about Gaza and Cæsarea, and from thence to the Arab's Tower (*Taposiris*), a line will show the extent to which the Delta has advanced; but this we must grant to be of a date very remote, for the currents which sweep along the north coast of Africa have prevented any rapid accession to the alluvial soil of the Egyptian shore. **Geology of the Delta.**

From Egypt, proceeding westward along the south margin of the Mediterranean Sea, we first reach the sterile and uninviting coast of the desert of Barkah, which extends to Razatin, or Ras-er-Tyn (*Cape Fig*); but its exact boundaries are very uncertain, neither the Pasha of Egypt nor the ruler of Tripoli having been able to tell me exactly where their respective dominions ended. Although the designation Libya was often applied to all Africa by the ancient Greeks, **Barkah.**

* The vast lakes recently explored by the French frigate commanded by Captain Bouët-Villaumez, within the Grand Bassam river, on the West Coast of Africa, are on the site of the deep inlet which appears on Fra Mauro's celebrated planisphere; it is there designated the Golden Gulf.

it was sometimes restricted to the sandy, waterless desert—(*sitientes arenas*)—between the Nile and the Cyrenaica, or country round Cyrene: this space was subdivided into Marmarica and Cyrenaica proper, the chief emporium of the former being *Paræetonium*, of which the site is traceable at the present Port Mohâderah (*Zygio*), point Ras al Harzeit being sometimes called Cape Baratún, a corruption of the ancient name. It is a curious coincidence, that Ptolemy's Katabathmos magnus and parva are now called by the Arabs 'Akabah-el-Kibír and 'Akabah-el-Soughaïr, the great and the little Descents:* the first being about 900 feet high, and the second 500 feet; and which some geographers mark as the separation between Asia and Africa, also as the western boundary of Marmarica. In the sea-board of this arid space there are the spacious harbours of Tebruk (*Anti Pyrgos* or *Tabraca*), and Bombah (*Bombæa* vel *Batrachus*), with several smaller havens for coasters: but not a vessel plied on those waters except foreign ones, and even those so seldom, that Tebruk and Bombah were unknown but by name when I first visited that coast. Indeed, between Alexandria and Benghází there was not, at that day, a single native boat, or any means of embarkation; a consequence of which was, that we found fish and seals in abundance.†

Paræto-nium.

Tebruk.

Bombah.

Passing the cove at Ras-er-Tyn (*Chersonesus*), between it and Benghází is the border of the mountainous tract called Jebel Akhdar, with the extensive remains of Grennah, or Kureïneh (*Cyrene*), the which, with the sea in front and sands in the rear, encourages the idea of its

Cyrene.

* Idrisi terms the first of these inclinations 'Akabah-el-Sollom, or staircase descent; whence the Port Sollom and Saloume of most of the early Portulani.

† I was told that the French Admiral, Ganthéaume, who possessed great acquaintance with the Levantine shores, saved his squadron from the British pursuers in 1801, by getting into Tebruk, a port of which our officers were utterly ignorant. Had we known it, there had been no escape for Ganthéaume!

having been once an island. It certainly differs in climate, aspect, wood, water, and resources, from all other districts between Syria and Tunis, well meriting its modern name, which expresses 'the green mountain.' On the margin of this space there are several small ports, but the only one resorted to is Dernah (*Darnis*), which however did not, in 1817, possess an embarcation of any sort; still vessels from Alexandria and Tripoli called there to embark honey, wool, wax, and butter. In the bight between the points Ras-el-Hilal (*Naustathmus*) and Cape Rasat, is Marsa Susah (*Apollonia*), a mere boat cove, though once the port of the potent city of Cyrene, formerly so celebrated for its knowledge, riches, and splendour. On the hills above, at the height of 1990 feet, its vestiges are to be seen from seaward; and from thence to Benghází are extensive ruins of the opulent cities of Dolmeïtah (*Ptolemais*), Taukrah (*Teuchira*), and other members of the *Pentapolis.*

Dernah.

Ports of the Pentapolis.

Between Cape Rasat (*Phycus prom.*) and Mesrátah (*Trierium* and *Cephalæ*), is the Gulf of Sidrah (*of the lotus*), the once dreaded *Syrtis major*, the navigation of which even Strabo thought it audacious to attempt: but native pilots confine the designation to the space within Ras Kharrah (*Zuca?*) and Ras Teyonas (*Borium prom.*) Our researches have deprived this extensive bight of its terrors, and shown that while it is comparatively free from danger, it is hardly worth the visits of shipping, there being in the whole space but one place deserving of being called a port, and even that is only fit for small vessels. Such is Benghází (*Hesperis* and *Berenice*), an insignificant fortified town, which yet derives a considerable trade by exporting cattle, dhurra (*holchus sorghum*), honey, wax, wool, and manteca, or coarse butter: to these may be added a little sulphur from the mines at the bottom of the gulf, the which has induced the Arabs to designate this Syrtis Joun al Kabrit. The castle holds the whole district in subjugation, though it is in such a dilapidated state, that

The Great Syrtis.

Benghází.

on my arrival there Halil Bey requested me to dispense with the usual salute, lest the concussion of the cannons should injure the walls; a fact already so well recorded by Captain Beechey, in his excellent account of the Expedition round the Syrtis (*page* 288).

Geological changes.

Syrtes was the name given by ancient geographers to these two great gulfs on the northern coast of Africa—the Sounds of Barbary; 'cursed and horrible places both,' says Philemon Holland, speaking English for Pliny. The Great Syrtis, that which we have now arrived at (ἡ μεγαλη Συρτις), had long been, from various obvious causes, unknown to navigation; insomuch that when I was first proceeding thither in 1816, the only information I could procure, even with the aid of the powerful Yússuf Báshá, and his admiral, Murád Reïs (a renegade Scotchman originally named Peter Lyell), proved that the gulf then was as great a bugbear as when the classical writers scared seamen about its fell shoals and whirlpools.* There can be no doubt, however, that it has changed its form, for it must once have penetrated further into the interior, and in a measure communicated with that great desert which separates the two potent human races—white and black—and gave birth to the fabled strife of Osiris and Typhon. Benghází was once

* Having placed some papers in the hands of that veteran and energetic geographer, Major Rennell, on my return to England, I cannot but record an extract of a letter which he wrote to me, dated Nassau-street, Jan. 19th, 1821, saying,—'My illness not having abated, I have not put pen to paper since I had the pleasure of seeing you till now. . . . . . The changes that have taken place (*in the Syrtis*) are nothing more than I supposed would take place at some period. Every flat coast or shoal is increasing; which can only be by a conversion of sand or gravel, &c. into firm land. This has taken place even on our own coast—40,000 acres of land have been accumulated in Romney Marsh, almost entirely by *sea* alluvion, which is marked by its having a slope inward from the coast. The surges, in tempests, have raised the sands of the Syrtis *too high* to be dissolved by the ordinary rise of the waters; which is exemplified by what Mr. Smeaton told me concerning his adventures on the Goodwin Sands. Landing, nearly about low water, the surface was so compact that he found a difficulty in inserting an iron crow to fasten the boat to: but on the rise of tide, it would hardly bear a man's weight.'

possessed of a large harbour, which may have communicated with the salt-water lake (*Tritonis?*) southward of the town. The deep inlet and quicksands where the *Philænorum Aræ* were erected, have disappeared, as has also the lake of Zuca, which Strabo mentions as disemboguing into the Syrtis: but the detritus of the Solocho Isles, off the western coast, may have formed the extensive bank of Isa, on which I more than once anchored, and found easy riding in rough winds. A little inside the beach here, a succession of large shallow marshes, where much salt is obtained in long blocks for commerce, may have been the inlet and naval station spoken of by Strabo.

From the Great to the Lesser Syrtis, now the Gulf of Khabs, the coast of Tripoli (*Oea*) offers little for intercourse **Tripoli.** by sea except the harbour of Tripoli itself, which lies nearly in mid-distance, and is the capital of the state of the same name; and the Regency having a sea front of upwards of 760 miles, from Dernah to Al Biban, is of some consideration as a maritime state of Barbary. The harbour is secured by a chain of rocks projecting from the north-east angle of the town, and a sand-bank off Point Tajúrah; and here are imported the woollens, cottons, muslins, hardware, arms, and ammunition of Europe; while the exports consist of **Exports.** cattle, leather, skins, soda, salt, natron, wax, saffron, senna, madder, oil, drugs, ostrich-feathers, gold-dust, ivory, gum, dates, and other articles of home produce, or commodities brought from Central Africa by the caravans. The coast, however, though for the most part low and shelving, is pretty bold-to, many parts affording good anchorage, as the north winds rarely blow home; and boats may generally find refuge in the little ports (*marsa*) Zoraik, Zilíten, Ugrah, at the mouth of the Wadi Khahan (*Cinyps*), Lebidah (*Leptis Magna*), Ligatah, Tripoli vecchio (*Sabrata*), Zoarah, al Biban (*the Gate of Pisida*), and Zarzîs: these creeks are mostly the result of the action of the sea and atmosphere on a friable shore, and the barriers are ridges of rock parallel to the line of the coast, which have withstood the attack. But neither the sea-board nor its details were known when I first visited it; insomuch that when I was with Lord Exmouth's squadron in 1816, he suddenly and hastily weighed from before Tripoli, and beat about in a northerly gale, expecting worse weather, without an idea of the excellent anchorage he might have taken up in the vicinity of the Lesser Syrtis.*

Jerbah. Jerbah (*Meninx* and *Lotophagitis*), an excellently cultivated and rich island, separated from the mainland by a basin with two straits forming entrances, is the commencement of the Regency of Tunis, which extends from thence as far west as La Cala, or El Kal'ah, near Bona, a littoral distance of more than 570 miles. The intervening country is greatly diversified with mountains and valleys, fertile plains and arid wastes; which are blest with one of the finest climates in the world, and a remarkably productive soil. Between Jerbah and the low group called the Karkenah (*Cercina*) Isles, lies the Gulf of Khabs, or the Little Syrtis (ἡ μικρα Συρτις), which may once have had a communication with Es Sibkhah, or the great salt plain of the interior, covered with water to the depth of three or four feet in the winter—a probable site of the *Tritonis Palus* of Herodotus. If so, a narrow strait among chains of eminences, and liable to very variable tides, may have confounded early navigation, and rendered it, as Scylax asserts, even more dangerous than the Great Syrtis. But the dread of the ancient mariners can now be but little understood, since from various changes the Syrtis Minor of the ancients is no longer recognisable: nearly the whole remaining space now affords good anchoring ground and smooth water, the

Little Syrtis.

Probable changes.

* On this occasion his lordship had embarked all the Christian captives save one, an Italian boy, then at the salterns of Zoara. Shortly afterwards I had the satisfaction of taking this boy to Malta, thereby carrying off the last Christian slave from Barbary. The principal of my proceedings on this coast, about that time, will be found in the Appendix.

bank of Karkenah preventing the sea from rolling home; and the shallows are indicated by the fishermen's palisades. In fact, from the unceasing operation of the sea in throwing up and depositing sand on a flat coast, where there is no river, torrent, or other back-water to sweep it away again, the shore must have had a continued augmentation.

The adjacent lands were abundantly fertile by nature, but, until the rule of the Carthaginians, were left without culture; for, in the words of Strabo, the ancient people of this country (*the Numides* or *Nomades*) abandoned their fields to savage beasts, to exhaust themselves by predatory warfare. The east coast of Tunis (*Byzacium*), though perhaps less cultivated than when it was regarded as a magazine of provisions, and dignified with the title 'Emporia,' is nevertheless abundantly fertile, and its tillage is very creditable to the Moors. The sea-board possesses several populous trading towns, and some excellent anchorages, to which numerous ships resort to take in the produce. Of these, the places in chief consideration—from Jerbah northwards—are Ghabs or Khabs (*Tacape?*) Sfákus (*Taphurah*), Mehadíyah or Afrikah (*Turris Hannibalis*), Lamta (*Leptis parva*), Monástir (*Hadrumetum*), Súsah (*Kabar Susis*), Ehrakliyah or Herkla (*Horrea Cœli*), Hammámét (*Aquæ calidæ*), Nabal (*Neapolis*), Khurbah, (*Curubis*), and Calibia or Iklíbiyah (*Clypea*). Some of these are of great consideration as towns in Barbary; but the principal is certainly the beautifully situated and opulent Sfákus, a place where I was most hospitably received before any Christian agent had ever been established there; and where my operations and journeys were viewed without that alarm, and inconvenient distrust, which were so often encountered on these shores at that time.

East coast of Tunis.

Towns.

Sfákus.

The Gulf of Tunis is a deep and safe bay, lying between Cape Bon (*Hermæum prom.*), the Rás-Adár of the natives, and Cape Farina (*Apollinis prom.*), which are thirteen leagues asunder. The site of the famous Carthage (*Car-*

Bay of Tunis.

Carthage.

chedon) is within and upon the promontory of that name on the west side of the bay, and the space from thence to Tunis still exhibits vestiges of the Tyrian mistress of three hundred African cities and towns (besides her power in Spain, Sardinia, Corsica, Sicily, and Italy itself) as well as of her Roman successor. Of these perhaps the most striking are the cisterns, the mole-basins inside of Cape Kamar, and the great aqueduct which conveyed water from the Jebel-ez-Zaghwán (*Zeugitanus mons*), a distance of fifty-two miles. But it must be conceded that the appearance of the land has greatly changed since the time of old Carthage, when the receding curvature of the beach threw the peninsula of the Byrsa more boldly out, so as to be all but insulated. South of the ruins of Carthage, and at the bottom of the bay, stands the city of Tunis (*Tunetum*), the metropolis of the Regency, with a population little short of 150,000. It maintains a busy trade, both of import and export; the produce and manufactures of the Beylik consisting of corn, oil, wool, hides, honey, wax, soap, silks, fine woollen cloths, shawls, fázes or scarlet skullcaps, burnuses, wrappers, indigo, madder-roots, orchilla, hennà, senna, dates, ivory, coral, sponge, pottery, tobacco, morocco leather, ostrich feathers, cattle, sheep, and other live stock. The city is separated from the bay by a shallow lake of intense saltness, occasioned by the powerful evaporation from a burning sun, and the aridity of the surrounding shores. This lake communicates with the sea by a narrow fortified channel, called by seamen the Goletta, but Halk-el-Wád by the Moors. Off the beach divided by the Goletta, the largest fleets may anchor in comfortable depths of water and good holding-ground. Indeed, when I joined Lord Exmouth's squadron before Tunis in 1816, I had been assured that it was considered a perfect anchorage under proper care; but the loss of the whole Tunisian fleet in March, 1820, and a heavy gale which I afterwards rode

Tunis.

Goletta.

out, induced me to reconsider that opinion; and on finding that large patches of the bottom consist of a hard clay which breaks short, I certainly cannot recommend it as a winter station where heavy ships are employed.

On the east entrance of the gulf, rise the Zembra or Zawámir (*Ægimurus*) Isles, of which the largest is 517 feet in height: and forty-two miles to the north of them are the dangerous rocks which have in recent ages been designated the Skerki, Squills, and Esquerques; and which appear to be the remains of the *Aræ* mentioned by Virgil, upon the *saxa latentia* of which, three ships of the Trojan fleet were said to be wrecked. They are the *Æginori* of Pliny, who observes that they lie opposite Carthage, and between Sicily and Sardinia—in his time more like rocks than islands, but recorded to have been inhabited, though afterwards to have sunk down. There was much doubt among our chart-compilers as to the existence of this reef, until public attention was unexpectedly aroused by the total and melancholy loss of the *Athénien*, of 64 guns, and most of her crew, in 1804.

Zembra. **Skerki.**

Besides the road of Tunis just mentioned, the northern coast of this state has the ports of Farina or Ghár-el-Milh (*Salt cave*), the sea-margin of Ouga (*Utica*), now fast filling up by the floods and alluvia of the river Majerdah (*Bagradas*); and Bizerta or Beni-zart (*Hippo Zarytus*)—the Venice of Barbary—with two interior lakes, in the inner one of which (*Sisaræ palus*) the water is fresh; the fisheries of these lakes are farmed at a high price, and are extremely profitable. Between this and the Algerine frontier, the only object of maritime interest is Tabarkah (*Tabraca*), formed by a fortified island and the river Ez-zeïne (*Rubricatus*) on the main. About twenty miles north-north-west of Rás-al-Manshar, or Cape Serrato, lies the uninhabited island of Galita (*Calathe*), west-south-west of which are two perilous sunken rocks, on which the

Port Farina. **Bizerta.** **Galita.**

Avenger frigate struck in December, 1847, when all hands but seven, and a boy, were drowned: the channels on either side of these rocks are both wide and safe.

Algeria.

From the Mazúlah hills, which bound the regency of Tunis, to the river Mulúwi on the west, an interval of 670 miles, the coast of the fine and fertile state of Algeria (*Mauretania Cæsariensis*) extends, the full capability of which has never, in modern times, been properly excited; and of late ages its Mohammedan rulers, termed Deys (*Dáïs*), preferred predatory warfare, which is always destructive of industry, agriculture, and commerce. Yet, though they preyed upon and braved the power of most of the Christian princes around the Mediterranean, their fleet—both vessels and crews—was always truly contemptible as an organized force; insomuch that the long sufferance of these barbarous and professed pirates is an anomalous phenomenon in the history of human polity. The origin of this degrading system of hostility, which, in despising the rules of civilization and the laws of nations, violated the rights of human nature, may be owing to the rancorous fanaticism of the Crusades; but its nearly uninterrupted continuation was a reproach to Christendom. It has, however, now passed away; and this country of physical beauty—though sapped by moral deformity—is now colonized by France; it will, therefore, inevitably advance in civilization, and, consequently, in the arts and pursuits of polished life. It is a fine stage for the exercise of philanthropy and commerce; for when I frequented the coast, even in the depth of its barbarism, there were, among the exports of its wilfully restricted traffic—corn, pulse, olive-oil, wax, honey, fruits, tobacco, kermes, live stock of all kinds, hides, wool, skins of wild beasts, coral, timber, charcoal, and ostrich feathers black, white, and grey.

Eastern ports.

The principal loading-places on the eastern coast of Algeria are—La Kalah (*Nalpotes*), Bastion of France (*ad Dianam*), Bona (*Hippo Regius*), Storah (*Rusicada*), in the gulf anciently termed the *Sinus Numidicus*, Kolah (*Culla* and *Collops Magnus*), the coves under Ras Sebah Rus (*Tretum prom.*) and at Wad al Kabir, or great river, Zergeli or Jijel (*Igilgilis*), Bujeïyah *Portus Saldæ*), Mersa Fahm or Zufún (*Audus*), Tedlez (*Rusucurrium*), and Marsa Zinet. Passing this, and rounding Cape Matafuz or Temedfús, we enter the great Bay of Algiers, or Al-Jezáirat, the characteristics of which are, bold shores, deep water, and excellent holding bottom; but it is not mentioned in early writings, the present name being derived from the islet before the town. On the western side of this bay, stands conspicuously the renowned city of that name, with its mole, forts, lighthouse, and Kasbah; surrounded by beautifully diversified hills, valleys, gardens, groves, and villas. Weighing from Algiers, and passing Cape Caxines or Ras Al-Kanátir (*Iomnium*), to stand to the westward, there is a rocky and precipitous coast, mostly bold-to; and we find in succession the ports and creeks named Sídí Ferej (*Via*), Tfesud (*Tipasa*), Nakous (*Cæsarea*), Shershél or Zerzahal (*Icosium*), Nakkous (*Iol* and *Julia Cæsaria*), Dniss or Tennez (*Cartenna*), Marsa Goleit, Musta-ganem (*Murustaga*), Arzaú (*Arsenaria* and *Deorum portus*), Wahrán or Oran (*Quiza*), Marsa Kibir (*Portus magnus*), within Ras al Harsbah (*Metagonium prom.*), and Ishgún (*Acra*). The energy of the French, and the use of steam, will doubtless increase the number of these ports as increase of trade shall require, for the whole coast of Algeria affords abundant materials for commercial enterprise.

Algiers.

Western ports.

We now approach the outermost of the Barbary States, the which has—from its having been formed by the union of various small kingdoms, or, rather, large provinces—been known as the *Empire* of Morocco, or *Mogh'rib-al-akzà*, the farthest west; being a remnant of the great African monarchies formed by the Saracens in Mauritania. Anarchy and intestine discords have reduced its boundaries, but it is still possessed of a surface equal to that of Spain; while

Morocco.

its Mediterranean coast—from Ceuta to the river Mulúwi—is 220 miles, which is not one-third of its sea-board. It is finely diversified with hills and valleys, a great part of which have never been visited by Europeans; and there are many rivers flowing from the great Atlas range of mountains which traverses the empire in its greatest length, and attains the snow-clad height of nearly 13,000 feet, modifying the aspect, soil, and climate of the whole region. Those rivers disembogue into the Mediterranean Sea and Atlantic Ocean, the large ones forming bar-harbours, which, though now so neglected as only to admit of small vessels entering them, may some day be converted into good stations for steamers. A ramification of the great mountain-range turns to the north, and is there known as the Lesser Atlas, of which Ape's Hill (*Abyla*), opposite Gibraltar, may be deemed the northern scarp. The climate of Morocco is at once mild and salubrious; and the soil, where cultivated, is in the highest degree fertile; but there are everywhere large tracts entirely uncultivated. Corn, dhurra, rice, maize, and pulse are extensively reared in most of the plain districts; there are raised and collected, oil, cotton, tobacco, indigo, sesamum, gum, honey, wax, fruits, horses, cattle, poultry, sheep, salt, saltpetre, hemp, saffron, and madder-roots; and they have also manufactories of linen, silk, háyiks, skullcaps, morocco-leather, slippers, barracans, burnuses, shawls, carpets, soap, earthenware, and hides. The declivities of the mountains are sprinkled with forests, in which the cedar, cork, ilex, carubba, walnut, acacia, and olive trees are prominent; and though iron, copper, lead, and antimony, as well as gold and silver, have been produced to a certain extent, the mineral wealth of these mountains may be said to be as yet unknown. It is truly a luxuriant yet indigent country, surpassingly favoured by nature, but blindly neglected by man.

Mount Atlas.

Produce.

The Muluwi.

The river Mulúwi, or Muluwyah (*Molochath*), which, as anciently, divides Algeria from Morocco (*Mauretania* from

Tingitana), and is therefore of political importance; it rises at or near the southern extremity of the lower chain of Atlas, and flowing through a diversified country as yet almost untrodden by Europeans, falls into the sea nearly in the middle of the bay at which we have arrived, the Gulf of Melîlah of our charts. About ten miles to the north-west of the mouth of the Muluwyah lies the Zaphran, or Ja'fereï group, consisting of three rocky uninhabited islets, the highest of which is upwards of 400 feet above the level of the sea; they afford good anchorage to ships taking refuge there in stormy weather, and, from the goodness of the ground, there is no danger of bringing home the anchors. About thirty miles distant from these rocks, on a north-west by west rhumb, is Cape Tres Forcas (*Mitagonitis prom.*) of the Spanish pilots, called Ras-ud-Dehir (Cape of the Monastery) by the natives; and in the bight formed between it and the Mulúwi, stands the Spanish penal fortress Melîlah (*Rusadir*), a Moor-bound space, with barely a pistol-shot range of territory. Deeper still in the bay is the great salt-lake Resífah, an excellent port till 1755, when an earthquake stopped up the entrance.

Zaphran isles.

Melilah.

In mid-distance between Cape Tres Forcas and the coast of Spain, lies the steep rocky islet Alboran, which has usually been assigned by geographers to Barbary; while some of the chart-compilers omit it altogether. Indeed, such was our ignorance till lately, that this sterile rock, with hardly anything of animal or vegetable life about it, has been more than once represented as a desirable place for a settlement; and so late as the year 1813, the *Naval Chronicle* published a view of it, and described the imaginary inhabitants as subsisting chiefly by fishing!

Alboran rock.

Westward of Cape Tres Forcas, which is the termination of an offset of the secondary chain of the mighty Atlas mountains, on passing Tiraka (*Tœnia Longa*), and standing across the bay of Mezemmah, or Al Buzema, we perceive a rock (*Sex insula*) on which the Spaniards possess a

Al Buzem

petty post, which is kept under greater restriction by the Moors, if possible, than Melîlah. About eight leagues further to the westward is another of the Spanish *presídios*, the fortress of Peñon de Velez (*Parietina*), an elevated islet surrounded by strong works; which, being nearly inaccessible, is therefore held to be impregnable. In these *presídios* the garrison and the *forzati*, or condemned elons, seem to be almost equal sufferers.

Peñon de Velez.

To the north-west of the Peñon, at the distance of about twenty-two leagues, is Ceuta, or Sebtah, the principal of the Spanish *presídios*, and the eastern extreme of the south shore of the Strait of Gibraltar. Though the whole of the intervening bight is called the Bay of Tetuán, that name is also applied in a restricted sense to the anchorage before the populous city of Titáwán—commonly, Tetuán (*Jagath*)—between the Capes Negro and Mazari, where our ships have often found shelter from south-west gales, and procured provisions and refreshments. In 1799, a fleet of seventeen sail of the line, under Lord Keith, watered there without any loss of time; but an unexpected impediment threatened the further supplies. Although so near Gibraltar, with whose merchants the Barbary Jews carry on a pretty considerable commerce, our admiral could not get fresh provisions and stock in exchange for his Government bills, and must have proceeded to the siege of Cadiz without refreshments, had not an English merchant happened to put into Tetuán for protection, with a few thousand Spanish dollars on board. Between Tetuán and the Peñon, the country is inhabited mostly by Moors; there is no town of any consequence upon the coast, and it is equally destitute of harbours, the only place resorted to by coasters being Mostaza, where grain, cattle, honey, wax, and other produce, are embarked, as well as camlets, barracans, mats, pottery, and the various articles of Tetuán and other native manufacture.

Tetuan Bay.

Mostaza.

Ceuta.

On the peninsula of Ceuta (*Exclissa* and *Septa*) is a

fortress opposite to Gibraltar, which seems, like it, to be impregnable by land. West of this are the fine cliffs of a mountain, 2200 feet high, known by us as Ape's Hill, the Sierra Bullones of the Spaniards, and Jebel Moúsa and Thâtúth of the Moors (*Mons Abyla*). From Ape's Hill to Tanjah, or Tangier (*Tingis*), a fortified town no longer a place of importance, the coast is broken by alternate cliffs and coves, some of which look tempting enough for landing at, but strangers are immediately fired upon by the Moors in ambush, disembarkation except at the regular towns being strictly prohibited. Leaving the anchorage at Tangier, and still standing to the westward, a bold shore presents itself under an uncultivated and arid aspect, as far as the fine headland called by the natives Ras-el-shukkár, or Red-flower point, and by us Cape Spartel (*Ampelusia*); which forms the north-west point of Morocco, and western entry of the Strait of Gibraltar.

M. Abyla.

Tangier.

From Cape Spartel to the south-south-west, as far as Arzílá (*Zilis*), the coast-line is a flat, sandy, and shingly beach, rising in the interior to a fine grazing country, but bearing a barren and deserted appearance. Off this part, and especially opposite the bight called Jeremiyah, there is good anchorage with easterly winds, to be chosen by the lead, there being no sea-danger; and the whole is safe and bold to Al Harátch, or Larâche (*Lixus*). But during those winds the water is smooth, and ships may keep under very easy canvas; as we experienced in the summer of 1811, when cruising off Cape Spartel with a squadron of four sail of the line and some smaller vessels, under the command of that excellent seaman, Rear-Admiral Sir Richard Goodwin Keats.

West coast of Morocco.

Such being the periphery of this not less interesting than extensive inland sheet of water, we must next proceed to consider its surface: though, before we entirely conclude our chorographical sketch, an account of the declared value

Mediterranean commerce.

of British produce and manufactures annually exported from the United Kingdom to the Mediterranean, may give a synthetical view of our commercial relations in that sea. Moreover, figures confer accuracy upon expression: 'to count,' observes Dr. Johnson, 'is a modern practice; the ancient method was to guess, and when numbers are guessed, they are always augmented.' The mean of the various obtainable returns for the years 1820 to 1824 was—

Spain and the Balearic islands	£582,891
Gibraltar	993,700
France and Corsica	312,866
Italy and the Italian islands	2,391,620
Malta	425,500
Ionian islands	323,650
Turkey and Continental Greece	989,260
Morea and Greek islands	32,000
Syria and Palestine	191,280
Egypt	257,760
Barbary and Morocco	51,600

English possessions.

For the English reader, it may be proper to add a few more statistical details respecting our own possessions on the shores of the Mediterranean, in order to prove their claim to the national regard. They are drawn up from various inquiries in proper quarters, and official returns procured for me by my friend the late estimable Mr. G. R. Porter, of the Board of Trade, reduced, as nearly as possible, to the close of the year 1824. In the following tables the islands of Malta and Gozo are included together, and the whole of the Septinsular Republic under the head of Corfu; a form adopted because the public returns are made from the chief city, or head-quarters of each garrison. And it will be borne in mind, that the respective forces (*Table II., page* 103) at each place are, of course, on the graduated peace establishment.

Here the mind is led to perceive how civilization enhances human enjoyment, by increasing the resources of a country. It is rather the industry exercised on the districts occupied, than their extent of area, which develops riches and power.

TABLE I. — STATISTICS.

NOTANDA.		GIBRALTAR.	MALTA.	CORFU.
Superficial area	*sq. geog. miles*	1½	125	1,059
Population	*males*	4,790	46,180	112,500
	females	5,560	49,300	98,240
	aliens	4,780	6,170	10,780
	total	15,130	101,650	221,520
Chief Town . . .	*name*	*Gibraltar.*	*Valetta.*	*Corfu.*
Inhabitants of do.	*number*	15,130	46,250	21,400
Public educational expense	£	740	2,090	6,880
Shipping from Great Britain	*number*	169	38	23
	tons	23,567	7,870	6,750
Ditto inwards .	*number*	37	26	53
	tons	6,500	4,805	7,930
Value of Commercial imports	£	1,041,600	300,700	43,300
Ditto exports . . .	£	*variable.*	200,000	510,000
Colonial Revenue . .	£	30,000	109,800	202,500
Expense to Great Britain,	£	146,000	100,000	82,000
Ground in crop . .	*acres*	70	53,670	271,890
Uncultivated land .	*acres*	750	47,350	219,440
Stock: Horses, mules, and asses .	*number*	*none*	4,910	13,810
Stock: Horned cattle	*number*	*reared*	5,560	11,200
Stock: Sheep . . .	*number*	*in the*	8,992	88,520
Stock: Goats . . .	*number*	*garrison.*	3,150	70,500
Property annually created	£	72,000	850,000	2,080,000
Do. moveable and irremoveable . . .	£	1,500,000	3,755,000	10,950,000
Acquisition of . .	*date*	A.D. 1704	A.D. 1800	A.D. 1815
Right of possession . .	*by*	*conquest.*	*treaty.*	*treaty.*

On showing the above tabulated items to Mr. Porter, who was then diligently occupied in compiling those admirable *Statistics of the Empire* which have since been published *in extenso* at the public expense, he kindly examined them, and made comparisons with documents in his office. He also supplied me with the Government returns of the agricultural produce of Malta and the Ionian Islands for the year 1839; assuring me, at the same time, that the document had his full confidence:

MALTA AND GOZO.

Description.	Area.		Produce.
Wheat	9,951 *acres*	...	17,453 *quarters*
Meslin	9,144 ,,	...	26,042 ,,
Barley	4,051 ,,	...	11,641 ,,
Pulse	3,206 ,,	...	7,614 ,,
Sesamum	493 ,,	...	488 ,,
Garden produce	4,345 ,,	...	125,816 *cwts.*
Cumin seeds	418 ,,	...	1,461 ,,
Cotton	10,898 ,,	...	32,602 ,,
Forage	7,594 ,,	...	208,778 *bushels.*
Pasture	4,670		
In crop	54,716		
Uncultivated	46,810		
Total	101,526 *acres.*		

THE IONIAN ISLANDS.

Description.	Area.		Produce.
Wheat	14,404 *acres*	...	47,266 *bushels*
Barley, maize, and meslin	24,471 ,,	...	115,997 ,,
Oats	4,474 ,,	...	18,651 ,
Currants	17,332 ,,	...	15,255,980 *lbs.*
Olive oil	94,038 ,,	...	75,005 *barrels*
Wine	61,267 ,,	...	209,270 ,,
Cotton	1,640½ ,,	...	45,620 *lbs.*
Flax	1,847 ,,	...	69,118 ,,
Pulse	4,676 ,,	...	13,125 *bushel*
Pasture	35,204		
Salt	(*extensive*)	...	194,000 *kilom.*
In crop	255,912½		
Uncultivated	228,949½		
Total	484,862 *acres.*		

In the two following Tables, the figures of Mr. Porter and mine differed a little, being drawn from different sources; but they substantially present the same view, which is the object of their insertion.

TABLE II. — GARRISONS.

Details.	Gibraltar.	Malta.	Corfu.
Field Officers	12	9	15
Captains	32	16	32
Lieutenants	44	26	45
Ensigns	24	17	29
Paymasters	5	3	5
Adjutants	4	3	4
Quartermasters	5	3	5
Medical Officers	8	5	10
Sergeants	149	89	170
Drummers	60	37	69
Rank and File	2987	2132	3506
Total	3330	2342	3890

TABLE III. — MARKET PRICES.

Articles (1824).		Gibraltar.	Malta.	Corfu.
		s. d.	*s. d.*	*s. d.*
Beef	*per lb.*	0 6	0 4	0 3
Mutton	*per lb.*	0 4½	0 4½	0 3½
Veal	*per lb.*	0 8	0 5½	0 6
Pork	*per lb.*	0 3½	0 3	o 3½
Ham	*per lb.*	0 6½	0 6	0 5
Supressada	*per lb.*	0 10	0 8½	0 9
Tunny salted	*per lb.*	0 5	0 4	0 5½
Turkeys	*each*	5 6	6 6	6 0
Geese	*each*	3 0	3 2	2 10
Ducks	*each*	1 4	1 6	1 2
Fowls	*each*	1 5	1 4	1 2
Eggs	*per dozen*	0 9	0 5½	0 7½
Butter	*per lb.*	0 10¼	0 8	1 0¾
Lard	*per lb.*	0 6	0 5	0 6
Cheese (*common*)	*per lb.*	0 4½	0 3¾	0 4
Bread	*per lb.*	0 1¼	0 1⅛	0 1¾
Flour	*per lb.*	0 3	0 2¼	0 2½
Rice	*per lb.*	0 2½	0 2	9 2½
Beans (*dry*)	*per bushel*	2 0	1 8	2 0
Wine	*per pint*	0 2	0 1¼	0 2
Oil	*per pint*	0 5	0 6¼	0 4¾
Milk	*per pint*	0 3	0 2¼	0 2
Common labour	*summer's day*	1 6	1 4	1 5
Common red wine	*bottle*	0 3	0 2	0 2½
Charcoal	*per* 100 *lbs.*	1 2	1 2	0 11
Firewood	*per* 100 *lbs.*	6 8	8 6	6 6
Fruit and vegetables		*cheap.*	*cheap.*	*cheap.*
Groceries and spices		*reasonable.*	*reasonable.*	*reasonable.*
Salt and tobacco		*gabelled.*	*gabelled.*	*gabelled.*

PART II.

OF THE CURRENTS, TIDES, AND WATERS OF THE MEDITERRANEAN SEA.

§ 1. Preliminary Matter.

Mediterranean level.

WE have seen the boundaries of the Mediterranean Sea, the surface of which, generally speaking, must have maintained nearly the same level for at least 2500 years; and as the low coasts are not liable to be overflowed, a comparative permanence of periphery may be expected. For although there are vestiges of submerged buildings in various places—as at Santo Stefano (*Portus Domitianus*) in Tuscany, Capo d' Anzio (*Antium*), Alexandria (or rather *Canopus*), and other places—still the same waters show the piers of Caligula's Bridge (despite the numerous mollusk or sea-worm borings, which indicate that it has not been left untouched the while) at almost the same height above the Bay of Puteoli, as they were upwards of 1800 years ago; and there are similar silent but undeniable witnesses in the marine works at Marseilles, Genoa, Civita Vecchia (*Centum-cellæ*), Navarin, Makri (*Telmissus*), and many ancient moles and littoral edifices which I have examined. It is therefore probable that occasional elevations and depressions of the bed of the sea have periodically compensated each other, as asserted so long since by Aristotle; nor, indeed, is it possible that any great difference can exist, since, under extraordinary evaporation, or suspension of supply from rivers, the equilibrium must be restored within a very few feet by the ceaseless flowing inwards of the oceanic waters.* In this action, both the greatness of the aqueous volume, and its peculiar properties as a liquid amenable to the power of gravitation in every particle, give the full power of reciprocation and constancy.

Geological remarks.

These remarks relate rather to the progressive changes which are clearly indicated as being in actual operation, than to the more archaic and violent convulsions which, by internal forces, have elevated vast continents, as appears to be shown by the remains of sea-life at great heights in the fossiliferous beds of the tertiary formation. Indeed, it is sufficiently evident that the energies of the volcano, the earthquake, and the torrent, have been for ages in mighty action on those shores; large tracts of which exhibit the effects of such subversive agents during a long series of ages. The present outer crust of the earth seems to repose on an unstable basis, and it is obvious that the outlines and consequent areas of land and sea are variable and displaceable, and that they actually have been displaced; but whether any momentous disproportion has occurred through unequal agencies during the process, is at present an undecided question. Much that is still at work depends on the slope with which the land and sea meet, the nature of the materials composing the coast, and the usual set of the local tides and currents; and the workings of those ought to be constantly attended to. The investigation into the causes of such changes belongs to the geologist; yet the geologist, in future, must receive his basis from the maritime surveyor. The treating of currents and tides may therefore be ushered in by such normal considerations as my inquiries

* Dolomieu thought the shores a foot lower near Alexandria, than in the time of the Ptolemies; but he must have overlooked a few circumstances respecting the rocks, the ruins, and the excavated baths, which strongly militate against his supposition. Indeed, relics under water, and adjacent ports full of sand, are not uncommon on these shores; and we are everywhere struck with evidences of places historically very ancient, but geologically recent.

and labours led me to observe, in order to make an approach to the actual depth, mean temperature, density, saltness, and specific gravity of the Mediterranean Sea. Such topics, by opening the arcana of the penetration of solar light and the propagation of heat, will, in the end, infallibly furnish materials for ascertaining the habitats of many tribes of marine animals, and the distribution of living beings in the sea; points so important to know, but till recently neglected, or very imperfectly regarded.

Phlegræan zone.

The great alternating body of force, in the agencies alluded to, has been the wonderful Phlegræan region observed to extend from the east about 1000 miles to the west; that is, from the borders of the Caspian Sea to the Azores, and perhaps from thence to Teneriffe. This zone is arbitrarily considered as about ten degrees in breadth, and certainly is well marked by points of eruption, earthquake-shaken districts, and other symptoms of igneous action throughout. This may be instanced on the north and centre of the zone by the hot springs and violent commotions at Tiflis, Ararat, Azof, Constantinople, Palestine, Smyrna, Santorin, Milo, Modon, Mount Majella, Vesuvius, Lipari, Stromboli, Etna, Sardinia, the Colombretes off Valencia, Olot in Catalonia, and Lisbon. On the south extreme the marks are fewer, but I found indications of volcanic action in the Ghâriyán hills to the south of Tripoli, and on my journey to Ghirzah passed over a black and forbidding tract called Ha'ráj. From Algiers through Morocco, severe earthquakes have been of frequent occurrence; and that which destroyed Oran, in 1790, was simultaneously felt in Tetuan and Tangier. Along the waters of this devoted space, the communication is also marked by the occasional protrusion of an islet, and frequent shocks of the *mare-moto* or sea-quake; which last was perhaps the reason why Neptune, a sea-god, should have been designated the *Earth-shaker* by Homer. Many of the igneous spots have been extinct for ages, and the forces are looked upon as having decreased in energy; but volcanic eruptions from the subterranean and subaqueous bed of heated crystalline rocks, though they may have diminished both in number and force since the earlier ages of the globe, are still in constant action.

Violent earthquakes.

Severe earthquakes are ever accompanied by an agitation of the neighbouring seas, as was specially noted during the tremendous calamity at Lisbon in 1755: and the fact was observed and recorded very early, for Herodotus (*Urania*, § 64) mentions a convulsion of the earth which was felt out at sea, by the fleet of Eurybiades; and I myself have felt such shocks on several occasions. Eruptions from the bottom of the sea, so far as I could learn, exhibit their phenomena exactly as from those subaërial vents which open at once into the atmosphere: subject only to the modifications produced by the greater density of the surrounding medium, and the greater external pressure caused by the weight of the overlying column of water, which then becomes an element of the repressive force. Professor Pallas mentions that in September, 1799, a submarine eruption took place in the Sea of Azof, 150 fathoms from the shore, opposite Temruck, accompanied with dreadful thundering, emissions of fire and smoke, and the throwing up of ashes and stones; after which an isle—'like a great sepulchral hillock'—rose from the bottom; but which sunk again, before he could visit it. In 1814, another new island was raised on that spot, by volcanic explosions: and both were accompanied by earthquakes in the vicinity. On the 13th of August, in 1822, when Aleppo was destroyed by a terrific earthquake, which instantly buried thousands of the inhabitants under the ruins of their houses, two rocks arose from the sea in the vicinity of Cyprus (*Journal of Science*, vol. xiv., page 450), an island ever partaking in the disasters of Syria.

Singular phenomenon.

A curious occurrence befel myself. On the 5th of February, 1820, being off the Lover's Leap at Leucadia, in the *Aid*, the weather became overcast, with variable winds and rain at midnight. During the middle watch, a dense cloud-bank was seen in the offing; and towards daylight there was some appearance of an island discernible in it, but being considered a mere Cape Flyaway, I was not called. Just before six o'clock, however, the morning being unusually dark and heavy, Mr. Skyring stated to Lieutenant Hose that he very distinctly perceived an island to the west, whereupon the ship was immediately put about, and after the sails were trimmed on the other tack, the bearings marked on the log-board were—'Cape Ducato south-east by east,—the Island, west.' At this moment I came upon deck, and was far from being satisfied that danger was sufficiently threatened for the ship to be put about without my orders; but the weather was very cloudy, and the wind fresh from south-east. On examining some of the men, and observing it was still very thick to windward, we braced up, and from thence till sunset worked to windward; but neither then, nor the next day, could we regain sight of the supposed island. The weather continuing very unsettled, I went into Port Bathi in Ithaca, where, on the 15th of the same month, we felt nine distinct shocks of *mare-moto*, during heavy rain and dark squally weather; giving the sensation of having grounded, although in fifteen fathoms water. From that day to the 6th of March, a succession of earthquakes occurred, and Santa Maura was particularly affected; nor should it be overlooked that in the same year, Zante was nearly destroyed.* In the following April, being at Corfu, Admiral Sir Anthony Maitland, then commanding the *Glasgow*

* It is advanced that earthquake shocks are not felt in the various Ionian Islands at the same instant of time. How this is ascertained we are not told; but when I was there, clocks and watches had a variance with each other of at least twenty minutes. It is clear that on these occasions, all the vicinity, and the ships in their ports, have unmistakeable notice, and apparently at the same moment.

frigate, informed me that a Greek vessel had arrived, with a report that they had passed heavy breakers off Santa Maura; on this I immediately weighed and ran down, placed sharp eyes at the mast-heads, sounded occasionally with 70 and 150 fathoms of line, but returned *re infectâ*. The myth became the 'rocky island' of report; and having gained a footing in the papers, was copied by Sir Charles Lyell from the *Allgemeine Zeitung* for 1820; and other authorities have recorded it. This may possibly have been an effect of submarine eruption; and it is not a little curious, that various old charts—as those of the Quarter Waggoner, Mount's and Page's Mediterranean Pilot of 1703, and others—have represented a small islet to seaward of Cephalonia. When I had written the above, it occurred to me that Lieutenant C. R. Malden, now the sole survivor of the *Aid's* gun-room officers, might have preserved a note on this subject, and I wrote to him for information on that point: the following is an extract from his answer, dated 4th August, 1852:—

> Away from home, I have not the means of verifying the date you mention, the 5th of February, 1820, of the discovery by the *Aid* of an island off Santa Maura. The fact is, however, fresh in my recollection: indeed, it is within the present year that I was talking of the cirumstance to some friends. I was not on deck, and did not see it, but I well remember the confidence with which Skyring and those on deck spoke of having seen an island, and the immediate inference that there must have been a shock of an earthquake at Santa Maura, and which I always understood we found to be the case on our arrival in port. Our subsequent cruise in search of the island was fruitless, but I well remember it.*

Experience has proved a constant relation between the earthquake and the interrupted activity of volcanoes, where the inner forces are not inert; and from the constantly-recurring mutability of the terrestrial crust, there can be little doubt of the existence of subterranean caverns communicating with active craters, however such commu- Volcanoes.

* The German naturalist, F. W. Sieber, experienced a severe *mare-moto* near this spot, on the 3rd of January, 1817, while on his passage from the Adriatic to Candia.—(See his *Travels in Crete*).

nications may be impeded, or even become closed, and then generate earthquake. This is not at all a new assumption, or Strabo seems to have considered such vents as safety-valves to their neighbouring districts; and, to be more exact, I found among the manuscripts of the Royal Society, a letter from their Secretary, the indefatigable Henry Oldenburg, addressed to Mr. George Cotton, at Rome, 30th of June, 1669, in which he says, 'And since the present eruptions of Mount Ætna are so considerable as that they fill the eares of all Europe with dreadfull reports, as well as they heape Sicily with great calamityes, my further request is, that you will please to send us the best observations that have been made of that horrible fire, and of all the circumstances and effects of it, and particularly of what kind of mineralls the fiery streams running through the valleys of Sicily did consist, as also what appearances there are in other neighbouring vulcans, as in the Strombylo and Vesuvius.'

Stromboli. In my account of 'Sicily and its Islands,' published in 1824, I mentioned that the crater of Stromboli has burnt without intermission from the earliest periods; appearing to be not only the vent of its own group, but to have a subterraneous communication also with Sicily and Italy, for, previous to a severe earthquake's taking place in those parts, Stromboli—after much internal rumbling, *rimbombi e mugghiti*—has been observed to be covered with dense clouds and smoke, and to emit, with increased activity, unusually ardent flames. And I said of the cove in which the cone rises abruptly from the sea, 'it is natural to imagine it would, from the constant action of the volcano, and the incessant discharge of matter for so many ages, be very shoal, or, at least, even allowing the stones to triturate, that a bank of sediment would have been deposited; the contrary, however, is the case, for I found gradual soundings of from four to twenty fathoms all round the coasts, even to the two points of Sciarazza Cove;

but immediately under the cone, as nearly as I could approach, and even within range of the ejected matter, there were forty-seven fathoms, and at the distance of a few yards, from sixty-five to ninety: an inspection of the chart will point this out more clearly. The circumstance is curious, and has not a little puzzled the sages of Stromboli, who at length, after serious deliberation, have decided, that a gulf at the base of the island, continually absorbs the ejections, and replenishes the volcano.' For particulars of the crater's action, as watched by myself from its margin, the reader is referred to the volume above mentioned, from which this passage is quoted.

In the case of subaqueous eruptions it is, of course, difficult to ascertain whether they take place from a new or an habitual vent, from an insulated or a parasitical cone. Von Buch, the geologist of the age, has given his opinion that, in submarine eruptions, the strata previously forming the bottom of the sea are uniformly elevated, and that positive eruptions do not take place from the vent until these strata have been raised above the level of the sea; but this postulate is far from being satisfactory. The protrusion of conical islets by the elastic force of volcanic action, of course lifts up the overlying horizontally-deposited strata; yet there is no reason *à priori* for supposing such an anomalous distinction between the mode of action of subaqueous and subaërial volcanoes. In the recent and well-observed cases of Sabrina Isle being thrown up off the Azores, and Graham Island in the Mediterranean, the substances which showed themselves above water were either ashes and cinders; or vesicular, lithoidal, and conglomerate lavas, the products of the eruption by which they were raised. Respecting this last-named volcano there are fortunately very detailed accounts, of which perhaps the most accurate is that by Dr. Davy, brother of the late Sir Humphry Davy. It seems that, as early as the 28th of June, 1831, Captain Swinburne, in passing nearly over the spot,

Submarine volcanoes.

felt several shocks of a sea-quake, proving that the cause was then in operation; but on the 19th of the following July the crater had accumulated to a few feet above the level of the sea, and was in great activity, emitting vast volumes of steam, ashes, and scoriæ. From that time it gradually increased in all its dimensions, till, towards the end of August, its circumference was about 3240 feet, and its height 107: then from October various changes took place, and it entirely disappeared in December. As there were certain mistakes propagated by the *Journal of the Geographical Society* and the *Quarterly Review*, as to this *Spiraglio*, and in order that physical inquiry might start fair, I addressed a note to the Royal Society, 'On an Error respecting the Site and Origin of Graham Island,' which is printed in the *Philosophical Transactions* for 1832.* I therefore need here only add what may be deemed the ultimate result, as contained in an extract from a letter written to me by Captain Graves, formerly of the *Adventure*, dated Malta, 20th June, 1846:—

> I have just returned in the hired vessel *Locust*, from a very pleasant cruise to Graham's Shoal, Girgenti, and Palermo; and I therefore lose no time in reporting to my old commander what I found upon his ground. Graham's Shoal I spent two days in examining, with your chart in hand. Since Elson (*late Master of the Adventure*) was there in 1841, the shoal itself has altered much both in depth and extent; it then had a sharp pinnacle, with one-and-a-half fathoms on it, and the water suddenly deepened all round, while the bottom was irregular, and composed of lava, cinders, &c. Now it has sunk down to a depth of thirty-five fathoms—as much under water as it was above it at its greatest recorded elevation—and as it descended it gradually spread out in extent, so that it now forms a flat bank, on which the sand and coral are already making a crust. Indeed, its present actual state is similar to those banks marked in your chart as Nerita, Triglia, Pinna-marina, &c., all which are probably extinct volcanoes.

Singular chasms.

The Italians have recorded instances of chasms suddenly formed by openings in the water, accompanied by discharges both of smoke and ashes; in one of which a vessel was lost

* On showing the proof-sheet of this to an excellent naval friend, he advised me to obtain permission from the Royal Society to reprint my letter in the Appendix to this work.

in 1813, in sight of the Neapolitan corvette *Stabia*, commanded by Captain Acton. The *Journal de Constantinople* also states that, on Sunday the 4th of April, 1847, one of these phenomena occurred in the Black Sea. An Austrian steamer of Lloyd's Company, the *Stamboul*, was proceeding to Constantinople in a calm state of the weather, and was within an hour's distance of Synope,—

When suddenly the sea opened under her; assuming the form of a vast funnel; the waves, in closing, covered her almost entirely, swept the deck, and did the most serious damage. The shock was so violent, that several leaks were sprung, and the vessel was some time in recovering herself from this terrible pressure, and getting fairly afloat again. She rose, however, after some pitching, but injured to such an extent, that if another shock had taken place, she would have inevitably been lost. It was with the greatest difficulty that she reached the port of Synope to refit, after which she proceeded to Constantinople, where she arrived safe last Tuesday. Those who were witnesses of this accident, thought at first it might have originated in an earthquake, but nothing of the sort has occurred elsewhere. It must be admitted that some submarine dislodgment opened under the vessel an abyss into which the waves rushed, and in this way they formed a gulf, in which she narrowly escaped being smashed and swallowed up.

Other causes of coast-changes.

The causes of successive alterations in the sea-board, other than volcanoes, are sufficiently obvious. Large quantities of detrital matter are at first carried with velocity down streams and torrents; but towards the mouths and deltas of large rivers, the action is so moderated that the waters bear a very diminished transport of terrene substances. Tidal sets and oceanic currents carry only the finer comminuted silts over the areas they traverse, except where narrow straits, projecting headlands, or other peculiar local features, interrupt them; or, not least of active causes, where breakers under the influence of prevailing winds exert both destructive and transporting powers. Gravel or sand contained in water having a specific gravity greater than that vehicle, can only be kept in suspension so long as the water is in a state of agitation; and as soon as such troubled water becomes quiet, the silt and other impurities gradually subside: nor is absolute repose necessary for this process; as the meeting of currents and the action of winds both affect the direction and accumulation of the deposit. But the cutting and scooping actions of streams, and their effect at great depths, cannot be so extraordinary as has been assumed by some writers, since the friction of the moving body is influential in an inverse ratio to the velocity of the super-current, so that the lower waters must inevitably be retarded: this seems to have been proved in a great measure by descents in a diving-bell. Hence it follows, that the bottom of very deep waters must remain comparatively undisturbed. Still the collective amount of soluble matter continually brought down by rivers and torrents, as well as what is even derived from rocks by the percolation of water through them—together with the action of frost, earthquake, and the undermining of the sea—must become sensible in the course of centuries—independently of the vast chemical changes which nature carries out, and upon which we have yet so much to learn.

Origin of the Mediterranean. But though the present intention is rather to point out a few palpable instances of what may be termed the secondary geological efforts, it may be necessary to mention that the formation of the expanse of the Mediterranean waters, commencing at such a mere frith as that of Gibraltar, and reaching as far as the sea of Azof, has been a subject of speculation from the earliest ages. The main features are sufficiently remarkable to awake inquiry; and the narrowness of the entrance, together with its local currents and tides, has excited conjectures both paradoxical and philosophical. A hasty glance at some of the most prominent will here suffice; only premising that we have no intention of entering upon a solution of the impossible problem of how and by what means the Mediterranean was brought into its present state, nor of either adopting or rejecting any of the several theories on the subject—mention is made solely for the purpose of aiding future examination.

By the earlier writers, poets, and mythologists, the two barrier mountains called the Columns of Hercules were said to have been united, till that wonder-working hero separated them by digging a communication between the Atlantic and Mediterranean seas. But among those who disregarded fiction, it was a prevailing notion that this opening had been forced by an impetuous onset of the inner or outer waters, as may be gathered from the allusions and testimonies of Strato Physicus, Aristotle, Diodorus Siculus, and Seneca: and what with the deluges of Deucalion and Ogyges, it may be inferred through all the subsequent poetical embellishments, that the shores of the Black Sea and Archipelago were actually twice devastated by sudden inundations of the sea, more than 3000 years ago. A vast *débâclage* followed the presumed disruption of the Euxine barrier, and the consequent mighty rush of water westwards.

Arabian opinions.

These opinions prevailed in different degrees, and swayed the minds of men for ages: but when the Arabs had substituted Mohammedan for Greek traditions, after the Tenth Century of our era, a new system was put forth. Thus, in the middle of the Twelfth Century, we hear Edrisi, or to give him his due, *Abu-Abdallah Mohammed ben Mohammed ben Abdallah ben Edris*—a noble Arab, born at Ceuta, and therefore probably well versed in the popular traditions of the Strait. He was the writer of a geographical work commonly known, since 1619, as the *Geographia Nubiensis* — from a false reading in the translator's manuscript, which made the author speak as a native of Nubia. It was, however, a description illustrative of a large silver terrestrial globe, constructed for his patron Count Roger, of Sicily, in the year 1153; his book was therefore known as *Ketáb Rujár* (Roger's book), though its true title was *Nuzhat al-mushták fi ikhtirák al-áfák* (the commencement of the journey of one who is desirous of travelling through the regions of the earth); it was also often cited as *Dhik al Memálik w-al Mesálek* (an account of kingdoms and countries). The complete work was long to be found only in the Bodleian Library at Oxford; but other copies having about twenty-five years ago been procured for the King's Library at Paris, M. Jaubert was persuaded to publish the translation of it, which appeared in two quarto volumes, in 1836, under the patronage of the French Geographical Society. Here we find that Edrisi terms the Mediterranean *Bahr-al-shám*, or Sea of Syria; and gives 1136 parasangs as its length from the commencement to its termination. He then continues, in his account of the Fourth Climate, § 1.—

The Syrian Sea was, as it is said, originally a lake enclosed on all sides, such as the Sea of T'aberistan is at present; its waters had no connexion with those of the neighbouring ocean. The inhabitants of Africa and Andalus (*Spain*), were constantly at war with each other, and mutually doing all the mischief they could, till the time of Alexander (*Iskender*). He came to Andalus, and the people of the country informed him of their contentions with the men of Sús (*Africa*). He therefore assembled labourers and engineers, and fixed upon the place of the Strait (*el-zokak*),—it was a hollow depression in the mountains. He then ordered the geometers to measure the ground, and take the level of the waters of these two seas; and he found that the level of the Great Sea was only a trifle above that of the Syrian Sea; he then commanded the soil to be dug away between the country of Tánjah (*Tangier*) and that of Andalus. The digging was continued as far as the lower part of the mountains of those countries. He then built upon them a mole of stones, twelve miles in length. He built, also, another opposite to it in the neighbourhood of Tánjah. Between these two moles there was a width of six miles. When the two bulwarks had been finished, the digging was continued till they reached the waters of the Great Sea, which rushed in with the greatest violence between these two moles, overwhelmed many cities on each side, drowned their inhabitants, and rose above the two moles nearly to eleven *statures* (eleven times a man's height). The mole which is near the land of Andalus appears very plainly when the water of the sea is smooth, near the place called *Es-safíhah* (the level); its length is in a straight line, and is a cubit in breadth, as measured by Al-Rabīī. I myself have seen it, and have sailed the whole length of this strait along this side of it. The people of the island (*Al-jezīrah*) call it the bridge; and the middle of it corresponds with the place where the stag's rock overhangs the sea. The other mole, which was near Tánjah, was carried away by the violence of the waters.

Remark on Edrisi.

Such is Edrisi's account; and it is not a little singular that where he places this mole, as appears from his own observation, I found that there is actually much less water than on either hand in the vicinity, as will presently be shown; and this is a fact which escaped notice from his time until my soundings were taken. It is also clear that his island is not Algeçiras but Tarifa, and he must have been conversant with the shoal off it, now called the Cabezos; for there can be no reasonable doubt but this is the Spanish portion of the mole which, according to Edrisi, 'appears very plainly when the water of the sea is smooth,' and which he himself—believing the legends and imagined aspect of the bottom of the sea there—had seen. It is moreover certain that the water on the African shore is the deepest; and the difference of level which he assigns between the two seas, is precisely what a 'geometer' would now find it—'only a trifle.' It is therefore, on the whole, a very curious description.

Nearly three centuries before Edrisi's time, according to Renaudot, the two Mohammedan travellers, or at least Abú Zeïd al Hasan, who wrote in A.D. 877, described the communication of the Mediterranean with the *eastern* ocean as a recent discovery. But as those early Arabian voyagers were better acquainted with India and China, than with the shores in question, they would not have been called in evidence but for a remarkable passage which proves—if Renaudot has rightly translated this sentence—that the Arabians must have adopted their geographical notions from the Greeks; for Abú Zeïd thought that the Indian ocean washed the coast of Tartary, fell into the Caspian Sea, and so by the Propontis into the Mediterranean. The story upon which this opinion was founded, runs thus:— Abú Zeïd al Hasan.

In our time, a discovery has been made of a circumstance quite new and unknown to those who lived before us. Nobody ever imagined that the great sea which extends from India to China had any communication with the sea of Syria; nor could any one apprehend the possibility of any such thing. Now behold what has come to pass in our days, according to what we have heard. In the Sea of Roum (*Mediterranean*) they found the wreck of an Arabian ship, which had been shattered by tempests; for all her crew had perished, and she being torn to pieces by the waves, her remains were driven by the winds and weather into the sea of the Khozars (*Euxine*), and from thence through the canal of Constantinople into the Mediterranean Sea, and were at last thrown on the coast of Syria. Hence it is evident that the sea surrounds all the country of China and Sila, the extremity of Turkistán, and the country of the Khozars: and that it passes through the strait till it washes the Syrian shore. This is proved by the structure of the vessel of which we are speaking, for the planks were not nailed, but joined in a peculiar manner, as if they were sewed together: all those built in the Mediterranean, or on the coast of Syria, are nailed together, and are not joined in any other way. None but the ships of Siráf are so fastened.

Now instead of the massoolah-boat suggested by a grave reasoner on this narrative, a supposition which is utterly inadmissible, it is within compass to suppose, that it was a wretched Sarmatian vessel from the Black Sea.

General idea.

Leaving this *non sequitur* to the Middle-Age geographer with whom it originated, it may be observed that tints of the Ogygian impressions still colour the mists of this question. Some of the most thinking among modern writers have supposed the Mediterranean to have been once a vast lake, the waters of which, having been suddenly increased by the irruption of those of the Black Sea, in consequence of some violent cataclysm, forced a passage through the Gut, and produced the awful inundation which submerged the island of Atlantis. This bursting of lakes has been a more favourite point with cosmogonists of former ages, than it is with present theorists. And with judgment: for nature adapts all her works most admirably to her designed end. No one ever yet saw a lake the barriers of which were suddenly incapable of resisting the contained water. It therefore follows that the flood, earthquakes, subsidence and upheaving of strata, must be the consequence of extraordinary volcanic action—one of the most powerful agents in changing the form of the earth's surface.

Buffon's theory.

The hypothesis of a rush of waters from the east was long a popular axiom, and held its sway until it was combated in form by the Count de Buffon. This eminent naturalist objected to the premises of his predecessors, on the ground that it is the Ocean which runs into the Mediterranean, and not the latter into the former. 'Cette opinion,' he says, 'ne peut se soutenir, dès qu'on est assuré que c'est l'Océan qui coule dans la Méditerranée, et non pas la Méditerranée dans l'Océan.' He further viewed the inner sea as having been a large lake, and considered that the Strait of Gibraltar was owing to a sudden disruption produced by some accidental cause, as an earthquake or a sinking of land, or otherwise by a violent effort of the ocean. This he strengthened and supported on the supposition that similar strata are observable at equal heights on both sides of the strait; and he concludes that when the ocean broke through this barrier, it rushed with overwhelming velocity into the lake. Here, aided by the former bursting of the Euxine through the Bosporus, the waters inundated the continent, transformed its marginal plains and valleys into sea-gulfs, and left only the eminences uncovered which now form Italy and the various islands. This idea, to which some of the ancient inferences closely approximated, was hardly originated by Buffon, though many of the details might be his own. Long before he took the field, the authors of the *Universal History* (vol. iii., p. 239, folio edition, 1744) had published these words:—'In the hypothesis of the ancients, the *Palus Mæotis*, the *Pontus Euxinus*, the *Propontis* and *Mediterranean*, were originally so many lakes, which, after having broke down, as it were, the dikes that parted them, with the impetuosity of their waters, opened themselves a passage between the mountains of Atlas and Calpe into the ocean. It is, perhaps, more likely that the ocean, having with the impetuosity of its waters dismembered the mountain of *Calpe* from the lands of Africa, poured itself into that vast space now called the Mediterranean, and, penetrating to the north, produced the *Propontis*, the *Pontus*, and the *Palus Mæotis*.'

So far one reverie was as good as another; and even the old Samothracian tradition relative to the bursting of the Bosporus—whether in consequence of volcanic action, of which vestiges are still observable, or from the pressure of accumulated waters—must not be lost sight of; especially as the forms of the islands, and their denudation, countenance the idea of their having been effected by a sudden and violent rush of waters. But Buffon, attached to theorems, became ambitious of taking the lead, by establishing his inference as a final authority, and a problem proved by facts: and, indeed, it was more palatable than his patronage of the dreams of spontaneous generation. The quidnuncs who criticised his speculation insisted that, if his opinion was of any value, the eastern waters of the Mediterranean would inevitably prove to be comparatively shallow. Whereupon his bosom friend, M. Sonnini de Manoncourt, asserted that he had himself found this confirmation of the theory, having, at Buffon's request, sounded the depth of the sea between Sicily and Malta, where he gained bottom in from twenty-five to thirty fathoms; and that in the middle of the channel, where the water is deepest, the depth never exceeded 100 fathoms; and further, that between Malta and Cape Bon in Africa there is still less water, the lead indicating no more than from twenty-five to thirty fathoms throughout the whole length of that passage. Now this, to use the softest term such an assertion admits of, is outrageously incorrect, for I can, from repeated trial, declare that thirty fathoms are only obtainable in-shore, till the experimenter arrives at the Adventure Bank; and there are parts to the north-west of Malta, right in the line for Cape Bon, where I have been unable to strike bottom with 500 fathoms of line. Again, we are assured that the same industrious explorer discovered that the Levant Basin is very shallow indeed, especially between the Archipelago and the shores of Libya: in this case, surely some astonishing change must have occurred, or how could I have found profound depths in the same space in less than half a century afterwards! I obtained from 70 to 90 fathoms, deepening to 150 and 250, at a small distance from the shore all along

Sonnini's assertions.

the Libyan coast; and tried in vain for soundings with 500 fathoms of line further out. At this time I was not aware of the Sonnini discoveries,* and the argument derived from them, or I would have attended more closely to the question. But on becoming acquainted with them afterwards, and to give that gentleman fair play, I wrote to my former assistant, Captain Graves, to try the northern side of the Levant Basin, in the neighbourhood of Cyprus. That active officer found 100 fathoms depth at about two miles from the island, and at about double that distance he had no bottom with 200: between the west end of Cyprus and the coast of Caramania, he struck ground in 650 fathoms; but about halfway between Cyprus and the coast of Egypt, he payed out 1000 fathoms of line without finding bottom. Hence we see that the expounders of Buffon's theory were all adrift; and also that the oracular prediction of Cyprus joining Asia Minor by the action and deposits of the Pyramus, is not likely to be yet fulfilled: but, in sober truth, it must be admitted that we are very little acquainted with the depth of the deeper parts of those waters.†

Captain Graves.

Quitting these portentous visions, however, which are supported by little more than doubtful conjectures, it is quite within the grasp of induction to investigate the effects and alterations which have resulted under a system of traceable causes, some extinct, but many still in ceaseless operation. Yet in stating the geological knowledge which we seem so recently to have obtained, it is impossible to forget that theoretically there is hardly anything absolutely new; for the system of

Ancient geological views.

* Sonnini was an oracle in some circles. He suggested that Candia was detached from Africa by an inundation of the low lands which formerly united them; an opinion, he says, which acquires an additional degree of probability when we direct our attention to the shallowness of the channel which separates them, 'whose bottom everywhere affords soundings!' For Sonnini's accuracy of observation, even above water, see page 70.

† For instance, while reading the proof of this sheet, a letter reaches me from Sir Francis Beaufort, informing me of a new search for the supposed rock 90 miles east of Malta, in which no bottom was gained with 2500 fathoms of line. We shall return to this.

Pythagoras. Pythagoras, as preserved by Ovid, is exceedingly applicable to our present notion of various phenomena in the inanimate world. Eratos-thenes. Eratosthenes (*apud Strabonem*) asked how is it that 'two or three hundred stadia inland there are still found numerous sea-shells, as well oysters as mussels, especially near the Temple of Jupiter Ammon, and all along the route leading to it?' And he cites with praise the Strato. Xanthus. names of Strato and Xanthus, as able geologists; of whom the latter, struck by the petrifactions which he saw far from the sea, boldly pronounced that Armenia, Media, and Phrygia were formerly under the sea. Much of this was Pliny. Lucan. adopted by Pliny; and perhaps inspired Lucan with his well-known prediction on the changes which have taken place in the Syrtes since his day; nor can this paragraph be better closed than by an extract from the Pythagorean philosophy (OVID'S *Metamorphoses*, book xv.), as rendered by Dryden:—

> The face of places, and their forms, decay;
> And that is solid earth that once was sea:
> Seas in their turn retreating from the shore,
> Make solid land what ocean was before;
> And far from strands are shells of fishes found,
> And rusty anchors fix'd on mountain ground.
> And what were fields before, now wash'd and worn
> By falling floods from high, to valleys turn,
> And crumbling still descend to level lands;
> And lakes, and trembling bogs, are barren sands.

---

§ 2. The Divisions, Temperature, Colour, Luminosity, and the various substances found in the waters of the Mediterranean Sea.

Mediterra-nean features. THE northern and southern shores of the Mediterranean are greatly contrasted in feature; the former expanding into peninsulas, isthmuses, sinuosities, and islands, while the configuration of the latter presents comparatively but little articulated variation of form. Exclusive of the Black Sea—which, however, must be considered as a part of it—

this sheet of waters is naturally divided into two vast basins; and these again are subdivided into particular portions. In the days of Strabo, this expanse was distinguished into three basins; the first comprehended the space between the columns of Hercules and Sicily—the second, that between Sicily and Rhodes—and the third, the sea between Rhodes and the shores of Syria.

The first great basin of the modern division extends from Gibraltar to Cape Bon and the Faro of Messina, washing the bases of the Pyrenees, the Alps, the Apennines, and the range of Mount Atlas; and it is again subdivided into two unequal parts by the islands of Corsica and Sardinia. The second, or inner grand basin, is of twice the area of the first; it extends from the coasts of Tunis and Sicily to those of Egypt and Syria, stretching on the north into two distinct and separate basins, known as the Adriatic and Archipelago; while on the south, the Gulf of Libya penetrates deeply into the African continent. The eastern portion of this basin is interrupted by Cyprus alone: it was anciently subdivided into the Pamphylian, Syrian, and Phœnician seas; but is now universally known as the Levant—a term, however, more proper for its coasts than its waters. **Basins**

Such are the commonly received designations; but geographers and pilots are rather vague in their denominations of the several subdivisions; and many portions being without physical boundaries, are only distinguished by epithets properly applicable rather to particular spots. Thus the space included between the Balearic Islands and the coast of Spain (*Mare Balearicum*), is called at times the sea of Majorca, and at others the sea of Valencia; to the west of Italy the waters are called the Tyrrhenian, the Ligustic, the Tuscan, or the Italian indifferently; and when viewed with regard to the Adriatic, they are often termed the Lower Sea, the two being the *Mare superum* and *Mare inferum* of ancient geography. The Sicilian sea **Subdivisions.**

washes the shores of the numerous isles which stud the centre of the Mediterranean; to the east of which—between Sicily and Western Greece—it joins the Ionian sea. From the south of Italy to the shores of Albania is the mouth of the Adriatic, or Gulf of Venice; and the space embraced by Greece and Asia Minor is the Ægean or Archipelago—the White Sea of the Turks. From this last a strait—the well-known Dardanelles (*Hellespontus*)—conducts us to the Sea of Marmora (*Propontis*); and another, now called the Strait of Constantinople (*Bosporus Thracius*) leads to the once-dreaded Black Sea (*Pontus Euxinus*). To the north-east of this breadth of waters is the Sea of Azof (*Palus Mæotis*), the utmost maritime limit in that quarter; though some insist that a strait once connected this part with the Caspian Sea, along the very peculiar and depressed plain beyond the Caucasus. The authorities are strangely discordant.

Temperature.

From the laws of gravitation it is inferred that the surface of the ocean is always at the same distance from the centre of the earth, and before the discovery that far above the freezing point the specific gravity again decreases, it was supposed that the temperature of its water decreases in proportion to its depth; whence some concluded that the profoundest gulfs must be coated with eternal ice. This theory is raked by fact. The result of my experiments leads to the conclusion that there actually exists a very sensible diminution between the surface temperature, and that obtained at great depths; and the difference may be roundly estimated as about one degree for every twenty fathoms of line near the surface, save where the agency of subterranean currents may be at work, for such streams are undoubtedly connected with oceanic influences: but below about 180 fathoms, to our utmost depths, the temperature varied but little from 42° or 43° of the Fahrenheit scale. We found that at equal depths the warmth is rather higher alongshore than in the offing, still no reliance can be placed here upon thermometrical indications of an approach to land, or a great bank, as taught in the Atlantic Ocean: and the supposed heating of the waves is a mistaken sensation produced by the cooling of the atmosphere in the mean time. The mere surface temperature is very variable, according to the weather and the altitude of the sun, differing at sunrise and in the afternoon by three or four degrees, and even more.

For my own experiments, I caused a hollow perforated cylinder to be made as non-conductive of caloric as the metal would permit, in which was placed one of the excellent self-registering thermometers invented by James Six. This was occasionally cast overboard to a depth of eight fathoms, and the mean results during several years give a difference from the temperature of the air of only 1·5 to 3·5; the greater variations being in the summer months. A comparison of my eight-fathom observations, with those of the ocean furnished by Mr. Purdy, led me to conclude that the Mediterranean waters average about 3·5 of Fahrenheit more heat than those of the western part of the Atlantic Ocean. Around Sicily I found the comparative summer temperature still higher, even at a greater depth, it being 10° or 12° warmer than the water is stated to be outside the Strait of Gibraltar; which accounts for a greater evaporation, and consequent effect on the currents. The surface temperature of the waters near Sicily was often higher than what was shown a few fathoms below it; but as that condition naturally depended in great measure on the state of the superincumbent atmosphere, and was a complicated condition, it was not included in the general estimate. Experiments.

The usual tint of the Mediterranean Sea, when undisturbed by accidental or local causes, is a bright and deep blue; but in the Adriatic a green tinge is prevalent; in the Levant Basin, it borders on purple; while the Euxine often has the dark aspect from which it derives its modern appellation. The clear ultramarine tint is the most general, and has been immemorially noticed, although the dia- Colour.

phanous translucence of the water almost justifies those who assert that it has no colour at all. But notwithstanding the fluid, when undefiled by impurities, seems in small quantities to be perfectly colourless, yet in large masses it assuredly exhibits tints of different intensities. That the sea has actually a fine blue colour at a distance from the land cannot well be contradicted; nor can such colour—however influential the sky is known to be in shifting tints—be considered as wholly due to reflection from the heavens, since it is often of a deeper hue than that of the sky,* both from the interception of solar light by the clouds, and the hues which they themselves take. This is difficult to account for satisfactorily, as no analysis has yet detected a sufficient quantity of colouring matter to tinge so immense a body of water: wherefore Sir Humphrey Davy's supposition of an admixture of iodine cannot be admitted, for its presence is barely traceable under the most careful analysis. Those who contend for there being no colour at all, may remind us that the blue rays are the most refrangible; and that being reflected in greatest quantity by the fluid (which, because of its density and depth, causes them to undergo a strong refraction) they cause a tint which is only apparent. Be that as it may, seamen admit of one conclusion—namely, that a green hue is a general indication of soundings, and indigo-blue of profound depth.

Colour.

Luminosity.

The peculiar occasional luminosity of this sea was particularly noticed by Pliny and many elders, and, in common with that of other waters, it has long been a subject of scientific inquiry, rational conjecture, and ignorant wonderment; and it is really as difficult of a full solution as it is superbly beautiful in effect. Every assignable

* When the surface of the sea reflects heavy clouds, so as to be apparently darker than usual, it is looked upon as the prognostic of bad weather.

cause has been advanced; putrescent fish, electricity, atomic friction, cosmical vortices, absorption and emission of solar beams, and what not, have all and severally been brought forward, and after various tilts of discussion, laid aside again. But most naturalists now impute this phosphorescent appearance partly to the decomposition of animal substances, and partly to the countless myriads of mollusca, crustacea, infusoria, and other animalcules which can voluntarily emit a luminous brilliance, the chemical nature of which is still unknown.

Regarding the constituents of the Mediterranean water, the analysts of former days were at considerable variance; for while some—observing the rapid desiccation of salt from its brine—maintained its muriate of soda to be considerably above that of the average of oceanic water, others would not admit that there existed any difference. Recent experiments, however, have shown that the water of the Mediterranean contains full four per cent. of salt, while that of the Black Sea has only a smaller proportion. M. Bouillon la Grange investigated the subject with great perseverance; and his conclusion is, that assuming the proportion of saline matter in the water of the Atlantic Ocean to be 38, that of the English Channel will be 36, and that of the Mediterranean 41. Components.

On returning to England in the winter of 1820, I became acquainted with the late Dr. Marcet, who was then studying the chemical composition of sea-water procured from different parts of the globe. Among other matters, he mentioned that Mr. Smithson Tennant and himself had been extremely desirous of getting specimens from great depths at and near the Strait of Gibraltar, in order to ascertain how the inner sea rids its waters of their excess of salt; and that having furnished Dr. Macmichael with an adapted machine, that gentleman procured them some water in the Strait at a depth of 250 fathoms, but found it fruitless to attempt to gain bottom, 'from the impossibility Dr. Marcet.

of reaching it, on account of the great depth at that spot.' Hereupon I assured him that bottom should be struck on my return thither, as I had then no notion of the difficulty to be encountered; for having found ground between Tarifa and Tangier in 160 fathoms, it was natural to conclude there would be no violent difference through the whole Strait. But we found to the east that no bottom was obtainable with 1000 fathoms of line, and that from the ship's drift, the line quickly formed a diagonal curve. By the employment of two vessels, however, so that the headmost could cast the line, and be met by the drift of the sternmost one while the weight was descending, we obtained soundings throughout this celebrated opening in 1824, for the first time since it was navigated. The depths varied from 700 and 750 to 950 fathoms, which last was struck as nearly as possible in a vertical line, on a bottom of gravel and sand, mixed with spoils of testaceous creatures and coralline fragments. Indeed, the whole Mediterranean Sea is so much deeper than would be expected from its proximity to surrounding lands, that it seems to be what, in geological dynamics, is termed a sunken basin.

Strait of Gibraltar.

But the revealment of this chasm was not more surprising than was the result obtained by the analysis of water procured in the immediate vicinity; for except in that instance, we detected no difference of density with increasing depths: and the whole is so remarkable, that I will give the account in Dr. William Hyde Wollaston's own words—it being the last document addressed by that philosopher to the Royal Society, before whom it was read on the 18th of December, 1828:—

Dr. Wollaston's analysis.

The object of the present communication is to do justice to the memory of my late friend, Dr. Marcet, by recording the result of one of his latest efforts in the cause of science.

In his examination of sea-water, of which he gave an account in the *Philosophical Transactions* for 1819, the specimens with which he had been supplied from different depths in the Mediterranean had not been sufficient to show what becomes of the vast amount of salt brought into that sea by the constant current which sets eastward through the Straits of Gibraltar.

For though the escape of the water of that current may be fully accounted for by its evaporation, which must be very rapid and copious on the sunny shallows of Africa, yet the salt which that water held in solution must remain in the basin of the Mediterranean, or escape by some hitherto unexplained means of exit.

In the hope of obtaining a more abundant supply of water from the greatest accessible depths, especially near the Straits, he begged assistance from Captain William Henry Smyth, R.N., who was engaged to make a survey of certain parts of that sea; and supplied that officer with the apparatus for raising water from great depths, which was contrived by Mr. Tennant, and is described in the communication already referred to.* Apparatus.

The zeal with which Dr. Marcet himself prosecuted his inquiries was so well known, that others were always willing to second his efforts, from a confidence that their labour would not be unprofitably wasted; and Captain Smyth did not fail to take every opportunity of collecting specimens in the course of his survey. But when he heard that Dr. Marcet was no more, not being aware of the interest with which the specimens would be received and examined by many surviving friends, he was unfortunately but too ready to oblige other persons with portions of his collection, which were afterwards applied by them to other objects.

Nevertheless, at the time I had the good fortune to be introduced to Captain Smyth, in the month of June, 1827, he still retained in his possession three bottles, the remainder of his stock, and at my request most obligingly sent them to me for examination. Specimens analyzed.

Happily, one of them is such as to accord in the most complete manner with the anticipation, that an accumulation of denser water might be found at great depths in the neighbourhood of the Straits, from which a counter-current beneath, though far less rapid, might carry westward into the Atlantic as much salt as enters with the eastward current near the surface from that ocean into the Mediterranean.

The evidence of this will be comprised, indeed, in very few words; for though the two first specimens, taken at distances of about 680 and 450 miles from the Straits, and at depths of 450 and 400 fathoms respectively, do not exceed in density that of many ordinary samples of sea-water, yet the last,

* This is a slight error. Dr. Marcet was kind enough to superintend the making of a water-bottle for me, by Thomas Jones, of Charing Cross. It consists of a thick bell-metal cylinder, about ten inches long and six in diameter, with strong caps on the ends, each having conical apertures in the same direction, through which passes a metal rod, having a conical projection at each end, both ends fitting exactly in the conical apertures in the caps of the cylinder. When in use, the piston-rod is lifted up, and held firmly by a spring, whereby the water can enter freely and pass upwards through the descending cylinder, which is closed at any required depth by letting a perforated iron ball slip down by the suspending line. This ball, on arriving, strikes the spring, when the bottle is instantly closed and forcibly locked up by the conical fittings, and water from the precise spot is obtained without a possibility of the intermediate fluid affecting it. The contents were then carefully emptied into bottles, corked, sealed, labelled, with all local particulars, and carefully placed in store.

which was taken up at about 50 miles within the Straits, and from a depth of 670 fathoms, has a density exceeding that of distilled water by more than four times the usual excess, and accordingly leaves, upon evaporation, more than four times the usual quantity of saline residuum.

Under-current.

Hence it is clear, that an under-current outward of such denser water, if of equal breadth and depth with the current inward near the surface, would carry out as much salt below as is brought in above, although it moved with less than one-fourth part of the velocity, and would thus prevent a perpetual increase of saltness in the Mediterranean Sea beyond that existing in the Atlantic.

On comparison of the relative specific gravities and quantities of salt, in the table subjoined to this paper, with those in Dr. Marcet's table, there may be remarked a want of accordance between the two experiments that will require to be explained.

This difference arises from the different temperatures at which his results and mine were dried. In his experiments the degree of heat chosen was 212°; in mine, the temperature was raised beyond 300°. In each case, it will be seen that the quantity of saline contents to be obtained may be estimated from the specific gravity, by multiplying the excess of density above that of distilled water by a certain factor, which will vary with the temperature that we may select for drying.

At 212° this factor is about ·144, and the product will then represent the saline contents + a quantity of water retained by the deliquescent salts. At 300°, and upwards, the factor is only ·134, on account of a nearer approach to perfect desiccation.

TABLE.

	Latitude.	Longitude.	Depth.	Sp. Gravity.	Salt pr. Cent.
No. 1	38° 30′	4° 30′ E.	450 fms.	1·0294	4·05
2	37° 30′	1° 0′ E.	400 ,,	1·0295	3·99
3	36° 0′	4° 40′ W.	670 ,,	1·1288	17·3
Gibraltar	36° 7′	5° 22′ W.			

Remarks on the analysis.

The high specific gravity here detected in No. 3, and the large amount of its saline contents, are so absolutely surprising, that I was in hopes that long ere this the matter would have had a fuller investigation; especially as Admiral Sir Edward Codrington promised both Dr. Wollaston and myself that he would attend to it, and for that purpose took out the machine I had used; but the Battle of Navarino clapped the stopper on. Meanwhile, my friend Sir Charles Lyell objects to the doctor's postulatum of carrying the salt out of the Mediterranean, on the inference that the dense water cannot possibly escape, because the bottom of the sea rises between Capes Spartel and Trafalgar, as appears 'by Captain Smyth's soundings, which Dr. Wollaston had not seen;' and he therefore concludes, that great quantities of salt would probably be deposited on the bed of that sea, in consequence of such a submarine barrier. Yet this ingenious theory can hardly be the true one, since the armings of our lead would have brought up salt from the deepest bottoms, instead of the mud, sand, and shells which we found. We still require much further information upon this subject; but in the mean time, I think cause will be shown why Dr. Wollaston cannot be right: indeed, it is not improbable that we might have struck upon a spring of brine. (*See page* 160.) Sir C. Lyell's objection.

As the question is of high import, it may be as well to compare the specific gravity of Mediterranean water in various parts of its extent, and at various depths, with that of the Atlantic Ocean, of which the mean is assumed as = 1·0283. The results obtained in the *Aid* were, by means of Clarke's hydrometer, so delicately adjusted, that when placed in distilled water, the mark of 100 grains exactly coincided with the surface line; and the experiments were made only in the finest weather:— Specific gravity.

Place.	Experimenter.	Depth. Fathoms.	Sp. Gravity.
The Strait of Gibraltar	*Marcet*	250	1·0301
Inside the Strait (50 miles)	*Wollaston*	670	1·1288
Off Marseilles	*Tennant*	(*surface*)	1·0273
Between Spain and the Baleares	*Smyth*	8	1·0270
Between Minorca and Barbary	*Wollaston*	450	1·0294
Between Carthagena and Oran	*Wollaston*	400	1·0295
Between Sardinia and Naples	*Smyth*	60	1·0285
In the mouth of the Adriatic	*Smyth*	45	1·0291
Between Malta and Cyrene	*Smyth*	60	1·0283
Entrance of the Hellespont	*Marcet*	34	1·0282
Mouth of the Bosphorus	*Marcet*	30	1·0144
The Black Sea	*Marcet*	(*surface*)	1·0141

Some of these and other sea-waters being submitted to analysis by Dr. Marcet, he obtained this final result of the Components

components precipitated by evaporation from 500 grains of the fluid; namely—

Muriate of soda	13·3
Sulphate of soda	2·33
Muriate of lime	0·975
Muriate of magnesia	4·955
	21·460*

While these crucial experiments were in hand, Dr. Wollaston put the question, as to whether it was not probable that traces of potash might be found in sea-water? Dr. Marcet instantly conceived its possibility, and begged Wollaston himself to test his own suggestion, which being complied with, the fact was soon established. (*Phil. Trans.* 1819, pp. 199—203). From the discovery of substances not previously known to exist in sea-water, Dr. Marcet much wished to repeat and correct his analyses; but he was not spared for that purpose, or he might have detected the two new elements—iodine and bromine—which have been faintly traced in oceanic fluid since his death. Perhaps the most perfect analysis hitherto made of Mediterranean water, is that of M. Laurens (*Journal de Pharmacie,* xxi. 93):—

Potash.

Iodine and bromine.

	Grains.
Water	959·06
Chloride of sodium	27·22
Chloride of magnesium	6·14
Sulphate of magnesia	7 02
Sulphate of lime	0·15
Carbonate of lime	0·09
Carbonate of magnesia	0·11
Carbonic acid	0·20
Potash	0·01
Iodine	*faint trace.*
Extractive matter	*a trace.*
	1000·00

Mediterranean physics.

Several elements of inquiry have, however, remained almost untouched, the principal of which relate to light,

* It should be remarked, that those ingredients which sea-water holds in a state of complete solution, are not united with it by any very intimate chemical combination, for they can be separated by distillation: yet the union is far from being simply mechanical.

heat, and the calorific effects actually in operation. It is admitted that the sea is impregnated with a mixture of gases, which especially affect the portion near its surface: yet M. Biot found that water which he drew up from a depth of 550 fathoms, yielded a mixture which contained no less than twenty-eight hundredths of respirable oxygen. 'But here,' he observes, 'several important questions in terrestrial physics present themselves, which cannot be solved by the apparatus I then employed. In proportion to the descent into the sea, does the pressure of the superior portion upon **Pressure.** the inferior become greater; and as a column of sea-water, eleven yards in height, is nearly of the same weight as a column of air of an equal base, extending from the surface of the earth to the limit of the atmosphere, it follows that, at a depth of 1100 yards, the water sustains a pressure of 100 atmospheres. How enormous, then, must this pressure be on beds still lower, if the mean depth of the sea, at a distance from the coasts, extends for several miles, as the laws of gravitation seem to indicate.' A question thence arises, as to the depth of water necessary to produce the liquefaction of those gases. Estimating the height of a column of water equal to the pressure of an atmosphere in the usual way, at thirty-four feet, and neglecting the saline contents of the sea, as well as the probable compression of water itself at vast depths, Dr. Faraday has shown (*Philosophical Transactions* for 1823) the pressure and temperature at which the gaseous substances below enumerated **Faraday.** become liquid in his experiments; and it results that those gases could not exist as such below the depths marked in feet on the last column.

							Feet.
Sulphurous acid gas liquefies under		2 atmospheres, at		45°	...	68
Cyanogen gas	,,	3·6	,,	45°	...	123
Chlorine gas	,,	4	,,	60°	...	136
Ammoniacal gas	,,	6·5	,,	50°	...	221
Sulphuretted hydrogen gas	,,	17	,,	50°	...	578
Carbonic acid gas	,,	36	,,	32°	...	1224
Muriatic acid gas	,,	40	,,	50°	...	1360
Nitrous oxide gas	,,	50	,,	45°	...	1700

Fundus maris.

The *fundus maris*, or bottom of the Mediterranean Sea, must, except inferentially, remain mostly unknown; but recent surveys—together with the labours of Count Marsigli off the coasts of Provence and Languedoc, and those of Dr. Donati in the Adriatic—go far to prove that this vast basin, at its creation, was composed of the same substances as the rest of the earth is; and that an artificial bottom of depositions and incrustations has filled all the interstices. In the forty-ninth volume of the *Philosophical Transactions*, Mr. Trembley gives a summary of Donati's contribution towards a 'Natural History of the Adriatic Sea,' the conclusions from which shall be subjoined: and although the Mediterranean basin is so extensive, and covered in most parts with an unfathomable depth of water, the observations of the Italian Professor on the available portion which he experimented upon, are of great value in an endeavour to form a judgment of the whole:—

Donati.

Bottom of the Adriatic.

His (*Donati's*) inquiries have enabled him to determine, upon his own knowledge, that there is very little difference between the bottom of the Adriatic Sea and the surface of the neighbouring countries. There are at the bottom of the water, mountains, plains, valleys, and caverns, just as upon the land. The soil consists of different strata, placed one upon another; and, for the most part, parallel and correspondent to those of the rocks, islands, and neighbouring continents. They contain stones of different sorts, minerals, metals, various petrified bodies, pummice-stones, and lavas formed by volcanoes.

Istria, Morlachia, Dalmatia, Albania, and some other adjacent countries, as well as the rocks, the islands, and the correspondent bottom of the Adriatic Sea, consist of a whitish marble, of an uniform grain, and of almost an equal hardness. It is of that kind of marble called by the Italians *marmo di Rovigno*, and known to the ancients by the name of *marmor Traguriense.*

This vast bed of marble, in many places under both the earth and sea, is interrupted by several other kinds of marble, and covered by a great variety of bodies. There are discovered there, for instance, gravel, sand, and earths more or less fat.

The variety of these soils under the sea is remarkable. It is to this that Dr. Donati ascribes the varieties observed with respect to the nature and quantity of plants and animals found at the bottom of the sea. Some places are inhabited by a great number of different species of plants and animals; in others, only some particular species are found; and lastly, there are other places, in which neither plants nor animals are to be met with.

The observations not only point out to us the affinity and resemblance between the surface of the earth and the bottom of the sea, but may likewise contribute to discover to us one cause of the varieties which are observed in the distribution of the marine fossils found in the earth.

Having had the advantage of the experience of Donati, Fortis, and De Luccio, I paid much attention to their results, and found the operation of the winds and currents of the Adriatic to be very uniform. The accumulation of matter on the western shore is readily accounted for by the constant set of the waters along the coasts of Albania, Dalmatia, and Istria; from whence they sweep along by Friuli, Venice, and Romagna, bearing their own silts, and carrying along the alluvial deposits of the rivers at the head of the Gulf: insomuch that the Venetian ports are encumbered, Ravenna is now high and dry inland; and from thence to the Isonzo, there is an uninterrupted series of terrene accessions. A singular effect is observed from the occasional strength of the river action over that of the current. Between the Malamocco and Parenzo, about the middle of the passage, there is a muddy bank resting on the solid limestone and concretions; it is about three miles in breadth, and in length extends to opposite Comacchio. In calms, the surface above it appears smooth and nearly stagnant, while the current which runs on each side, being weakened by diffusion over it, deposits matter in the centre; and thus, in the lapse of ages, an island may be formed. The coast just to the north of it, having its rivers under tidal action, though small, is broken into estuaries; but as that action weakens, it enables the Po to form a delta. The bottom, however, of that part of the Adriatic between the Po and Trieste, being everywhere of moderate depths, forms a submarine plateau, which must be considered as only the continuation of the great plains of Lombardy and Friuli.

Adriatic actions.

Singular effect.

A study of the motions in the Adriatic waters affords a tolerable clue to those of the Mediterranean in general. When fresh winds, by their friction, for any considerable time, force the surface waters home in some given direction against the coast, the movement is quite sufficient to carry any mechanically-suspended substances to distances proportionate to the strength and duration of the cause: and when a current of water freighted with matter passes a projecting point, or flows through a narrow channel, so that by pressure and resistance its rapidity is increased, and passes from thence into a bay or opening, where its force is weakened by diffusion, it will deposit the chief part of its burden at the bottom of that wider space. It is thus by natural means, in ceaseless action, that constant cargoes of detritus and matter in solution are borne to the sea, and there committed to the currents by rivers and torrents; and to these are added the heavy occasional falls of the cinders, ashes, and lapilli, thrown out from volcanoes. It is not, therefore, matter of surprise, that the sedimentary accumulations should have formed a thick coating over the whole bottom.

Basins.

Perhaps the most remarkable submarine feature of the Mediterranean is its perfect hydrographic division into two great basins by the form of its bottom; thus confirming the allotment made by geographers from a study of the form of its shores. The barrier at the entrance of the Straits marks the commencement of the western basin, which descends to an abysmal profundity, and extends as far as the central part of this sea, where it flows over another barrier, and again falls into the as yet unfathomed depths of the Levant basin. My means were not equal to my wishes in examining this surprising fact; but after fixing (*or rather discovering*) the subaqueous bank to which I gave the name of Adventure, I got occasional moderate soundings nearly across from Sicily to Tunis, in a winding line of connexion crowned by the Skerki rocks—doubtless, as already stated, the now-abraded Aræ of Virgil (*see page* 93). Yet in making an occult line on my chart, to indicate this rise from a depth at present beyond estimate, I by no

Adventure Bank.

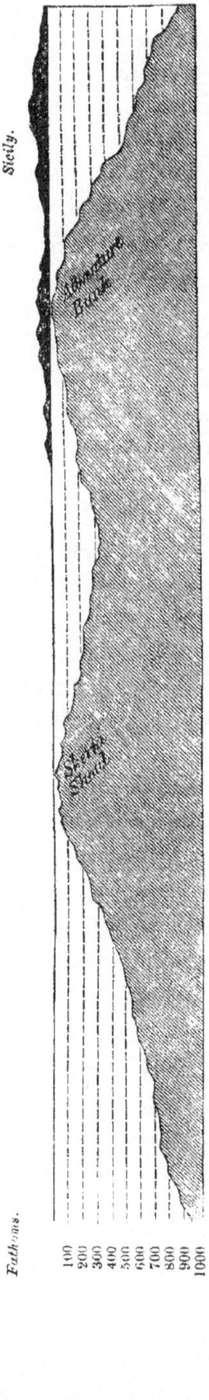

means meant it as a mark for shoal water in which navigation is concerned; for though I found occasional spots of from 30 to 90 fathoms, and still less near the central reefs, there were 140, 157, and 260 fathoms on either side, as also places where 190 and 230 fathoms of line were run out without striking bottom. The Adventure Bank is developed from this deep ridge, and is a comparatively shallow plateau, affording good anchorage in many parts; and it is much frequented by fish. A section through these basins carried across Palestine into that chasm, the Valley of the Jordan and the Dead Sea—which last is 350 fathoms deep, with a surface 256 fathoms below the level of the Mediterranean — would unfold a wonderful geological structure in the eastern boundary. Another section through the central part, in a line of 240 miles north-west and south-east, through the Skerki and Adventure banks, under a proportion of thirty in length to one in depth, is given in the annexed diagram.

Submarine plateau.

Dead Sea.

While Sicily is thus shown to be a continental island, there is that prodigious depth of water around Corsica and Sardinia which marks them at once as pelagic: and the Mediterranean in general is so much deeper than analogy and the proximity of lands would lead us to expect, as to countenance the idea of its sunken basins being partly formed by volcanic agency.

Continental & Pelagic islands.

Coast contour.

On glancing over the general chart, we cannot but be struck with the marked difference between the northern and southern distribution of the great terrestrial masses which give form and feature to the coasts: yet, however accidental that contrariety may appear to be, it offers a characteristic development in geological chronology, since there could not have been any depression or elevation of the included waters, without a sensible alteration of littoral contour, such constituting the true line of contact between the land and the sea surface. Such a datum is absolutely necessary to further inquiry, for the whole shores are as remarkable for difference of altitude as for variety of outline.

Omar el-Aalem.

In giving an opinion on the general permanence of the sea-level, I am aware that many men, and some of them of respectable authority, have entertained a different view; but modern accuracy has dissipated most of the classic and mediæval visions on this point. In the tenth century, *Omar el-Aalem* (Omar the Wise), attacked the question with laborious industry, and produced his work on the *Ebb* (el jezr) *of the Sea.* On comparing the documents of his own time with some said to have been written 2000 years before, he became convinced that important changes had taken place from the subsidence of the waters; and he considered his opinion to be confirmed by the numerous salt-ponds in the interior of Asia, a conclusion also recently arrived at by Pallas, the eminent Prussian traveller. Even had Omar been misled, as hath been suggested, by the features and phenomena of the Caspian Sea, he might have recanted his error, without fear of the banishment, or impeachment of his understanding, which he actually incurred. But his intention is not very clear; and possibly his work refers only to the effect of tides observable in the Persian Gulf, Red Sea, and Indian Ocean.

§ 3. The Extent, Supply, and Evaporation of the Mediterranean Sea.

THE headlands, bays, and sinuosities in the margin of this aqueous expanse—so many centuries acted upon by the united effects of breakers, wind-wave action, tidal streams, and terrene displacement—render it difficult to state dimensions with positive accuracy in a few words; and even the entering into comparatively exact details would be tedious and indistinct. The substantial elements, however, shall be given, and for minute particulars the chart may be consulted.

Extent.

The Mediterranean Sea extends from the longitude of 6° west to 36° east of Greenwich, while the extreme limits of its latitudes are from 30° to 46° north; and, in round numbers, its length, from Gibraltar to its farthest extremity in Syria, is about 2000 miles, with a breadth varying from 80 to 500 miles, and, including the Black Sea, a line of shore of 4500 leagues. The ancients, who considered this sea a very large portion of the terrestrial globe—although it turns out to be but equal to about one-seventieth part of the Pacific Ocean—assigned to it a much greater length, as will be shown in Part IV., § 1; but Strabo seems to have flattened-in largely, since his principal distances for establishing its length were:—

Strabo's dimensions.

	Stadia.		Nautic miles.
From the Columns of Hercules to the Straits of Sicily	12,000	...	1028
Cape Pachynum to west end of Crete	4500	...	380
East end of Crete to Alexandria	3000	...	257
Alexandria to Rhodes	3600	...	308
From Rhodes to Issus	5000	...	429

Measures so strangely distorted by Ptolemy!

The superficial extent of the Mediterranean Sea, as intercepted between its various limits, and calculated by the parallels of the latitudes and the longitudes of those limits, from thence deducing the areas in square statute miles (which are used by Halley), may be thus tabulated:—

The Western Basin	325,272
The Adriatic	52,819
The Levant Basin	518,755
The Archipelago	75,291
The Sea of Marmora	4,644
The Black Sea	159,431
The Sea of Azof	13,075
Total surface	1,149,287

Rivers.

Including the Black Sea as a branch of the Mediterranean, the chief feeders of this vast sheet of waters are the great rivers Nile, Danube, Dnieper, Po, Ebro, and Rhone; there are also the secondaries, Var, Magra, Arno, Ombrone, Tiber, Adige, Isonzo, Tagliamento, Lo Drino, Samana, Achelous, Alpheius, Meander, Pyramus, Bagrada, and Mulvia; besides which, the smaller streams and streamlets, with their tributaries, are almost innumerable. Contrary, therefore, to the long-received opinion respecting the scanty supply of water poured into this sea by rivers, the quantity which it is constantly receiving forms an important integer; although a number of these streams may not run above a hundred miles, and the drainage of others may be comparatively small in proportion to the surface through which they pass.

Fresh-water springs.

Moreover, fresh-water springs exist in the sea, near the shore, which are more or less copious according to circumstances; but those of Stamfane rock and Syracuse are popularly held to proceed from the Alpheius by submarine communication.* In the Gulf of Spezzia there is a spring

* In my Memoir of Sicily (page 171), it is mentioned that in the harbour of Syracuse, opposite the fountain of Arethusa—and probably from the same source—a copious spring of fresh water rises from the bottom, without intermingling with the brine. It is called *Occhio della Zilica*, or Alpheius, which Moschus (*Idyllium*, viii.) represents as bearing leaves and sacred dust from Elis.

which constantly discharges a very considerable body of water, rising with such force as to produce a slight convexity on the surface; this stream is probably derived from a system of cavernous passages in the neighbouring limestone rocks, but its place, as marked on my plan of the gulf, has been immemorially the same. In the *Mare piccolo*, or great port of Taranto, and at some distance from the mouth of the Galesus, fresh water springs up in such force and abundance that it may be taken up without the least brackish mixture; and in the briny lagoon of Thau, at Cette, there is a deep spot called the *Avysse*, from which rushes up a column of potable water, with such force as even to make waves. Near Ragusa, the Kalamota Channel terminates in the port called Val d'Ombla, which is watered by the Ariona, a subterranean river bursting up with amazing volume and force from the foot of Mount Bergatz; fresh-water springs are also copious in the gulfs of Cattaro and Aulona. At Agio Janni, below Parga, between the mouths of the Acheron and Thyamis, is a circular space of fresh water, about forty feet in diameter, rising through the sea with great activity; this is probably the ascending spring alluded to by Pausanias, (*Arcad.* vii.) Off the little desert islet, Ruad, near Tortosa on the coast of Syria, a spring of fresh water gushes up in the sea in such volume, that it may be skimmed off without the slightest impregnation of salt. 'You may draw up potable water,' says Pliny, 'out of the sea about the Chelidoniæ islands and at Aradus;' and there must be many unrecorded jets of the same nature, mingling with the sea unnoticed.

Spezzia. Taranto. Thau. Ariona. Agio Janni. Ruad. Aradus.

These may seem but insignificant addenda to the supplies of the Mediterranean, in the opinion of the brine-theorists; but in the aggregate they form a goodly volume, and have all, perhaps, exerted their influence for many ages. The duration of some is matter of record. In the Argolicus Sinus, between Kivéri and Astros, is the Anàvolo (*Deine*), a copious spring of fresh water, rising with consi-

Anàvolo.

derable strength through the sea, at the distance of about a quarter of a mile from the shore. If this can be reconciled to the rather vague early notices with which it substantially corresponds, it must have been thus in action for nearly 1700 years at least. From the account of Pausanias, Deine appears to be the emissary of the Zarethra, which drained the plain of Argos (Ἀργὸν, *inert*); and it is thus described by

Col. Leake.

my friend Colonel Leake:—'The body of fresh water appears to be not less than fifty feet in diameter. The weather being very calm this morning, I perceive that it rises with such force as to form a convex surface, and it disturbs the sea for several hundred feet around. In short, it is evidently the exit of a subterranean river of some magnitude.'—(*Travels in the Morea*, vol. ii. p. 480.)

Percolation.

The general percolation is also very great: Pliny the Younger, giving a description of his villa near Ostia, to Gallus, mentions the wells in his garden, adding, 'And indeed the quality of this coast is rather remarkable; for in whatever part soever you dig, you meet, upon the first turning up of the ground, with a spring of pure water, not in the least salt, though so near the sea.' I have also

Sea beaches.

noticed the same on the beaches of Calabria, the Terra di Bari, and the Capitanata; and in my account of Sicily, I mentioned the well of good fresh water at Milazzo, which, though several feet below the level of the sea, is so near, that it is only sheltered from the surf by a wall. In the same work, I also state, that on both sides of the Faro of Messina, 'pure, though rather hard, fresh water is procured, by digging a hole in the sand, within two or three feet of the margin of the sea; this is occasioned by the filtering and percolation of the fiumare (*torrents*), which, though apparently dry, are never actually so; and this accounts, in some measure, for the malaria arising on their banks.'*

---

* Though not exactly in the same line of argument, perhaps I ought to mention those ebullitions near the volcanic regions, which arise from an

It may be that, without the constant supplies from the Straits of Gibraltar and the Black Sea, the Mediterranean would not receive an equivalent to the loss by evaporation from its rivers and atmospheric precipitation: but from what is here advanced, it is evident that conclusions have been made *per saltum*, and that the question requires a more assiduous attention than it has yet received. It must be acknowledged, however, that great strides towards a fuller knowledge have been made; and the intelligent Berghaus presents the following data for the condition and extent of drainage by the larger rivers:— **River drainage.** **Berghaus.**

MEDITERRANEAN FLUVIAL SYSTEM.

Rivers.	Basins in square miles.	Direct length in geographical miles.	Development in geographical miles.	Extent of windings.	Ratio of windings to direct length.
Nile	520,200	1320	2240	920	0·7
Danube	234,080	880	1496	616	0·7
Dnieper	169,680	548	1080	532	1·0
Don	168,420	408	960	552	1·3
Po	29,950	232	352	120	0·5
Rhone	28,160	248	560	352	1·4
Ebro	25,100	268	420	152	0·5
Dniester	23,040	360	440	80	0·2

abundant disengagement of carbonic acid gas, sulphuretted hydrogen, and other hot vapours from subaqueous vents; for some of them being solvent, and others capable of decomposing rocks, they cannot be without effects. One of these, near Panaria, in the Æolian islands, is thus described in my account of Sicily (page 260): 'In this strait, a strong smell of sulphur is perceptible; and in two places, near the north extremity, are springs emitting sulphureous gas, the bubbles of which rise in quick and constant succession to the surface, where they have been known to flame on bursting in the atmospheric air. Wishing to ascertain something respecting this indication, I submerged a thermometer in a bottle, which I found gave 97° of Fahrenheit, in 21 feet of water; but not satisfied with the result, I had a tin tube made for me by an ingenious mechanic of Messina, with a valve at each end, which, as it descended, allowed a free passage to the water; but on being drawn up, closed at both ends by the pressure, and contained a sufficient quantity of water to keep the thermometer to the heat of the depth to which it was lowered. The result obtained was 105° in 22 feet of water, while at the surface it was 84°, and at a mile distant the temperature of the sea was 76½°, that of the atmosphere at the same time being 71°.' This was on the 22nd of April, 1815.

Supply and loss.

Such is the tabulated view of a general system; but in the river with which I am personally best acquainted, the Po, the numbers are small, for that *rex fluviorum* of Italian streams, and its tributaries, assuredly drain a basin, the area of which cannot be much less than 40,000 square miles. In further assumptions on the dimensions and velocities of the principal effluents, M. Berghaus shows, that by taking the running waters of Europe as unity, or 1·00, the quantity discharged into the Black Sea will be as high as 0·27 parts, while the Mediterranean receives only 0·14; the former ingulfing nearly one-third part of all the running water of Europe. As these supplies must be deemed far too little to compensate the loss in vapour of so great a surface under a powerful and often cloudless sun and hot winds—where the air is proved by hygrometrical registers to have only half the moisture of the English atmosphere—the oceanic influx through the Straits, together with the surplus of the Euxine, constantly flowing through the Dardanelles, may be cited as making up the deficiency. Yet as these grand affluents produce no perceptible increase in the height of the internal waters, that circumstance has attracted the attention of philosophers; but trustworthy evidence is still a desideratum. Theorists, indeed, threaten us with the filling up of the Black Sea, in a handful of ages, by which the shallower parts of the Levant Basin may become exposed, and a new adjustment of levels take place: yet even under this condition, this sea would conform itself to the requisite balance, by means of its free communication with the ocean, between Spain and Africa, where the stream, as Horace said of his imaginary river, ceaseless flows and must for ever flow—

Proportion of river water.

Steadiness of level.

———————— At ille
Labitur, et labetur in omne volubilis ævum.

Halley's theory.

This brings us to the celebrated Halleian theory, which is still the *ne plus ultra* of a numerous series of Mediterranean inquirers; and which, having been *proved* by experiments at once plausible and ingenious, demands a brief recapitulation in this section.

Halley—in whatever light we consider his extraordinary powers of mind and wonderful diversity of knowledge—was one of the most remarkable of a galaxy of giants in science and literature, who illustrated the close of the seventeenth century, and therefore must be approached with deference and respect, even where there may exist a difference of opinion from him: at all events, he is entitled to the full merit of originality, nor would it be prudent to differ from him without strong reasons. His essays on the quantity of vapour raised out of the sea by the heat of the sun, are printed in the *Philosophical Transactions*, and in the first volume of the *Miscellanea Curiosa.* Upon certain assumptions derived from experiment, Halley holds that 'every ten square inches of the surface of the water, yields in vapour, per diem, a cube-inch of water; and each square foot half a wine-pint; every space of four foot, a gallon; a mile square, 6914 tons; a square degree, supposed of sixty-nine English miles, will evaporate thirty-three millions of tons: and if the Mediterranean be estimated at forty degrees long, and four broad, allowances being made for the places where it is broader by those where it is narrower (and I am sure I ghess (*sic*) at the least), there will be 160 square degrees (761,760 *statute miles*) of sea; and consequently the whole Mediterranean must lose in vapour, in a summer's day, at least 5280 millions of tons.' This he esteems to be a vast quantity, though as little as can be concluded from the trials he made, adding—'And yet there remains another cause, which cannot be reduced to rule, I mean the winds, whereby the surface of the water is licked up sometimes faster than it exhales by the heat of the sun; as is well known to those that have considered those drying winds which blow sometimes.' Our philosopher then proceeds very methodically to show, and gives figures for it, that little more than one-third of this is returned by the

Halley.

Loss by evaporation.

nine great rivers—'the Iberus, the Rhone, the Tiber, the Po, the Danube, the Neister, the Borysthenes, the Tanais, and the Nile, all the rest being of no great note, and their quantity of water inconsiderable.' Under this impression, by a laboured estimate founded on a calculation of the waters of the Thames at Kingston Bridge, he concludes that the nine rivers contribute only 1827 millions of tons in a day.

Remarks on the theory.

Now it is inconceivable that one of such accurate powers of calculation as Halley unquestionably possessed, should have established so plausible a theory on such very imperfect data; and it is still more inconceivable that it should have stood so long unshaken, although his own argument proves at once the fallacy of its premises, and consequently the untenableness of its result. With all the deference just mentioned, and esteeming *Captain* Halley as a brother naval officer, as well as a brother-surveyor, I cannot look upon the keeping a small vessel of water, by means of a pan of coals, for several hours at a summer-heat in this country, and measuring the decreasing weight of water in a given time, to arrive at the amount of evaporation, as at all meriting confidence; especially as he has not given us the degree of 'summer heat,' although so main a point, and one of which he records that the thermometer showed it nicely. However, by the method pursued, he found that a depth of 0·1 inch, from a surface of eight inches in diameter, was vaporized in twelve hours; and lumping together the summer and winter, as well as nights and days, of the thus assumed Mediterranean, he reckoned that the same depth, 0·1 inch, may on the average be evaporated every twenty-four hours. Starting with this very arbitrary and doubtful quantity from his 'little pan,' in which even the salted water was artificial, he obtained the normal numbers just cited, by 'exact calculus.' *Aliquando bonus dormitat Homerus:* besides the jumble of inferentials brought into play, one of the most obvious and tangible conditions of the theory is more than one-third of its

Halley's method.

whole amount in error; for the surface of the Mediterranean, by recent measurements, has been shown to be 1,149,287 square statute miles, instead of 761,760: so that the proportionate quantity of evaporation—or property by which water has the power of emitting vapour of an elastic force proportioned to its temperature—by Halley's own rule would be 7966 millions of tons, instead of 5280 millions, per diem. And it should have been recollected, as a peculiarity of inland seas, that their shores in summer are of a higher temperature than water, and hence the aërial dryness already alluded to: it therefore follows, that the vaporization over such places will be much greater than that of the ocean, in the same parallels, where the air, saturated with aqueous vapour, continues at the same heat for several days successively. Substantial error.

But as the Halleian Theory had become a received postulate in Mediterranean physics, it struck me that it would be as well to re-examine the whole argument; and in order to test its merits, with better materials than Halley's, the latest surveys, adjusted to the points determined by Captain Gauttier and myself, were brought to bear on the question. The superficial dimensions were established by rectangular sections of each of the chart divisions being neatly cut to the limiting parallels of latitude and longitude, and then carefully weighed with a delicate balance: but as the evaporation from the sea must be in proportion to the quantity of surface presented by the water to the evaporative influence, it follows that, from the interposition of islands and promontories, the quantity would be very unequal at different places,—a point which was not considered in Halley's computations. Moreover, another quantity in the inquiry, depth, is still beset with uncertainty,—for though I sounded beyond 1000 fathoms without striking the bottom, further experiments to ascertain the greatest profundity were inconsistent with my means and time, and therefore, in some of the deductions which will Re-examination of the experiment.

Assumed depth.

follow, I resorted to a geographical mile as a unit of depth, which inference fully warrants. After this operation, tables were drawn up, of which that on the opposite page is an abstract: in which, as the third column of numbers is, from what I have just stated, but a guess, the cubic contents will necessarily be vague and inconclusive, being only intended as a mere assumption. From this remark, however, the Sea of Azof may be excepted, since a fair approximation to the amount of its contents can be obtained. This sea, which —if the assertion of Herodotus (*Melpomene*, lxxxvi.) has any value—must have greatly contracted its boundaries even within the historical period, having been well sounded throughout, allowed our weighing process to be conducted with such nicety as to be within 0.1 grain = about seven square miles. The mean depth was ascertained by crossing the sea in seven different directions, noting the soundings at short intervals along each line, and then taking an average of the whole. This may answer well enough in the present state of the question; but greater accuracy might have been obtained by priming the chart paper with linseed-oil, and some preparation of lead, to increase its weight. An example of the treatment may suffice:—

Sea of Azof.

Basin.	*Measured weight of the sections.*	*Log.*	*Surface of sections as compared above.*	*Log.*	*Measured weight of sea only.*	*Log.*	*Surface of sea deduced.*
	Grains.		Statute miles.		Grains.		Sq. stat. miles.
Western	518·57 ..	2·7148074 ..	331,257 ..	5·5201651 ..	509·20 ..	2·7068884 ..	325,272
Adriatic	91·30 ..	1·9604708 ..	54,147 ..	4·7335744 ..	89·06 ..	1·9496827 ..	52,819

Other treatment of the question.

Some of the methods of bringing out the deductions about to be given, are considerably altered from those used by Halley, because—I. A day is too short a period: a whole year should be taken as a cycle, in which all the varying temperatures of the different seasons complete their rounds, and become equalized. II. As the initial or starting point is 0·1 inch evaporated in twenty-four hours, there was no occasion for leaving this linear measure, and going to measures of capacity and weight—as wine-pints and tons;

more especially as all the other quantities are in measures of length: and by keeping to one quantity throughout, the subsequent correction of the result is so much the more easy, should more correct experimental data be afterwards procured. Taking, therefore, 0·1 inch in twenty-four hours to be ·000001515150 of an English statute mile in the same length of time, and ·0005533973 of a mile in one year, the quantity evaporated is here given in cubic miles. III. The evaporation about Alexandria must be so very much greater than in the Sea of Azof, that in stating the amount for each basin, some modification of the mean quantity adopted for the whole Mediterranean, became necessary. Considering, therefore, that under the equator and a vertical sun the vaporization would be a maximum, and that under the perpetual ice at the pole it would be nothing, the amount may safely be assumed to vary as the cosine of the latitude—the 0·1 inch in twenty-four hours being considered the quantity at 40° of polar altitude. On that principle, therefore, the evaporated quantities of each basin have been modified according to their distance above or below 40 degrees of latitude; and the following are the results— Conditions and data.

Division.	*Mean latitude.*	*Area in square miles.*	*Depth in miles?*	*Cubic contents in cubic miles?*	*Annl. evap. in cubic inches.*
The Western Basin	39° 00′ ...	325,272 ...	·9 ...	292,744 ...	180·66
The Adriatic Sea .	42 30 ...	52,819 ...	·1 ...	5,282 ...	28·13
The Levant Basin .	34 30 ...	518,755 ...	·6 ...	311,253 ...	308·84
The Archipelago .	37 45 ...	75,291 ...	·1 ...	7,529 ...	43·01
The Sea of Marmora	40 40 ...	4,644 ...	·05 ...	232 ...	2·54
The Black Sea . .	43 45 ...	159,431 ...	·107 ...	17.059 ...	83·20
The Sea of Azof .	46 15 ...	13,075 ...	·0079 ...	102·9 ...	6·53
Mediterranean total . .		1,149,287		634,201·9	652·91

In conclusion:—Halley also attempted to get at the quantity returned in the form of showers. This he hoped to obtain by calculating the tons of water brought down by the various Mediterranean rivers; of which, taking about half a dozen, and estimating that each brings down ten times as much water as the Thames, he finds the evapora- Evaporation.

tion more than sufficient to meet the supply. Hence, some have imagined danger, from an inevitable and constant concentration of brine.

River proportions.

Now these assumptions are desperately inaccurate, since his stated capacity of the rivers is involved in error; the Nile alone being considered to deliver annually into the Mediterranean a body of water about 250 times that which flows out of the Thames; and the estimated lengths of the two principal streams, compared with the Thames as unity, are Danube 7, and Nile 12¾ times. Starting with his diminutive quantity of 0·1 inch in twenty-four hours, we may put that down as equivalent to 36·523 inches in the course of a year, a normal point in Halley's computations, and more than the quantity now assigned for the mean fall of rain for the whole temperate region of the old world, which is 34 inches. Yet, although the fall of rain differs widely in the various Mediterranean countries—especially as regards the vicinity of the Atlantic on the west, or the arid shores to the south—the average annual quantity, carefully estimated from the evidence of many registers, is under twenty inches per annum; so that the evaporated water, after having returned twenty inches by precipitation to whence it was extracted, has 16,523 inches to spare for distribution over a space of land of equal extent with the sea. And when we augment the 16·523 of Mediterranean rain, on account of the powerful energy of atmospheric precipitation, that must be derived from the vast volumes of vapour ever rolling in from the Atlantic Ocean, there is evidently an abundant supply, even on the Halleian data, for the whole of those countries which drain into the Inner Sea.

Rain.

Remark in conclusion.

Still, assuming—as we have reason to do—that the main exit of the surplus water of this sea is by evaporation, then to get the actual quantity evaporated in one year, we ought to measure all the water that falls into it at the mouths of all the rivers, and at the Straits of Gibraltar and the Dardanelles; and if to this we were to add the real contemporaneous fall of rain, then, and then only, would the required quantity be obtained. Meantime it may be implicitly relied upon, that all is right, for it is evident that Nature comprehends the exquisite system of compensation, and knows no waste.

---

§ 4. On the Currents of the Mediterranean.

BY the term 'currents' is understood those progressive movements of the water by which vessels, or anything floating upon it, are carried in their direction, and precisely with their own velocity, when no wind prevails. Currents differ essentially from tides, in deriving their motion from other causes than solar and lunar attraction; and in their constant circulation they traverse extensive regions, where they necessarily emit or imbibe heat. But though it is inferred that currents may extend to a vast depth, our exact acquaintance with them is nearly confined to the supernatant effects only. They appear to be in continual motion in a certain direction; yet their course must be treated with relation to the points of divergence and convergence of their route, for it is well known that irregularities of outline in the shore, without any reference to elevation or depression, have very considerable consequences in modifying the action of the sea, by turning the course of both current and tidal streams. It may readily be inferred that currents perform important offices in the grand economy of nature, disturbing the general hydrostatic pressure, rendering the fluid favourable to submarine vegetation and piscatory life, and preventing stagnation by agitating the waters: but we are still in comparative ignorance of them, for their extent, direction, depth, strength, and temperature are very various, and often fluctuating. Currents are always named after the points of the compass towards which they run or

Currents.

Motion and course.

set; being therein exactly the reverse of wind, which is designated according to the point from which it blows.

That progressive movement of the waters in the Mediterranean which is independent of tide, and constitutes a true current, is more remarkable for constancy than strength, except in places where local peculiarities exert a peculiar influence, and prevailing winds occasion a difference of level. We have just seen that evaporation has so powerful an action as probably to cause a general proportional depression of surface, and thus give rise to the principal phenomena already mentioned. From obvious causes, this Inner Sea is, for the greater part of its extent, warmer, both in summer and winter, than the Atlantic, which therefore flows into it; at the same time, the Black Sea is somewhat colder than the Mediterranean, and consequently flows into it also.

Relative heights of seas.

In all ages, wherever there are two neighbouring seas, it has been customary to consider that one was more elevated than the other; and till very lately, the operations of modern inquirers countenanced the time-honoured opinion. Thus the early philosophers were borne out by Toaldo, in their notion that the Mediterranean was much higher than the Atlantic—thus Count Marsigli, the voluminous historian of the Danube, showed that the ancients were justified in asserting that the Euxine was thirty or forty feet above the ordinary level of the Ægean—thus M. Fauvel confirmed the opinion given by the engineers of old to Demetrius Poliorcetes, that so great a difference in height existed between the gulfs of Egina and Corinth, that it would be dangerous to cut across the isthmus which divides them—and thus the observations of the French Egyptian Institute were supposed to prove that the surface of the Red Sea is neither more nor less than twenty-eight feet higher than the Levant Basin; whence it followed that the ancients were right.

But what broken reeds are occasionally trusted to!

Maraldi and Cassini pronounced the Mediterranean, as ascertained by them, to be exactly one toise higher than the Atlantic: the hall of the Paris observatory being forty-six toises above the ocean, and forty-five above the Inner Sea. Shortly afterwards, Count Morozzo showed that the Adriatic must be more elevated than the Mediterranean (*Memoirs of the Academy of Sciences at Turin*, 1788); and recently the still more precise observations of Delambre, Méchain, Gauttier, myself, and MM. Coraboeuf, Peytier, and Bourdaloue, have proved, by successive reductions of height, that in all and each of those places, the waters in repose have surfaces of so nearly the same level, that the differences are but barely ascertainable by our present improved instruments and methods of determination.* We must therefore look to all other probable causes for currents, besides a difference of level; and however narrow the communication between the two seas may be, the Mediterranean is still part and parcel of the one vast expanse called the ocean:† and whatever may permanently affect the level of the one, must eventually affect the surface of the other also. The hypothesis, therefore, as to a durable depression of an actual branch of that ocean, will not stand the test of sound inquiry: were the Mediterranean always much lower than the Atlantic, it would be impossible for the current ever to set out of it, agreeably to the laws of hydrostatics, unless the body of waters should be influenced by winds or the attractions which cause tides. Observers.

Besides the Halleian doctrine of evaporation, which evidently solves much of the theorem, we must now bring forward the argument of other writers, that an under- Supposed under-current.

---

* By General Monteith's experiments with boiling water at the mouth of the Kalla (*Journal of the Royal Geographical Society*, vol. iii. p. 37), he inferred that the Black Sea was precisely of the same level as the ocean, since the point of ebullition was exactly equal to 212° of Fahrenheit's scale.

† Eustathius must have had currents in view when he derived the word Ocean from *ὠκεως ναειν, to slide swiftly.*

current—or one running counter to that at the surface—exists; which is presumed to carry a vast volume of fluid back to the great ocean. I shall presently advance a fact or two which may present obstacles to this, but in the mean time a fair hearing shall be given to the hitherto received statements: and it must be recollected, that to establish a counter-current setting outwards below, unless a greater gravity be conceded, it is necessary that the Mediterranean water be of a lower temperature than that of the Atlantic, for otherwise it must run out at the surface, and the supply be received underneath. This is well known not to be the case; yet, in order to avoid prejudgment, the instances usually brought forward shall be duly cited, for, without denying the subcurrent assumption, I merely insist that its existence is not yet proven.

Dr. Smith. The first formal paper on the subject, that I am aware of, was read before the Oxford Society on the 21st of December, 1683, and is printed in the fourteenth volume of the *Philosophical Transactions*. Dr. Smith mentions the vast draught of water poured continually in, and says:

> I here omit to speak of the several hypotheses, which have been invented to solve this difficulty: such as subterranean vents, cavities, and indraughts, exhalations by the sunbeams, the running out of the water on the African side, as if there were a kind of circular motion of the water, and it only flowed in upon the Christian shore: which latter I look upon as a mere fancy, and contrary to all observation. My conjecture is, that there is an undercurrent, whereby as great a quantity of water is carried out as comes flowing in.

This shows, as Dr. Smith speaks of the *several* hypotheses, that the phenomenon had been under discussion, and that the supposed existence of super and sub-currents has long been received. But one of the most notable instances in support of that opinion was afterwards brought before the Royal Society, and published in the thirty-third volume of its *Transactions*. It is there stated that, in

Captain de l'Aigle. the year 1712, M. de l'Aigle, 'that fortunate and generous commander of the privateer called the *Phœnix*, of Marseille,' gave chase to a Dutch ship near Ceuta. On coming up with her, he sunk her with a broadside, in the middle of the Gut between Tarifa and Tangier. A few days after, the foundered ship, with her cargo of brandy and oil, floated upon the shore near Tangier, at least four leagues to the west of the spot where she had been sunk, and in direct opposition to the surface-current. The fact is thus vouched by Dr. Hudson, the communicant:— Dutch ship sunk.

I was at Gibraltar when this happened, where I saw above a hundred of the butts of that cargo of brandy, which were sent thither from Tangier; I likewise spoke with the captain of the Dutch ship, who told the governor, myself, and many others, where his vessel sunk; and her rising afterwards at Tangier appeared very unaccountable to us, as it does to me to this day: for there is no doubt but the ship sunk where the Dutchman told us, since the Spaniards from the land who saw it confirmed it to us. The water in the Gut must be very deep, several of the commanders of our ships of war having attempted to sound it with the longest lines they could contrive, but never could find any bottom.

This very circumstantial evidence appeared to establish the conclusion that there exists a recurrency in the deep water in the middle of the strait, and certainly, to some of the smaller philosophers, afforded as satisfactory a solution of the problem as that of the unequal effect of evaporation, the which must be an ever-varying operation of nature. There are, however, two or three points of the deposition on which we could have wished the Dutchman to have been cross-examined, for we are left in the dark as to the why and wherefore a merchantman should have incurred so spiteful a broadside, how her people were saved, whether she was water-logged instead of sunk, &c. &c.: and, by the way, we regret that while Sir John Jennings had the combined English and Dutch fleets inside the strait, and Vice-Admiral Baker's squadron was just outside, a French privateer should have been permitted thus to lord it in the Gut. For a ship to founder with her cargo in a medium incapable of supporting the load, and then rise again without being specifically lighter, is contrary to hydrostatic laws. A sinking vessel actually heavier than water, must go to the bottom; but if, from her cargo being washed out, she Remarks on this story.

is rendered lighter than the fluid in which she is immersed, then she would float to the surface, and be amenable to the laws of tides, winds, and gravitation, combined with local circumstances. A ship cannot conveniently alter her gravity so as to sink or float merely to confirm a paradox; and in the case before us, the Dutch vessel must have been water-logged within the influence of the LATERAL set at the surface.

The *Hermenegildo.*

In like manner, the wreck of the Spanish ship *Hermenegildo*, of 112 guns, which was blown up in action with the squadron of Sir James Saumarez, in June, 1801, floated into Tangier Bay three days after the explosion, with one man still alive on board. On this occasion there appeared to be some striking anomalies between the ebb and flood, which so awakened the attention of Don Vicente Tofiño, that in the ensuing October, when peace had been ratified, he sent his nephew, Captain Tofiño, to gather information at Tangier, where he was assiduously aided by Mr. Salmon; but I was unable to learn the result.

The Patton experiment.

It is recorded, and much stress is laid on the fact, that when the late Admiral Philip Patton, who died in 1815, was Lieutenant of the *Emerald*, a 32-gun frigate, that ship was overtaken by a gale of wind in approaching Gibraltar; and at night was hove to, nearly in the middle of the strait. Before daybreak she struck at the back of the rock, where it was presumed that a counter current had carried her. Here she had to ride it out within half a cable's length of the breakers, and no room for even freshening the nip in the hawse; whence her destruction was most probable. This narrow escape induced the Lieutenant to study the currents of the strait with serious attention; and while upon this inquiry, he endeavoured to ascertain how far the theory of upper and under streams could be sustained by experiment. On the ground that when two fluids of specific gravities meet in a narrow channel, the heavier will run out below at exactly the same rate as the lighter will flow in from above, a number of bottles were filled with the water from the Atlantic at a distance from all land, and another set of bottles were filled with water from the inner part of the Mediterranean. Upon as accurate a method of weighing as he could command, a flask containing one pound, six ounces, and five drachms of the ocean fluid, was considered to be thirteen grains lighter than the same flask filled with an equal quantity of the inner water. He also filled two decanters, of equal size, with the respective fluids, one being slightly tinged with ink, with their necks placed in a luting of putty; when the whole was held horizontally, the interchange of the heavy water displacing the lighter, was thought sufficiently sensible to justify an inference that the two liquids were of unequal densities. From these experiments, which were rather pains-taking than philosophical, Lieut. Patton came to the conclusion, that the Mediterranean surcharge was prevented by an ever-flowing undercurrent into the ocean.*

Having here given the plainest of the many propositions which have been advanced respecting this doubtful subject, and, without presuming in the present state of our knowledge of the question, to attempt casting the die, I shall now continue my essay without any longer running on the Scylla or Charybdis of this controversy. Locke, in addressing the understanding, insists that 'doubtful positions relied upon as unquestionable maxims, keep those in the dark from truth who build on them; and to be indifferent which of two opinions is true, is the right temper of mind that preserves it from being imposed upon, and disposes it to examine with that indifferency until it has done its best to find out the truth,'—this sound axiom we commend to the future investigator of any two theories. But it will be remembered, that the notion of upper and under Remark.

* I made an unsuccessful endeavour to repeat the decanter experiment, but with water taken from the surface, and at fifty fathoms of depth. The fluid was vexatiously sluggish.

currents, has descended to us for many ages; and if we accept the reasonings of Lucretius (*De Rerum Natura*, lib. v.) it was not opposed to the cosmogony of Epicurus. It was, perhaps, a knowledge of this which led Count Rumford to demonstrate, by direct experiment, that fluids of all kinds, when heated to different temperatures in different parts of their volume, must necessarily have an opposition of currents: the warmer, from its rarefaction and specific levity, occupying the upper, and the colder portion the lower part.

Count Rumford.

Mediterranean indraught.

The Strait of Gibraltar is so remarkable, both to the navigator as well as to the geologist, that it becomes necessary to treat it with greater detail than will be requisite in other cases: and, although the *Fretum Herculeum* was applied to the space between Cape Trafalgar and Spartel on the west, and from Gibraltar to Ceuta on the east, a more enlarged hydrographical view authorizes us to extend the western mouth of that magnificent ocean-channel to Cape St. Vincent on the north, and to Cape Cantin on the south. This assumed breadth of entrance is the more necessary, since the whole of its waters are affected by the draught into the Mediterranean. On this head, my venerable friend, the late Major Rennell, entertained an opinion that there is a general tendency of the Atlantic waters between 30° and 45° of north latitude, and from 100 to 130 leagues off the land, to move towards the Strait of Gibraltar at a rate of not less than from fourteen to seventeen miles in twenty-four hours. Now, though extreme cases might occur during a long prevalence of particular winds, wherein such an indraught would be appreciable to nice experiment, and outward-bound vessels from the English channel might find themselves rather to the east of their reckoning, the Major's assumption must be received *cum grano salis;* especially if depth be admitted as a condition of these 400,000 square miles. Another friend, enlarging upon Rennell, considers the gulf-stream as the primary

Major Rennell.

mover of the Strait current; but when he reflects that the Gulf stream is strongest under easterly gales, and is generally much weaker in winter than in summer, he must perceive that his position is unstable, and that the inflow is occasioned by some local cause exclusively connected with the Mediterranean Sea.

Gibraltar Strait.

Such being the natural entrance of this Strait, the true boundaries of its narrows, designated the Gut, are between the capes of Trafalgar and Spartel, which are twenty-two miles apart—the isle of Tarifa and Alcazar point, nine-and-a-quarter miles—and Gibraltar and Ceuta, which are twelve miles distant from each other; the whole occupying a length of about thirty-five miles. These three form the principal stations for averaging a breadth; and the local peculiarities are necessary to a full inquiry as to how much the current owes to differences in the specific gravity of the contiguous waters—how much to the depth and form of bottom—what to the density of the several media—and what to the fluctuations of atmospheric pressure. Such inquiries were of course beyond my time and means, for it must be borne in mind that I was only correcting a nautical chart for the use of navigators: but, in the hope that my observations may aid philosophical researches, I may state that, in a line between the two first-named points, the body of the stream is of much less depth than it is to the eastward, as it carries but from twenty to seventy fathoms to half the distance across from Spain: and even the deepest part between that seventy fathoms and Cape Spartel, is but 220 fathoms. A few miles more within, the channel has not above 160 fathoms at the greatest; but between Tarifa and Alcazar point, it deepens to 500, and immediately beyond, gets to 700. This depth rapidly increases towards the Mediterranean Sea, and is 950 fathoms in mid-distance between Gibraltar and Ceuta; and as there is no bottom with 1000 fathoms of line up-and-down (*upwards of 1300 payed out*), a little farther to the eastward, it is

Depths of the Strait.

clear that the bottom, from the meridian of Cape la Plata, forms an inclined plane, through which a mid-channel section of eleven lengths to one depth, appears thus:—

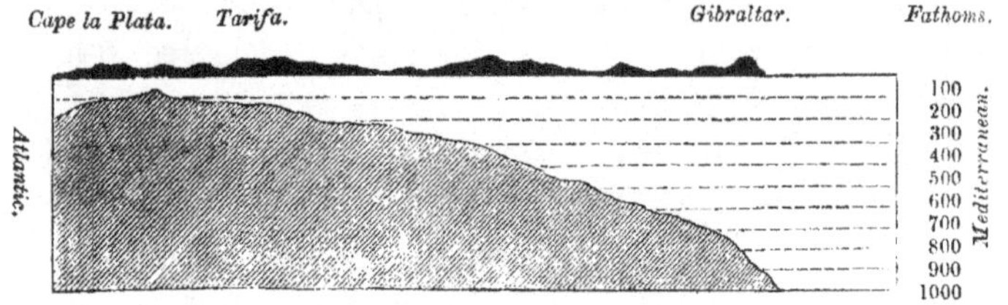

Remark.

A glance at this singular formation at once throws doubt on Dr. Wollaston's position, insomuch, that on hearing of the conclusion he had formed, I wrote a particular statement for him; but when it was received, he was upon his deathbed. His executors, therefore, returned the paper to me, and I forthwith sent it to Sir Charles Lyell, who was then compiling his well-known work—the *Principles of Geology*. Hence the conclusion on page 131.

Velocity of the current.

When to this underwater abyss we add the roots of the mighty Atlas chain on one side, and the elevated tabled barrier formed by Spain on the other, the feature of the strait is still more surprising. In its narrowest part, the central stream may be about four miles in width, but, of course, with variable limits; and its average rate of flowing is from two to three miles per hour; but Gibraltar pilots have assured me, that they have known it to run, under special circumstances, at upwards of five knots; which being without proof, is rather assertion than fact. Its course sets so constantly to the eastward, that a temporary surface-current towards the Atlantic — which sometimes, though very rarely, is known to occur — can hardly be deemed an exception to the general rule: such an action may result from westerly gales causing a partial elevation of the oceanic waters, and consequent resurge on regaining their usual level, or it may be effected by a superficial rise from strong levanters, which, in either case, is entirely local.

The solution of one part of the difficulty seems to be, that by an extraordinary natural effort, there are two returning or counter-lateral streams, one on each shore, so that a very complex motion is constantly observable; and these remarkable streams being governed by lunar influence, shall presently be treated as tides. But the phenomenon of a strong middle-current setting inwards, while only two feeble streams return in the opposite direction at given times—a tidal reflux, far inferior to the quantity flowing in—may be attributed to the pressure of a greater fluid mass on a small body of water; a pressure which, from the force of its impulsion, must necessarily displace the upper strata of the smaller mass.

Effects of the current.

The central current being established, and running from west to east, it consequently follows that the action must be felt on and in the neighbouring waters; which it assuredly is, though only to such a degree, and in the more distant parts, as to be hardly appreciable in practical navigation. The influence of the stream is sensibly experienced inside as far as Cape de Gata, a distance of 150 miles, but it gradually diminishes, being more diffused: yet it then takes a direction not only according to the curves of the coast, but also from the winds, especially those from the sea. About the vernal equinox, with winds between west-south-west and north-west, we found that the current ran, along the coast of Granada, at the rate of one knot an hour eastwards; after passing Cape Palos, it set to the east-south-east; and when we neared the Balearic Islands, it flowed very gently to the north-east. In a word, the reaction of the streams against the coast, with the operation of winds, together with the variable currents induced by the smaller straits in this sea, occasion lateral and adverse 'sets' in all directions. But under ordinary circumstances, and in settled weather, when the great Atlantic stream has its usual course into the Mediterranean, the current moves strongly east along the African coast, and across the bay of

Coast of Spain.

Tunis, to the coast of Sicily: hence we may see why a ship, sailing eastwards along the African shore, should be generally ahead of her reckoning.

Remarks. These *sea-motions*, as our early hydrographers named them, are everywhere sensibly affected by the prevalent winds; as is strongly instanced in bights, inlets, and channels—of which the Gulf of Lyons, the Riviera of Genoa, the Faro of Messina, the upper portion of the Adriatic, the Gulf of Corinth, the Euripus, the Syrian sea, and the two Syrtes, are examples. Strong ripples resembling breakers are frequently caused in the vicinity of the larger islands, by the waves of one particular division meeting those of another: often breaking in so confused a manner as to account for many of the ideal shoals which find places on charts, to the confusion of navigators.

Sets of current. After flowing along the shores of North Africa eastwards—with occasional interruptions—the general current sweeps by Syria and Karamania, and returning westwards, sets out along the coasts of France and Spain, though in many parts it is so sluggish, as to be almost imperceptible. Strong winds from the north-west reverse this order of things; for then the stream sets in along the same coasts, at times making strongly round the Gulf of Lyons, and varying its course conformably to the contour of the coasts of Provence, Languedoc, and Catalonia. Off Toulon the easterly set was sometimes so strong after levanters, that the inshore ships of our blockading fleet had considerable difficulty in keeping their stations. In the sea of Tuscany, the south-west winds occasion the greatest elevation of the waters; and a continuance of *labeschades* (e *libeccio*), or gusty gales from that quarter, have been known to raise the height of the surface no less than twelve feet above its ordinary level. In the phrase of the pilots, the waters are then up, and consequently occasion a strong surface-drift through the Strait of Bonifaccio.

A curious feature of these shores, and one not unconnected with currents, is the changeable nature of the beaches, especially at the mouth of the Var, and at Nice, where the sea-margin is seen alternately consisting of large shingle, fine sand, or small gravel, and, a few days afterwards, coarse shingle again. This must be a consequence of the fluent and refluent action of the surf upon the materials composing the beach, according to the force of the surges; but the cause of that force has not yet been satisfactorily investigated. M. Risso, the Savoyard naturalist, had, even in fine weather, very frequently observed the swell of the surf tumbling in something like the rollers of the Atlantic. For this he could assign no cause, but he assured me that the phenomenon was generally higher after heavy rains in the Alps and Apennines—producing the rise in the rivers called 'freshes'—than at any other time; and therefore an unequal atmospheric pressure might contribute to the effect, by causing a circulation of the waters below; for as a surf sets, it acts from this cause on the ground at some distance in the offing; but its returning action having a tendency to restore the level by reverberation, is almost confined to the surface, and has no effect on the bottom. The sea-motions are certainly sensible at a depth of twenty-five fathoms.

Beaches.

M. Risso.

Though inferior in importance, the Strait of Messina has occupied the attention of philosophers for as many ages as that of Gibraltar; and the laws of its current are still among the desiderata of physical inquiry. While most of the ancient reasoners on the subject only gave us terrible pictures of the dangers of navigating the *Mamertinum fretum*, Eratosthenes ascribed the cause of the bewildering currents and counter-currents to a difference of level in the vicinity, especially insisting that the *descending* waters flowed from the higher Tyrrhenian sea: and Aristotle follows on the same ground. But as the effect has been proved to be influenced by the attractions of both sun and moon, the subject will be resumed in the next section.

Faro of Messina.

Here, however, we will step for a moment to the opposite coast of Sicily, and introduce a strange current, thus recorded in my account of Sicily, *page* 224:—

Marobia.

The Marobia is an extraordinary phenomenon, most probably deriving its name from *Mare ubbriaco*, or drunken sea, as its movement is apparently very inconsistent; it occurs principally on the southern coast of Sicily, and is generally found to happen in calm weather, but is considered as the certain precursor of a gale. The Marobia is felt with the greatest violence at Mazzara, perhaps from the contour of the coast. Its approach is announced by a stillness in the atmosphere, and a lurid sky; when suddenly the water rises nearly two feet above its usual level, and rushes into the creeks with amazing rapidity, but in a few minutes recedes again with equal velocity, disturbing the mud, tearing up the sea-weed, and occasioning a noisome effluvia; during its continuance the fish float quite helpless on the turbid surface, and are easily taken. These rapid changes (as capricious in their nature as those of the Euripus), generally continue from thirty minutes to upwards of two hours; and are succeeded by a breeze from the southward, which quickly increases to heavy gusts. This phenomenon may be occasioned by a westerly wind blowing, at some distance in the offing, towards the north coast of Sicily, and a south-east wind, at the same time, in the channel of Malta, the meeting of which would take place between Trapani and Cape San Marco. I advance this idea, because the westerly wind most usually precedes, and the south-east succeeds, the Marobia.

In addition to this I ought to have added, that it was during a turbulent marobia that H.M. ship *Raven*, of 18 guns, was lost on Cape Granitola, on the 6th of January, 1804: and this was the 'unusual current' of Captain Swaine's defence, at the consequent court-martial of inquiry into the loss. When very violent, its effects of action and reaction are felt even on the opposite coast of Barbary.

Central currents.

This central and important portion of the Mediterranean forms the passage or channel of communication between the Western and the Eastern Basins; and with respect to the prevalent 'sea-motions' by which it is affected, I have only to repeat what was published in my account of Sicily (*page* 184), so long ago as 1824. It is there stated that the currents

arising from the constant evaporation and the action of the winds, observe no regularity, rising a foot or two, according to the weather and the peculiarities of locality and depth; thus the north-west wind, raking the shores, promotes a strong set to the south-east; while the south-western, which is here very sensibly felt during the vernal equinox, causes strong counter-currents; and at length, on its changing to the opposite quarter, the whole body of water rushes to the westward with considerable velocity . . . . . . In long-settled weather, the currents between Sicily and the Barbary shore, and from thence to the westward of Galita, run to the eastward at the rate of from half a mile to a mile an hour. In the channel of Malta, the south-east current has occasionally been so strong, that ships have found it difficult to beat up to Maritimo; while others, driven to leeward of Malta, have been obliged to carry a press of sail in order to hold their own, until a change of wind enabled them to make the island again. Another proof of the influence of this current is, that ships stretching over from Cape Passaro to Valetta, with a northerly wind, usually keep a point higher, to ensure fetching it.

Between Malta and Tripoli, the current generally sets to the southwards and eastwards; but between Malta and Tunis, a prevalence of south-east winds throws the waters upwards to the barrier formed by Adventure Bank and the Skerkis, where, beside that impediment, meeting the general easterly set from Gibraltar, the current sweeps away northwards, at the rate of about a knot and a half per hour, while at other times the set is southerly.

Adriatic currents.

The operation of the winds and waters in the Adriatic, is more uniform than in the parts just treated of. The current usually sets in along the Albanian and eastern shores, sweeps round the head of the gulf from Trieste to Venice—often at the rate of a knot an hour—by the Romagna, and thence out again along the Italian shores, with a somewhat diminished force; but the Bora makes a surcharge of a foot or two on these latter coasts. This general action is accompanied by a sufficient tidal influence to cause a variety of local sets, called *ligazzi*, some of which prevail right across, a natural consequence of the contour of this sea, and the islands which stud it; but these variable streams are neither rapid nor dangerous. Much has been written on the subject by the Venetian pilot, Vicenzo di Luccio; and he has not only described currents for the different months in the year, but has gone so far as to give almost an hourly course and velocity for them. When in the 'City of the Sea,' I made inquiry for this Signor di Luccio, but without effect, for as his details—however particular—have the air of mere arbitrary assumption, I was desirous of a *viva voce* explanation.

Ionian currents.

Although the Ionian Sea feels the general set of the main current, there is sometimes a surface-run to the southwards, which is retarded or increased according to the nature and degree of the offing gales. A stream is generally running through the fine channel of Corfu, which is remarkably influenced by the wind; when it blows pretty strong from the north, the waters set to the southward at the rate of 1½ or 2 knots an hour, and it occasions a fall in the level of from three to four feet: a southerly wind raises it to about the same height, and the current then sets northwards. But this is not confined to the channel, although it is there the most marked, for over the whole Ionian Sea, southerly winds cause an extraordinary rise of about a foot, and northerly ones a fall of about the same amount; but if they are strong and continuous, the elevation and depression are naturally greater. Still the traces of tidal action are extremely faint; for even the remarkable ingress and egress from this sea into the Gulf of Arta (the commercial value of which is detailed in a memoir written by General Vaudoncourt; their supposed full-and-change days have been stated by others) can hardly, from present data, be yet considered as a regular tide, since the sets are known to be more influenced by the winds than by our satellite. A stream runs into the Gulf with the sea-breeze by day, and in the night, when the land-winds prevail, the water returns outwards. Tidal action is more decidedly marked just below, in the Gulf of Corinth, although the current movements are not dissimilar in cause and effect from those of Arta, for the strength of the set and the height of rise depend on the direction and force of the wind, the current running most strongly when it is blowing down the Gulf, and often taking a direction against the wind. On many occasions, the meeting of the waters of Patras and Corinth, under the influence of the offing and gulf winds, causes a broken foam across the narrow channel at the entrance of the Gulf of Lepanto, and a considerable swell. In running through this, the agitation reminded me of that well-known and often dangerous spot called the bridge, between Drake's Isle and Mount Edgecumbe, at Plymouth, though the apparent causes differ widely.

Archipelagan currents.

In approaching the Archipelago, and from thence the coasts of Asia Minor and Syria, many peculiarities are observable in the currents, of which the principal is the action of the waters descending from the Euxine, through the numerous inter-insular channels of the Cyclades, upon the main current which sets along those coasts westwards. On the north coast of Candia, it is observed that, with the wind blowing strong from the west for any continuance, the waters rise two or three feet above their common level; and with the wind from north or east, they fall two feet below that level, the effect of the westerly set acting on the usual conditions of surcharge and discharge. The whole of the Archipelago, however, is affected from the north-east; for the Black Sea, receiving a greater accumulation of water from its tributary rivers than is withdrawn from it by evaporation, pours out a constant and copious stream through the Sheïtan akindí-sí, or Satan's current, into the sea of Marmora, whence—an expansive surface being offered to exhalation—the discharge through the Hellespont, though still considerable, is perceptibly less rapid, but very constant.

Currents in the Black Sea.

The water of the Black Sea has a lower specific gravity (1·01418) than that of the other Mediterranean basins, a fact which proves that it is not liable to much evaporation. The overflowing current just alluded to, which, especially from the mouth of the Dnieper and the Danube, runs rapidly through the entrance of the Thracian Bosphorus, the rate being estimated at from three to five knots per hour according to the prevalent direction and force of the winds, makes counter-currents and eddies along the sinuosities and points by which it is diverted in its course. From the relatively small amount of salt in these waters, the shallower parts of the Euxine are sometimes frozen; and the Sea of Azof, into which the inundating Don (*Tanais*) and the many branches of the Kuban discharge themselves, is frozen over during three or four months of the year, so that laden sledges and troops of people pass and repass upon it.

The Levant. It has been pretty fairly established, that owing to the action of the main or general current, a set constantly runs by Cyprus and along the coast of Karamania to the north and west: whence, a ship leaving Malta, and bound to Smyrna or the Dardanelles, on meeting a strong north-easter off Cerigo, as is so often the case, instead of beating against the drain of current then setting down from the Dardanelles, would, at no loss of time, stretch away to the south-east, as far as Alexandria, nearly with an easterly current, and so along the coast of Syria with the northerly set. At times between Rhodes and the mainland, in consequence again of the effect of a prevailing north-east wind sweeping the whole surface of the other parts of the Archipelago for nearly two-thirds of the year, the current is liable to run like a sluice; insomuch that in a calm, a ship may be carried up to the north, by carefully looking out for eddies, and keeping within the islands near and after passing Rhodes. Sir Francis Beaufort, when Captain of the *Frederiksteen* frigate, made some very judicious observations on the currents of this part of the sea: he also made an experiment on the under-currents, which I regret not having heard of before I quitted the station, or that simple and ingenious operation should have been repeated in other places. His words are,—

Sir F. Beaufort.

From Syria to the Archipelago there is a constant current to the westward, slightly felt at sea, but very perceptible near the shore, along this part of which it runs with considerable but irregular velocity: between Adratchan Cape and the small adjacent island, we found it set one day almost three miles an hour; and the next, without any assignable cause for such a change, not half that quantity. The configuration of the coast will perhaps account for the superior strength of the current about here: the great body of water, as it moves to the westward, is intercepted by the western coast of the Gulf of Adalia; thus pent up and accumulated, it rushes with augmented violence toward Cape Khelidonia, where, diffusing itself in the open sea, it again becomes equalized.

The cause, the progress, and the termination of this current would form an interesting subject for future investigation. To trace its connexion with the volume of water which enters by the Straits of Gibraltar, with the influx of the currents from the Euxine, and with the effect of the Nile, and of the numerous though small rivers of Asia Minor, will require a series of corresponding observations on both sides of the Mediterranean. The counter-currents, or those which return beneath the surface of the water, are also very remarkable; in some parts of the Archipelago they are at times so strong as to prevent the steering of the ship; and in one instance, on sinking the lead when the sea was calm and clear, with shreds of bunting of various colours attached to every yard of the line, they pointed in different directions all round the compass.

Currents along North Africa.

The main current, as already said, sweeps from Gibraltar along the African shores, modified by the several sinuosities; but regaining its regular course along the coast of Lybia, it flows by Alexandria, and, trending north-eastwards, makes for the shore of Syria, and in its advance seems to acquire new strength. There is frequently a strong outset from Abúkír Bay, and variable flaws off Damietta; but the grand outlet of the Nile has great influence around. The northerly winds which prevail in summer, carry with them the vapours raised from the Mediterranean—though without forming regular clouds—over the valley and low ranges of the Egyptian hills, as far as the Abyssinian Alps and the lofty mountains beyond; where, being cooled and condensed, they fall in rain, and are in some measure carried back to their native sea by the periodical inundations of the Nile. The overflowing generally begins at the end of June, sometimes from a fortnight to a month later, and continues for above two months, after which it gradually subsides. The river rises from fourteen to twenty-three feet in vertical height, and the volume of water which it carries into the sea is twenty times greater in its latter than in its former state; insomuch that during the full surcharge—as before stated—potable water may be baled on the surface of the Mediterranean, even out of sight of land. Here the current exerts itself on the large quantity of alluvial substances brought down by the Nile, and drifting the sediment eastwards, exerts its silt-depositing property (before alluded to in page 8) with such effect, that rapid accretions along the Syrian shores, thereby leaving Tyre and Sidon inland, are directly traceable to it. Indeed, this is so palpable as scarcely to require the eye of the geologist, for I have seen the waters discoloured with impurities for many leagues; and in 1801, a rather alarming phenomenon was encountered here by H. M. frigate *Romulus*, commanded by Captain Culverhouse, on her passage from Acre to Abukir Bay. It is thus related by Dr. E. D. Clarke, who was then a passenger on board:—

H. M. ship *Romulus.*

July 26th.—This day, being Sunday, we accompanied Captain Culverhouse to the gun-room, to dine with his officers, according to his weekly custom. As we were sitting down to dinner the voice of a sailor employed in heaving the lead was suddenly heard calling '*half four!*' The captain, starting up, reached the deck in an instant, and almost as quickly putting the ship in stays, she went about. Every seaman on board thought she would be stranded. As she came about, all the surface of the water exhibited a thick black mud: this extended so widely, that the appearance resembled an island. At the same time, no land was really visible, not even from the masthead, nor was there any notice of such a shallow in any chart on board. The fact is, as we learned afterwards, that a stratum of mud, extending for many leagues off the mouths of the Nile, exists in a moveable deposit near the coast of Egypt; and when recently shifted by currents, it sometimes reaches quite to the surface, so as to alarm mariners with sudden shallows where the charts of the Mediterranean promise a considerable depth of water. These, however, are not in the slightest degree dangerous. Vessels no sooner touch them than they become dispersed; and a frigate may ride secure, where the soundings would induce an inexperienced pilot to believe her nearly aground.—(*Clarke's Travels*, vol. iii. p. 13.)

The circular motion of the current round the Mediterranean, shown in the preceding remarks, appears to have been first observed, or at least described, by the celebrated Geminiano Montanari, in the year 1681,—the same philosopher to whom is attributed the discovery of the method of determining the heights of mountains by means of the barometer. It is therefore to be wished he had been the first to detect that the rise and fall of waters—under either

Montanari.

tides or surcharges—are also shown by that truly philosophical instrument; it being low water when the barometer is highest, and *vice versâ*.

---

§ 5. On the Tides of the Mediterranean.

THE word 'tide' signifies properly the body of the oscillation, and comprehends the difference between high and low water; tidal motion being rather the elevation of a wave than an absolute transfer of water. The tide-wave differs from the wind-wave, because it is the result of forces acting both parallel to the surface and perpendicularly on the surface of the sea; whereas common waves are all occasioned by lateral disturbances of wind, current, and terrestrial modification. It therefore follows that, so far as their primary causes are concerned, tides may be considered merely as alternate elevations and depressions of the water, without any necessary transfer from place to place; but the whole being produced by an undulating motion, in which the surface swings upon certain average curves, recalls the trite appearance of the waves over a field of corn in a gale of wind. Nevertheless, although astronomical demonstration is strong upon this point, practical observation of the phenomena has shown that there is often a positive transfer of water from one place to another; and all waves which are produced by causes acting near the surface of the water—as in the case of a shelving or gradually inclined shore—are in so far impelled in a lateral direction, and the waves then are consequently a propagation of motion through that water. The above remarks must also be qualified by recollecting that in the case of comparatively shallow water, such as all seas may be called, the forces parallel to the surface produce the greatest part of the effect: in a word, that the horizontal transfer *must* considerably exceed the vertical movement.

Tides.

Tides and currents are so similar in movement and effect, and so constant in their operation, as to be in many cases difficult to distinguish; yet they are so distinct in cause, that a discrimination is here attempted, even where those agents are difficult to investigate from want of action.

Anomalies.

So many causes contribute to the varied courses of the waters, and so many interfere with the very slight indications of Mediterranean tides, that we are obliged to infer rather than assert results of direct observation; in this state of knowledge it is therefore impossible to give any general rule for the observable effect. My own time and attention were necessarily more devoted to fixing latitudes and longitudes, and delineating coasts and harbours, than to studying the physics of this sea; but I made a few experiments, which I hope may render the subject a peculiar object of attention to some who have better means and more leisure. Indeed, I have little doubt that a day will arrive when it shall be proved that these inner-sea motions—except the extreme local ones—are actually connected with those of the great oceanic waters. The closely following up a few apparent anomalies in this beautiful department, added to the crowning tidal knowledge which may yet be expected from our explorers in the Polar Sea—for which theoretical science yearns—must inevitably lead to a clear perception of all the phenomena presented by tides.

Mediterranean tides.

The Mediterranean, though poetically termed a 'tideless sea,' is far from being so in reality; for accurate observation detects a sensible elevation and depression of its waters —independent of currents, surface drift, or wind-raised swells. This, if not wholly, is partly ascribable to the lunar sympathy, as manifested by the alternate changing of the stream, and a periodical rise and fall, somewhat coincident with the oscillations of the Torricellian tube; the lowest surface accompanying a high barometer, and *vice versâ*. But as yet these are hardly admissible terms, for though there are places—as Venice and Jerbah—where the fact of a tide is shown in the amount and periodicity of its recur-

Barometer.

rence, and others where it is obvious from not immediately mingling with water differing in temperature, set, and velocity, still the tides over most part of this sea are so feeble and irregular as to be difficult to ascertain. Hence it has been asked, if these motions are attributable solely to the attraction of heavenly bodies and centrifugal force, how is it that the moon, which is acknowledged to have an attractive power sufficient to move such vast bodies of water as the Atlantic and Pacific oceans, should exert its influence so slightly over the inner sea, that many will hardly believe there are any tides in it? To this the Newtonian answers, The strait by which it communicates with the ocean is so narrow, that it cannot in so limited a time receive or discharge sufficient water to alter the elevation of the whole surface sensibly: and he moreover insists, that instead of the faintness of the Mediterranean tides being an objection to the theory of planetary attractions, it is a fair proof in its favour. For herein, the moon acting at the same moment in all parts, diminishes the gravity of the mass, while the difference of atmospheric pressure upon such a sea may tend to obliterate any slight appearance of tide that would occur if the pressure were uniform over the whole surface. Over a large space the air is increased in bulk, and consequently diminished in weight, by an almost tropical heat, thereby occasioning mobility and alternation. Yet there being little or no neighbouring water to move forward and increase the liquid elevation—which is produced in other cases less by a vertical rise of the waters attracted than by a lateral flowing of adjacent waters by virtue of their greater density—there consequently can be but weak tides in small seas, especially when the entrances are comparatively narrow and shallow, and face the west, a direction opposite to the general movement of the great mundane tidal wave.

Paradox.

Answer.

Still, although the Mediterranean tides are irregular, in many parts scarcely perceptible, and mostly so inconsiderable in a nautical point of view, that with a few exceptions

Opinions of the ancients.

they are scarcely worth appreciating, they are unquestionably interesting when physically considered, as exponents of a general cause; nor will it be forgotten that the theory of tides was first studied on those very shores, even from the time of Pytheas. Posidonius, who measured an arc of the meridian, explained the ebbing and flowing of the sea from the motion of the moon; and he seems to have been the earliest who declared the law of these phenomena, although Cæsar nearly at the same time (*De Bello Gallico*, lib. iv.) alluded to the nature of spring tides, as being connected with the moon's age. But assuredly Pliny advanced on this subject almost all that was possible for human sagacity, before Sir Isaac Newton unveiled the great law of the universe, and demonstrated that the same force which guides the planets in their courses causes the waters to rise and fall. Now Pliny had formally said, that the cause of the phenomena is in the sun and moon—*verum causa in sole lunâque*—adding the remarkable assertion, that the moon exerts her power as well under the earth as when she is seen aloft.

Tides at Gibraltar. Among the most palpable of the Mediterranean tideways are those in the Strait of Gibraltar, where various anomalous eccentricities are found, in consequence of its being the avenue between that sea and the ocean. While in command of a gun-boat at the siege of Cadiz, I found the tide-hour* in the bay to happen at two o'clock, or no

* In the present rage for foisting new terms into the *vernacular* technicals of a working profession, both meaning and brevity should be sought: I am therefore glad that Lieutenant Raper, in his 'Practical Navigation,' has adopted this appropriate phrase (*Ora del Porto* of the Italians) for the *High water on Full and Change days* of erst, and its recent substitute *Establishment of the Port.* The word *Pharonology* has been introduced to teach us where lighthouses stand; and an attempt is in hand to supersede the time-honoured and appropriate term, *Variation* of the needle, by the equivocal word *Declination,* which latter has been so long held by seamen as belonging to the sun. *Watershed* is absurdly forced into geography, to denote the highest ridges bounding the valleys of a country. It is a sort of echo of the German word *Wasserscheidung* (Water separation); but *shed,* either as a verb or a noun, has no such sense in English, and is almost exclusively appropriated to the falling or dropping of tears or blood. 'Culminating divisions' would better express what is meant.

less than two hours and a half sooner than all the tables in 1810 gave it to be; in consequence of which the movements of our flotilla at the siege were at first often embarrassed. By my own observation, the full and change at Gibraltar occurs at $12^h\ 50^m$; the rise at the former place ranging from eight to twelve feet, and at the latter, as shown in the boat-camber, from three to five feet. But between these two stations, I was assured by Don Felipe Bauza, the hydrographer of the Spanish navy, that it is high water at Tarifa at $11^h\ 15^m$, and under Cape Trafalgar at $5^h\ 40^m$; from which it would seem that from Europa Point the flood sets round Cape Carnero, and passing that headland flows to Tarifa, in the vicinity of which it meets the tide coming from the west off Cape Trafalgar, where it is low ebb when it is high water at Tarifa. This is singular, but Bauza was satisfied of the substantial truth of the facts. Captain Bauza.

On the southern coast of the strait, another tide runs alongshore from Ceuta—where the tide-hour is at $1^h\ 45^m$—by Tangier, where it is flood at 12^h, to Cape Spartel and its offing. These lateral streams average a distance from the respective shores of more than a couple of miles, and their rate of velocity varies from two to four knots per hour, their regularity being interrupted by the prevalent direction and force of the winds; and their action in impinging on the central current, occasions eddies and whirls in the most prominent parts of the strait. But these repercussions are so very transitory and changeable, often not occurring at all, that, on being consulted, I could not approve of General Don's allowing Ignazio Reiner, his pilot at the Rock, to insert them in a chart for publication; and the same of the very useless tabulated floods and ebbs, making time and tide rather more synonymous than, with all their strong points of resemblance, we find them to be. Ceuta and Tangier.

From what is thus advanced, it will be evident to the seaman that, with a moderate wind, there will be no difficulty, by watching the tides, in beating to the westward Proof of the tide's uses.

through the strait. And upon such conviction I acted; for during the investment of Tarifa by the royalist general, O'Donnell, in August, 1824, the besieged constitutionalists were ill-advised enough to fire at an English merchant-ship which was passing, whereby she incurred a detrimental delay, and had she had any munitions of war on board, would have been plundered, though the threat was softened by a promise of *bills* in payment. On learning this, being the senior naval officer at Gibraltar, I instantly despatched

H. M. sloop *Pandora*.

the *Pandora*, sloop of war, Captain William Gordon, to expostulate with General Valdez, the commander of the rebel garrison; and I moreover directed Lieut. M'Causland, in the mortar-boat *Hamoaze*, to lead the insulted trader through the straits. The wind was then westerly, blowing fresh at intervals; but I assured both these officers that by making short boards with the flood tide on the Spanish shore, the passage would readily be effected. It being a point of strict service in which promptness was requisite, my wishes were cordially seconded: the *Pandora* quickly brought an ample apology from the unhappy constitutionalists—numbers of whom were destined to be shot in cold blood a few days afterwards—and the *Hamoaze* succeeded in beating through with the heavily-laden merchantman in tow. This mortar-boat was but a tub of a vessel at best, yet she thus performed a nautical feat, so far as I know, then unprecedented.

Tides along the coast of Spain.

On the Spanish coast inside the Mediterranean, the tides are certainly of the most moderate order; and during some long spells in Port Mahon, when our Toulon blockading fleet used to winter there, I found, after numerous trials, that that fine harbour was barely affected, the ebbing or flowing a foot or two being irregular, and evidently more ascribable to winds than to lunar attraction. This was also the opinion of Mr. Gaze, the master of the fleet; who told me, however, that a regular rise and fall had been detected at Malaga, where it was flood at about 12 hours. This, in consequence of strong sea-winds while there, I was unable to prove; but the assertion was in a measure corroborated by the captain of the port, although his notions as to the distinction between a current and a tide were not of a very definite character.

Tide at Carthagena.

There was another point which gave me more trouble than the allegation respecting Malaga, and it was this: Polybius, who is usually very exact as to what he personally knew, says that, at the siege of Carthagena by the Romans, Scipio observed that a certain part of the walls was left undefended when the tide fell; as the besieged judged the sea to be a sufficient barrier on that side. Now I diligently attended to the historian's statement, because it involves a greater rise and fall than is known along this coast; but no present evidence, either ocular or oral, would lead to Scipio's conclusion. My experiments were made in the inner floating-harbour, which appears to occupy the site of the cothon which occasioned Doria's aphorism, that June, July, and Carthagena were the best ports in the Mediterranean. Here a fairly-placed tide-pole only announced an alternation of about sixteen inches; and the pilots and fishermen of the spot knew of but little variation from this amount, except in offing gales. But another assertion offers a still greater puzzle; for Polybius (*lib.* x.) pointedly boasts that he can speak of Carthagena with assurance, inasmuch as he takes his account, not from hearsay, but from what he had himself seen and examined. In this spirit he writes: 'The whole of this gulf takes the character of a perfect harbour. For an island lies in its mouth, and leaves on either side a narrow entrance; as it receives all the force of the swell from the sea, the whole gulf remains entirely calm.' Now, as the term gulf cannot allude to the cothon, or to the marshes then existing to the north of it, this island can be no other than the bold and rocky Scombrera; but instead of being in front of the gulf, it lies quite over on the south-east side, with the open bay

Scombrera.

on the west, and a boat-passage between it and the main. Polybius, however, might have viewed it from one of the eastern eminences, whence it was apparently brought to bear more centrally. This was a point which the late Dr.

Dr. Arnold. Thomas Arnold—although he admitted the general accuracy of Polybius—told me he would endeavour to ascertain in his next vacation, as it was a part of Spain he should like to visit: that vacation he never saw, for, within a fortnight after he wrote to me, he was suddenly seized with *angina pectoris*, which carried him off in a few hours, on the 12th of June, 1842.

Along the shores of France and Italy.

Round the Mediterranean shores of France and Italy the tides are of little moment, the most exact observations giving only a foot or two of rise from that cause; but though this may be accepted roundly, I am not inclined to assign much weight to the tide-hours at Toulon, Spezzia, and Naples, which are respectively given as $3^h\ 30^m$, $1^h\ 45^m$ and $11^h\ 20^m$, because there was considerable inconsistency in the accounts placed before me by General Visconti, who assured me that Sir Charles Blagdon's time for full and change at Naples—between the hours of nine and ten (*Philosophical Transactions* for 1793)—is erroneous. The tides, however, and the currents caused by them, in the

Faro of Messina.

beautiful *Stretto Mamertino*, or Faro of Messina, demand an express mention; I shall therefore repeat what was published, for the most part, nearly thirty years ago, in my account of Sicily and its Islands, especially as the appended details are not irrelevant to this inquiry:

Scylla.

As the breadth across this celebrated strait has been so often disputed, I particularly state that the Faro tower is exactly 6047 English yards from that classical bugbear, the Rock of Scylla, which, by poetical fiction, has been depicted in such terrific colours, and to describe the horrors of which, Phalerian, a painter, famous for his nervous representation of the awful and the tremendous, exerted his whole talent. But the flights of poetry can seldom bear to be shackled by homely truth; and if we are to receive the fine imagery that places the summit of this rock in clouds brooding eternal mists and tempests—that represents it as inaccessible, even to a man provided with twenty hands and twenty feet, and immerses its base among ravenous sea-dogs;—why not also receive the whole circle of mythological dogmas of Homer, who, though so frequently dragged forth as an authority in history, theology, surgery, and geography, ought, in justice, to be read only as a poet. In the writings of so exquisite a bard, we must not expect to find all his representations strictly confined to a mere accurate narration of facts. Moderns of intelligence, on visiting this spot, have gratified their imaginations, already heated by such descriptions as the escape of the Argonauts, and the disasters of Ulysses, with fancying it the scourge of seamen, and that in a gale its caverns 'roar like dogs;' but I, as a sailor, never perceived any difference between the effect of the surges here, and on any other coast, yet I have frequently watched it closely in bad weather. It is now, as I presume it ever was within the reach of history, a common rock, of bold approach, a little worn at its base, and surmounted by a castle, with a sandy bay on each side. The one on the south side is memorable for the disaster that happened there during the dreadful earthquake of 1783, when an overwhelming wave (supposed to have been occasioned by the fall of a part of the promontory into the sea) rushed up the beach, and, in its retreat, bore away with it upwards of 2000 people, whose cries, if they uttered any in the suddenness of their awful fate, were not heard by the agonized spectators around. * * *

On the whole, from the adhesive quality of the sands, and a strict examination of the various localities, particularly the lighthouse of the Faro point, which was constructed more than 200 years ago on the ruins of an ancient tower (then, as now, on the margin of the sea), I do not believe that the channel has widened; indeed, it is not clear to me, that this part was not originally wider, and that the two lakes have been gained from it; the story related by Hesiod and Diodorus, of the sea being broad here, until Orion raised the promontory of Pelorus to place a temple on, though not a confirmation, gives some colour to the supposition. **Faro Point.**

The four principal stations of the distances across, in my trigonometrical operations, by theodolite angles from a base line on that part of the beach near Messina called Mare Grosso, are—from Faro point to Scylla castle, 6047 yards, as before stated; from Ganziri village to Point Pezzo, 3971 yards; from Messina lighthouse to Point del Orso, 5427 yards; and from Messina lighthouse to the cathedral of Reggio, 13,187 yards. * * * **Breadth of the Faro.**

The currents in the Faro are so numerous, and so varied, with respect to their duration and direction, that I found it very difficult to ascertain anything with precision, as one series of observations seldom agreed with another; but I have generally found the statements of the most experienced pilots, after making due allowance for localities and weather, approximate very near to each other. In settled seasons there is a central stream running north and south, at the rate of from two to five miles an hour, and which though, properly speaking, only a current, when uninfluenced by strong winds, is guided by the moon. On each shore there is the *refluo*, a counter or returning set, at uncertain distances from the beach, often forming eddies to the central current;* but, in very fresh breezes, the lateral tides are scarcely perceptible, while the main increases so as to send, at intervals, slight whirlpools to each shore. There is, in general, an uncertain rise and fall of a few inches; but before the vernal equinox, when the sun is nearest **Currents and tides.**

* With a descending current, the *Refoli*, or contrary sets, occur on the Sicilian shore; with an ascending one, they are near Calabria.

the earth, and the moon in her perigee, they rise to 18 or 20 inches. When the *Rema montante*, or main current, runs to the northward, it is called the ascending or flood,—and the contrary, the *Rema scendente*, the descending or ebb; and this has obtained, perhaps, even from the time of Eratosthenes. There is usually an interval of from fifteen to sixty minutes between the changes; and the tide runs six hours each way, though I have known it, during a south-east gale (which has the greatest influence), flow to the northward upwards of eight hours. By the most precise observations I have been able to make, it is high-water on the days of full and change of the moon, off the Faro point at *6h. 56m.*; and in the harbour of Messina at *8h. 10m.*, or rather later. But these times are in themselves irregular and uncertain, owing to the great waves without, and contingent agencies which are not amenable to such calculations as mine were. A descending current makes the strait the roughest.

Navigation of the Faro.

The Faro channel is entered from the north on passing the lighthouse on the point; and though, from the nature of its winds and currents, it has long been clothed with imaginary terrors, yet as the Athenians and Syracusans, and the Locrians and Rhegians fought there, it could not have been considered so fearfully horrible by ancient sailors as by ancient poets; and the language of the former would probably have borne a tenour very different from the romantic embellishments of the latter, notwithstanding the passage through it might have been an affair of some moment with their small vessels and inexperienced seamen. But we have been gravely assured in a recent publication, that this strait is still extremely dangerous, and—forgetful of the memorable names of Loria, and Byng, and Walton*—it is added that Nelson was the first who ventured through with a squadron of men-of-war: while, on the contrary, it has always been used as an expeditious route by those, bound to the south-eastward, who have not been accustomed to a 'coat of terrors and a cap of fear:' and I am convinced that no persons well acquainted with this channel, will think it hazardous, especially if they have been in the habit of keeping well over to the Sicilian shore.

From the baffling winds to be expected, however, it certainly requires caution, though, except the set of the current towards the rocks under the Torre di Cavallo (a situation extremely disagreeable at night in bad weather), the beaches are so steep, that the stream enables vessels to glide safely along them. In light breezes, the current may be stronger than the ship's effort, and by turning her round, often alarms a person unacquainted with the phenomenon, although there is no actual danger: and the losses there, during my residence in the island, were certainly not more than would have been the case in any other part frequented by an equal number of vessels.

---

* This was the officer who, after the action between Byng and Castaneta, being detached in pursuit of six sail of the line and as many smaller ships that had escaped, reported his complete success to the Admiral in the following laconic terms:—

'Sir,—We have destroyed all the enemy's ships and vessels on the coast, as per margin.—I am, &c.,

'George Walton.

'Canterbury, off Syracuse,
16th August, 1718.'

I would not, indeed, advise a stranger to push through in the night, unless with a fine free wind, as the light at Messina is so indifferent, that it cannot be distinguished among the numerous torches of the fishermen, who, every tranquil night, cover the strait with their boats. Precautions should also be taken against the heavy gusts which, at times, from the mountainous nature of the coasts, vehemently rush down the fiumare (*torrent-beds*), and are dangerous to small vessels. I have twice, with grief, seen the neglect of them prove fatal; one of these circumstances occurred in the Sicilian flotilla, to which I was then attached; a fine barge, with eighteen of the best sailors we had, in attendance upon Colonel Caffiero, one of our officers, had been on constant duty in this strait for several years; when, in the early part of 1815, having carried the Princess of Hesse Philipstadt on board a vessel bound to Palermo, the barge was assailed by so sudden a squall on her return, that they could not lower the mainsail, and she instantly overset; the bodies of the unfortunate men were picked up the next day, between Scaletta and Taormina, about twenty miles to the southward. It is remarkable that there has been found in Messina a Greek inscription to the memory of thirty-seven youths of Cyzicus, who met a similar fate in the Faro; and in honour of whom, as many statues—the workmanship of Calion—were erected with a suitable inscription. Caution.

My description of Charybdis must follow that of Scylla:—

Outside the tongue of land, or Braccio di Santo Rainiere, that forms the harbour of Messina, we see the Galofaro, or celebrated vortex of Charybdis, which has, with more reason than Scylla, been clothed with terrors by the writers of antiquity. To the undecked boats of the Rhegians, Locrians, Zancleans, and Greeks, it must have been formidable; for, even in the present day, small craft are sometimes endangered by it, and I have seen several men-of-war, and even a seventy-four gun ship (*the Queen, bearing the flag of Rear-Admiral Sir Charles Penrose*), whirled round on its surface; but by using due caution, there is generally very little danger or inconvenience to be apprehended. The Galofaro appears to be an agitated water, of from 70 to 90 fathoms in depth, circling in quick eddies; but rather an incessant undulation than a whirlpool, and the cases are only extreme when any vortiginous ripples threaten danger by absorption to laden boats. It is owing, probably, to the meeting of the harbour and lateral currents with the main one, the latter being forced over in this direction by the opposite point of Pezzo. This agrees in some measure with the relation of Thucydides, who calls it a violent reciprocation of the Tyrrhene and Sicilian seas; and he is the only writer of remote antiquity I remember to have read, who has assigned this danger its true situation, and not exaggerated its effects. Many wonderful stories are told respecting this vortex, particularly some said to have been related by the celebrated diver,* Colas, who at last lost Charybdis.

* A diver on my establishment, named Dionisio Ninfo, had been brought up at the Faro, who, though an elderly man, could descend in seven or eight fathoms water, and there remain a minute and a half. My servant having accidentally thrown some spoons overboard at Milazzo, in six fathoms

his life here. I have never found reason, however, during my examination of this spot, to believe one of them.

Galofaro. The formation of the Tangdora shoals, stretching out on each side of the little kind of bay, off which the Galofaro is situated, is probably owing to the eddies of Charybdis; and the sand, being united by the bituminous particles before mentioned, is as hard as a rock. I first surveyed these shoals, and supplied the Senate of Messina with a large plan of them. To strangers entering the harbour at night, they are dangerous, as ships are apt to close the light too much; and if the vessel grounds, the rapidity of the stream, and great depth of the water outside, are obstacles to getting off again. To prevent the repetition of an accident, not unfrequent, I recommended a smaller light to be placed between the established one and Fort Salvador, which has since been adopted, and must prove of infinite service.

Taranto. The Captain of the Port at Taranto, by General Visconti's desire, promised to send me a set of observations on tidal phenomena in that gulf, which he pronounced to be 'molto singolare;' but I never received his document, and from my own notice detected nothing along those shores, but the usual atmospheric influence in the Mediterranean.

Adriatic tides. In the Adriatic Sea, the tides, in most parts, are so weak as not to be easily recognised; yet that they exist throughout it, has been ably shown by Professor Toaldo, of Padua, in an essay intituled *De reciproco Æstu Maris Veneti*, of which a copious abstract appears in the *Philosophical Transactions* for 1777. The head or upper part of the Gulf of Venice (often termed the bottom) has a very notable rise, ranging from one foot to nearly four in springs, and according to the prevalence of winds up or down. The times of high water before the moon's passing the meridian, are considered as fairly noted in the following table for Venice. Venice and Chioggia, which was forwarded to me by Colonel Campana—at those two places the rising and setting of the moon is the time of low water for that day, and about 1^h 30^m before the moon reaches the meridian, is the time of high water.

---

depth, Dionisio was overboard in a trice, and recovered them, to the surprise and amusement of some officers who had breakfasted with me, and who could watch his movements in the clear water.

	New Moon.		Full Moon.	
	Day.	*Night.*	*Day.*	*Night.*
	H. M.	H. M.	H. M.	H. M.
January	2 40	1 40	2 41	0 56
February	2 8	1 57	2 13	0 57
March	2 5	2 5	2 27	1 11
April	2 18	1 19	0 58	0 58
May	0 30	0 8	0 40	1 25
June	1 2	2 47	0 15	2 45
July	0 38	0 53	0 23	1 22
August	0 3	0 9	0 31	2 1
September	0 54	1 39	0 47	0 47
October	1 40	0 35	1 47	0 47
November	1 56	0 41	2 29	1 0
December	1 25	1 11	2 45	1 0

Effect of winds on these tides.

By this table, it seems that our own conclusion of the tide-hour falling at or about ten on full and change days, is not alarmingly in error. But though the elevation of the tidal waters is stated above, it should be added, that northerly winds lessen this amount, in neaps, most disagreeably to the olfactories; while those from the south throw in a surcharge which sometimes raises the surface to five or six feet above the general level, inundating all the lagoon marshes. Towards the end of the month of December, 1821, after a continuance of fresh south-east winds for several days, the sea was raised to an extraordinary height; so much so, that Venice appeared like one extensive lake during the whole of Christmas-day and the 26th. On this occasion the gondolas were plying in the *Piazza di San Marco;* and from the evidence of records and votive pictures, this is not at all a solitary case.

Professor Toaldo.

According to Professor Toaldo, a few days about every new and full moon, the tide is higher than ordinary; and by means of these spring tides only is it that the larger ships are carried in and out of port. He also found that, of two daily tides, the one is higher and of longer duration than the other; and that the greatest spring tide scarcely ever happens on the very day of the syzigies, but either on, before, or after it, by one, two, three, or sometimes four days. Toaldo likewise saw reason for assuming that the height of the springs at Venice is above what it was formerly, because he ascertained that the tides now really flow to places considerably above what they reached in ancient times; and certainly a comparison of his own mean heights—taken less than a century ago—gives on the average, less than those of our own day. But the instances are too unstable for building upon; exact registers have never been kept there, and we must remember it is on record, that tide-mills were established at Venice so far back as the year A.D. 1078.

Coast of Istria.

In H.M.S. *Aid* we found the tides off Istria set against the north-east wind at the rate of nearly a knot an hour, and then return to its south-east course; and, at times, the effect of the ebb was to cause an apparent stand-still of the offing and central waters. The gale called Bora certainly occasions a surcharge along the coast of Italy, but at Barletta, Bari, Monopli, and Brindisi, the sailors insist on experiencing a tidal action, ranging from a few inches to three feet. Our operations were not sufficiently nice for confirming this assumption, nor does it rest on very strong grounds, being due more to a transient notice than to direct experiment. Neither could we detect positive indications of a regular tide in the Ionian Sea,* except in one instance; although I have known sets of stream running at more than a knot, with a rise and fall of nearly two feet, generally corresponding with the Adriatic movements. The exception alluded to is Patras, at the entrance of the Gulf of Corinth, where we established the mean tide-hour at $6^h\ 54^m$, with a range of about two feet and a half. This was obvious; whence we must conclude that the lunar influence is manifest on the neighbouring shores—a fact,

Ionian Sea.

* There was much talk about a current in Port Argostoli, which the Cephaloniots believed to flow uniformly against the wind, 'owing to subterraneous caverns.' It is, however, but an effect of the form and contour of the harbour and its vicinity, as acted on by winds in heaping and heading up the waters of one arm, and draining them off by the other.

however, which it was out of our power to ascertain; for the rise and fall observed at Lepanto, Galaxidi, Corinth, and Vostitsa—though covering and leaving dry alternately a considerable extent of shore—were evidently so dependent on winds, that as yet we must set them down as currents. But there is a singular periodic motion prevailing in the waters of this gulf, independent of whatever flux and reflux there may be; it generally takes place twice in twenty-four hours, when unopposed by fresh gales, the inset being termed *embasmos*, and the outset *eugalmos*. A knowledge of this peculiarity aids the arrivals and departures of shipping, and facilitates local intercourse. **Gulf of Corinth.**

The Archipelago presents, with certain anomalies arising from its shape and its islands, the general aqueous motions of other portions of the Mediterranean Sea, and is more amenable to currents than tides. But, ages ago, Herodotus spoke of the ebbing and flowing of the sea in the Gulf of Milis, before Thermopylæ, which, he asserts, 'may be seen every day' (*Polymnia*, § 198); and in the Euripus, or Strait of Negropont—which in the gorge is only forty yards wide — a very remarkable phenomenon of reciprocated motion in the waters is observed. During the first quarter of the moon, as well as from the 14th to the 20th of its age, and also for the last three days of the lunation, the tide ebbs and flows regularly four times in the twenty-four hours; while during each of the other days, it ebbs and flows, with the great force of five or six knots, from eleven to fourteen times per diem, though the difference of elevation rarely exceeds two feet. This is, to a degree, accounted for by assuming that a change of wind in the Gulf of Volo, or Ægean Sea, produces also a change in the relative levels of waters around, and a stream consequently flows through the bridge at Egripos to restore the equilibrium; and it is well known to those along the shore, on account of the contrary motions given to mill-wheels, that a southerly wind produces a strong set to the northward, while a **Archipelago.** **Euripus.**

northerly wind is accompanied by a southerly current. These tidal irregularities, and their residual phenomena, have attracted attention for ages; and there is an idle story that Aristotle drowned himself here, because he was unable to explain the cause. Persecution and banishment after he had enjoyed power, more probably shortened the life of the philosopher, than any dissatisfaction from an intelligent failure; especially since he had always affected to scorn suicide as dastardly and disgraceful.

Aristotle's death.

Smyrna.

At Smyrna, it is said that the tide flows on full and change days, when regular, from three to four o'clock, with a rise of two feet; but it is added, that the neaps are always irregular. This tide-hour, however, is not recognised by Sir Francis Beaufort. 'Neither on this coast' (*Karamania*), says he, 'nor in the Gulf of Smyrna, where the *Fredericksteen* was stationed for some months, could it be perceived that the direction of the current, or the rise and fall of the water, were influenced by the moon. The depth of the water does, indeed, frequently vary, but this effect is produced by the direction of the winds; those from the south and west universally raising it, in some cases even two feet, and those from the opposite quarters depressing it in an equal degree.' Recent observations attempt to show that in the port of Mermerícheh, the tide flows on full and change at $9^h\ 30^m$, and rises eight or ten inches; but, unless with a tide-gauge under able inspection, I should very much doubt the absolute accuracy of such a determination.

North Africa.

Along the African shore the tides are distinctly traceable in some places, though imperceptible in others. Thus, at the mouth of the Tetuan river, the water rises nearly four feet on full and change, at $1^h\ 30^m$, and is hardly discernible a little farther to the east. The often-cited flux and reflux at Bizerta—noticed by the younger Pliny,* in his strange

Bizerta.

* This is a point upon which the *ipsissima verba* of the Younger Pliny

story of the dolphin—is caused by vaporization, the action of winds and rains, and the consequent effect on the sea; which rendered it, in Dr. Shaw's opinion, a miniature of the Strait of Gibraltar. At the Goletta of Tunis, there is a rise and fall of nearly three feet, which is so variable in its times of recurrence, that it must be ascribed to local rather than lunar causes; but towards the Lesser Syrtis, the moon's influence becomes less equivocal. Along the Karkenah and Sfákus channel the tides are fairly developed, running about a couple of knots, a rate often increasing to upwards of three, as they gather round the Gulf of Khabs; until, on passing Jerbah, they flow away to the eastward, and weaken by diffusion. This is one of the greatest latitudinal distances from Venice, and therefore increased tidal vigour might have been expected; but I was amused rather than vexed, on finding our boats lying high and dry nearly a mile from the *Burj-er-Rús*, a pyramid of human skulls just outside the castle of Jerbah, under which we had landed two or three hours before. The Mediterranean station had made us overlook the wholesome nautical rule of keeping the boats afloat; but, happily, our being left aground by a receding tide entailed none of the dreadful disasters which befel the Spaniards under Lacerda and Doria, in 1561, when the slaughter took place that supplied the Christian heads with which the Burj-er-Rús (*Tower of Heads*) is built. The tide rises till about $3^h\ 10^m$, ranging from four to six feet, and at times even to eight; the waters around must consequently be affected in some degree by its action. Still, the great bank formed between Jerbah and Lampedusa shelves so gradually, that the great sea-swells roll in and disperse without breaking. I therefore, on several occasions, when the wind was dead towards the shore, and the waves rising,

Goletta.

Lesser Syrtis.

Rise and fall at Jerbah.

---

should be cited: 'Adjacet ei (*Hippo*) navigabile stagnum, ex quo in modum fluminis æstuarium emergit, quod vice alterna, prout æstus aut repressit aut impulsit, nunc infertur mari, nunc redditur stagno.'—(*Lib.* ix. *Ep.* 33.)

ran the *Adventure* to leeward out of the sea-swell, till we found a convenient depth for anchoring in smooth water. Had I known of this in 1816, it would have saved Lord Exmouth a world of hurry and anxiety, as well as the loss of a few anchors, when his squadron was caught by a northerly gale in Tripoli roads.

Lord Exmouth.

It was then that a circumstance which I have elsewhere related, took place: the wind came on suddenly while the Admiral and most of his captains were on shore negotiating the treaty for the suppression of slavery. By his lordship's desire I mounted to the terraced roof of the consulate, which overlooked the anchorage; and perceiving that some of the ships were driving, and others heeling over prodigiously, my report quickly brought up Lord Exmouth and his flag-captain, Sir James Brisbane, the others all hurrying down to their boats. On the Admiral's nearly gaining the terrace, I called down the stairs, 'My lord, the *Montagu* is under sail.' 'Oh! as for Heywood,' he replied, 'no fear of him; what are the others doing?' By this time he had gained the summit of the house; he looked at the ships, and was off in an instant.

Beyond Tripoli, between Mesrátah and Grennah, or Kirenneh (*Cyrene*), is the wide and open gulf which the ancients called the Greater Syrtis, mentioned in the first chapter. Of this once-dreaded spot, the dangers to navigation are said to have been occasioned by the frequent occurrence of banks and shallows formed by the flux and reflux* of the sea, and still more by these movements themselves. Now, as we found only the slightest possible indications of tide here, this flux and reflux can only apply to the indraught which follows the sea-winds, and the reaction of the body of waters when the opposite ones

The Greater Syrtis.

* Pliny the Elder speaks of both the Syrtes as being 'vadoso ac reciproco mari diros,' in a passage which Philemon Holland renders—'The third gulfe is parted into twaine, cursed horrible places both, for the ebbing and flowing of the sea, and the shelves betweene the two Syrtes.'

prevail. Captain Beechey, then a lieutenant of the *Adventure*, in charge of the party which journeyed round its shores by land, while the ship examined the coast by sea, mentions many parts where he perceived the effect of very violent surges; but, on the whole, he concluded that the land had advanced upon the sea in those regions, 'since we find their ancient ports now filled up with sand, their lakes to have taken the character of marshes, and their quicksands (if ever they had any) to have become solid and firm.' The result of our operation is, that navigation, if necessary, can look the Syrtis in the face, for the whole is now proved to be approachable in cases of necessity; but without such necessity, no vessel, especially of the smaller sort, ought to get embayed there; for northerly winds have a long and uninterrupted fetch. But the difference of impression between the present time and that which prevailed when I first went thither, in 1816, is remarkable, for all the local seamen then spoke of it with dread: yet I could find no one—not even in the Báshá of Tripoli's squadron —who had any personal experience in the matter, except one Monsieur Lautier, whose relation I could not rely upon, as I soon found that he carried too much canvas for his ballast. It may therefore illustrate the matter, if part of a letter which I addressed to Baron de Zach, and which he published in his *Correspondance Astronomique* for 1822, be here subjoined:—

Captain Beechey.

M. Lautier.

All the world know that the two Syrtes are the great gulfs on the northern coast of Africa, between Carthage and Cyrene, and that they were the terror of the ancient mariners: so it is reported by Herodotus, Scylax, Diodorus Siculus, Pomponius Mela, Edrisi, and many other historians, geographers, and poets; among the last, Lucan and Apollonius Rhodius, &c. The lines of the latter represent the general notions on the subject a century and a half before our era:—

Apollonius Rhodius.

'Near the fell Syrtis is the vessel borne.
There shifting sands the labouring bark embay;
Thence never crew pursued the homeward way.
A hideous tract the slimy marshes spread;
The putrid waves are motionless and dead:

A treacherous depth of seeming land is seen,
Devouring water, clothed in fraudful green.
Along the brine a spume corrupted lies,
And pestilential vapours load the skies.
Inhospitable rise the sandy heaps;
No bird has dwelling there; no thing that creeps.'

It was with descriptions so terrible and alarming, that I attacked this classical bugbear. I entered by Mesrátah, and by the flat shore of Isa: my expectations were realized. I did not find it a coast desolate, monotonous, and melancholy—without form, and so low as to justify the character which has been given by the old navigators, that it is 'neither land nor sea.' We did not see submerged plains, or drowned lands; but we saw distinctly how the waves, which are perpetually breaking against the shore, wash and leave the rocks uncovered which abound on this coast, and which are also strewed with the remains of many wrecks. Horrible swamps, however, seem to extend over a superfices of nearly 200 miles, and are so perfectly level, that

Wrecks.

they appear rather like a sea than a shore. The wrecks are, without doubt, those of ships which have deviated, or been driven, from their proper routes, being misled during the night, or during thick fogs, which are common here. On other parts of the coast there are few or no dangers, excepting several little heads of rocks, scattered about different points. The tides are insignificant. With the hand-lead going, a vessel may approach all parts of the coast between Mesrátah and Cape Razat, which is thirty-five leagues beyond Benghazi. This is a singular contradiction to the reports of the difficulties that the ancient mariners pretended to have found; but it must be acknowledged that vessels should not enter into this gulf, unless chased by tempests which it is impossible for them to resist; for, in rough weather, the sea rises here to a prodigious height. It must also be considered, that the navigators of those times were always at a loss in estimating their reckonings; they were also troubled by the terrors which their imagination created, knowing that they should experience no mercy from the wandering and barbarous tribes inhabiting this coast.

But of what utility can it be to enter here? there being but one place in the whole gulf worthy of being called a port,—and even that a poor one.

Lucan's prediction.

Having alluded (*ante*, *p.* 122) to Lucan's prediction respecting this gulf, it may be here further noticed. In the ninth book of the *Pharsalia*, he speaks of nature as having left the Syrtis a mingled and useless mass of drowned land, stagnant pools, waters, and swelling tides; after which (*voce Rowe*) he thus perorates:—

Perhaps, in distant ages, 'twill be found,
When future suns have run the burning round,
These Syrts shall all be dry and solid ground;
Small are the depths their scanty waves retain,
And earth grows daily on the yielding main.

LUCAN, ix. 539.

Stability of the Mediterranean.

Such are the boundaries and such the contents of the Mediterranean; but, even at the risk of repetition, a remark or two is called for before noticing the inhabitants of these waters. After what has been advanced, there appears to be no sufficient reason for supposing that there is any positive diminution of waters in this Inner Sea, or any alteration of its general level; by which term is meant the actual line coincident, or nearly so, with the mean point of time between its greatest and least height—a *datum*, however, which in these regions it is hardly necessary to discuss. On a due consideration of all the circumstances, the problem resolves itself into this: whether the level has actually been raised a few feet in the course of twenty centuries, or whether the ground has subsided to a similar amount in consequence of unknown causes. Although, as I have already stated, many submerged ruins are met with, an examination of various cothons, moles, and other sea-works, proves the permanency of the level of the Mediterranean from a remote historical period; and, on the whole, my impression is, that its coasts have gained at least as much on one hand as they have lost on the other.

This never-ceasing reciprocity bears out the 'velut paria secum faciente natura (*velut sua damna compensante Natura*): quaqua hauserit hiatus, alio loco reddente,' of the Sage. The question forms the point to which modern geological inquiry has now arrived, for it is clearly discernible that the effects of such alternations must be connected with the history of life upon our globe. But precisely the same idea was distinctly started by Strabo, nearly 1900 years ago. 'It is not,' says that philosophical geographer, 'because the lands covered by the sea were originally at different altitudes, that the waters have risen, subsided, or receded from some parts and inundated others; but the reason is, that the same land is sometimes raised up, and sometimes depressed, and the sea also is simultaneously raised and depressed, so that it either

Strabo's geological view.

overflows, or returns into its own place again.' Would any of our most practical geologists express any other opinion at the present moment?

§ 6. Mediterranean Ichthyology.

THIS is a most interesting field for inquiry, especially as connected with the general physics of the Inner Sea; and though the full development of the Mighty Deep may, probably, never be vouchsafed to the means or curiosity of man, much additional information will, no doubt, reward perseverance. And again must I express regret at the imperfect tenour of my researches, for professional duties left no choice, and the flight of time was ever defeating inclination.

Penetration of light.

The solar rays, we are told, only penetrate to the depth of twenty-five fathoms, below which the sea receives no light, and consequently little or no direct heat from the sun; but this hypothesis is assuredly lame, since I have sunk a plate, and then viewed it with the marine-tube hereafter described, at that depth; which is a very different operation from the penetrating power of the solar beams. Indeed, as water is pervious to light, it must necessarily pierce the limpidity to a vast depth—under the obvious conditions as to intensity of the rays and smoothness of the surface—before these rays are intercepted. It is analogically presumed that air is disseminated through the waters, without which marine creatures could not exist, as they would be incapable of decomposing the fluid for the purpose of procuring the oxygen that may be really necessary for them. At present it is impossible to pronounce how far all marine life may require light, for our knowledge of the pelagians inhabiting great depths is necessarily limited, and, as was just said, we know little respecting the penetration of solar rays into the deeper abysses. In such a medium, we might anticipate some modification in the organs of vision of those mollusks which possess eyes; and naturalists have found instances of such adaptation, not only in the eyes of the lower marine inhabitants, but also in the very fitting of the air-bladder. A very remarkable instance is said to be afforded in the *Pomatomus telescopus*,* a creature found in the depths of the Mediterranean, which is furnished with remarkably large eyes, so formed as to gather every ray of light which can illume the darkness of its abode. **Pomatomus telescopus.**

Pressure. The real amount of pressure borne by animal life in profound depths, is truly an interesting element for consideration and experiment. At 16 fathoms, a living creature would have to sustain only about 60 pounds to the square inch, and at 60 fathoms as much as 180 pounds. At 100 fathoms depth the pressure would amount to 285 pounds, and at 700 fathoms the creature must bear with impunity a quantity equal to 1830 pounds upon the square inch; while the pressure of 1000 fathoms of superincumbent water on the same area considerably exceeds a ton. Now, I have drawn up star-fish alive through 170 fathoms, but since then Professor E. Forbes has nearly doubled that depth

---

* Not having met with this fish either in markets or books, I applied to my friend, Professor Edward Forbes, to hunt it up for me: his answer was —'I have at length succeeded in tracing him to his lair. The *Pomatomus telescopus* is a sea-fish, one of the true *Percedæ*. It is remarkable for its enormous eye, and is very rare indeed.' Risso describes these eyes, the opercula of which are in three pieces. He mentions a pelagian, which he names *Alepocephalus rostratus*, with still larger eyes, saying,—'C'est un phénomène très digne de toute attention des Ichthyologistes, que les poissons les plus remarquables des bords de l'Europe méridionale, qui habitent à deux milles pieds et plus de profondeur, ont leurs écailles adherentes très faiblement à la peau, et les organes de la vue d'une grandeur disproportionnée à l'ensemble de leur corps; que leur vessie natatoire est si vaste, que leurs cæcunes sont si nombreux, et que les teintes qui les colorent réfléchissent si peu de nuances. Quant à leurs habitudes, elles resteront encore pour les naturalistes long temps ensevelies dans les profondes régions des mers.'—*Hist. Nat. de l'Europe mérid.*, t. iii. p. 449.

with success; and I understand that M. Biot has made captures from still deeper water—his own expression being, that they existed '*dans les grandes profondeurs des mers.*' Of course these animals are properly fitted for such an extraordinary condition of existence; but the pressure of the sea on inanimate bodies, and at comparatively no great depth, is sufficiently obvious. I have twice found that the cylindrical copper air-tube, under the vane attached to Massey's ingenious patent sounding-lead, was unable to stand; for it collapsed at little more than 200 fathoms' depth in the first instance, and in the second was crushed flat under a pressure of about 300 fathoms. Moreover, a claret-bottle filled with air and well corked, burst on its descent to 400 fathoms with the brass Marcet cylinder, and others broke at little more than half that depth. We also found that bottles filled with fresh water—and we even wasted wine on some occasions—and corked, had the cork usually forced in at about from 150 to 180 fathoms below the surface. In these cases the fluid sent down is exuded, and the vessel returned full of sea-water; the cork which had been forced in, is sometimes inverted within the neck of the bottle.

Effect on copper cylinders.

Bottles.

Animal life in the sea.

To return. It is impossible to overlook the teeming animal life in the 'vasty deep;' which not only affords subsistence by one marine race feeding on the other, but gives ducks, divers, gulls, shags, petrels, tern, and all sorts of aquatic birds—as well as turtles, seals, and other amphibia—a constant supply of food, and also adds abundantly to the sustenance and traffic of man. From the wondrous productive power of fishes, their numbers are incalculable; yet so numerous are their enemies, that it has been questioned whether any of them die a natural death. Still, in the brief span between the ova and the end, all and each of the constituent individuals of those myriads, together with all the subaqueous vegetable tribes, have their allotted portions in the universal economy; they aid in giving circulation to the waters, and thereby tempering the climates of the globe; for even those mollusks which of themselves seem hardly capable of locomotion, qualify, though in a degree almost infinitesimal, the equilibrium by their secretions.

Habits of fishes.

There is no doubt that marine animals strictly seek those districts and depths in each of which their respective foods are found; and herein is an extraordinary adaptation of means to the end, insomuch that fishes swimming near the surface and those a few fathoms below them differ, and these, again, are found to be different from those with habitats at greater depths. Yet, whatever profundity the fishes may inhabit—and pelagians are presumed to frequent the profoundest—as their respiratory organs and specific gravity seem to be admirably adapted to the nature of water, they can live and breathe with ease at every inch throughout. The distribution of mollusca, radiata, and others of the lower organization, is also palpably arranged for the fore-mentioned ends, although greatly dependent on local conditions. Professor Forbes, who was for eighteen months in the Ægean with Captain Graves, divides that portion of the sea to which his inquiries were directed into eight regions of depth, each characterized by its peculiar fauna. 'Certain species,' he says, 'in each are found in no other, several are found in one region which do not range into the next above, whilst they extend to that below, or *vice versâ*. Certain species have their maximum of development in each zone, being most prolific in individuals at that zone in which is their maximum, and of which they may be regarded as especially characteristic. Mingled with the true natives of every zone are stragglers, owing their presence to the secondary influences which modify distribution. Every zone has also a more or less general mineral character, the sea not being equally variable in each, and becoming more and more uniform as we descend.' (See his *Report on Ægean Invertebrata*, 1843.)

Professor E. Forbes.

Whales. Instances have occurred through successive ages, of the larger cetaceous animals having made their appearance occasionally in the Mediterranean waters. Various individuals have even been captured; as a pike-headed whale (*Balæna boöps*), upwards of 100 feet long, off Corsica, in the year 1620; a fin-fish (*Balæna physalis*), near Barcelona, in 1744; another fin-fish near Tunis, in 1787; a round-nosed whale (*Balæna musculus*), killed on the coast of Provence in 1790; and two or three cases of the common whale (*Balæna mysticetus*) being beached. But not being natives of this sea, they must only be regarded as stragglers adrift. The *Orca* that grounded in the port of Ostia, which Pliny saw so gallantly attacked by the Prætorian guards, was no doubt a stranded whale. Much discussion has

Sharks. arisen about Jonah's 'great fish,'* which custom has recognised as a whale; but others consider the large basking shark (*Squalus maximus*) to have been the creature in question, although it is the tamest and most harmless of the ichthyological races, feeding mostly on medusæ, small crustacea, and sea-plants. The Lamia (*Squalus carcharias*), or white shark, the most voracious of human food of all fishes, has a better claim to have been the 'great fish' that swallowed the prophet, since he can readily ingulf a man whole; and it has therefore sometimes been designated *Jonæ piscis*.

Fisheries. Though many of the most valuable species of fish are abundant in the Mediterranean, the quarantine regulations, arbitrary exactions, and deficient enterprise of most of the

* The English authorized version of the Book of Jonah (i. 17 and ii. 1, 10), is literal and exact. No epithet is used except 'great,' and the Hebrew word *dag* is a common term signifying fish: that which swallowed Jonah is not specifically named in the Hebrew Bible, but in the New Testament (*Matthew* xii. 40), the word is rendered κῆτος in Greek, which usually signifies a whale, but is also taken for any very large fish. The noted idol Dag-on (1 *Sam.* v. 1, 2) was represented as half man, half-fish, and has therefore been taken for the Assyrian monster Derceto (*Diodorus Siculus*, ii. 4), the original mermaid, but without reason.

people who inhabit the coasts, have combined to prevent the fisheries from being carried to the desired extent as an object of external commerce, most of the produce being consumed at home. From this, however, we must except the tunny, the sword-fish, the anchovy, and the sprat, the capture and curing of which are carried on with great spirit; while the coral fisheries form an important branch of industry, though often far from being highly remunerative. In an economic view of the central parts of this sea, perhaps the tunny is the most important fish; and I have already described the method of taking it, and other particulars, in my accounts of Sicily and Sardinia. I also alluded to the migratory visits of this fish having become more capricious of late than formerly, insomuch that sometimes the produce of the *tonnare* barely repays the expenses of their establishment. This may arise from accidental obstructions to their course, a point on which they are said to be very sensitive; and they are extremely gregarious. The shoal enters the Mediterranean from the ocean in spring, passes along the European shores into the Black Sea, where they are supposed to spawn, and returns along the African shore to the ocean in the fall of the year. But in the Black Sea it has been noted to enter along the coast of Asia, and return along that of Europe; a peculiarity which Pliny, following Aristotle, accounts for by supposing the fish to see better with the right eye than with the left. The more natural opinion is, that the prevailing winds are the cause, those of summer being chiefly from the south, and those of the later seasons from the north; the fish, therefore, may be presumed to prefer the smooth water under the weather shore. This is not said to impugn the merit of those writers, for they—together with Archestratus, Ælian, Ovid, Oppian, Isodorus, Athenæus, and Ausonius—have recorded such numerous interesting and instructive facts relative to the customs and instincts of Mediterranean fishes, that we almost overlook their neglect of specific

Tunnies, sword-fish, &c.

Migratory fishes.

Ancient writers

differences. Recent inquiry has, indeed, confirmed the truth of many of their statements which had for ages been stigmatized as fabulous.

Aristotle. But among all inquirers into marine zoology, none can claim a footing on the same plinth with Aristotle; the generalizations of whose admirable researches (περι ζώων ἱστορία) in these waters, remain to this hour unshaken.

Remark. But a study of the habits of the finny tribes, with their affinities and analogies, though so interesting in itself, is not the legitimate object of these pages, the intention here being merely to give a glimpse of the Mediterranean ichthyology, by appending a list of the inhabitants of its central waters — a region chosen as probably affording specimens of the whole sea. In this enumeration—which is necessarily deficient—I have taken every pains to arrive at a sound general view, having attended the fisheries on the coasts of Tunis, Sardinia, Sicily, and Calabria, as well as the various markets in those places. There has been no small trouble in drawing it up, for the trivial name of many a fish differed in the course of a few miles; and difficulties occurred in reconciling the scientific synonymes of various species, especially where the differences are barely demonstrable, or where personal names of the naturalist's friends are foisted in. My course was, therefore, to cling towards the Linnæan classification—which I had already followed in the memoirs of Sicily and Sardinia—because it is best suited to the degree of my knowledge in that department, and enables me to keep clear of the more intricate systems recently introduced. Among the Sicilian names a few are added in the Sardinian vernacular, the two being sufficiently in alliance to make but little difference; and an Italian will find no difficulty in understanding them at sight. I will therefore proceed to the enumeration, merely remarking by the way, that though mostly handsomer than British fishes, those of the Mediterranean Sea are, in general, not to be compared with them in flavour.

I. THE PRINCIPAL MEDITERRANEAN FISHES.

Latin Names.	Sicilian.	English.
Accipenser huso	*Beluga*	Great sturgeon.
——— sturio	*Storiunu*	Sturgeon.
Ammodytes argenteus	*Lussi*	White sand-eel.
——— lancea	*Agugliattu*	Riggle.
——— tobianus	*Aguglia*	Sand-lance, or hornel.
Anarrhichas lupus	*Pisci lupu*	Sea-wolf.
——— strigonus	*Sarpananza*	Sea-cat.
Argentina aphya	*Nunnatu*	Argentine.
——— sphyræna	*Curunedda*	Spit-fish.
Atherina hepsetus	*Pisci virgatu*	Mediterranean smelt.
——— menidia	*Trotischeddu*	Grey atherine.
——— presbyter	*Majetica*	Sand-smelt.
Balistes lunulatus	*Fanfra*	Crescent balistes.
——— scolopax	*Pesce balestra*	File-fish.
——— vetula	*Pesce sozzu*	Old-wife.
Blennius alauda	*Durgannu*	Sea-lark.
——— cornutus	*Mustia 'mperiali*	Horned blenny.
——— galerita	*Bavusa cu tuppè*	Crested blenny.
——— gattorugine	*Patuvanu*	Tom-pot.
——— gibbosus	*Tombarella*	Butter-fish.
——— gunellus	*Gurgiuni*	Gunnel.
——— labrus	*Tordu bavusuni*	Guffer.
——— mustela	*Bausedda*	Weasel blenny.
——— ocellaris	*Mesoro*	Sea butterfly.
——— pholis	*Missuru*	Shan, or shanny.
——— physis	*Bavusuni*	Forked hake.
——— tentacularis	*Bausa ucchiuta*	Tentaculated blenny.
——— viviparus	*Gurgiuneddu*	Eel pout, or green bone.
Callionymus dracunculus	*Velleiu*	Gowdie.
——— lyra	*Dragone marinu*	Skulpin.
——— pusillus	*Ampisciu*	Small skulpin.
Centriscus scolopax	*Trumbina*	Sea snipe, or bellows-fish.
Cepola marginata	*Spirdottu*	Tape-fish.
——— rubescens	*Signu di Salomone*	Red snake-fish.
——— tænia	*Pisci bannera*	Ribbon-fish.
Chætodon paru	*Muolla*	Square chætodon.
——— vetula	*Ogiusa*	Sea-rabbit.
Clupea alosa	*Saboga*	Shad.
——— amara	*Aleccia*	Gipsy herring.
——— encrasicolus	*Anciova*, or *alici*	Anchovy.
——— pilchardus	*Saraca*	Pilchard.
——— siculus	*Cicirelli*	Sicilian whitebait.
——— sprattus	*Sardella*, or *Sardina*	Sprat.
Coryphæna hippurus	*Capuni*	Dolphin of seamen.
——— imperialis	*Pettinu 'mperiale*	Dorado.
——— novacula	*Pettinu*	Razor-fish.

LATIN NAMES.	SICILIAN.	ENGLISH.
Coryphæna pompilus ...	*Lampuca*	Striped coryphene.
Cottus cataphractus......	*Pogge*	Mailed bullhead.
——— dracunculus	*Mustuzola*	Tommy Logge.
——— gobio	*Capo grosso*............	Miller's thumb.
——— scorpius	*Pisci capone*	Sea scorpion.
Cyprinus alburnus	*Donzella*	Bleak.
——— auratus	*Pesci di oru*	Gold-fish.
——— barbus	*Barbio*..................	Barbel.
——— brama	*Mutzula*	Bream.
——— carpio	*Carpiuni*...............	Carp.
——— erythropthalmus	*Laccia*..................	Rudd, or red-eye.
——— gobio	*Ghiuzzu*	Gudgeon.
——— jeses	*Capitano*...............	Chub, or jantling.
——— leuciscus	*Albula*..................	Dace, dare, or dart.
——— phoxinus	*Pisciulinu*	Minnow, or pink.
——— rutilus	*Pisci duci*	Roach.
——— tinca	*Cheppia*	Tench.
Delphinus delphis.........	*Delfinu*	Dolphin.
——— orca	*Cetaceo*	Grampus.
——— phocæna	*Pisci porcu*............	Porpoise.
Echineis cidaris	*Ampiscia*	Sea turban.
——— naucrates	*Sussapega*	Long sucking-fish.
——— remora	*Pisci 'ntoppu*	Sucking-fish.
Esox acus	*Cavanucci*	Lax.
—— belone	*Aguggħia*	Gar-fish.
—— lucius	*Cane di sciumi*	Pike, Jack, or Luce.
—— saurus	*Sauru*	Skipper.
—— sphyræna	*Aluzzaru*	Sea pike.
—— stomias	*Stomica*	Piper-mouthed pike.
—— synodus...............	*Fra di mari*	West India pike.
Exocætus exiliens	*Ancileddu 'mperiali*	Swallow flying-fish.
——— volitans.........	*Saltatore*	Flying-fish.
Gadus æglefinus	*Baccalà friscu*	Haddock.
——— asellus mollis ...	*Moncaru*..............	Groundling.
——— asellus varius ...	*Asnelli*	Bibb.
——— barbatus	*Tavila*	Whiting pout.
——— blennoides.........	*Mirruzzu duci*	Dorse.
——— carbonarius	*Ciaula*..................	Coal-fish.
——— lota	*Concunieddu*	Burbot.
——— luscus	*Munaceddu*	Miller's thumb.
——— Mediterraneus ...	*Sazzaluga di mare*...	Mediterranean cod.
——— merlangus.........	*Merlangu jancu*......	Whiting.
——— merlucius	*Mirruzzu*	Hake.
——— minutus............	*Pesci ficu*	Capelin, or Poore.
——— molva	*Muncaru*	Rock ling.
——— mustela	*Mustia*	Five-bearded cod.
——— pollachius	*Vacchetta*	Whiting pollack.
——— punctatus	*Asnellu*	Whistle gade.
Gasterosteus aculeatus ...	*Maccionu*	Banstickle.
——— ductor	*Capitanu*	Pilot-fish.

Latin Names.	Sicilian.	English.
Gasterosteus pungitius ...	*Spinarola*	Lesser stickle-back.
——— spinachia ...	*Ispriotta*	Thorny stickle-back.
Gobius aphya	*Gurgiuneddu*	Spotted goby.
——— bicolor	*Teurrazza*	Black-and-brown goby.
——— joso	*Gobbiu jancu*	White goby.
——— melanurus.........	*Gobbiu pureddu*......	Sea gudgeon.
——— minutus...........	*Urgiuni di fangu* ...	Polewig.
——— niger	*Urgiuni niuru*	Rock-fish.
——— paganellus.........	*Gorgionu*..............	Brown goby.
Gymnotus acus	*Anciddazza*	Naked gymnote.
——— electricus	*Diavulicchiu*	Cramp-fish.
Labrus Adriaticus........	*Perciudda*	Basse.
——— anthias	*Munacedda*	Holy basse, or barber.
——— cappa..............	*Lappanu*..............	Gold sinny.
——— Cretensis	*Zigarella*..............	Cretan basse.
——— cynædus	*Pizza di Ré*	Yellow basse.
——— donzella....	*Dunzedda*	Bergil.
——— fuscus	*Iodiolu*	Tawny basse.
——— guttatus	*Turdu stizziatu*	Comber.
——— hepatus	*Lappanu saragu*......	Liver basse.
——— Julis	*Arusa*, or *Marabut*...	Rainbow fish.
——— maculatus	*Menduredda*	Spotted wrasse.
——— merula	*Turdu d'Arca*	Black labrus.
——— olivaceus	*Pettineddu*	Sea-wife.
——— pavo	*Lappanu beddu*	Peacock labrus.
——— psittachus	*Rucchia*	Parrot wrasse.
——— reticulatus	*Turdu arrocali*	Reticulated wrasse.
——— scarus	*Bricchese*	The scare labrus.
——— tinca	*Verdaliddu*	Golden maid.
——— turdus	*Turdu*	Sea tench.
——— venosus	*Serra*	Bloated basse.
——— vetula	*Zittu*	Little sea-wife.
——— viridis	*Virdu*.................	Green labrus.
Lophius Europeus.........	*Rannu di mari*	Toad-fish, or sea-frog.
——— piscatorius	*Piscadrixi*	Angler, or sea-devil.
Mugil auratus	*Daurinu*..............	Gold-headed mugil.
——— cephalus	*Malettu o cefalu*......	Common mullet.
——— labrosus	*Labronu*	Thick-lipped mullet.
——— saliens..............	*Flavetoni*..............	Leaping mugil.
Mullus apogon	*Trigghia svarvata* ...	Bearded mullet.
——— imberbis	*Re di trigghia*	Beardless mullet.
——— ruber	*Trigghia mangiadori*	Red mullet.
——— surmuletus	*Trigghia di solu*......	Sur mullet.
Muræna anguilla	*Muragliunu*	Sharp-nosed eel.
——— catenata	*Murena ficu*	Chain-striped murena.
——— conger	*Anguidda grongu* ...	Conger eel.
Muræna Helena...........	*Murena nera*	Roman eel.
——— marina	*Anguidda di mari*...	Grig.
——— myrus	*Smiru*	Sea-snake.
——— punctata	*Gargiuni*..............	Murey.

LATIN NAMES.	SICILIAN.	ENGLISH.
Ophidium aculeatum	*Nasoni*	Snout-fish.
——— barbatum	*Calagneris o lissa*	Bearded ophidion.
——— hydrophis	*Bandiera niuri*	Water serpent.
——— imberbe	*Culuri di mari*	Beardless ophidion.
Osmerus eperlanus	*Tarantula*	Smelt.
——— saurus	*Tammurru*	Lizard smelt.
Ostracion gibbosus	*Pesce luna*	Oyster-fish.
——— hystrix	*Rizza*	Porcupine-fish.
——— mola	*Papa tundo*	Large sun-fish.
——— nasus	*Pesce soddu*	Trunk-fish.
Perca asper	*Serraina*	Yellow perch.
—— cabrilla	*Cabrilliu*	Smooth serranus.
—— cernua	*Pizzuni*	Ruffe, or pope.
—— fluviatilis	*Ragnu vuraci*	Perch.
—— giber	*Boragie*	Hunchback perch.
—— labrax	*Spigula*	Wolf perch, or basse.
—— lucio	*Percia stizzata*	Spotted perch.
—— marina	*Percia grossa*	Bergylt.
—— punctata	*Spinula*	Thorny perch.
—— pusilla	*Conaditu*	Dwarf perch.
—— sacer	*Tumulu*	Holy perch.
—— scriba	*Mulassu*	Learned perch.
—— telescopus	*Occhi grosso*	Large-eyed serranus.
—— umbra	*Umbrinu*	Dusky serranus.
Petromyzon branchialis	*Lampernu*	Pride.
——— fluviatilis	*Alampria*	Nine-eyed eel.
——— marina	*Papa pixi*	Lamprey.
Pleuronectes flesus	*Pisci passera*	Flounder, or flook.
——— hippoglossus	*Stocapisci 'mperiali*	Holibut.
——— limanda	*Palaja di arena*	Dab, or saltie.
——— maximus	*Rumulu 'mperiali*	Turbot.
——— passer	*Passera picciula*	Whiff.
——— platessa	*Palaja*	Plaice.
——— rhombus	*Lupiddu*	Kitt, or pearl-fish.
——— solea	*Linguata*	Sole.
Raia altavela	*Amiema*	Finless ray.
—— aquila	*Pisci aquila*	Sea eagle, or whip ray.
—— aspera	*Pesci lepre*	Shagreen ray.
—— batis	*Cappuccina*	Skate, or maid.
—— bicolor	*Razza*	Trygon, or brett.
—— clavata	*Picara pitrusa*	Thornback.
—— lævis	*Liscia*	Slippery ray.
—— marginata	*Miragliettu*	Small-eyed ray.
—— miraletus	*Quattro occhi*	Homelyn.
—— oculata	*Occhiateddu*	Mirror ray.
—— oxyrynchus	*Farassa*	Sharp-nosed ray.
—— pastinaca	*Cadairu*	Sting ray, or flaire.
—— radiata	*Pigara scappucina*	Starry ray.
—— rubus	*Pigara spinusa*	Rough ray.
—— torpedo	*Pisci diavulu*	Torpedo, or cramp-fish.

LATIN NAMES.	SICILIAN.	ENGLISH.
Salmo albula	*Cefalu*	Phinock, or whitling.
—— eperlanus	*Sazzaluga*	White smelt.
—— fario	*Troucia*	Trout.
—— saurus	*Tammurru*	Sea-lizard.
—— thymallus	*Ombrina*	Grayling.
—— trutta	*Trota russigna*	Sea-trout.
Sciæna aquila	*Feguro*	Stone basse.
—— cappa	*Tiligugu*	Maigre.
—— cirrosa	*Umbrina 'mperiali*	Hairy sea-hog.
—— lineata	*Spatula*	Streaked sea-hog.
—— nigra	*Umbrina niura*	Black umbra.
—— umbra	*Tristareddu*	Sea-crow.
Scomber aculeatis	*Serviola*	Cross spine.
—— alalunga	*Alalungu*	Albicore.
—— colias	*Scurmu 'mperiali*	Spanish mackerel.
—— ductor	*Capitanu*	Little pilot-fish.
—— glaucus	*Savrella*	Sea-green mackerel.
—— pelamis	*Palamitu 'mperiali*	Bonito.
—— scomber	*Scurmu*	Mackerel.
—— thynnus	*Tunnu*	Tunny.
—— trachurus	*Sureddu*	Horse-mackerel, or scad.
Scorpæna lutea	*Scrofaneddu*	Yellow sea-scorpion.
—— porcus	*Scrofanu*	Porcine scorpæna.
—— pristis	*Capuluzzu*	Sea-scorpion.
—— scorpius	*Mazzuni*	Father lasher.
—— scrofa	*Cepola capuni*	Sow-scorpion.
Silurus electricus	*Babbauru*	Sheath-fish.
—— glanis	*Glannu*	Sly silurus.
Sparus annularis	*Lappanu spareddu*	Grey pickerel.
—— aurata	*Canina 'ndorata*	Gilt head.
—— boöps	*Vuorpa*	Bull-eyed sparus.
—— cantharus	*Ciuciastra*	Brown bull-fish.
—— chromis	*Monacedda*	Maroon spare.
—— dentex	*Dentici*	Four-toothed spare.
—— erythrinus	*Pagedda lavaru*	Spanish bream, or rotchet.
—— hurta	*Prau 'mperiali*	Fork-tailed spare.
—— mæna	*Minnula*	Cockerel.
—— melanurus	*Macchiettu*	Black-tailed spare.
—— mormyrus	*Ajula 'mperiali*	Mormyre.
—— pagrus	*Pagru*	Red gilt-head.
—— salpa	*Scilpa*	Braize.
—— sargus	*Saracu*, or *murruda*	Egyptian spare.
—— saxatilis	*Sparagghiuni*	Black rock-fish.
—— smaris	*Minnula 'mperiali*	Smare.
—— sparus	*Spargu*	Becker.
—— vetula	*Varatulu*	Black bream, or old wife.
—— vulgaris	*Gujicidduzzu*	Braize.
Squalus acanthias	*Pisci scioccu*	Picked dog-fish.
—— canicula	*Pisci cani*	Morgay, or cott-fish.
—— carcharias	*Canuzzu*	White shark, or lamia.

Latin Names.	Sicilian.	English.
Squalus catulus	*Rusetta*	Hound-fish.
—— centrina	*Gattu di mari*	Brown shark.
—— galeus	*Nocivolo*	Tope, or miller's dog.
—— glaucus	*Lupu di mari*	Blue shark.
—— maximus	*Grossu cani di mari*	Basking shark, or sail-fish.
—— mustelus	*Pisci palummu*	Smooth hound-fish.
—— pristis	*Sia*, or *Sega*	Saw-fish.
—— spinax	*Chelpu*	Lesser picked dog-fish.
—— squatina	*Squadru*	Monk, or angel-fish.
—— stellaris	*Pisci tigrinu*	Spotted shark.
—— tiburio	*Magnusa*	Rock shark.
—— vulpes	*Gaddolu*	Thresher, or sea-fox.
—— zygæna	*Marteddu*	Hammer-headed shark.
Stromateus argenteus	*Lampuga*	Pampus.
—— fiatola	*Fiatula 'mperiali*	Striped stromat.
Syngnathus acus	*Agujeddu*	Pipe-fish, or sea-adder.
—— hippocampus	*Cavaddu santu*	Sea-horse.
—— marinus	*Trumbettina*	Little pipe-fish.
—— ophidion	*Cavanu*	Sea-snake.
—— typhle	*Pisci tialu*	Needle-fish.
Tetrodon hispidus	*Luna di mari*	Sea-globe.
—— mola	*Pisci tammurru*	Sun-fish.
—— truncatus	*Pisci tundu*	Oblong sun-fish.
Trachinus draco	*Traccina*	Sea-dragon, or sting-bull.
—— jugulares	*Majaru la rocca*	Weever.
—— vipera	*Aragnu*	Otter pike.
Trigla cataphracta	*Pisci curruda*	Sea-rocket.
—— cuculus	*Labbru russignu*	Red cuckoo gurnard.
—— gurnardus	*Gurnardu*	Nowd, or grey gurnard.
—— hirundo	*Fagiani 'mperiali*	Tub-fish.
—— lineata	*Belunganu*	Streaked, or rock gurnard.
—— lucerna	*Tigiega*	Lantern gurnard.
—— lyra	*Gaddinettu*	Piper.
—— milvus	*Tavia*	Yillock.
—— volitans	*Pisci volatori*	Flying gurnard.
Uranoscopus cocius	*Cocciu 'mperiali*	Little star-gazer.
—— scaber	*Papa cucculo*	Bearded star-gazer.
Xiphias gladius	*Pisci spata*	Sword-fish.
—— platypterus	*Macairu*	Broad-backed sword-fish.
Zeus aper	*Pisci tariolu*	Boar-fish.
—— faber	*Pisci di Pedru*	John Dory.
—— gallus	*Gaddu*	Silver-fish.
—— luna	*Cetola*	Opah, or king-fish.

II. THE PRINCIPAL CRUSTACEA, TESTACEA, AND MOLLUSKS.

Latin Names.	Sicilian.	English.
Acalephæ (varieties)	*Attaccaticciu marinu*	Sea-jellies.
Actinia (varieties)	*Sciuri di mari*	Sea-anemonies.
Alcyonium bursa	*Borza marina*	Sea-apple.
——— digitatum	*Cinque dita*	Dead-man's hand.
——— epipatrum	*Penna marina*	Sea-pen.
——— ficus	*Fichi di mari*	Sea-lungs.
——— lyncurium	*Arancia di mari*	Sea-orange.
Anomia caput-serpentis	*Capo di serpe*	Terebratula.
——— ephippium	*Matriperna fausa*	Saddle anomia.
——— vitrea	*Terra bratula*	Palermo terebratula.
Aplysia depilans	*Leporina*	Coarse sponge.
Arca barbata	*Sponguli pilusi*	Bearded ark.
—— navicularis	*Luntra*	Boat ark.
—— Noae	*Spongulu*	Noah's ark.
—— pilosa	*Nuci pilusa*	Hairy ark.
Argonauta argo	*Todari*	Paper nautilus.
——— calcar	*Nautiliu shperoni*	Spur nautilus.
——— carinaria	*Firola*	Keel-edged nautilus.
——— scafa	*Navicella*	Boat-shaped nautilus.
Asteria (varieties)	*Stiddi di mari*	Star-fishes.
Asterias aranciaca	*Ragnatelu*	Butt-horn.
——— caput-Medusæ	*Stidda de Medusa*	Shetland argus.
——— ophiusa	*Stidda serpentara*	Sand-star.
——— rubens	*Stidda russigna*	Cross-fish.
Buccinum echinophorum	*Castagna di mari*	Purple whelk.
——— galea	*Brognu*, or *Vrognu*	Helmet-shell.
——— gibbosulum	*Gobbo di mari*	Hunchback.
——— hæmostoma	*Vocca 'nsanguinata*	Red-lipped whelk.
——— sabarun	*Vrognu d'arina*	Grey casket.
——— Tyrrhenum	*Vrognu 'mperiali*	Purple whelk. Burret.
Bulla ampulla	*Gunfiata*	Obtuse dipper.
—— carnea	*Vessica di mari*	Ovula, or egg.
—— Cypræa	*Velidda di mari*	Common cowry.
—— hydatis	*Orecchiu*	Pillar-lip.
—— lepida	*Squamosa*	Orange-coloured dipper.
Cancer arctus	*Cicala di mari*	Broad lobster.
——— astacus	*Gammaru di sciumi*	Cray-fish.
——— bernardus	*Diavulicchiu di mari*	Soldier crab.
——— crangón	*Granciulinu*	Shrimp.
——— depurator	*Granciu di fangu*	Cleanser crab.
——— gammarus	*Granciu*	Lobster.
——— locusta	*Alausta*	Spiny lobster.
——— mænas	*Granciu di rina*	Common crab.
——— pagurus	*Granciu fudduni*	Hermit.
——— squilla	*Gammaru*	Prawn.

LATIN NAMES.	SICILIAN.	ENGLISH.
Cardium aculeatum	*Galli spinusi*	Prickly cockle.
——— edule..............	*Chiocchiolu a mangi à*	Common cockle.
——— tuberculatum ...	*Frutti d'arena*	Sand cockle.
——— unedo	*Crocchiula 'ncanalata*	Ribbed cockle.
Cellepora spongites	*Spongia vitrosa*	Fragile hydra.
Chama antiquata	*Nuci di mari*	Sea-nut.
——— bicornis	*Ostrica monaca*	Sea-cabbage leaf.
——— calyculata	*Chiocciola spinusa* ...	Scaly clamp.
——— cor	*Coru di voi*...........	Bull's heart.
——— gryphoides.........	*Ostrica russigna* ...	Rock clamp.
Chiton aculeatis	*Scaglia spinusa*	Prickly coat-of-mail.
——— fulvus..............	*Scaglia gialliccia* ...	Tawny coat-of-mail.
Conus Mediterraneus ...	*Ammiraglio*	Lake cone.
—— monachus	*Cappuccinu*	Crown-shell.
—— rusticus..............	*Ammiraglio giallo* ...	Olive.
—— siculus	*Cappuccinu beddu* ...	Volute cone.
Corallina acetabulum ...	*Sertolariu*	Sea-parasol.
———— fragilissima ...	*Muscu marinu*	Milk-white coralline.
———— officinalis	*Alga viva*	Vermifuge grass.
———— opuntia	*Scuteddu di mari* ...	Sea-kidney.
Cypræa lota	*Ciprignu*...............	White tooth-shell.
——— lurida	*Surriceddu*...........	Sea-mouse.
——— moneta	*Ciprignedda janca*...	Black-man's tooth.
——— pantherina	*Ciprigna stizzata* ...	Spotted cowry.
——— spurca	*Ciprigneddu*	Sea-louse.
Dentalium artalis	*Occhi duru*	Lake tooth-shell.
Donax irus..................	*Arceddu di scogliu*...	Rock Venus.
——— scripta	*Arceddu stizziatu* ...	Solen, or razor-fish.
——— trunculus	*Arceddu giarnusu*...	Sea-wedge.
Doris argo	*Carciofu di mari* ...	Sea-lemon.
—— stellata	*Carciofulu*	Speckled sea-lemon.
Echinus cidaris	*Rizza a sfera*	Turbaned sea-urchin.
——— esculentus.........	*Rizza carisa*	Sea-egg.
——— purpureus.........	*Ficu d'India di mari*	Grey urchin.
——— spatagus	*Rizza spatagu*.........	Hairy sea-egg.
Flustra hispida	*Escara securu*.........	Sea-mat.
——— pilosa...............	*Milleporu*	White flustra.
——— truncata	*Cervunu*	Foliaceous polype.
Gorgonia antipathes......	*Curaddu niuri*	Black coral.
———— coralloides......	*Curaddu giallu*	Yellow gorgon.
———— flabellum	*Albero di mari*	Branched gorgon.
———— mollis............	*Gramegna*	Coriaceous gorgon.
———— nobilis	*Curaddu veru*.........	True red coral.
———— patula	*Curaddu schiacciatu*	Horny gorgon.
———— verrucosa	*Curaddu puorrosa*...	Sea-fan.
———— verticillaris ...	*C. Spezzatu*...........	Sea-feather.
———— viminalis	*C. Salciosu*...........	Isis polype.
Haliotis bistreata	*Pateddu a doppie righe*	Ovate ear.
——— lamellosa	*P. Sfogliatu*	Smooth ear.
——— striata	*P. Strisciatu*	Wrinkled ear.

LATIN NAMES.	SICILIAN.	ENGLISH.
Haliotis tuberculata	*Pateddu reali*	Common sea-ear.
Helix decollata	*Lumaca scapezzata* ...	Sea-slug.
—— lacuna	*Lumaca surcata*	Whorl.
—— limax	*Lippariddu*	Sea-snail.
Holothuriæ (varieties) ...	*Citriolu marinu*	Sea-cucumbers.
Holothuria physalis	*Aretusa*	Portuguese man-of-war.
——— tremula	*Tremante*	Fistularia.
Lepas anatifera	*Summuzzaroli*	Duck barnacle.
—— anserifera	*Conca pedata*	Goose barnacle.
—— balanus	*Ghiannaru di mari*	Acorn shell.
—— costata	*Ghiannaru surcatu*	Ribbed barnacle.
—— pollicipes	*Ghiannaru murtipedi*	Cornucopiæ.
—— rugosa	*Ghiannaru grinzosu*	Wrinkled barnacle.
—— tintinnabulum	*Ghiannaru sonante*	Bell acorn-shell.
Mactra corallina	*Truogolu curaddusu*	Smooth mactra.
——— solida	*Truogolu marmoreu*	Ribbed mactra.
——— stultorum	*Truogolu di pazzi* ...	Gaping tethys.
Madrepora ananas	*Matripora ananosa*	Starry madrepore.
——— anthophyllum	*Matripora frondosa*	Simple medusa.
——— cerebrum	*Piedra cervulosa* ...	Brain-stone.
——— cyathus	*Tazza di nettuno* ...	Saucer madrepore.
——— verrucaria	*Matripora caccia porru*	Little medusa.
——— virginea	*Matripora ianca*	White finger.
Medusæ (varieties)	*Ortica marina*	Sea-nettles.
Medusa aurita	*Campanulu*	Sea-umbrella.
——— cruciata	*Medusa a croce*	White-cross medusa.
——— infundilatum ...	*Medusa a imbutu* ...	Sea-blubber.
——— marsupialis	*Medusa a bursa* ...	Sea-purse.
——— noctiluca	*Ogghiu a mari*	Sea-lanthorn.
——— pilearis	*Medusa pilusu*	Hairy blubber.
——— pulmo	*Purmonariu*	Eight-arm medusa.
——— vetella	*Escariunu*	Naked-eyed medusa.
Millepora aspera	*Idra rozza*	Erect hydra.
——— carduncalus ...	*Cardonu di mari* ...	Sea-thistle.
——— cellulosa	*Idra a merlettu*	Lace polype.
——— miniacea	*Millepora vermiglia*	Red hydra.
——— pumila	*Picciuna di mari* ...	Shell-sucker.
——— reticulata	*Millepora a rete*	Porous millepore.
——— tubulosa	*Idra maccaronaja* ...	Parasite hydra.
Murex cutaceus	*Buccinu pellicciatu*	Coated murex.
——— gyrinus	*Ranocchieddu*	Rock frog.
——— melongena	*Pirulu*	Pear-shaped murex.
——— olearium	*Ranellu*	Oil-jar.
——— purpura	*Buccinu purpureu* ...	Purple whelk.
——— puso	*Vessigutu*	Wreath rock.
——— Syracusanus	*Buccinu Sicilianu* ...	Keeled rock.
——— trunculus	*B. Truncatu*	Knotty rock.
Mya arenaria	*Cardinu*	Sand gaper.
—— truncata	*Ascidiu*	Toothed gaper.
Mytilus barbatus	*Nicchia varvata*	Bearded mussel.

Latin Names.	Sicilian.	English.
Mytilis bidens	*Nicchia a dui renti*	Double-toothed mussel.
—— hirundo	*Rondinellu*	Swallow mussel.
—— lithophagus	*Percia-pietra*	Burrowing mussel.
—— rugosus	*Modiuli*	Furrowed mussel.
—— vilelia	*Nicchia a vela*	Sea-nettle.
—— unguis	*Unghianatu*	Claw mussel.
Nerita glaucina	*Naticao*	Blind nerite.
—— officinalis	*Valvatu*	Snail nerite.
—— viridis	*Concha nivea*	Neritina.
Ostrea crenulata	*Ostreca intaccata*	Little oyster.
—— edulis	*Crocchiuli*	Common oyster.
—— lima	*Ostreca raschia*	Imbricate oyster.
—— maxima	*Pettenu*	Scallop.
—— pes felis	*Pettencuru*	Striated oyster.
—— pusio	*Picciridda*	Long oyster.
—— plicatula	*Ostreca torciuta*	Grey oyster.
Patella atra	*Patedda niuri*	Black limpet.
—— cærulea	*Patedda turchina*	Blue limpet.
—— crepidula	*Pianeddu*	Oval limpet.
—— flaviola	*Patedda gialliccia*	Yellow slipper.
—— lacustris	*Patedda di lacu*	Ancylus, or bonnet.
—— mammillaris	*Frutta di mari*	Striate limpet.
—— nimbosa	*Fizzureddu*	Ovate limpet.
—— oculus	*Patedda ucchiatu*	Goat's-eye limpet.
—— pectinata	*Patedda erpicata*	Wrinkled limpet.
Pennatula antennina	*Pennuzzu*	Spotted sea-pen.
—— grisea	*Lucioleddu*	Shining sea-pen.
—— mirabilis	*Pennuzzu filatu*	Filiform sea-pen.
—— rubra	*Pennuzzu russignu*	Variegated sea-pen.
Pholas candida	*Dattoli janchi*	Piercer.
—— dactylus	*Dattoli di mari*	Piddock.
—— striata	*D. Strisciatu*	Ovate pholas.
Pinna marina	*Lana conca*	Wing shell.
—— muricata	*Lana spinusa*	Prickly nacre.
—— nobilis	*Pinnula*	Pearly nacre.
—— squamosa	*Madre-perna scagliata*	Scaly nacre.
—— sacata	*Saccone*	Sea satchel.
Sepia loligo	*Calamaru*	Ink-fish.
—— octapus	*Ottapedia*	Long-armed cuttle.
—— officinalis	*Siccia*	Cuttle-fish.
—— sepiola	*Calamareddu*	Sea-pulp.
Serpula echinata	*Verme spinosu*	Glabrous sea-worm.
—— glomerata	*Agghiuommeratu*	Winding sea-worm.
Sertularia abietina	*Pigna di mari*	Sea-fig.
—— halecina	*Cornu di Bove*	Horny polype.
—— misenensis	*Acciu di mari*	Sea-thread, or bristle.
—— myriophyllum	*Musca maritima*	Leafy polype.
—— pennaria	*Sciuru marinu*	Sea-tuft.
—— thuja	*Cellulariu*	Bottle brush.
Solen cultellus	*Cannulicchiu*	Sheath, or razor-shell.

Latin Names.	Sicilian.	English.
Solen ensis	*Cannulicchiu stortu*	Scimitar.
—— siliqua	*Conca niura*	Pod.
Spondylus gæderopus	*Ostreca spinusa*	Prickly oyster.
Spongia fasciculata	*Fasteddu di mare*	Sea-bunch.
—— ficiformis	*Ficu spognusu*	Top-shaped sponge.
—— infundiliformis	*Imbuto di spogna*	Sea-funnel.
—— officinalis	*Spogna comune*	Common sponge.
—— tomentosa	*Artica marina*	Stinging sponge.
Strombus clavus	*Brogniuni*	Trumpet shell.
—— pes pelicani	*Conca piegaru*	Cormorant's foot.
—— tuberculatus	*Conca torta*	Sea-screw.
Tellina cornea	*Foglia dura*	Pandora.
—— digitaria	*Arceddu*	Lucina.
—— donacina	*Faccia di rosa*	Rayed tellen.
—— gargadia	*Rematoru*	Toothed tellen.
—— leporina	*Fimbriu*	Thetis.
Teredo clava	*Verme di legnami*	Clavated borer.
—— navalis	*Vergale marina*	Ship-worm.
Trochus conulus	*Cunieddu*	Top-shell.
—— divaricatus	*Stregone di mari*	Camisole.
—— perversus	*Cunieddu a manu manca*	Left-handed top.
—— striatus	*Guscio di mari*	Channelled camisole.
Tubipora flabellaris	*Nodo di mari*	Depressed nereis.
—— pinnata	*Alcyone di mari*	Erect nereis.
—— serpens	*Pietra sertolaria*	Tubular coral.
Tubularia cornucopiæ	*Penna di mari*	Tubular coralline.
—— fistulosa	*Salce di mari*	Bugle coralline.
—— indivisa	*Alga di vermi*	Grey tubularia.
Turbo clathrus	*Curnicchi di mari*	Wreathed turban.
—— littoreus	*Lumaceddu*	Periwinkle.
—— rugosus	*Occhi di S. Lucia*	Screw-winkle.
—— sanguineus	*Lumacedda russigna*	Purple wreath.
—— terebra	*Curnicchiuli*	Auger turban.
—— turritella	*Turbu stortu*	Staircase-shell.
Venus exoleta	*Bagatteddu*	Zigzag Venus.
—— tigerina	*Conca bedda*	Tropical Venus.
—— verrucosa	*Vongulu*	Rough Venus.
Voluta mitra	*Turricula granulata*	Mitral volute.
—— oliva	*Ruolo oliva*	Olive-shell.
—— rustica	*La Trenga*	Cylinder.
—— tornatilis	*Tornuteddu*	Creeping olive.
Zoanthus (varieties)	*Sciuri vivi*	Animal flowers.

PART III.

OF THE MEDITERRANEAN WINDS, WEATHER, AND ATMOSPHERICAL PHENOMENA.

§ 1. Climate of the Mediterranean.

Ancient climates.

WE are not here going upon the old geographical parallels and longest-day divisions, by which the inner sea occupied nearly five of the ancient climates, but shall adhere to the general sense of the word climate as at present used,—namely, to denote the state of the country as to changes in heat, moisture, winds, and other agents which sensibly affect our organs, promote the development of plants, and thus render the land fit for animal and vegetable life. Yet the Mediterranean is a large and varied space to be thus included under one head; since, besides the greatness of its longitudinal extent, it includes a latitudinal space of sixteen degrees—from 30° to 46°. So extensive a region is liable to almost numberless variations; for while on its northern shores the vicissitudes are sudden and violent, even hyperborean cold existing at certain seasons, the heat is all but inter-tropical in the south; and though, as will presently be shown, its general salubrity is deteriorated by malaria, this sea has ever—both from its atmospheric and geographical position—borne a high reputation for a temperate and healthful climate.

Meteorology.

The vital importance of meteorology, as well to the landsman as the sailor, has been acknowledged for ages; yet the laws which govern atmospherical changes have not received the full attention of investigation which human interests require: before it has been more completely reduced to a demonstrable theory, it cannot be ranked among the positive physical sciences; and for this it yet demands a large accumulation of accurate and well-arranged facts. We are not, however, on this account, to suppose that we know nothing about the rare and elastic medium in which we live and move; on the contrary, while Philosophy so long neglected her duty, Knowledge had been sufficiently alert to render many persons to whom it was necessary, in a degree, weather-wise. It was therefore—besides our own general use of the observations—with a hope of aiding further inquiry into the properties of our wondrous envelope, that I diligently registered the fluctuating phenomena as they occurred. These were my views upwards of forty years ago, when the inconstant and uncertain nature of wind made it seem impossible to reduce it to any certain law, much less to connect it with the general movement of the atmosphere over the whole earth. In the more fickle latitudes this may be a labour of much time, since the difficulties appear almost insuperable; but the powerful light thrown on the inquiry by monsoons, trade-winds, and hurricanes, places the further explanation of apparent anomalies within hope. The laws of atmospheric phenomena near the equator have recently been sufficiently made out to afford reason to hope that the theory will, ere long, be more nearly approached: the laws already ascertained are such an unexpected advance, and display such a regularity and order, as could scarcely have been looked for in our times.

On my means of observation.

With an earnest desire to strengthen the grand outline then so indistinctly traced, I availed myself of the opportunity before me of substituting for prevalent impressions and erroneous notions suggested by the senses, the more exact and secure method of observation now so happily applied. But though all possible pains were taken, our instruments were too few and deficient for the attainment of that accuracy which science now demands: indeed, what chance had I in 1812 of purchasing perfect barometers, thermometers, and hygrometers, when, in this very year of our Lord 1852, Mr. Glaisher, of the Royal Observatory at Greenwich, in a lecture read to the Society of Arts, on the probable effect of the Great Exhibition in the Crystal Palace in improving our mechanics, complained that his exertions in the cause of meteorology had been checked by the glaring deficiencies in the atmospheric and eudiometric instruments, and the present impracticability of procuring better? In the course of his address, this able meteorologist quoted a communication made to him by the able Professor, W. H. Miller, of Cambridge, which is so much in point, that an extract must be here given:—

State of instruments.

Respecting barometers, Schouw discovered a remarkable law between latitude and barometric pressure; but nearly every one of the English observations were doubtful, on account of the badness of instruments and neglect of data for reducing the observations, many of the observers having used worthless instruments, in ignorance that better were in existence. I tried to verify this law of Schouw's by using various English observations.

Six years observations in the Mediterranean, by Captain Smyth, I reduced as far as I could; but the labour was thrown away, because the instruments did not admit of determining the errors,—that is, the error was not constant.

Professor Chevallier, of Durham, had observed with a high-priced barometer for nine years, and that an observation should not be lost, had instructed the ladies of his family to observe. He tried to obtain the constant error by comparison with a barometer of my own, of Bunton's construction. The error was extremely variable; he anatomized his barometer, and found that the error mainly depended on the hygrometric state of the atmosphere.

Observations made at Madras, for twenty-three years, by Mr. Goldingham, and printed in the East India Company's costly volume, are, for the same reasons, worth less than nothing. Lieutenant Sullivan, R.N., made numerous observations at the Falkland Islands, which, for the same reasons, are worthless. The same observation applies to Captain Fitzroy's observations at an important meteorological station—the neighbourhood of Cape Horn.

Object of my register.

Still, for a right comprehension of the subject before us, it should be borne in mind that the material uses to which I applied the barometer and thermometer—namely, for correcting the refraction of the heavenly bodies in altitude, and for watching atmospherical changes—were very fairly answered; and though we shall return to the subject, it may at once here be emphatically stated, that the marine barometer—with all its alleged defects—is one of the most valuable of the boons which science has given to navigation. But I certainly had entertained hopes of being more useful to abstract philosophy, being for some time unaware of the imperfect graduation and other defects of the means requisite for pursuing that object. Sicily, Sardinia, Malta, the Ionian Islands, and Tripoli in Africa, were the points of my chief experience; and I considered that careful meteorological registers made at all the places visited, would have afforded some normal points of utility to inquiry. When my friend, Professor Miller, returned the six years' observations which he had kindly undertaken the drudgery of sifting, I was less surprised than disappointed at the severe sentence above expressed; and he added,—'I intended to compare your observations in the Channel with the Royal Society observations contemporaneously made, but was deterred by the account Mr. Hudson (*the Assistant-Secretary*) gave of the careless manner in which the latter were made at that time.'

Professor Miller's discussion.

Yet, although defeated, my efforts were not abandoned; and as that *whole* range of observations had failed, I determined to re-examine those which were made after I returned to England in 1820, and procured a fresh supply of instruments. In the usual way of obtaining the mean height of the barometer, it is indispensable to the accuracy of the calculation, that an equal number of observations be obtained corresponding to winds from opposite directions. Now, this being circumstantially precluded, I proceeded after Professor Miller's own method of treatment, upon my register from leaving Sheerness on the 7th of August, 1821, to August the 2nd, 1824; by which process a more

My own discussion.

satisfactory conclusion than his first was obtained. It is as follows:—

The sum of 1049 readings of the barometer between the above-mentioned dates, reduced to the freezing point, is 31333·6 inches, and the sum of the latitudes of the places of observation is 38578°; hence, dividing by 1049, we have—

Mean barometer reading, 29·870; and mean latitude, 36° 46′ 33″.

The observations were taken at 8h. A.M., to accord with the ship avocations. Now, according to Forbes, the diurnal oscillation is, in inches, = 0·1193 (cos. mean lat.) $\frac{5}{2}$ 0·0149 inches; and the pressure is at a maximum at 9h. A.M.; hence one hour before the maximum being = one hour after it, there remain two hours to noon, or $\frac{1}{12}$th of the daily oscillation to be applied.

Cos. lat.	36, 46, 33	....	log. 9·903629
$\frac{1}{12}$th of ·1193 . . =	·0099	....	log. 8·000000
Nat. Num. . . .	·008	=	log. 7·903629
Mean barometer . .	29·870		
Corrected to noon =	29·862		

To correct this reading to the standard of the instruments in the Paris observatory, we take the mean of 28 observations made on board my ship, the *Adventure*, at Sheerness, and on the way from thence to Falmouth, at 0° C, = 29·826; and the mean of 28 observations made at Paris at 9h. A.M. on the same days, at 0° C, = 29·788. Now the pressure at the level of the sea in the latitude at Paris, *exceeds* that at the level of the sea in the mean latitude of the track from Sheerness to Falmouth by 0·015 inches, according to Schouw; which, deducted from 29·788, leaves 29·773; and on the other hand, the pressure at the Observatory at Paris being less than at the level of the sea in the same latitude, we must add,—

By Schouw .	0·213	29·986	which, deducted from 29·826, leaves the various deficiencies of . .	·160	hence, the mean barom. in Mediterranean = 29·861 corrected by Paris bar.	= 30·022
By Ramond .	·270	30·043		·217		·079
By levels . .	·222	29·995		·169		·031
By other levels	·282	30·055		·229		·091

Conclusion thereupon.

As this process yields a mean = 30·056 nearly, I trust that general comparisons are afforded by my barometric conclusions: and they were constantly compared and marked on diagrams with the oscillations of the sympiesometer, a very portable instrument, contrived, by means of hydrogen gas and oil, to indicate most sensitively the changes of the pressure of the atmosphere. The scales of the thermometers, though not perfect, were very fairly divided, and we had the best that Dollond and T. Jones could furnish. Our weakest point was in the hygrometer, one of De Luc's, for Mr. Daniell—my regretted friend and successor as Foreign Secretary to the Royal Society, when I was called into Wales in 1839—had not then promulgated his beautiful experiments on wet bulbs and dew-points. Aware, therefore, that my means were short of the desired excellence, and relying on the diligence exerted in watching them, I still think my meteorological remarks will prove trustworthy to navigators, and perhaps to others; more especially as, in drawing up my general deductions, I had the great advantage of access to the registers kept by my excellent friend, the Abbate Piazzi, from 1791 to 1815, in the Royal Observatory at Palermo — near the mid-latitude which I have assumed. **Mr. Daniell.**

In my account of Sicily, published in 1824, I remarked that 'The medium height of the thermometer is 62·5°, of the barometer 29·80 inches, and the annual amount of the pluviometer 26 inches. The thermometer in the hottest days rises as high as 90° or 92° (in the shade), and very seldom falls lower than 36°, even in the depth of winter. The highest barometer index I have observed in very serene* weather, and light westerly airs, was 30·47 inches; and the lowest, in gloomy weather and south-east gales, 29·13 inches. In the year 1814, there were 121 overcast and cloudy days, on 83 of which rain fell; 36 misty days; 49 of very variable weather, and 159 fine bright days.' **Sicily.**

Of Sardinia, my sketch of which was published in 1828, it is stated (*page* 79) that 'the island lies between the 39th and 41st degrees of north latitude, and though the thermometer ranges from 34° to 90°, I found its mean temperature, by a register of Six's thermometer, 61·7°; but this being the average only in my cabin in the various ports and bays, I tried that of a very deep and limpid spring (44 *feet*) near Porto Conte, in a cavern 120 feet below the surface of the **Sardinia.**

---

* This word, in my *Sicily* (page 4), is unfortunately printed *severe*, but the context will point out the error. Being absent from England during the time it was printing, I was prevented from seeing the proof-sheets.

earth, and found it to be 60·2°. The medium height of the barometer appears to be about 29·69, the highest point I have known in it being 30·40, and the lowest 29·20.' A further scrutiny of my register leads to the general conclusion that the prevalent winds are from the west to the north, and the next prevalent are from east-south-east to south; and also that the spring season is usually mild and balmy, with frequent showers—the summers are sultry, with occasional thunder-storms—autumn is warm and genial, with occasional rain—and winter, though fine at intervals, is at times rainy, tempestuous, and humid. Such, indeed, may be assumed as the climate for the mid-latitude of the Mediterranean; and the view will be more complete by showing the monthly variation of the temperature for the mean parallel—an isotherm for comparison—as nearly as my tables enable me to carry it:—

Conclusions.

	Max.	Min.		Max.	Min.
January	50·1	44·3	July	79·9	74·1
February	51·5	46·0	August	81·7	76·0
March	58·8	50·7	September	80·1	73·5
April	63·6	61·0	October	77·4	65·4
May	68·7	64·0	November	69·3	58·9
June	77·9	67·6	December	60·5	49·7

In addition to these deductions, it may aid the inquirer to see what may be deemed an approximation to the actual climate condition of several stations of the Mediterranean Sea, at very slight elevations above its level:—

PLACES.	Barometer.		Thermometer.		Rain. Inches.
	Max.	Min.	Max.	Min.	
Gibraltar	30·90	28·62	85·0°	46·8°	31·1
Marseilles	30·55	29·04	79·2	40·5	26·8
Sardinia	30·40	29·20	90·0	34·0	27·5
Rome	30·28	28·73	82·5	41·0	30·4
Sicily	30·47	29·13	91·0	36·0	26·0
Malta	30·39	28·80	90·2	46·4	15·0
Cephalonia	30·32	29·07	90·5	43·5	21·9
Constantinople	30·38	29·16	90·0	53·4	31·6
Alexandria	30·16	29·42	91·5	51·7	7·6
Tripoli	30·25	29·50	92·5	51·2	10·0
Algiers	30·28	28·99	86·8	41·5	25·6

These figures afford a pretty fair estimate of the atmospheric pressure and temperature of the Inner Sea, under the unequal action of all the disturbing causes; and in the latter table, where I have availed myself of the contributions of friends, I have carefully eschewed the wondrous but impotent conclusions derived from observations of a thermometer placed, as it is called, in the sun—a practice on which some of our countrymen abroad are apt to rely, not being aware that the heat thus marked by the scale is nothing more than that of the instrument itself, thus exposed to the continued action of the solar rays. Nothing, however, in meteorology, presents a greater uncertainty, with the means as yet contrived, than the attempt at measuring the annual fall of rain in a given spot; since so much depends upon the nature of the gauge, the ability and industry of the observer, and the height and exposure of the place. Of this, a crucial example is afforded by the registers carefully kept at Gibraltar for nearly half a century, wherein a range of the mean fall yearly is presented, varying from 14·16 to 62·87 inches in the least and greatest quantities; and we are assured that in the year 1796, when the latter fall took place at Gibraltar, at Madrid—the centre of an elevated plateau in the middle of Spain—the yearly rain measured only ten inches. Such differences in places comparatively contiguous,* are not, as might be expected, at all uncommon. The mural semicircle formed by the Alps on the north of Italy, encloses a basin into which the warm southerly winds blow, and the effect is such, that while on the northern foot of those mountains there are but thirty-five inches of rain, there falls an average of fifty-eight inches upon the southern foot; and at Tolmezzo in Friuli, in the south-east part of that curve, forming an

Remarks on the results.

Annual fall of rain.

* The greatest annual fall on record is that at Sierra Leone, on the west coast of Africa, which amounts to no less than 400 inches! Yet some of the adjacent parts are comparatively dry.

angle where the vapours accumulate, a mean of twenty-two years' registration gives ninety inches as the annual quantity, whilst the fall at Venice—only sixty-five miles to the south-south-west—is but thirty inches. A similar relation exists with respect to the other mountain-chains of the Mediterranean, where like causes produce, at certain places, the fall of twice, or even a greater excess over the mean annual quantity.

Stability of temperature.

Imperfect as the above observations must necessarily be, they will be undoubtedly useful in physical geography; and had Pliny given us similar numbers in his 'Cyclopædia,' we should have had a range of 1800 years upon which to discuss the *questio vexata*, as to an alteration of the mean heat on the shores of the Mediterranean. It is considered as absolutely shown, that little or no change of the ordinary temperature can have taken place in Syria during 3000 years; because the Israelites found the date and the vine flourishing in Canaan, and they exist there still. My good correspondent, M. Arago, insists that a trifling alteration of temperature would have destroyed one or the other of these fruit-bearing trees, since the vine will not ripen where the mean temperature of the year is higher than 84°, or the date flourish where it sinks below that degree. This argument would be beautiful, as appealing to Nature herself where instrumental means are wanting, but his numbers are assuredly too high, since I found the date-tree flourishing exuberantly around Tripoli, under a mean thermometer of 75·35°; and its luxuriance in the neighbourhood of Cairo is well known. I have no register for Cairo; but on the hypothesis that the climate of Lower Egypt may be nearly inferred from its parallel of latitude, since there are but few disturbing causes acting on the atmosphere in that valley, it may be considered as low as 70·56°, on Sir David

Brewster's formula.

Brewster's formula, T=81·5 (*inferred mean at the equator*) cos. lat. No one will hesitate to admit that the date-tree and the vine were simultaneously cultivated in the valley of the Jordan; but assuredly no country is more broken by mountain and plain. The date is not recorded as an eatable fruit in Scripture, nor can it be proved that ripe dates were gathered and wine pressed on the same spot.

It is now impossible to say how far we might have agreed with the ancients, or differed from them, had they transmitted quantitative determinations to us, derived from a systematic registration and discussion of the meteorological facts; and in the absence of these, it would be very unsafe to admit the vegetation of Italy as a natural thermometer, into rigorous argument, until we are positively agreed as to the identity and place of the trees adduced. The hypothesis, it is true, has recently been so ably treated by my friend, Dr. Rothman, as to be almost convincing; but the unanimous testimony of a host of classical writers, the aspect of various localities, the greater mildness of recent winters, and the harvests being now somewhat earlier than formerly, certainly lead to the inference that the south of Europe is much warmer than when Cæsar so carefully noted the changes of season. Of course the meteorologist is aware, from the recent brilliant researches into the condition of the earth as connected with the length of day, that it is concluded the mean heat of the globe cannot have diminished as much as $\frac{1}{300}$th part of a degree of Fahrenheit during a period of 2000 years; yet it cannot be decided how far local modifications may exercise local influence. Therefore, without entering into an elaborate discussion as to the grounds on which the above opinion was formed, I may at once state, that such is the impression left by my inquiries; and my views were strongly corroborated by the late Dr. Arnold, as reconciling some ancient and modern statements respecting Greece. The facts advanced by history are substantial, the objections to them all but mere inference: numerous tracts which were abundantly fertile are now almost irremediably dry and barren, and—whether from neglect of drainage and tillage,

Absence of authority.

Rothman.

Dr. Arnold.

or other causes—some districts which were tolerably healthy of old, have become so pestiferous as to be uninhabitable. A contrary opinion to this has been received, under the

Apennines.

assumption that by stripping the Apennines of wood, the degree of cold had been sensibly increased; but a comparison of recent registers with those of the Accademia del Cimento, proves that the temperature has scarcely varied from the time of Galileo,—that is, prior to the denudation of those mountains.*

Ancient remarks.

The ancients distinctly speak of effects produced by the cold of winter in Italy, Greece, and Asia Minor, which have been in later ages unknown; and the almost insupportable severity of the climate between the Euxine and

Herodotus. Ovid.

Gaul is historically infamous, from the days of Herodotus to those of poor Ovid, whose dismal lamentations over his sufferings at Tomos are well corroborated by the *septem assurgit in ulnas,* the travelling on ice recorded by Strabo, and the pointed assertions of other writers. Both Herodo-

Cæsar.

tus and Julius Cæsar, however deficient in minute geographical accuracy, agreed with regard to the very material and striking fact — of which there were then thousands of living witnesses—that the winters in Gaul were most rigorous, and that the rivers were at such seasons nearly all

Columella.

frozen. Columella, the agricultural writer, who flourished under Claudius, and is the first author who speaks of vines in Gaul, has this remarkable passage: 'I find that it is the opinion of many respectable authors, that the quality and state of the atmosphere have been altered in the course of a long series of ages; for Saserna, in the work which he has left on agriculture, infers that the state of the atmosphere is changed, because certain districts, which formerly were incapable of producing vines and olives, on account

* Means of ascertaining the temperature by instruments are, however, truly modern, for the thermometer was only reduced, by the skill of Fahrenheit, to a correct standard in 1724, although it had been invented 135 years earlier.

of the continued severity of the winter, now yield abundant vintages and plenty of oil, by the climate having become milder and warmer.' Whether this was the result of embanking rivers, draining morasses, cultivation of the soil, extirpation of forests, or of accidental causes, we know not; but it may be considered as a substantial fact.

Our purpose, however, being to tell what can be gathered as to any alteration in the Mediterranean climate, it may be as well to assume Rome as a starting-point; and here we have the testimony of both naturalists and poets concurring upon the sharpness of their winters. Pliny the Elder—*De Naturâ Cœli ad Arbores* (*Nat. Hist.*, *lib.* xvii. *cap.* 2)—says that he who wishes well to his trees and corn will desire that the snows may remain long on the ground: 'Alioqui vota arborum frugumque communia sunt, nives diutinas sedere.' A passage in Pliny the Younger (*lib.* v. *ep.* 6), which aided Arago in estimating a mean temperature for Rome, shows that sometimes the bay-tree (*Professor Martyn's rendering of* '*laurus*' *) was killed by the cold of Italy; writing from his Tusculan villa, he says, 'The winters here are severe and cold, so that myrtles, olives, and those trees which require continued warmth, will not flourish here (*aspernatur ac respuit*); but it produces the bay (*laurus*) in great perfection: yet sometimes, though not oftener than in Rome, they are killed by the severity of the seasons.' This passage is noted, though somewhat incorrectly, by the Hon. Daines Barrington, in an investigation of this question which appears in the *Philosophical Transactions* for 1768; and he adduces Ælian's chapter of instructions (*lib.* xiv. *cap.* 29) how to catch eels whilst the water is covered with ice, as the strongest proof of the Italian rivers being constantly frozen over: 'Now, if we may believe the concurrent accounts of

Roman climate. **Pliny I.** **Pliny II.** **Ælian.**

* The Laurel of the ancients was certainly the *Laurus nobilis* of Linnæus: our bay-tree, or common laurel, being a plum—*Prunus Lauro-cerasus.*

modern travellers, it would be almost as ridiculous to advise a method of catching fish in the rivers of Italy, which depended entirely upon their being frozen over, as it would be to give such directions to an inhabitant of Jamaica.' And certainly, from the result of my local inquiries, in the more temperate winters now experienced, the Tiber is never frozen over, and when snow falls, it lies on the ground but for a few hours; a duration of two days is held to be a rigorous visitation.

Poetical evidence. In now summoning a poetical witness or two, we cannot but note that the opponents of this view object to the evidence of such authorities, and impute their expressions to the exaggerations of a glowing fancy. Allowing some weight to the observation, it must still be admitted that there must have been ground for assertions as to local facts and customs. Would Milton have made every shepherd 'tell his tale,' had it not been customary to count sheep? or would Gay have written *Trivia* without encountering annoyances in walking the streets of London? There can be no sound reason why 'glacie currus frænaret aquarum' of Virgil's application to the river Galesus, or the 'fractâ glacie' for the matutinal immersions of the Isis-stricken ladies of Rome in the Tiber, should be rejected as merely poetical fictions, since they are collaterally corroborated. Now, Virgil is quoted as an authority in matters of husbandry, and he is constantly advising the farmers, throughout the *Georgics*, to guard their flocks and herds against the ice and snow of the winters; advice which no native poet of the present time would dream of insisting upon. Horace, who, it must be admitted, hated residing in Rome, has various passages which allude to the streets of that capital as disagreeably impure * from snow and ice, as well as from

Virgil.

Horace.

* Dr. Hawkins (*Medical Statistics*), on the authority of ancient documents, tells us that the mean term of life among the Romans was 30 years: among the easy classes in England it is 50 years.

noise and smoke. Nor must the testimony of Martial (*lib.* viii. *ep.* 14 & 68) be omitted, since, in advising the protecting of plants against wintry frosts by placing them in conservatories—for such must be his 'specularia,' which 'puros admittunt soles, et sine fæce diem'—stamps a fact as to climate, and also intimates an early invention in the arts. Martial.

On the whole, I cannot but repeat my belief, that the climate of the zone described may have become somewhat more temperate than it was of old, from some undiscovered or local cause; but both it, and the amount of its effect, are, in the utter absence of better data, more difficult to account for, than it would be to explain why the abject subjects of PIO NONO differ so greatly from the energetic Romans of the times of Scipio and Cæsar. Such partial changes of climate are not out of record; for in the *Philosophical Transactions* for 1766, page 230, Mr. Bowles, Director-general of the Mines in Spain, says, 'Eight leagues square of this Upper Montana (*near the source of the Ebro*) is the highest land in Spain; the mountains rise in the atmosphere to the line of congelation; I see snow from my window (*at Reynosa*), the 4th of August, 1766, as I am writing this letter. Some years ago there used to fall so much snow, that the people were forced to dig lines through it, to go to church, in the winter; but there has fallen little snow since the earthquake at Lisbon, and some years none at all. I am persuaded it changed the climate of many parts of Spain; for no man living saw, nor heard his father say he saw, snow fall in or about the city of Seville, until the year 1756.' In a word, a few days of frost constitutes the severest winter near the isothermal line which I propose to mark in the mid-latitude of the space here described; and snow thaws along the coast in a few hours. It is true that there are records of the Adriatic's having been sometimes frozen over at distant intervals, and of prodigious falls of snow north of the assumed line of heat;

Considera-tion of the evidence.

Upper Montana.

Adriatic frozen.

but occasional incidents are only exceptions to the general rule, and the climate must be pronounced to be generally delightful and salubrious.

Deterioration of climate.

The last word introduces another and a serious consideration. Whatever doubts may be thrown on a variation of the thermal force in these regions, there appears to be very little, among the native physicians, as to a gradual deterioration of the healthfulness of the climate: even here, however, I shall take the middle course, for the essence of a recent Italian work is to swallow every possible statement which favours the author's hypothesis, and reject whatever disagrees with it; and even in begging the question, a *circulus vitiosus* damages the author's argument throughout. In his effort to prove the salubrity of ancient Rome, a somewhat triumphant citation is made from the newly-discovered

Cicero.

fragments of Cicero *De Republicâ*. Speaking of the happy choice which Romulus made of a site for his new city, the orator says, *Locum delegit in regione pestilenti salubrem*, from which it is inferred, that whatever ailed Latium, the

Strabo.

air at Rome was always good. Now Strabo, a man less likely to be influenced by his imagination than this writer, assures us that the situation of the Eternal City was fixed by necessity, and not by choice; and however much healthier it may have been than it is known now to be, it is impossible to note the 'regione pestilenti' above quoted, the complaints of old writers, the recorded plagues, and the numerous temples, altars, statues, and medals to the honour of Apollo, Esculapius, Salus, and Hygeia, without entertaining strong misgivings as to its former wholesomeness. As to the Campagna around, the pestilent nature of the autumnal air in those days is notorious from the writings of Strabo, Martial, Cato, Seneca, Galen; and, among others, Varro may be particularly instanced, from his having advised the proprietor of an unhealthy farm either to sell it at any price, or else to abandon it.

Malaria.

In the first chapter of this work, mention was made of

the pestiferous nature of the air in many places; and other situations afflicted with a summer fever of such a distinctly remittent type as to give plain evidence of the effects of malaria, are pointed out. Yet the subject is of such frequent occurrence in treating of the Mediterranean climate, that it is necessary, even at the risk of repetition, to return to it. And in order that my opinions may obtain attention, I can here state, that although by a singular blessing—considering my continual exposure—I never was a day in the doctor's list while afloat, yet, from constantly attending to the surgeon's daily reports on the sick, examining into particular cases during my official visits to naval hospitals, and generally keeping my weather-eye open, I was enabled to procure such an insight into these matters as a careful officer ought to possess. By this knowledge, and a proper attention to clothing, habits, food, and employment, men may often be preserved from illness by their commander; though, when disease has actually made its appearance, the treatment must rest entirely with the doctor. Among the inquiries which ought to engage the attention of those who are officially placed in charge of many people, surely those which tend to the preservation of human life, and the increase of human enjoyment by health, may justly claim pre-eminence;* it therefore follows, that while the captain's influence should be con-

Duty of officers.

* On publishing my descriptions of Sicily and Sardinia, I appended tables relating to the air of all the towns and villages in those islands; and in quoting the following eulogy on them from Dr. Macculloch's *Essay on Malaria*, 1827, page 368, I am prompted by the hope of inciting attention to so important a point, and not by the personal feeling with which they cannot but impress me:—'Let me only further add,' he says, 'in gratitude to a person without whose assistance I could not even have written what I have, that I am indebted to Captain Smyth for nearly the whole of that topographical information which relates to the shores of the Mediterranean; while they who may choose to abstract that portion, will see that it forms the greater part of the subsequent details. He is not a physician, it is true: yet if but one physician out of a thousand had observed as well, the entire geography of Malaria would not be now to write, and physic would be relieved from a heavy disgrace which it deservedly endures for this neglect.'

stantly employed in the adoption of preventive means, the surgeon should as diligently apply his knowledge of the curative treatment required. And as it has been admitted by medical men, that marsh fevers carry off or disable one-fifth of the dwellers along the Mediterranean shores, our attention to such a scourge cannot be misplaced.

Nature of Malaria.

Although the *effects* of malaria are at last pretty well known to our medical corps, its mysterious nature and *origin* have never been hitherto unravelled; and whether it be an atmospheric agent, whether sulphuretted or carburetted hydrogen gas, or whether it be a material or an aëriform substance, is still unknown. Notwithstanding it is notorious for infesting rice and flax grounds, morasses, and stagnant waters, the febrific tendencies of which are too well known, malaria exists independent of marshes and rank vegetation, in barren and apparently arid places. It may possibly be influenced by the drying power of the atmosphere, or the energy of evaporation under local causes, as expressed by the relation which the dew-point bears to the temperature of the atmosphere. The effluvia from marshes is presumed to indicate its presence; but even where a warning smell may exist, its cause should be sought for, since it is as yet doubtful whether curing an effluvium of its scent also destroys its hurtful quality. Future discoveries may unveil the matter.

Singular effects.

That the vagaries of malaria, as evinced in effects which display but little affinity with each other, are at present almost inscrutable, is no reason for a neglect of observation and inquiry. We are told of towns where one side of a street is infected, and the other not; of streets in which some houses alone escape; and even of barracks in which some divisions are healthy and others filled with sick men. But in these cases the anomaly is rather apparent than real; it may be owing to exposure of site here, or the prevalence of contagious disorders there. Sometimes the miasma has been known to rise from its marshy bed along the nearest side of the adjacent uplands, infecting all that it passes over, though becoming so rarified and dispersed as to lose its malignity at the summit. Over valleys the action is somewhat different; during the summer nights, the exhalations of the day are partly precipitated, and meeting those which for some time after sunset try to ascend, the two baneful gases concentrate. This is malaria of a most malignant type, and will be more or less pernicious according to the season of the year, and the predisposition of bodies exposed to its influence. This unseen enemy is also wafted by the winds to a considerable distance, contaminating the air of places not otherwise unhealthy, the virulence depending on local and aërial circumstances. As to season, it may be considered to prevail from the summer solstice to the autumnal equinox, when fevers, visceral complaints, and general bodily derangement, mark its presence; and it is even advanced by native physicians, that at such times fatal epidemics among men, and epizootics among cattle, display, on dissection, the same appearances of inflammatory affection. The operation of malaria is not much dreaded during day-time, since all the emanations are dissipated by the solar beams; evening causes more cases of fever even than midnight, when the poisonous exhalation is completely condensed upon the soil; and hence those who sleep in the upper stories of houses are less liable to disease, and take it in a milder form, than those whose beds are on the ground-floor. The time of sleep seems to be the moment of attack, as the debility of the body, and the peculiar state of the local night-air, combine to aid the effective reception of miasma: foulness of stomach excites redundant bile, and consequently lays the trap for fever, but the remote cause assuredly exists in some of the volatile bodies in the atmosphere.

Time of operation.

Plague, or pestilence, being but a capricious visitor to the Mediterranean, is only here named because its effects have been confounded in history with the ravages of

Plague.

malaria; yet, should a stray reader desire my opinion on contagion, non-contagion, and quarantine, he will find it already expressed in the *United Service Journal*, Nos. 49 and 51. The misnamed plague which afflicted the Syracusan and Roman armies (LIVY, xxv. 26), when the former, according to the historian, perished to a man, was the autumnal

Malaria mistaken for plague.

malaria which affects the fatal plain of the Anapus; it was as deadly to the Athenians before Himilco's time, as I have since known it to be to some obstinate sporting English officers. When the French, under the luckless Viscount de Lautrec, had overrun all the kingdom of Naples, except the capital and Gaeta, by injudiciously encamping near Baiæ—a neighbourhood where causes of endemic fevers are never wanting—their army was reduced from 28,500 to 4100 men. A melancholy retreat was the consequence; Marshal de Lautrec, the Prince of Vaudemont, and many other persons of eminence, being among the victims. So after the battle of Tchesmè (*the Franco-Russ spelling of Chesmeh*, i.e. *spring*), in 1770, when the Russians were masters of the Archipelago, and might therefore have chosen any port therein, Count Orloff, in opposition to all advice, persisted in making Port Naussa, in Paros, the cantonment and depôt of his forces. And bitterly he paid for his obstinacy, in the death of the greater part of his soldiers and seamen, and the sickening of nearly all the rest; by which the objects of that campaign were frustrated.*

Consequence of such error.

Under the colloquial term plague the cause of illness was mystified, and men's minds were so misdirected, that the real enemy was utterly slighted till nearly the present

---

* Instead of the Empress adding Tchesmè to Orloff's name,—the which was rather due to Lieutenant Dugdale,—she should have stamped the brute with Tarrakanoff as an agnomen. To show how little Orloff's disaster weighed with Hygeian tacticians, I will just mention that so late as the year 1809, I served on the grand and powerful expedition to Walcheren, where, from precisely the same causes, we lost 10,000 men, inflicted thousands of others with pertinacious ailments, and utterly wasted twenty millions of money.

day. In 1812, during our occupation of Sicily, three or four times, and by means of as many successive parties, was it determined to occupy a point between Cape Rasaculmo and the unhealthy village of Spadafora, in the telegraph line between Messina and Milazzo; nor was the intention abandoned until thirty men had been destroyed by malaria, against which the natives had warned the officious staff-officer then holding his brief authority. Such instances are truly deplorable, especially as I can distinctly state that most of these places infected with malaria are well known to their respective neighbours, and by them at once pointed out to strangers. Yet I have known of both naval and military officers treating such admonitions with incredulity, and even contempt; and loss of life, or lingering illnesses, have been the consequence. It is to be hoped that the hour of scepticism has passed, and that our authorities have begun to learn that there actually are pestilential spots which should be carefully avoided. At least, if a commander who is made acquainted therewith allows his want of vigilance, or contempt of means, to cause a loss of life, that commander is not only recklessly neglectful of his duty, but is also morally guilty of culpable homicide. Captain's responsibility.

Still, with the exception of these scourges, the Mediterranean climate must be considered as highly salubrious; and although, from possessing a more humid atmosphere than is generally inferred, it may not be so good for pulmonary disease as medical men have supposed it to be, and its vicissitudes are trying to invalids, it is gratifying to know, by evidence which cannot be disputed, that the highest degree of health is enjoyed by the British fleet on this station. In proof of this I shall here submit a table which was kindly furnished to me by Sir William Burnett, himself formerly Physician of the Mediterranean fleet, and now Director-general of the Medical Department of the Navy. It shows the sanatory condition, from official General salubrity of the Mediterranean.

returns, the total number of cases of principal diseases and injuries, with the ratio of each per 1000 of mean strength for seven years, from 1830 to 1836; and is a tabular view of the clear and elaborate report on the 'Health of the Navy,' compiled by Dr. Wilson.

Wilson's report.

PRINCIPAL DISEASES.	Total number of cases in Seven Years.	Annual ratio attacked per 1000 of mean strength.	Total died in Seven Years.	Annual ratio died per 1000 of mean strength.
Fevers	4,681	84·	98	1·8
Organic diseases of the brain	113	2·	42	·8
Inflammation of the lungs	1,742	31·3	54	1·
Inflammation of the liver	403	7·2	12	·2
Inflammation of the digestive organs . .	142	2·5	13	·2
Consumptive diseases of the lungs . . .	285	5·1	105	1·9
Expectoration of blood	147	2·6	3	—
Dysentery	742	13·3	18	·3
Malignant cholera . .	96	1·7	22	·4
Delirium tremens . .	64	1·1	6	·1
Syphilis	2,771	49·9	—	—
Gonorrhœa	1,451	26·	—	—
Ulcer	3,969	71·2	6	·1
Wounds and accidents	12,415	222·9	101	1·8

§ 2. Winds, Weather, and other Atmospherical Phenomena.

Nature of the atmospheres.

WITHOUT attempting a set discussion upon winds—or air in action—it will not be out of place to state, that the recent meteorological investigations of Dove, Daniell, and Howard, have led to the conclusion that the earth is surrounded by two atmospheres, the relations of which to heat are different, and their states of equilibrium, from the unequal temperature of the sphere which they envelope, incompatible with each other. Hence arises a system of antagonist currents, gaseous evolution, and action and reaction from the consequent different densities and temperatures; and while they are perpetually pressing in opposition, the silent processes of evaporation, condensation, and precipitation, tend to equalize the temperature, and govern the weather. In admiration of this development of design in such an imperfectly known department of natural knowledge, the intelligent Mr. Daniell offers his humble tribute of gratitude to a beneficent Providence; adding, 'By gradual but almost insensible expansions, the equipoised currents of the atmosphere are disturbed, the stormy winds arise, and the waves of the sea are lifted up; and that stagnation of air and water is prevented, which would be fatal to animal existence. But the force which operates is calculated and proportioned; the very agent which causes the disturbance bears with it a self-controlling power; and the storm, as it vents its force, is itself setting the bounds of its own fury.' On this principle, it must be remembered that the winds, which are so sensible to us, are acted upon by agents on or near the earth's surface; which are usually parts of minor systems of compensation as compared with the grand movement of the atmospheric ocean in which 'we live, and move, and have our being.' **Daniell.**

These are now the accepted premises, and from thence viewing the diversities and vicissitudes of weather as being in a degree amenable to the conformation and orographical mass of the countries surrounding the Mediterranean Sea, a few words on the subject may present, as it were, a working diagram of the principal mountain aspects to the mind's eye. **Remark.**

On the west of the Mediterranean are the lofty mountains of the Pyrenees and Granada, with the generally elevated land of Spain, having obvious influence on the continuity of the aërial currents which flow over them; especially as the gigantic range of Mount Atlas is so near on the south, that meteorologically it may be deemed a continuation; while the high summits of Majorca, Corsica, and Sardinia, aid the consequent action. Under so wide **Western division.**

Central portion.

a scope, the centre of this sea presents a singular feature, in that the Alps and Apennines, as well as the branches of Hæmus and Pindus, must all be considered, *de facto*, to be parts of one and the same mountain range. The chain may be assumed as beginning in Calabria, whence it passes through Italy, crosses the states of Genoa, and sweeping round Piedmont and the basin of the Po, extends down the other side of the Adriatic, and so onwards till one branch terminates in the marble cliff at Cape Sunium, and the other in the Balkán, on the shores of the Black Sea. The aërial connexion is doubtless carried over the high lands of the Morea, which end in the abrupt steeps of Tænarus.

Eastern division.

On the east side similar influences are exerted by the Caucasus and Taurus ranges, with Mounts Athos, Ida, Takhtahlu, and Lebanon; together with the elevated peaks of Candia, Cyprus, and other islands of the Levant. Now the great bases of climature are formed by latitude and local elevation above the sea. The determination, therefore, of these two elements will go far towards answering these questions. It is fully established that the decrement of heat increases as we rise in the atmosphere; and that it augments gradually with an accelerating progression, till the commencement of congelation in each latitude. By a geometrical investigation of the difference of density produced by the changes in the air's temperature, the gradation of effect is clearly traceable; and it follows that, as the mountain-chains on each side of the Mediterranean lie between the latitudes of 30° and 47° north, the snow-line—or lower limit of perpetual congelation—can vary only from about 11,000 to 7,000 feet—a conclusion which, though deduced theoretically, is found to coincide with the result of actual observations wherever they can be obtained.

Result.

Such being the broad feature, some of the resulting meteorological changes may be accounted for, or even determined, however remote we may yet be from a power of fully investigating these causes and effects. One fact, however, has accrued from inquiry—namely, that the temperature is more equable on the shores of the Mediterranean than further inland; and some of the prevalent winds may almost be prognosticated. Still, such is their fickleness as concerns direction, force, change, and temperature, that a complete cognizance of the laws which regulate their course might be despaired of, but for the conviction of there being nothing fortuitous in physics; this, added to the splendid results reaped by inquiry into the equatorial winds, leads to a hope of our one day knowing something more about the differences of aërial temperature beyond the tropics, and whether the winds in the Mediterranean zone are the results only of currents thrown off from warmer regions, not subject to any especial law.

Leaving these matters to time, we will proceed to state a few of the acknowledged facts; from which it will be seen that the caution and watchfulness of the sailor are as necessary in the Mediterranean as in other seas. It is true that fine weather is there predominant; but, fortunately for the welfare of the surrounding population, that sea is not quite the placid inland lake lauded by the strains of certain poetasters. Those who expect to find it constantly serene will therefore be disappointed, and, as they ought to have known better, will meet with no commiseration. Remark.

The most prevalent winds in the Mediterranean are those which blow between west round northwards to north-east, as they occur with little intermission for nearly two-thirds of the year; and pretty constantly so in summer. In February, March, and April, the south-east and south-west winds prevail, but their character varies greatly with the locality, and even their true course and velocity have hitherto been registered with palpable laxity; and in speaking of leading winds it is not unusual, instead of naming a rhumb, to term proceeding from Gibraltar eastwards, *going up* the Mediterranean, and from the Levant to Gibraltar *going down*. In this region, although the Prevalent winds.

general characters are not dissimilar, the winds are certainly more diversified in circumstance, consequence, and local peculiarities than in the more northern parts of Europe; they may, therefore, have led to the ancient notion that climates—strictly, with old geographers, a tract of the earth definitely-bounded in latitude—were also amenable to the influence and power of a consequent atmospheric temperature. As they rule so largely the navigation of the inner sea, I endeavoured some years ago to call the attention of seamen to the importance of studying their fluctuations, as indicated by the marine barometer. In the abstract which I then made from my memorandums for insertion in our professional work, the *United Service Journal*, there might be little perhaps that was positively new; but it was given out as a link in the chain of evidence to show the reliance which may be placed on that unerring monitor. The following account will therefore be, in some degree, a mere expansion of that essay both in style and matter:—

Ancient climates.

Winds outside the Straits.

Between the Capes of St. Vincent and Spartel, the south-west winds are the most disagreeable; a violent one is sure to be denoted by a depression of the quicksilver in the barometer. These gales were greatly dreaded in the late Spanish wars by inexperienced navigators, who, from not knowing how they came on, frequently fell into difficulties. They are always precursed by a long hollow swell, generally commence with a breeze between south and south-south-west, from which point it continues to blow for five or six hours, although the sea sets from the westward. It was too common for the cruizers before Cadiz, unaccustomed to that bight, to have their minds impressed with the danger of the shoals of San Lucar, which were then exaggerated as very alarming. Under this apprehension they were induced to haul their starboard tacks on board, and push for the Strait of Gibraltar, whereas the real danger lies at the entrance of this Strait, and consists of shelves and reefs with soundings so uncertain as not to be depended upon at all. On the other hand, by standing to the westward on the *port* tack at the announcement of the gale by the mercury, while the wind is from the southward, a gaining westerly board may be made, lee-way included.

A solano at Cadiz.

The outer harbour of Cadiz, where the allied squadrons rode during the siege of that city by Marshal Victor, is greatly exposed to the waves thrown in by the westerly winds. But the hardest gale of the neighbourhood is the Solano, or Levanter of the Gibraltar pilots; which, although it comes over the land, is so violent as to justify the Portuguese proverb which makes the gravel fly before it—*Quando con Levante chiove, las pedras muove.* This wind is preceded by a peculiar haziness and clammy humidity, as if owing to a diminished atmospheric electricity; the air is encumbered with the cirro-strativeness of the wane-cloud; and the mercury in the tube gradually sinks. Meanwhile, parasitic clouds, as they are termed by meteorologists, cap the hills of Medina Sidonia, and the atmosphere becomes raw and bleak to the feelings. The apparently stationary clouds are, in fact, the result of a descending storm of dispersion, for instead of being stationary on the mountains, they are formed and redissolved every instant; the vapour being precipitated by the arriving current, and dissipated in the departing one. The solano now sets in from the east-south-east to the south-south-east, for it is not the true Levanter of Mediterranean seamen; the one so named, inside the Strait, blows directly from the east, freshens as the sun rises, and lulls as he declines—being generally at the maximum about noon. A very notable solano occurred on the 27th of March, 1811, when the *Milford's* barge, of which I was accidentally in charge, unable to face it even with her well-appointed crew, was only rescued from being driven off to sea, by passing astern of the *Undaunted* frigate, the outer ship in the bay, where her commander, the present Vice-Admiral Richard Thomas, had veered out hawsers—per signal from Sir Richard Keats—for us to get hold of. On the morning of the 28th, the bay presented a singular scene of tumult and devastation, with signals of distress flying in all parts; spars and merchandize floating in every direction. It was found that fifty-three trading vessels were wrecked on the rocks and under the walls of Cadiz during the night, and that upwards of one hundred more were damaged—crews for the most part saved. Had not this gale been prognosticated, and in some measure provided against by striking lower yards and topmasts, bracing to the wind, freshening hawse, and getting everything snug, the consequences might have been serious to the men-of-war as well as to the merchantmen; but except the sinking of four gun-boats inshore, and the driving of two or three vessels to sea, the English squadron sustained no injury. One of the vessels thus forced from her anchors into the offing was the *Basilisk*, a gun-brig, to which I was taking aid in the barge, though unable to fetch her, from the combined force of the wind and sea, as above described.

Effects of the Solano.

Strait of Gibraltar.

That the winds in the Strait of Gibraltar blow either from the east or west points of the horizon (technically termed *down* or *up*) in general, has been immemorially remarked; and the conformation of its coasts on both sides renders the reason palpable. Of these winds, the east is the worst and most violent, being often the cause of much inconvenience in the bay, from its gusty flaws and eddies, besides its always being found raw and disagreeable on shore: hence, Señor Ayala, the Spanish historian of Gibraltar, terms the east wind the 'Tyrant of the Straits,' and the west their 'Liberator.' A strong Levanter, in December, 1796, fell heavily on the British fleet at Gibraltar, not only rendering them powerless spectators of Villeneuve's squadron running through the Strait to the westward in safety; but it was also nearly fatal to the *Gibraltar*, 80, and *Culloden*, 74, while the *Courageux* was driven from her anchors, and the wind increasing to a perfect hurricane, with a dense fog, she was forced against Ape's Hill, on the Barbary shore, where she was dashed to pieces, and 465 of her crew perished. Many remarkable gales, productive of great wear and tear of ground tackle, occurred while I was on the station, of which, perhaps, one of the most mischievous was that which happened early in 1822, when upwards of forty vessels were driven on shore, and the new mole at Rosia Bay—constructed at a great cost—was nearly washed away. In this storm many lives were lost at Leghorn, and the harbour and piers of Genoa were seriously damaged.

Loss of the *Courageux*.

Within the Mediterranean, the predominant breezes are, as above said, from the north and west quarters, except in the spring, when south-east and south-west ones prevail; but their duration and strength are extremely uncertain about the period of the equinoxes, at which times the wind seldom changes suddenly without an accompanying fall of rain, or, at all events, the formation of rain-clouds; for it rarely happens that the new wind is of the same degree of heat as the one it has superseded. Such changes are frequent in the spring; and the local pilots entertain a notion, that vernal storms which commence in the daytime are more violent, and of longer duration, than those which spring up in the night. Be this as it may, I can, from a long and careful practice, assure the circumspect navigator, that no very perilous weather is likely to assail him without his being sufficiently warned; yet, as the barometer does not usually vary more than a few lines, even to pretty sharp gales, careful attention is required to mark its indications. It may, however, be laid down as a general rule, that whenever the mercury subsides so low as 29·40 inches, severe weather may be looked for; especially if accompanied by dark globular clouds in masses, or when a gloomy haze

Weather within the Straits.

Prognoses.

encumbers the sky after serenity.* So far, in fact, did this conviction aid me, that during the last three years of my commanding a ship in that sea, by attending to my silent monitor, and arranging accordingly, I never once had occasion to turn up the hands in the night: besides the comfort and regularity thus bestowed on the crew, those who have toiled in contrary and vacillating winds will readily estimate the confidence with which an officer is inspired when watches are relieved at their regular hours, recruited both in strength and animation, and cheerfully ready for whatever may betide. But the ease thus afforded to the people is not the only benefit of importance rendered by the marine barometer, for its saving to government in expenses of wear and tear is great; insomuch that, in reporting my arrival at Spithead, in October, 1824, to the Admiralty, I expressly said,—'It is with great pleasure I am able to add, that though, from the very nature of my mission, I have been obliged to hang on lee-shores, and shoals, and coasts little known, and therefore avoided by other navigators, this service has been effected, not only

H.M. ship *Adventure.*

* Among Mediterranean prognostics worthy of being rescued from contempt, two or three may be cited:—Small clouds increasing prove that their weight prevents their rising in the air, and therefore denote rain; while large clouds decreasing, being obviously under dissipation by solar heat, or winds, assure us of fine weather. Therefore, as their ragged aspect shows the process of condensation, the *cirro-stratus* and *nimbus* clouds invariably announce rain; an uncommon twinkling of the stars denotes humidity; a steadiness of the stars, and patches of haze, dryness. The rising or setting sun tinging the air with yellow, indicates vapour; and the atmosphere assuming a reddish tint, serenity. A lunar halo, coloured near her, is significant of great humidity; and a cloudless night, unaccompanied by heavy dew, betokens fine but sultry weather. Small masses of *Cumuli*, with detached flaky clouds, mark settled weather and warm winds; the elegant *cirrus* shows approaching change; while *Cumulo-stratus*, with detached blackish and irregular clouds, precurse variable weather and cold winds. Lightning near the horizon without thunder indicates wind from the opposite quarter; and the same from high clouds announces fine weather. The water in port being unusually clear, so that the bottom is seen in several fathoms, prognosticates the approach of a hard gale; as does also a diaphanous atmosphere. It is, however, difficult to catch the characteristics without experience in observation.

without the *Adventure's* having touched the ground, but without the loss of a spar or sail, or cable or anchor.'

Weather on the coast of Spain.

Proceeding up the Mediterranean by the coast of Spain, we find a climate which in summer is usually fine and dry, with the advantage of freedom from rain and humidity—if that can be deemed advantageous, where a cloudless sky occasions scorching drought, to the injury of men, animals, and vegetation. In the winter, the flaws and gusts of wind from the mountain ranges are often furious : and this impetuosity is severely felt in the vicinity of the Pyrenees, at the eastern bases of which I have observed some very remarkable weather-worn rocky steeps. The south wind seldom blows on these shores, except in the winter season ; at which time the south-west gales, called *birazones*, send in a great sea along the coasts of Andalusia and Granada, where it blows dead on the shore. But a singular change is known to occur here ; for frequently, in arriving at the coast abutment called Cape San Martino, which divides Murcia from Valencia, a ship running free before a westerly wind there encounters another from the north or north-east, often blowing fresh. Along these shores, and especially those of Catalonia, the 'sea-fret,' or dense mist generated on the ocean, precurses the easterly winds which drive it in, and occasions lassitude both to animal and vegetable life. When this first appears, ships at anchor should look to their ground-tackle, and those under sail should gain an offing.

Gulf of Valencia.

Gales from the north-east, easterly to south-south-east, are teasing along the shores of Valencia and Catalonia; though by attention it may be foretold when they are coming on, sufficiently so for the adoption of precautionary measures. Thus, for example, when, with a sluggish barometer, the horizon is overcast in those quarters with thick, whitish clouds—partaking of both *cirrus* and *cirrostratus*, and passing eventually to the *nimbus* form—it may be considered that the wind is about to blow from those points of the compass : and it may be recollected, as a scale, that towards the mean Mediterranean latitude the heavy summer clouds are from about 500 to 700 feet above the sea. The commencement of such a breeze is usually from the eastward, and moderate, but it freshens as the wind draws round to the south-east, and then blows very violently, with a heavy sea rolling on-shore, insomuch that an embayed ship would find it difficult to claw off. A specimen of such a gale upon this coast may be given from the journal of the present Rear-Admiral Lovell, who was a lieutenant on board H. M. S. *Melpomene*, of 44 guns. This ship was despatched from off Toulon, in company with the *Orion* 74, the *Endymion* frigate, and the *Weazle* sloop-of-war, in quest of a squadron of frigates under the command of Jerome Buonaparte, which was reported to have left Genoa for some Spanish port. Our force parted from the fleet under Lord Collingwood on the 8th of December, 1805, and off Barcelona, on

H M. ship *Melpomene.*

Sunday, the 15*th,* at 9 P.M., came on a most tremendous squall, with thunder, lightning, rain, and sleet; clewed all up. At about 9·15, the main-mast was struck by lightning; the fluid exploded by the pumps, and hurt an officer (*Mr. Lovell, then Badcock*), and a sailor.

Monday, the 16*th,* wind more moderate and steady; examined the main-mast, found it severely splintered in many places, particularly about the hoops and in the wake of the trusses, where copper had been nailed on. Stood towards Barcelona, in hopes of rejoining the *Orion*.

Tuesday, the 17*th,* at 9 A.M., wind in heavy gusts from the north, which veered round with much fury to east-north-east, the sea rising all round us, and striking hard under the counter, with water-spouts and flashes of lightning in every direction; furled all the sails, and prepared for another gale; at 11, a very heavy sea pooped us, stove in the dead-lights, and filled the cabin with water; P.M., the wind increased to a perfect hurricane; at 1, the ship was struck by lightning, and the main-mast much hurt; at 2, most tremendous squalls, with continued rain, thunder, and lightning; the storm stay-sails blew to atoms, the ship entirely unmanagable, and whole seas breaking over her. The rudder-head gave way, chocked the rudder, and secured it with the pendants. At 3·30, the main-topmast went in three pieces; and at 4, both the rudder-chains gave way. At 6, a man fell from the fore-yard on the best bower-anchor, but was not killed. All the pumps obliged to be kept constantly going.

Wednesday, the 18*th,* wind veering in gusts from north-east to east-south-east; the quarter-boats were stove; found the rudder gone from the stern-post. At 10 A.M., the carpenter reported the main-mast sprung a few feet above the quarter-deck. P.M., the sea mountains high;

got a cable from the stern with hawsers, &c., and struck the mizen-topmast, but found it impossible to wear the ship.

Thursday, the 19th, more moderate, with a heavy swell; employed in making a Pakenham's rudder. Saw the *Colombretes*, two points on the lee bow, distant five leagues. Made all sail on the fore-mast, in hopes of wearing, as we were drifting bodily down on those rocky uninhabited islands. Finding she would not wear, anchored with a spring on the cable, in sixty fathoms. At midnight, tremendous squalls, with thunder, lightning, and rain.

Friday, the 20th, at 1 A.M., found the ship driving, cut the cable and spring, set the storm stay-sails and foresail; saw the islands west-south-west; the ship would lie no higher. No chance remained of saving a single life, when the wind shifted in a dreadful squall, and allowed her to lie up south-east for forty minutes, which put us clear of danger. P.M., succeeded in shipping the rudder, and found, to our great joy, the ship once more under command.

The coast of France forms a deep bight between the Pyrenees and the Alps, which from its gusty turbulence, even in the summer months, has been immemorially designated the Gulf of Lyons. Here, when a breeze springs up in the afternoon, and freshens as the sun goes down, it may be expected to blow strong at midnight. Hard gales are sometimes preceded by a heavy swell and surf, in character not unlike the rollers of the South Atlantic Ocean, though of inferior volume. In this notorious gulf, so proverbial for the treachery of sudden anemological changes, I have weathered some tough gales; and can therefore render personal testimony to the violence of its squalls, and the furious ebullition of its waters: of which recorded instances are numerous. In March, 1795, a French man-of-war, having received rough treatment in the conflict, or rather affair, with Hotham, off Genoa, parted company from Mons. Marten, and falling into the Gulf of Lyons in a violent gale of wind from the south-east, which chopped round to west-south-west, was quickly dismasted, and nearly torn to pieces; here she lay so utterly prostrate for six days, that, had one fallen in with her, she must have surrendered to any of our attendant frigates or sloops—nay, even the *Fox* cutter would then have been an annoyance. It was in this gale that we lost the *Illustrious*, a fine 74, which had received great

Coast of France.

French man-of-war.

The *Illustrious*.

damage in the recent battle in the Gulf of Genoa; for having struck the shore, and there being no hope of saving her, she was abandoned and burnt. Every seaman will recollect that on the 22nd of May, 1798, Nelson was assailed by a sudden storm in this gulf, which carried away all the *Vanguard's* topmasts, broke the foremast into three pieces; sprung the bowsprit; washed a man overboard, killed a midshipman and a seaman, and wounded several more. This ship, which acted her name at the Nile only two months afterwards, rolled and laboured so dreadfully, and was in such distress, that Nelson himself declared, 'the meanest frigate out of France would have been a very unwelcome guest!' And in the winter of 1808, when his true and tried associate, Lord Collingwood, was blockading Toulon, with his flag flying on board the *Ocean*, a roomy new 98-gun ship, he was assailed by a succession of hard gales. In one of these gales, that noble three-decker was terribly crippled, and so nearly lost, that I here give the words of a spectator, Captain Fead, who thus wrote to me in August, 1845:—

The *Vanguard.*

The *Ocean.*

I was standing on the *Royal Sovereign's* forecastle, and at the same time looking at the *Ocean*, which was then about half-a-mile on our lee bow, and on the starboard tack. At that moment she was struck by a very heavy sea, which threw her nearly on her beam-ends, so much so, that several of our men called out, 'The Admiral's gone down!' But in a few seconds I had the pleasure to see her right again. We understood afterwards, that the blow completely disabled her; and that nearly all the bolts of her iron knees were broken. It was the most awfully terrific scene I ever beheld. Lord Collingwood told Admiral Thornborough, a short time after, that he thought the top-sides were actually parting from the lower frame of the ship; and that the heavy guns were suspended so nearly vertical, that the effect was alarming. This happened in December, and we must have been about the middle of the Gulf of Lyons, with the wind at north-west.

Wind and sea.

One of the peculiarities of this gulf is the sudden rising of its waves, and their attaining a size not at all proportionate to the strength of the winds. Both their amplitude and elevation are greater than would be considered to result only from the action of the wind on the aqueous particles; and their increase under a gale cannot be regarded as uniform. The absolute height of these waves, from the trough to the crest, in severe weather, cannot be much less than thirty feet, even close to Provence, where Count Marsigli pronounced that, 'in a very violent tempest,' they only rose to seven feet above the natural level of the sea. The waves of the Mediterranean, in general gales, may be estimated between fourteen and eighteen feet in height, and are often, from want of range in some parts of the short seas, called 'chopping.'

Height of waves.

Towards the close of the war, many ships of our Toulon fleet were struck by the electric fluid, while cruising off Cape Sicie; among which those fine three-deckers, the *Hibernia*, the *Ville-de-Paris*, the *San Josef*, the *Union*, the *Ocean*, the *Barfleur*, and the *Royal George;* which, together with several of the two-deckers, besides having men hurt, received very considerable damage in gear and spars, between 1811 and 1814. In the beginning of September, 1813, Sir Edward Pellew anchored off the mouth of the Rhone with thirteen sail of the line, and there watered; but they had to ride out a very heavy gale of wind, with two cables an end and topmasts struck. Of this blockading force, about one-half were damaged by lightning, and at least five ships were obliged to shift their topmasts. This gale commenced from the south, and sent in a heavy sea, but on the 10th it blew violently from the north, and then the water was comparatively smooth; so far, therefore, it may have been considered a veering storm; but though some other Mediterranean gales within my experience approximate even nearer than this towards the rotatory theory, the subject has not yet been properly discussed. That able and active meteorologist, Colonel Sir William Reid, the present Governor of Malta, is, however, at his post, and an extract from a letter which I received from him, dated Valetta, 8th of January, 1853, is satisfactory:—

Fleet off Cape Sicie.

Rotatory symptoms.

Sir William Reid.

I have not been altogether neglecting observations on the winds since I have been in Malta, but I am obliged to give my mind to my official duties almost exclusively. I shall send some of the local newspapers, to show you that they now report the weather daily.

There is, doubtless, much unknown to us; but I have seen enough to satisfy me, that the storms here are progressive and revolving, as in corresponding latitudes elsewhere. On the 1st of February, 1851, there was a whirlwind storm of vast diameter, extending from Sardinia to Syria, which moved towards east-north-east, and, I suppose, came from Africa.

My second work, entitled *The Progress of the Development of the Law of Storms*, is being translated into Italian, and is now half printed. When published, I hope we shall have many Mediterranean observers of the winds; but the translation is found to be extremely difficult, on account of the many nautical terms, and I fear it will be a year yet before it can be finished.

The *Rodney*.

In January, 1812, the *Rodney*, a superb new 74-gun ship, commanded by the present Admiral Sir E. D. King, on board of which I was then serving, was so torn and disabled by the united violence of wind and wave, that Sir Edward Pellew was obliged to send her to England in the following autumn, although thereby lessening his effective force in time of need. Noble, however, as this ship appeared on the waters, it must be admitted that she was one of that hastily-built batch of men-of-war sarcastically termed the *Forty Thieves.*

The *Aid*.

I afterwards visited the gulf in very passable weather; but on the 3rd of October, 1820, while standing for Marseilles in H.M. ship *Aid*, the atmosphere became so very transparent as in itself to be of a suspicious character; yet the peculiar beauty of the romantic hills before us, the glorious sun above, and the smooth, glistening fluid around, conspired to lull apprehension. But when, about three o'clock in the afternoon, the lieutenant of the watch entered the cabin with—'Sir, a light breeze is springing up; shall we set top-gallant studding-sails?'—I, having that moment looked at the barometer, and found that it had suddenly fallen three-tenths of an inch, with a surface still extremely concave, replied, 'No; turn the hands up to shorten sail, and we'll get the top-gallant yards on deck?' This answer surprised him; but all the officers being well acquainted with the reliance which, both from experience and reason, I placed on my marine monitor, the preparations were briskly executed to the desired extent, although there was no other discernible aspect of mischief. Scarcely, however, was the canvas reduced, and the ship under command with close-reefed topsails, before a gust rushed so furiously upon us, that had we made sail instead of shortening it, the masts must have been carried away, if that were the least accident. As usual with northerly gales in this gulf, great numbers of birds were blown off, which, though of very opposite characters, were all subdued out of their several instincts, and laboured to find a common shelter on the decks. That same night we lay-to, with the sea occasionally making fair breaches over us; but, from the premonition thus obtained, excepting a boat washed from the quarter-davits, a jib-boom sprung, and the weather-bulwarks stove in, we sustained scarcely any damage. **A gulf gale.**

Among the severe atmospheric visitations on the otherwise charming shores of Languedoc and Provence, must be enumerated the chilly and searching northerly wind called the *mistral* or *mistraou*, the *bize*, *la grippe*, and one of *les fléaux de la Provence.** This wind, by which all these parts of France are so much visited, after getting chilled in passing the high Alps and their extensive snows and glaciers, takes its course with increasing violence towards the warm atmosphere of the Mediterranean, and is very impetuous in coming down the valley formed by the Rhone. Diodorus Siculus, Pliny, Strabo, and other ancient writers, appear to have been well aware of the properties of the mistral; and it has been customary to consider it the same wind with the *circius* of Lucan, to which Augustus—'dum **The Mistral or Bize.**

* This wind is not only disagreeable to the human feelings, but is injurious to the young fruits and vegetables, and all trees exposed to it become bent: in allusion to which, the Provençaux couplet runs—

La Cour de Parlement, le Mistral, et la Durance,
Sont les trois fléaux de la Provence.

Vent de cers.

in Galliâ moraretur'—erected and dedicated a temple (SENECA, *Nat. Quæst., lib.* v. *cap.* 17); but there can be no reasonable doubt of the *vent de cers*—a boisterous wind from the heights of Cevennes, in Languedoc—having the better claim. The piercing cold complained of by the natives of the south of France, during the continuance of the mistral, is owing to the immediate transition from a high temperature to a lower one, as well as to its actual frigidity; for I have myself experienced very chilling sensations in this part, with the thermometer at 50°. The winter climate of Nice is excellent, with bright skies and pure air; but the spring is often unpleasant by reason of the great inconstancy of the weather, and the violence with which winds from the mountains sweep its valley or basin, while it is ever liable to be scourged by the mistral. Although to valetudinary persons, with whom a clear, dry atmosphere agrees, the city and its suburbs form a desirable residence, there are serious drawbacks in the remarkable alternations of temperature, the dirt of the dwellings, and the offensive treatment of grounds, in consequence of there being little or no cattle or other stock. A Provençal proverb is repeated as a warning against night promenades:—

Weather at Nice.

Que lou sol y la sereine
Fan veni la gent mouraine.

On the coast of Piedmont, and from thence to Tuscany, the summers are fine; though the *labeschades*, or south-west gusts, drive home on the shores, load the atmosphere with humidity, and raise the water to a high level. The winter is ushered in by *ouragans*, or violent storms of lightning and rain, with occasional hail; but the northerly winds always clear the air.

Ouragans.

Prognostic.

The Tyrrhenian Sea is greatly agitated by south-west gales of wind; and those from the westward are sometimes known to be on their way, by a peculiar cloud in that quarter, after the manner of the Harmattan, on the west coast of Africa, yet not so regular or so striking. Virgil, though somewhat deficient in accuracy as a navigator (see his description of the departure of his hero from Carthage), has, at the opening of the fifth Æneis, marked the prognostic of the change to a western wind with the discrimination of an observer. Dryden, exercising his usual licence, gives the passage thus :—

> But soon the heavens with shadows were o'erspread;
> A swelling cloud hung hovering o'er their head:
> Livid it looked, the threatening of a storm;
> Then night and horror, ocean's face deform.
> The pilot, Palinurus, cried aloud,
> 'What gusts of weather from that gathering cloud
> My thoughts presage!'

Land squalls.

In those bays which bound the ravines and valleys of the higher grounds, the *raggiature*, or land squalls, are violent, though not of very extensive action ; for I have felt these descending easterly gusts in shore, though the breeze was fresh from south-west in the offing of the Gulf of Gioja, and, standing with this into the Faro of Messina, have there met with a steady south-east wind. The more notable *burrasche*, or mountain storms of Calabria, commence with massy clouds coursing, displacing, and effacing each other like oceanic waves, in rapid grandeur ; the tempest then rages, but its energy is soon exhausted.

Calms.

Along the whole coast, there is a time between sea and land winds which is calm, and called by the Italian seamen *bonaccia* as being unaccompanied by danger ; but their more sturdy Roman predecessors designated it *malaccia*, from being a cause of detention. It commences as the land winds die away, and lasts till nine or ten o'clock, by which time the solar rays are sufficiently effective to act ; then the sea-breeze generally increases till about two hours after the sun has culminated, and at or near sunset subsides. During this time, the ascending current of rarified air seems to have a considerable effect on the clouds which it meets with, even on those in the zenith, sometimes changing *cumuli* into *cirri* with magical celerity.

Raffiche. The summer serenity of the Corsican waters is frequently disturbed by boisterous gusts called *raffiche,* which blow off the hills. Speaking generally, the mountains of the Mediterranean may be said to supply cool air to the valleys; but as these winds occasionally rush out seawards in descending blasts, necessity dictated the use of lateen, or triangular, sails attached to yards that can instantly be let go by the run, for the xebecs, polacres, feluccas, and other craft which coast the shores within their influence. In winter there are strong gales from December to March, which often occasion considerable damage: and the north-westers send in a heavy swell upon the exposed shores of Corsica. The *Berwick.* In January, 1797, the *Berwick,* of 74 guns, was riding in San Fiorenzo Bay, under refit, with her lower masts stripped of their rigging; yet such was the force of the swell, that she rolled all three clean over the side. This unlucky vessel was captured by the French a few weeks afterwards, under jury-masts, and was one of the very few of her class won from us by the enemy during that eventful war: she was, however, retaken at Trafalgar, after a very gallant defence.

In and around Sardinia, the most prevalent winds are from west-north-west to north, and from the eastward; the proportion of days being for the first—which is the healthiest quarter—210, and the latter 145: these may be respectively Maestrale. termed the dry and the humid. The prevalent *Maestrale,* or north-west breeze, brings in a long swell from seaward; and it acts with such violence over the Nurra districts, that the trees exposed to it are bent nearly horizontal into the opposite direction, and so they grow. The west wind seldom blows without bringing rain; still it is always welcomed on the coast, on account of its favouring the arrival of the tunnies; when it veers to south-west it is injurious where it rakes. The south wind rarely occurs but as a stormy winter visitor, and is annoying in the exposed bays; in February, 1793, when Sardinia was invaded by a French force under Admiral Truguet, a gale from this quarter occasioned the loss of the *Leopard*, a fine 80-gun ship, with several smaller vessels, in the bay of Cagliari, besides greatly damaging the whole fleet. The *gregale*, or north-east wind, is called double-faced, from being very squally and inconstant, with heavy rains; and the east wind, or *bentu de soli* (the coming of which is indicated by parasitic clouds on the mountains), is usually accompanied by very vivid lightning, and, from its being loaded with vapours, becomes extremely disagreeable after a long continuance. The *maledetto levante*, so complained of by the natives for its debilitating effect, is actually a south-east wind, the scirocco of Sicily and Italy, and the 'plumbeus auster' of Horace: so great are its effects in driving up the hygrometer towards the damp point, that it is termed *mollezza*; whereas the healthy and agreeable *tramontana*, or north wind, from its opposite quality is called *gli secchi*, or dry. But Sardinia has much very fine weather, and the calms of the summer months are harvest-times to the fishers. In settled seasons the *imbattu*, or sea-breeze, sets in about ten o'clock, A.M., keeping on till about two P.M., and is exceedingly refreshing during the heat of the day; it then weakens, and falls calm as the sun goes down, and is succeeded in the evening by the *rampinu*, or land wind, which holds through the greater part of the night.

The *Leopard*. **Bentu de soli.** **Levante.** **Imbattu.** **Rampinu.**

The island of Sicily occupies the central station of the Mediterranean sea, and may be said to enjoy the average means of its winds and weather. Whilst the sun is in the northern signs, the sky, although it seldom assumes the deep blue tint of the tropics, is, nevertheless, beautifully clear and serene; then after the autumnal equinox, the winds become boisterous, and the atmosphere comparatively dense; the dews and fogs increase, particularly on the coasts, and the rain falls in frequent and heavy showers. In summer it is generally calm in the morning, but a sea-breeze springs up about nine or ten o'clock, freshens until

Sicilian weather.

two or three, and gradually subsides again into a calm towards evening. The winds are variable both in their force and direction. The most prevalent are the northerly and westerly, which are dry and salubrious, producing, with the clearest sky, the most agreeable sensations; and a modifi-

Mamatili.

cation of the maestrale, called *mamatili*, is enjoyed by the Palermitans as a most refreshing sea-breeze. Those from the east round to southerly are heavy, and loaded with an unwholesome mist, often accompanied by heavy rain, thunder, and lightning; storms in which I have seen vessels struck by the electric fluid, and in one of these I was a witness of the all but destruction of Scylla castle, in the spring of 1815.

Æolian islands.

On the north of Sicily are the Æolian isles, the fabled residence of the god of winds; and whether from the heat of the water by volcanic springs, the steam of Vulcanella, the incessant hot ejections from Stromboli, or all of them added to the general temperature—it is certain that there are more frequent atmospherical changes among this group, than in the neighbourhood. These extend their influence

Faro of Messina.

to the Faro of Messina, but are there modified under local conditions. Thus when a northerly wind blows through the Strait, and meets a southerly one some twenty miles below it, or a wind from the Adriatic off Cape Spartivento, it is the occasion of much aërial commotion; especially in the offing between Taormina and Mascali, where the weather is then called *Del Golfo di Cantara.* Another singularity of the Faro is *La Lispa,* a calm in the Strait, with masses of super-impending clouds, though blowing fresh outside: this continues till the next *taglio di rema* (See page 180) of the descending current, when, as soon as this gush of water is established, the wind bursts in with squally gusts and accelerating force.

Malta and Goza.

South of Sicily lie Malta and Goza, its geographical though not political dependants; and although these islands are blessed with the steadiest climate in Europe, there are occasional and beneficial disturbances to its serenity. Sometimes winds of a very boisterous character rage, accompanied by rains of tropical profusion; the winters consequently were much dreaded by the galleys of the Order, and their opponents in Barbary were equally influenced. Under such induction, with all their known and acknowledged expertness at sea, the Maltese sailors were absolutely astonished at the blockade of their ports by Sir Alexander Ball, observing that 'English ships could winter without harbours.' The most violent gales they experience, are those from the north-east, the dreaded *Gregale* (from *Greco*), which rakes the harbours of Valetta, sends in a prodigious swell, and has often caused serious damage, as well on shore as among the shipping. The south-wester is the hottest of the summer breezes; it is much disliked by the Maltese, and even in the spring of 1816, I saw the fields on the neighbouring isle of Lampedusa so burnt and parched by it, as to blight all hope of a harvest. From the heat imbibed by the calcareous surface of Malta, the sultry nights which follow the *festa di San Lorenzo*, in August, and continue till after the autumnal equinox, are sometimes very distressing to strangers, the warmth being of that oppressive degree termed *implacable*. The gregale.

But the most annoying visitor of the regions around is the *scirocco*, or south-easter, a wind detested equally by the ancients and moderns; being, no doubt, the evil vapour of Homer (*Iliad* v.) into which Mars retreated when wounded by Minerva. This debilitating breeze—the dreaded *samiel* of Egypt—sweeping over the parched deserts of Arabia and Africa, where the hottest summer climate in the world is to be found, is moderated by its passage over the sea, to a tolerable degree of temperature; and on the east coast of Sicily, where it first arrives, its effects are inconsiderable; but seeming to acquire additional heat in its progress over the land, becomes a serious inconvenience as it advances. At its commencement, the air is dense and hazy, with long white clouds settling a little below the summits of the The scirocco.

Indications of a scirocco. mountains, and at sea, floating just above the horizon, in a direction parallel to it: it often terminates by a rapid lull, which is succeeded by a north-west breeze. The thermometer does not, at first, experience any very sensible change, though it slowly rises with the continuance of the scirocco to 90° and 95°, which last is the highest I have observed, though the feelings—which are certainly a very inaccurate measure of actual heat—seem to indicate a much higher temperature; but the hygrometer shows increased atmospheric humidity, and the barometer gradually sinks to about 29·60 inches. This wind generally continues three or four days, during which period, such is its influence, that wine cannot be well fined, or meat effectually salted; oil-paint laid on while it continues will seldom take or harden: and while from seeming dryness it rives unseasoned wood, and snaps harp strings, it makes metals oxydize more readily, mildews clothes, and renders everything clammy. We are told, however, that dough can be raised with half the usual quantity of leaven; and though blighting in its general effects during summer, it has been known to favour the corn-harvest, and the growth of several useful herbs and plants in winter.*

Scirocco at Palermo. This wind is peculiarly disagreeable at Palermo, although situated in the north-west part of Sicily; but the plain is surrounded on the land-side by mountains, which collect the solar rays as if to a focus. Although somewhat inured to the heat of the East and West Indies, and the sands of Arabia and Africa, I always felt, during a scirocco here, more incommoded by an oppressive dejection and lassitude than in those countries; and it matters little to the person attacked, whether the sensation is attributable to the immediate parching of the skin and the absorption of his

* The Scirocco appears to be so modified by its transit across the sea, or else is such a contrast to the Bize, that when it visits the shores of Provence, it is welcomed as a very *alizé Méditerranée.*

electricity, or to a positive increase of temperature. At such times the streets of Palermo are silent and deserted, for the natives can scarcely be prevailed upon to move out while it lasts, and they carefully close every window and door of their houses to exclude it. Still the scirocco does not appear to be actively prejudicial to human health, though it is said that, if it be of long continuance, wounds are sometimes attacked with erysipelatous inflammation, and it often is troublesome to people of a plethoric habit. It is more frequent in the spring and autumn than in the summer, and in winter possesses no disagreeable qualities, except to invalids; many persons refuse medicine during its continuance, but whether right or wrong, deponent knoweth not. Queen Caroline of Naples said, in a note to an English lady, that she had risen—*en déshabille*—from a marble floor to write, and must throw herself down again, in order to alleviate the oppression she felt: such was the inconvenience endured even by royalty—by the daughter of Maria Theresa—in the otherwise enchanting valley of the Conca d'Oro. One of our generals held a levée during a scirocco, and however booted and belted the smaller fry came, he himself exhibited the happy ease of undress: and the late Lord Holland, fainting under the oppressive heat, passionately invoked the colder breezes, breaking forth with—

Queen of Naples.

Lord Holland.

Oh! my soul's panting wish in mid-day dreams!
Oh! native soil! Oh! verdure, woods, and streams,
Where are ye? And thou, lovely Redlynch! where
Thy grassy prospects, and thy vernal air?
Oh! send thy spacious waters to my aid,
Lend me thy lofty elms' protecting shade;
Henceforth within thy limits let me live,
Oh England! injured climate! I forgive
Thy spleen-inflicting mists.

Remark.

And, indeed, when the sultry and withering blaze of heat, the earthquakes, hurricanes, diseases, misery, personal insecurity, reptiles, mosquitoes, flies, fleas, and other major and minor evils are recollected, the pleasure of visiting warm climates is considerably alloyed.

Weather in the Adriatic. The navigation of the Adriatic is rather dangerous, except under the careful attention of a good officer, from the liability of being caught without sea-room in extremity. The winds generally draw or incline up and down its length, seldom blowing right athwart; during the summer months they are light and variable, with frequent calms* and occasional squalls, with the usual accompaniments from the northward; these gales are, however, of short continuance. Winds from the south-east bring in a high sea, with fog and rain, but they are usually steady, and not unfrequently succeeded by a fresh north-west breeze. The south-wester, or *Siffanto*, is vehement, but short-lived, and often draws round to the south or south-east, when it is succeeded, near the vicinity of the Po, by the gale and sea called *Furiani*. The entrance of this sea is liable to sudden gusts, which do not always give warning of their approach, and when it continues to blow hard there, the waves are tumbling and confused; subsiding, however, with the weather. Towards the centre of the gulf, the winds are steadier than at the mouth; though at the upper part they are still more variable. From a comparison made by the late Sir William Hoste—who had much experience of this sea, and being one of those useful officers who both fight and write, drew up some excellent sailing directions for navigators—it was found that the ships off the Po, those before Trieste, and those in the Quarnero, had usually different winds at the same times. From the votive offerings of seamen in the churches of the—on this side—'harbourless shore of Italy,' such mutable weather must long have scourged the coasting traders, before the few places of refuge were constructed; for Dante, when in the ninth bolgia of Hell, in alluding to the atrocious throwing overboard of the two citizens of Fano, off

Siffanto.

Furiani.

Conflicting winds.

* The air of the Adriatic in easterly breezes, as in most parts of the Mediterranean, has the pernicious quality of mildewing sails: seamen will therefore be careful to air them in north and west winds.

Cattolica, says it was so managed that against the winds of Mount Focara, it were needless for them to offer up a vow, or to pray— Dante.

> Poi farà sì, ch'al vento di Focara,
> Non farà lor mestier voto, nè preco.

To prevent the mischiefs of private cupidity, the laws of the mediæval times forbade merchant-vessels from putting to sea in the bad season; and so late as 1569, Venice prohibited her vessels, under heavy penalties, from attempting to return home between the 15th of November and the 20th of January. But this was a great improvement in bold navigation, as compared with dicta of the thirteenth century, which assigns the winter to fools only:— Aphorism on the weather.

> Tempo di navigare—d'April dei cominciare:
> E poi securo gire—finche vedrai finire
> Di Settembre lo mese—che l'altro a folli imprese.

Off Croatia, and indeed generally from the Gulf of Trieste to the Mouths of Cattaro, the weather is notoriously unstable; calms, thunder, water-spouts, and the hot wind called *youg* by the Sclavonians, being frequent all the summer; and heavy northerly blasts called *Boras*, the *Sebenzanas* of Dalmatia, with fogs and hard squalls during the winter. Nor are these variations confined to seasons. Obviously Bora appears to be a mere corruption of Boreas, though said to be derived from a Sclavonic term for furious tempest. It is greatly dreaded in the upper part of the Gulf of Venice, particularly in the Canale di Maltempo, and other channels of the Quarnero and Quarnerolo, where it rushes down from the whole line of the Julian Alps with such irresistible fury, that not only numbers of vessels are sacrificed, but it ravages the shore also, being feared as much for the suddenness of its attack as for its violence. From this cause, the emporium of Fiume is nearly confined to a summer intercourse in trade, and the otherwise eligible haven of Porto Ré is useless as a government arsenal; there are also districts which are rendered nearly uninhabitable

The youg. The bora. Effects of the bora.

by it. As the maritime cliffs and surfaces of those shores which are most exposed to the bora are well marked—for not a bush nor a blade of grass can grow on them—the local craft usually anchor opposite the parts where vegetation is most abundant.

Progress of the bora.

The coming on of this wind may fortunately be known some hours beforehand, by a dense dark cloud on the horizon, with light fleecy clouds above it, a rather lurid sky, and it is immediately preceded by a breathless, but speaking stillness. Its general source is between north and north-east, and its most usual continuance about fifteen or twenty hours, with heavy squalls, and terrible thunder, lightning, and rain, at intervals: but the bora most feared, and with justice, is that which blows in sudden gusts for three days, subsides, and then resumes its former force for three days more. Ships caught by it generally let fly everything to receive the first blast; then immediately bear up to the southward to seek safety in any port they can fetch, or remain under bare poles till it is exhausted. We lost many prizes during the late war, by these impetuous winds acting on vessels, the rig of which was new to the young prize-masters, and even some of our cruisers, when caught unawares, have been nearly thrown on their beam-ends. In December, 1811, the French frigate *Flora*, of 44 guns and 340 men, was surprised by a bora, on her passage from Trieste to Venice, which threw her on the coast near Chiozza, where the captain and two-thirds of her people perished; in 1815, two merchantmen, which had anchored off the mole of Trieste with the intention of entering the following morning, were assailed in the night, and foundered with all hands; and in 1820, the *Monte Cuculi*, a fine Austrian corvette of 20 guns, was met by a bora while under all sail, and instantly went down, with the whole of her passengers and crew.

The *Flora*.

The *Monte Cuculi*.

These boras, however, as I have already hinted, give sufficient notice of their approach to an attentive observer;

although *Borinos*, or strong squalls, from the same quarter, of short duration, may sometimes be encountered without much barometrical indication. A very hard summer bora, which I experienced in Lissa harbour, on the 13th of July, 1819, occasioned a fall in the mercury from 30·15 inches to 29·77; it was precursed by the usual denseness near the horizon, with a fresh south-east wind; and during the two preceding nights—although the weather was fine—there was much lightning in a vast cloud-bank which had formed. On the third evening, this bank spread over the sky to the zenith, and the coruscations became incessant; whereupon, as we were lying at single anchor, prepared for going to sea, we dropped the best bower, braced the yards to the wind, and took measures for the safety of our observatory, tents, and instruments on Hoste's Isle: these had been left to the eleventh hour, in order to watch a new and brilliant comet which was then following Capella, and standing towards Dubhe. In the midst of this aërial commotion, at about one in the morning, the gale suddenly chopped round from south-south-east to north-north-east, with such fury, as to make the ship heel over in an extraordinary degree; and the cables were veered out until she was uncomfortably close to the marina. It was fortunate that we were in so excellent a port, for the sudden shift of wind must have done injury to any vessel under sail, however well prepared. In about an hour, the acme of its force somewhat abated, rain fell in large drops, and for two days afterwards we had cool breezes from the north, and clear weather.

Summer bora.

Shortly afterwards, we underwent another of these blasts, of which I particularly noted the advent, progress, and termination. On the 9th of August of the same year, while moored with the stream and small bower in the perfectly land-locked harbour of Lossin Piccolo, the morning was suspiciously cloudy, although the preceding evening had been remarkably clear over-head; insomuch as to allow of my making some satisfactory observations in the observatory-

Lossin Piccolo.

tent, and also showing Saturn's ring to the magnates of the town, it having just then become again visible after its temporary disappearance. On the morning stated, the wind was in the south-west quarter, the clouds lurid, the atmosphere dark, and the whole celestial aspect so singular

Indications.

and threatening, that, notwithstanding our apparent security, I ordered the top-gallant yards and royal-masts on deck, top-gallant masts to be struck, the best bower to be ranged, and the sheet cable bent. In the afternoon, the horizon, from north-west to north, was as black as possible, and the gloominess of its appearance was contrasted by a bed of white fleecy clouds which rose immediately above it, and soared rapidly till they joined a series of waved distinct streaks overhead; forming an immense arch from west-south-west to east-north-east, with a deep blue sky on each side. In a few minutes a strong wind had evidently arisen in the north-west, as it blew the clouds right and left, though we still felt the south-wester even stronger than in the morning.

The bora rages.

The scene was now awfully grand; masses of cloud were in motion from the zenith downwards, excluding by degrees the brassy sky, while a momentary stillness was but a presage of the coming storm. At this time all the fishermen were making for the shore, and the whole marina resounded with the shouts of people endeavouring to rowce up their vessels on the strand. At length huge drops of rain plashed down, and the whole atmosphere seemed to resolve itself into black smoke, while the north wind was seen approaching, by the eddies of sand which it threw up before it. The gust now reached the ship, roaring tremendously, with such force that both our cables were snapped like twine, and before we could bring up with the best bower and sheet anchors, veer to forty fathoms, and brace the yards by—which was effected with a celerity that delighted me—the ship was nearly thrown upon the quay. The rain now poured a deluge, and the apparent mill-pond of a harbour was soon covered with long rolling waves, the crests of which were cut off in foam. Every boat in the port was either swamped or capsized; oars, rudders, and thwarts were floating on every side, and the vessels along the marina were driven one upon the other. Such a gust, if it had continued, must have destroyed the place; but providentially, its excess of violence lasted only a few minutes, and in less than an hour all was restored to comparative tranquillity. Among other disasters, we noticed the destruction of a trabaccolo astern of us; she had escaped the first blast with being merely thrown on the mud, but after she was aground, the rain falling on her cargo of unslacked lime occasioned her conflagration, and loss of sight to some of her crew. The mischief done on shore was much greater than that afloat: numbers of trees were torn up by the roots, the roofs of houses blew away like chaff, windows and doors were forced in, and even floors were displaced by the wind getting into the lower stories.

Effects of the bora.

Accident.

The crews of two of our boats—the gig and cutter, under the charge of the able master, Mr. Elson—which were capsized outside the harbour at the very commencement of the bora, though within a few feet of the land, were obliged to lie along the ground on gaining the shore, and grasp the brushwood while the main force passed over them: the masts, oars, sails, and arms of these boats were lost, together with some of the surveying instruments. In the morning the barometer stood at 30·05, and after the rain at 29·91 inches: this bora, though a summer one, was pronounced to be the severest which had happened in the memory of the 'oldest inhabitant!*

Mr. Elson.

The *Bora* is much modified in the immediate vicinity of

* Captain Cosulich, who published a portulano of this sea—*a spese sue*—in 1848, in speaking of the dangers of this vicinity, advises vessels not to venture into the Quarnero if Mount Velebich should be capped with white clouds; adding, 'Lo parlo per esperienza, perchè nacqui sull' isola Lussini.'

Cattaro and Ragusa, but between those places and Monte Gargano, I have experienced very fresh weather. A curious phenomenon occurs among the cliffs of Montenegro: in the midst of the most steady season of the year, in the finest day, of the purest atmosphere, when not a speck of cloud is perceptible, thunder is heard to roll with loud repercussions among the mountains; and it is remarked, that at these times the springs of the neighbourhood gush up with increased force.

Curious incident.

Ionian winds.

In the Ionian Sea, the prevalent winter winds are from south-south-west to east-south-east, and those of summer from north to east-north-east; but in general, among the islands, rarefaction commences soon after sunrise, and continues to increase with the solar force till noon, during which interval there is not a breath of air in the valleys. About mid-day the rarefied air begins to ascend rapidly, and, agreeably to statics, a cooler and denser air rushes in to supply its place, and restore the equilibrium. Inside the islands the winds are variable to an extreme, insomuch that a ship may be seen coming in at Corfu through the north channel, and another through the south, both before the wind; while in mid-channel it is either calm, or the wind is veering to all points of the compass. That these are mostly mere surface currents of wind, is shown in the fact that the courses may be asleep while the royals are flapping to their masts; and coasters often heel to the breeze, while the citadel flag, about 130 feet above them, hangs motionless on its staff. On the 2nd of August, 1818, after an agreeable interchange of civilities and survey-communications, the French corvette *Chevrette*, commanded by Captain Gauttier, and the *Aid*, got under weigh; myself being bound into the Adriatic, and my friend to the Archipelago, when we both had a fair wind! Eddying breezes often blow violently from the mountains of Epirus, which are at times, from their force and coldness, very unwelcome; and hard gales in the Corfu channel are

The Corfu channel.

occasionally preceded by a fitful roar on the waters—(*spaventosa mugghito*)—which was described to me as really awful by Capt. Kirkness, whose packet, *the Countess of Chichester*, only escaped being wrecked in the south entrance by sound seamanship. Besides this *mugghito*, seamen are warned by another local phenomenon. In the northern part of the Corfu channel rises the steep and rocky Pantokrator, or Table-mountain of Salvatore, marked by a conical summit at each extremity; these are usually enveloped in dense white clouds previous to the approach of bad weather.

Warning noise.

Salvatore.

In the Gulf of Arta, the winds, when regular and not stormy, follow the sun's diurnal course, commencing with light morning airs from the eastward, veering round southerly till about an hour before noon, when a fresh westerly wind sets in, which dies away at sunset; and this is the simple fact of the 'alternating winds' so much marvelled at by travellers. The Gulf of Corinth, as might be expected, is extremely subject to *raffiche*, or sudden squalls from the mountains, which whiten its surface with foam; outside similar gusts blacken the aspect of the waters. The warm and disagreeable easterly wind—called *Vento del Golfo* by the Ionians—commences a little after midnight, and continues till nearly mid-day; the westerly breeze sets in soon after noon, and lasts till nearly midnight; but the Greek pilots say, that from spring to winter, however strong the wind may have blown during the day, it almost constantly moderates at sunset. In the winter the north-east winds are prevalent and strong, especially along the Roumelia shore; and their meeting with the southwesters in the offing, is often the cause of the commotion which affects the Ionian islands, where the descending winds from the hills are sometimes absolutely furious. The *Myrmidon*, commanded by the Hon. Robert Spencer, with Sir Charles Penrose on board, when at anchor in Koinos bay, near Port Bathi in Ithaca, heeled so deeply and so repeatedly

Gulf of Arta.

Wind at Ithaca.

to the blast from the mountains, that the Admiral assured me the wind must have been strong as a West Indian hurricane while it lasted. It is to be regretted that we have not yet arrived at an anemometer for use on sea-board; for till we have such an instrument, force and direction can only be inferred. Nor is it easy to obtain accuracy at such times with respect to the course of the gales, for the air is wafted over the high and precipitous lands of Acarnania, Epirus, and the islands, each giving a direction to it; and this direction naturally varies, according to the angle of incidence corresponding to the surfaces against which they strike in their progress.

Want of exactness.

Morea.

The climate of the Morea differs more with its localities than its area would lead one to suppose; but the aspect of its mountains and valleys, with the varied exposure to sea-winds, accounts for the difference between the amenity of its maritime situations and that of the rugged mountains of Arcadia, where the atmosphere is more keen and cold, besides being occasionally very foggy. The north-east wind is clear and sharp, and is generally attended with fine weather; but at times blows with great violence, and is severely felt as the *Gregale* in Malta.

Electric agency.

The whole of the Ionian Sea is subject to intense lightning, especially in the neighbourhood of Corfu, where the Acroceraunian 'infames scopuli' sufficiently prove the justice of that classic designation. The production of free electricity during the conversion of water into steam is well known to be rapid and abundant: in like manner, while the solar heat is converting into vapour the moisture of the earth, electricity is largely disengaged during the process. Slightly-liberated electricity produces lambent or phosphoric flames, which are unattended with danger; but when an overloaded atmosphere is animated by opposite powers and driven by antagonistic currents, the ferment and explosion of the elements are exhibited in fury, and the coruscations of fierce fluid matter are energetic. These lightnings are of the kinds called sheet and forked, and when vivid are awful as well as beautiful; the first, in noiseless, far-spread forms, momentarily illuminating every object, and then leaving an indescribable gloom. At times the flashes follow each other in such rapid succession as to appear almost incessant; so that a military wag, at a mess-table at which I was sitting, proposed to put out the candles and dine by lightning. This buffoonery would not have deserved repetition, but that it conveys an idea of the powerful glare which must have prevailed to call forth the jest. Lightning at Corfu

It is not uncommon, especially in and near the middle-latitude zone of the Mediterranean, to experience typhoons (τυφῶν), or whirlwinds, of which some of the most obvious instances that have passed under my notice are in the vorticular columns of sand in the deserts of north Africa. From such currents of air rushing through the atmosphere, and along the surface of the sea, with an impetuous spiral rotation, there very frequently result, in the warm months, those extraordinary phenomena somewhat inappropriately named waterspouts, since they are owing to a commotion of rarefied air only: of these syphons I have frequently seen several at once, of various magnitudes, round the ship. In round terms, they may be described as trumpet-shaped cones descending from a dense cloud, with the small end downwards, beneath which the surface of the sea becomes agitated and whirled round, and the water, converted into vapour, ascends with a spiral impulse, till a junction is effected with the cone proceeding from the cloud; frequently, however, they disperse before the union takes place, especially when the action of the winds drives them out of their perpendicular position. There can be little doubt that the Franklinian theory is substantially right, and that, from the vapour being evidently drawn or forced upwards, waterspouts are the consequence of a previous whirlwind; a point cavilled at by some recent lecturers. Must it not be conceded, that, from the equal Typhoons. Waterspouts.

Rotation is necessary. distribution of the atmosphere, it follows that no extraordinary movement can take place in any of its parts, except by means of a positive rotation? Yet a vortex will not be regularly formed nor continue in action, without the aid of an external propelling force, and a constant discharge from that spiral extremity of its axis towards which the motion tends; both of which conditions appear to be fulfilled in the object before us, although the collision of such masses of air may render the effect both excentric and brief. In addition to the operation of wind, atmospheric electricity and its opposite may be also found to exert influence; but Dr. Franklin's argument will here suffice, for the upper air is rarer than at the base, and the syphon itself is mechanically elevated by the centrifugal effect of

Gyrations. its own whirling motion. The gyrations in this sea are thought to be in accordance with the hands of a watch, but their revolving spirally makes this rather difficult to establish; and there may exist a great disparity in their temperature, humidity, and substance.

From the earliest times navigators have always, and very naturally, entertained great apprehensions of this

The Prester. phenomenon, the noted Prester (πρηστὴρ) of the Greeks, the destroyer of those at sea; of which Lucretius (*lib.* vi. *v.* 422, *&c.*) gives so terrific a description. But though most sailors still believe it to be dreadfully dangerous, and small craft have been known to founder immediately on being struck, in most cases it would probably be productive of no serious injury to a vessel of any tolerable size; nor do I believe that a well-authenticated disaster occasioned by these waterspouts to a well-found man-of-war is on record. I had indeed been informed of the staving in of the quarter-deck of an 80-gun ship, the *Tonnant*, and of the expression of Sir

Sir J. Gore. John Gore, that 'for the first time in his life he was alarmed:' but on my asking particulars of that officer, he neither recollected the accident, nor the exclamation—so uncertain is hear-say evidence. Yet careful seamen should avoid this phenomenon, and as it is moved in space by the prevailing wind, which is acting equally on the ship, it may be made to pass, by skilful manœuvre. I think it improbable, however, that, with sails taken in and hatches battened down, the consequences would be very serious to the hull, although from being more active aloft than below, the upper spars might suffer. Still I must own to having felt more comfortable on board an English man-of-war than in a Sicilian gun-boat of paranzello rig, when in presence of these most curious visitors, for whose advent, agreeably to my experience, the barometer does not prepare us.

Precautions necessary.

During the formation of a water-spout, the winds around are generally light and variable, with frequent whirling cat's-paws and calms; but the weather heavy, with clouds of small dimensions and flaky, in very slow progression over a deep blue sky. At length one of them enlarges, takes a position, becomes elongated, and sends forth a syphon, which finally reaches and agitates the sea; but the moment of contact is not readily made out, for the effect is manifest in the ebullition on the surface before the extremity of the spout has visibly approached it. The base of the column, which may be from 50 to even 100 feet in diameter —enclosing a smaller, more transparent, and apparently hollow cylinder—is first seen darkening the agitated area beneath it, as well as a wider circle of deep-blue water around; and afterwards it discharges a volume of vapour upwards, with an audibly whizzing noise, into the column of the protuberant cloud above it—a fact of which Pliny seemed to be aware. The dispersion commences with the vertiginous point, which becomes broken, less defined, and shrinking, as it were, upwards; the syphon often appearing to be suspended to the cloud for some time afterwards; and though other spouts may then be forming, I never noticed the production of a second one from the same cloud. The duration is from two or three to ten minutes, or even longer; and their dispersion is frequently owing to the springing up of a breeze. Such

Formation of a spout.

Sometimes suddenly formed.

is the most frequent line of action; but I have also known them to form suddenly in squally weather, on the chopping round of the wind, or where two winds meet; and they are seen both before and after heavy rains, frequently attended with thunder and lightning. When they appear to be approaching a ship, it is not unusual to fire a gun at them, which, by the concussion of the air, may scatter them; but where the experiment is tried at all, it should be well done, and I have been assured that the vibration caused by firing several guns in a salvo, infallibly makes the column separate and dissipate in heavy rain, accompanied by local lightning and hail. This process I never tried; but on one occasion, off Maretimo, a fine columnar one of 1300 or 1400 feet in height being within a mile was about to be thus operated upon, when it suddenly passed a-head of us while we were gazing in admiration of the magnificent phenomenon. The water which falls on such an occasion is, of course, perfectly fresh, but so instantaneous a chemical process cannot be sufficiently considered. Dante, Camoens, Thomson, and other poets, have described this phenomenon as poets are wont to do; and even Falconer, that truly nautical bard, gives rather a more terrific account of the dispersion of a water-spout than would suit the staid sobriety of prose:

Fresh water.

The horrid apparition still draws nigh,
And white with foam the whirling billows fly.
The guns were primed, the vessel northward veers,
Till her black battery on the column bears:
The nitre fired, and while the dreadful sound
Convulsive shook the slumbering air around,
The watery volume, trembling to the sky,
Burst down a dreadful deluge from on high!
Th' expanding ocean trembled as it fell,
And felt with swift recoil her surges swell!

Evolution of electric fluid.

In furtherance of the cursory allusion I have made to the probability of electric agency as the cause of water-spouts, the reader may be reminded that there is a rapid and profuse evolution of electric fluid in the process of evaporation. The presence of a surcharge of this fluid is established by the great frequency of noiseless sheet-lightning over the surface of the waters, and also by the appearance and play of that lambent flame about the mast-heads of ships, known to seamen as the *compazant* (a corruption of Corpo Santo). It was the Dioscuri of classic times, and its remarkable appearance is noticed by Cæsar (*De Bello Africano*), on which occasion it settled on the points of the spears belonging to the fifth legion. This harmless meteor is also hailed in the Mediterranean with the appellation of the fire of Sant Elmo, or San Pietro and San Niccolo; in either case under similar notions to those which inspired the ancients on the appearance of their Castor and Pollux. It is a beautiful meteor which usually occurs at the close of squally weather, and in nights of intense darkness; it reveals itself as a pale blaze of phosphoric light,* hanging on the trucks in the form of a sea-medusa, to a depth of two or three feet down the mast, with gentle scintillating flittings such as might be represented in shaking a large jelly. Its duration varies from five or six minutes to nearly a quarter of an hour in vigour, when it gradually dies off, and is generally succeeded by fine weather; nor is this so much a matter of marvel as the native pilots wish it to be thought, for if the compazant is the effect of a mild or diluted electric

St. Elmo's fire.

Castor and Pollux.

---

* A curious instance of this meteor occurred in my own knowledge, in the Pacific Ocean, when serving on board the *Cornwallis* frigate, in 1807. We were working out of the Gulf of Panama towards Acapulco, in dark, squally weather, and the log entry runs—'*Tuesday, September 29th*, at sunset it fell calm, with such heavy rain, thunder, and lightning, as are seldom surpassed. The corpo-santo uncommonly vivid.' This, however, is not all; for I well remember the first impression the light gave was that a lanthorn had been taken aloft, but increasing brilliance soon revealed its nature. Meantime a spirited young main-top-man shinned the royal-mast, to break off the spindle round which it was resting on the truck, without any pendant parts. On touching it, the fluid ran down his arm, and from him overboard, and all was instantly pitch dark: he arrived on deck rather terrified, as he told me, from the 'queer numbness' it gave him. It is not a little singular, that forty-five years after its occurrence, the captain of that ship, the present Vice-Admiral Charles James Johnston, should have told this anecdote to some members of my family, at his hospitable mansion near Dumfries, in August, 1852; adding, that its brilliance at one time was so great, 'that they could see each other's faces on deck.'

fluid, it is but natural that the storm which is caused by the same should cease when the electricity becomes no longer visible in its dazzling state. These luminous appearances are esteemed ominous when a single one is seen fleeting down the masts; and this must be the inauspicious flame pointed out by Falconer, who, both a seaman and a poet, thus shows it:—

Ill omen.

> High on the masts with pale and livid rays,
> Amid the gloom portentous meteors blaze.

Ages, however, before Falconer's time, Pliny (*Nat. Hist. lib.* ii. *cap.* xxxvii.) had described these lambent stars, and his description is thus rendered by Philemon Holland:—

> I have seene myselfe in the campe, from the soldiers sentinels in the night watch, the resemblance of lightning to sticke fast upon the speares and pikes set before the rampier. They settle also upon the crosse sail yards and other parts of the ship, as men do saile in the sea, making a kind of vocall sound, leaping to and fro, and shifting their places as birds do which fly from bough to bough. Dangerous they be and unlucky when they come one by one without a companion; and they drown those ships on which they light, and threaten shipwrack, yea, and they set them on fire if haply they fall upon the bottome of the keele (*si in carinæ ima, should have been rendered 'in her hold'*). But if they appear two and two together, they bring comfort with them, and foretell a prosperous course in the voiage, as by whose coming, they say, that dreadfull, cursed, and threatening meteor called Helena (*the single one*) is chased and driven away. And hereupon it is that men assigne this mighty power to Castor and Pollux, and invocate them at sea no lesse than gods.

Helena.

Balls of electric fire.

These, as well as the singular balls of electric fire sometimes seen gliding on the surface of the sea, are classed as *glow discharges*, in contradistinction to the violent form of lightning called the *disruptive discharge.* The fire-balls are mischievous (*see the Philosophical Transactions*, 1750, *for the Montague's case*), but the compazant is deemed harmless. Even now, when there are two or more, for they are not unfrequently at each mast-head, they are hailed with great pleasure both by the local and foreign seamen; more especially when they remain stationary for some time, and then gradually disappear. So favourable a representation of the elegant Ariel was not lost by the master-mind of Shakspeare (*Tempest, Act* 1, *Scene* 2), who, recognising the then popular notions of the 'Fire Spirits' of the storm, makes the active sprite say to Prospero :— Ariel.

> I boarded the king's ship; now on the beak,
> Now in the waist, the deck, in every cabin,
> I flam'd amazement: sometimes, I'd divide,
> And burn in many places; on the topmast,
> The yards and bowsprit, would I flame distinctly.

The Archipelago—or sea of seas as smatterers think its name imports—is perhaps the most interesting spot in the world, to the eye of the poet, the artist, the scholar, and the accomplished tourist; and it has really been the scene of such grand and heart-stirring events, that even denying its inherent claims to regard as the cradle of genius, taste, philosophy, and the arts, it is hardly possible to eschew enthusiasm when writing upon it. The duty, however, of a sailor is merely to treat of it as regarding navigation and climate; and though that restriction will be adhered to, as far as mere classical recollections are concerned, an occasional reference to the early mariners and meteorologists must be made. Archipelago.

The climate of Attica, the diadem of Greece, is in general dry and serene; during the summer months the prevalent winds from north-east to east-north-east, rarely blow hard for more than two or three days; and from thence to the winter, nothing can surpass the delicious temperature of favourable seasons. Here the east wind (ἀπηλιώτης) so detested on other shores, is esteemed brisk, pleasant, and refreshing, both to animal and vegetable life. In the winter the weather may sometimes be sharp, but the severe Bœotian winters of Hesiod, and the ice—κρύσταλλος—of Thucydides (*lib.* 3 § 4, *Platœa*) are not common in that latitude of late, since the thermometer rarely descends to the freezing point. The air of Attica was always esteemed the purest in Greece, and is still the best; and such is its extreme dryness, that Sig. Lusieri, Lord Elgin's artist—whose house was on the site of the Prytaneum—told me that he could leave a sheet of paper on the open ground all Attica.

night, and write or draw upon it on the following morning. This freedom from atmospheric moisture has, no doubt, greatly contributed to the admirable preservation of the Athenian structures. Such is the climate of the country before us; but the neighbouring Ægean sea is broken by so many headlands and isles, that its air is less genial, being liable to sudden squalls, accompanied by rain, thunder, lightning, and hail.

Etesiæ.

In settled weather, the customary Etesian gales, or *meltem* (calm weather) of the Turks, predominate; they blow from the north-east nearly through the summer months, though their constancy is considered certain only for forty days. Being equally dry and wholesome, they attemper the general atmosphere, and relieve the crassitude of the air in the valleys. The name of these winds is derived from ἔτος, year, as they occur annually about the same season, and though from custom it is principally understood to mean the Hellespontic, or north-east wind of the Archipelago, it is not strictly confined to any particular direction, but is frequently applied to such as blow at stated seasons from any point of the compass. The true Etesiæ (ἐτήσιαι αὖραι, i. e. *annual breezes*), however, commence about the middle of July, rising at 9 *a. m.*, and continuing during the day-time only. The direction of this current of air is from north-east to south-west; and it is probably caused by the rarefaction of the atmosphere nearly under the tropic of Cancer, in consequence of the solar heat at that season. From Aristotle and Theophrastus down to Des Cartes and others still more recent, a theory has obtained which amounts to the same thing, namely, that the Etesiæ derive their origin from the melting of the snows and ice of the polar regions, and the consequent southerly elemental rush; assigning as a reason for their blowing strongest in the day, that the snow ceases to melt in the cold of the night. But there is no end of the names by which these winds have been known: from intermitting at night and rising with the

Supposed origin of the Etesiæ.

sun, they were called *venti delicati*, and *venti somniculares*; yet none of them blow exactly from the north. Pliny has pretty well described these breezes, and their *prodromi* (forerunners), the light north-east airs by which they are for eight or ten days preceded : but his speculation thereupon is rather amusing. 'The sun's heat,' he observes, 'being redoubled by that of Sirius, is thought to be attenuated by the Etesiæ, and no winds are more constant, nor keep their times better.' Cicero remarks, that they moderate the violent heat of the weather during the dog-days; and he has been confirmed in the present day by Baron Theotoki, of Corfu. But it should be observed, that in the Gulf of Egina the north-east winds are extremely sultry to the feelings; although in the month of July, I found the range of the thermometer during the day was but between 75° and 86° of Fahrenheit. Here the land-breeze generally begins in the evening, and continues till near seven o'clock on the following morning, when it frequently falls calm till eleven or twelve, and is then succeeded by the sea-breeze.

Pliny's account.

Cicero.

Theotoki.

The north-east and north-west winds blowing almost constantly during the summer, may—*sic parvis componere magna solebam*—be termed the monsoons of the Levant, and to them the Grecian coast owes many of its advantages both of climate and intercourse. With every due respect to the sagacity of the ancients, the cause may be thus approached. When the sun, on advancing towards the north, has begun to rarify the atmosphere of southern Europe, the general Etesiæ of spring commence in the Mediterranean sea; these, as was recorded by the elder meteorologists, blow in Italy during the months of March and April, and were called by the Romans *favonii*. Their influence is at first but slightly felt, but so soon as the earth becomes considerably warmer than the sea, the current of air advances towards the land, and produces the western breezes. In the autumn, the winds alter to variable, sometimes blowing from

Monsoons of the Levant.

Favonii.

Alternating winds. the sea towards the coast, and at other times in a contrary direction, from the sudden alteration in the temperature of the two elements; for as the sun regularly declines towards the equinoctial, the earth, both on the continent of Europe to the northward, and that of Africa to the southward, gradually cools again, subject for some time to slight variations, either on the land or water, which must necessarily produce changeable winds in the Mediterranean, until some weeks after the autumnal equinox. In round terms, we may say for the Archipelago, that north-westerly breezes often usher in fine weather, and are extensively favourable in cooling the air, and dissipating unwholesome moisture; while the contrary may be expected from opposite quarters. The spring winds of record are those which blow in the first days of March, and which, from periodically bringing Ornithii. flights of birds of passage, were termed the *Ornithii;* whence, when the Bœotian in Aristophanes is enumerating the daws, ducks, and coots he has brought to Athens for sale, Dicæpolis exclaims—'Why, you come to market driving all before you, like the bird-storm!'

The regular north-easter, or far-famed Etesian wind, is Venti Stati. one of the *Venti Stati* of Bacon's *Historia Ventorum*, by which he means 'stayed winds,' or such as do not blow alike in Egypt, Greece, and Italy. This wind was thought by the Greeks to draw clouds to it—*cæcias nubes ad se trahere*, whence their proverb compared it to usurers, who by laying out money do swallow it up (idem. *Qualitates et Potestates Ventorum*, § 32). From the descending, rising, Bacon's metaphor. and progressive motions of clouds, Bacon derived his curious but correct dancing metaphor—*cum enim* (the winds) *choreas ducant, ordinem saltationis nosse jucundum fuerit* (idem. *Topica particularia*, § 18). The Etesian winds bear the vapours of the Mediterranean into the Sahara Desert, and are there dissipated; but the south-westers are arrested by the Alps and the Apennines, and robbed of their contents.

A northerly wind suddenly blowing on a summer's day, is held by the Greek boatmen to presage a fine night; whilst on the other hand, as gales in that season are sometimes preceded by a dead and glassy calm, the mariner is warned. Thus in the storm of August, 480 years B.C., which wrecked nearly 500 of the ships of Xerxes, and was otherwise so disastrous to him, we learn that the sea and sky were previously serene; but when the furious levanter (*apeliotes*) came on, his fleet were on a dead and iron-bound lee shore, which would be as trying to laden transports at present as it was to his crowded vessels, although navigation was then so imperfect, that Euripides gave, as an expressive figure—'The Oar, the sovereign of the Seas!' About the time of the solstices, or longest and shortest days, the south-east and south-west winds blow with great force; but the brumal northers are still more dreaded, since they are often accompanied with storms of hail, sleet, and snow, insomuch that the navigation amongst so many islands becomes extremely dangerous to a stranger. During one of these occurrences, in the year 1771, a Russian three-decker, of the noted Orloff's fleet, was driven from her anchorage at 'Psara, and thrown upon the Kalogero rocks, where every man perished; a Turkish 64-gun ship shared a similar fate a few years afterwards, and the disasters to smaller vessels in the north wind are both numerous and distressing. This same *tramontana*, or north wind, is a deviation of the Etesiæ, sometimes blowing with great violence even in the summer months; and though generally held to be an auspicious harbinger of a change for the better, it is mostly cold and injurious to vegetation, obscuring the horizon to a remarkable degree. After its continuance for only a few hours, the mountain summits of Albania and Greece are covered with snow: and the clearing off of the clouds rendering this visible, with strong solar beams and large blue patches appearing in the sky, indicate the moderating of the *tramontana*,—

Fleet of Xerxes.

Northers.

Disasters.

Tramontana.

Though the bold seaman's firmer soul
Views unappall'd the billowy mountains roll,
Yet still along the murky sky,
Anxious he throws th' inquiring eye,
If haply through the gloom that round him low'rs,
Shoots one refulgent ray, prelude of happy hours.

Winter in the Archipelago.

Winter in general is a trying time to the navigator; for the Archipelago is liable to violent gusts of wind, nearly equal to those of a hurricane, though, fortunately for all concerned, more transient. They are perhaps the same formerly dreaded under the name of Schiron: these are not only preceded by an agitated barometer, but often afford a timely warning of their approach by dense lowering clouds, vivid lightning, and crashing peals of thunder. Yet although forewarned, the mariner cannot always reap the full advantage of pre-monition in such a hampered sea, for ships are too often caught where the exercise of nautical skill is paralyzed by their peculiar position. It was thus that the *Phœnix*, a frigate of 36 guns, ably commanded by the late Admiral C. J. Austin, was totally wrecked on the shores of Tchesmè Bay, in February, 1816, the wind for the time blowing a perfect hurricane. The ship's company were all saved: but in another such storm, on the 9th of January, 1826, the loss of life was more severe. It appears that the *Revenge* 74, bearing the flag of Sir H. B. Neale, the *Cambrian* frigate, and the *Algerine* sloop of war, weighed from Garden Bay, at Hydra, at 5*h*. 30*m*. P. M. on that day, with light southerly winds. About three hours afterwards, a gale suddenly arose, after much painfully bright lightning to windward. The ships were standing towards Cape Colonna; but at ten the *Revenge* lowered her topsails to reef them, when, in a furious squall from south-south-west, she carried away her fore and cross-jack yards, split every sail, and was nearly driving on shore. This blast was fatal to the poor *Algerine*, for at that very moment she must have been overpowered by the elements, and foun-

H. M. S. *Phœnix*.

H. M. Sloop *Algerine*.

dered: nor were Commander Wemyss, the officers, the crew, or a vestige of the vessel, ever heard of after.

The south wind, even in summer, is also disagreeable on account of the sudden changes to which it is notoriously liable; and still deserves the description of *pollens fulminibus* given by the ancients, for it is potent in thunder and lightning, though, as recorded by Sophocles, of short duration. I was myself once off Milo, standing for Attica with a leading southerly breeze and fine weather, when unexpectedly the wind shifted smack to the northward in a heavy squall, by which the sea was thrown into an up-and-down agitation, the crests of the old waves being cast over us in foaming spray. As this subsided, the wind with us still at north, a vessel was seen in the east, descending the Archipelago before a brisk easterly breeze. Such baffling instability often keeps the mariner's nerves on the full stretch, and besides the losses, close shaves and touch-and-go incidents are every-day matters hereabouts. Captain John Stewart, of the *Sea-Horse* frigate, another of those useful sailors who could both write and fight, drew up some excellent directions for the navigation of this sea, which have long been in use. This gallant and regretted officer made it a general rule, while cruising in the 'Arches' during the unsettled months, to anchor under the lee of any land when the winds were from the north, since they usually subside so gradually as to afford sufficient time to weigh; whereas those from the southward, by yawing in all directions, or chopping about at once, are not to be trusted with ground-tackle.

A sudden squall.

Captain Stewart's rule.

We have sufficient evidence that the ancients dreaded the stormy season in the Archipelago, and were the prototypes of the Venetians in legislating thereon. It is from their records that the difference between this sea and all others has been known for so many ages; and though the ancients do not seem to have understood how much the changes of weather are affected by the sun's place in the ecliptic,

Stormy season.

Laws for shipping. they took note of the perennial winds. Some of their laws were expressly designed to curtail such litigations as impeded commerce; and since merchant ships kept the sea only between the months *munychion* and *boedromion* (from the beginning of April to the end of September), all such causes were but to be heard during the time those vessels were in port. Corinth was then the emporium of Greece, and the mart of Asia and Europe; the merchandise of Italy, Sicily, and all that was known of the western world, was brought up the Gulf of Corinth to Lechæum, on the north side of the isthmus; and that from the Ægean islands, Asia Minor, Phœnicia, Egypt, and Lybia, to the port of Cenchreæ on the south. In the coasting voyages of those times, Corinth necessarily became the centre of trade; the circumnavigation of the Peloponnesus was considered both tedious and uncertain; and mariners were so little inclined to brave the stormy sea between Crete and Laconia, that a proverb was current, saying that the man who doubled Cape Malea 'should forget all he held dearest in the world.' A notion was entertained that the rising of the star Capella was inauspicious to seamen; and its two dependants, ζ and η, Boötis—the Ἔριφοι, Hædi—were emphatically styled the *horrida et insana sydera.* Arcturus—of ἄρκτου and οὐρὰ (*Bear's tail*)—was also noted by the early seamen for its ungenial influence; and among other prepossessions against it, we learn from Demosthenes that a sum of money

Bottomry. was lent at Athens on bottomry, upon a vessel going to the Thracian Chersonese (*Krim Tartary*) and back, at 22½ per cent. on the voyage out and home; but unless they returned before the rising of Arcturus, 30 per cent. was to be paid. A meteorologist has recently insisted that this star has still a malign influence on the weather, and he quotes Gadbury in proof of the assertion; but the authority of John Gadbury, and the value of the conclusion, deserve exactly the same degree of respect. Nevertheless, the *season* of its rising may support the old prejudice without a reference to astrology; for it is pretty certain, that if a long double stratum of clouds appears just above the horizon at that time, a gale may be expected. Modern seamen may rest assured, that a rising sea, attended by a sinking of the mercury in the tube, is an infallible prognostic of a storm. **Prognostics.**

Some of the Greeks of the present day affect to be wonderfully weather-wise, and give all sorts of gratuitous advice about arrivals, departures, anchorages, and all that. But although navigation took its rise in the Mediterranean, it is there, even now, in comparative infancy; and from its climate, and the ignorance of its seamen, is likely long to remain the theatre of well-inclined but mere fine-weather sailors. On the appearance of foul winds they seek shelter under the lee of some headland or island, or bear up for the nearest port—with too great a deference for the elements to think of contending against them. They study omens of all descriptions, of which I procured a rich assortment from Kampse, a Greek pilot who served me for upwards of three years, and was well versed in such matters. Among the best-established tokens is that derived from the first appearance of the egg-plant (*solanum melongena*), which is believed by the native seamen to be constantly followed by a north-easter of some continuance; and therefore ships bound for the Black Sea sail before this harbinger of foul wind makes its appearance. This, at least, indicates the time of the apprehended change. **Greek practices.** **Egg-plant.**

In summing up this brief sketch of the Ægean winds, it may assist our inquiry to give the more ancient notions on this subject. Homer only mentions the four cardinal winds expressly—viz., ΒΟΡΕΑΣ, ΕΥΡΟΣ, ΝΟΤΟΣ, ΖΕΦΥΡΟΣ—though intermediate ones are inferred: but it must be confessed that the early notions are not clearly expressed, for even the Iliad and the Odyssey are at variance respecting the properties of the gentle Zephyros, while the troublous Euros is sometimes represented as serene, and Achilles is made to invoke Boreas at the funeral pile of Patroclus. **The winds.**

Cardinal points.

Aristotle, Timosthenes, and others, enlarge 'the rose of winds;' but the exact gradation between the above-named points, and the twenty-four of Vitruvius, cannot easily be attained. Fortunately, however, the tower erected by the astronomical architect, Andronicus Cyrrhesthes, at Athens, has survived the storms and revolutions of many ages, and not only gives us the eight points of the compass then recognised, but also the reputed quality of the winds from those quarters in the meridian* of Attica, by express symbols. Now, as the same meteorological causes must have operated through all time, this interesting structure affords an admirable record of ancient observations; and it proves that more than 2000 years ago the characteristics were the same as at present. Indeed, simple but accurate and close observation carried the ancients much further on the road to truth than some moderns admit. Having studied Vitruvius (*lib.* i., *cap.* 6), Stuart and Revett, Choiseul Gouffier, and many other authorities on this head, one of my first visits in Athens, with Signor Lusieri, was to this temple; with which I was delighted, notwithstanding its having been degraded to a tekkiyeh, or chapel for the dances and frenzies of the howling dervishes.

Andronicus Cyrrhestes.

Tekkiyeh.

The Tower of the Winds is an octangular marble edifice, which, in 1820, was in very tolerable preservation, being entire, with the exception of the moveable brazen triton which surmounted it, and pointed with a wand to the quarter from which the wind was blowing. On the upper story of each side of the tower is excellently sculptured a large winged figure in relief; those which represent cold weather are mature old men, full-clothed and bearded, in

Brazen triton.

---

* As the east dial is only the west dial reversed, and as the noon-day line in the south dial is perpendicular to the correspondent hour lines, it is evident that Andronicus sought the true meridian. From inference, this was probably 150 years B.C.; but the silence of Pausanias is unfavourable to the supposition. This author has, however, carefully recorded an altar of the Winds near Sicyon (*Lib.* ii., *Corinthiacs, cap.* xii.), with four caves (βόθρους) or pits, for the purpose of assuaging storms.

a style which the Athenians chose to call barbarian; and the milder winds are personated by youthful figures, more lightly clad. Above them their names appear in uncial characters; and they are divided below by a cornice from large dials constructed and accommodated for each face; those for the verticals of the cardinal points being regular, and their intermediates declining. It appears truly admirable for its object as an indicator of weather and time to the Athenians, though, from its proximity to the Acropolis, it was badly placed for the vane-triton's showing the true line of all the winds, since it could not be free from eddies. Over the door appears Schiron, the representative of north-west winds; he is robust and bearded, with warm robes and boots, and, though mostly a dry wind, to show that he occasionally brings rain, he is scattering water from a vase. Zephyros, the soft and benign western breeze, is a lightly-clad, bare-legged youth, gliding slowly along with a pleasing countenance, and bearing flowers and blossoms somewhat significant of Ζωὴν φέρω (*I bring life*), in allusion to his genial influence in gardens. Boreas, the impersonation of the fierce and piercing north wind, is a bearded old man, warmly clothed, but without a water-vase; and he is so much affected with cold, that he guards his nose and mouth with his mantle—an action which has been mistaken for blowing the flabra, or wreathed conch-shell. Kaïkias, or the north-east wind, which in winter is the coldest in Attica, is represented as an elderly man spilling olives off a charger, to denote his being unfavourable to the fruits of the earth, and especially to olives, with which the plain of Athens abounds: Stuart, however, insists that instead of fruit he is holding hailstones in a shield. Apeliotes, who represents the east wind, is a handsome youth, indicating gentle motion, and bearing various fruits in his mantle, together with a honeycomb and wheat-ears, in token of his being favourable to orchards. Eurus, the south-east wind, so often accompanied by tempestuous weather, is represented as a morose old fellow,

Seasons marked.

Schiron.

Zephyros.

Boreas.

Kaïkias.

Apeliotes.

Eurus.

nearly naked, the agitation of whose drapery implies occasional violence. Libs, the south-west wind and the *traversia* of the Peiræus, a robust, stern-looking man, bearing the aplustre of a ship, which he seems to push before him. The Romans, who usually copied the Greeks, gave dusky pinions to Libs, in allusion to its changeful energies, being by turns hot, cold, dry, rainy, serene, and stormy, insomuch that it was reckoned unfavourable for ships to sail from the Athenian ports while the weather hung in the south-west.

Libs.

Notos.

Notos, the south wind, has a sickly aspect and clouded head, significant of unwholesome heat and dampness; and he is emptying a water-jar, as the dispenser of heavy showers in sultry weather. On the whole, these weather *influences* agree remarkably well with those of the same winds for our own climate.

Weather in the channels.

The winds of the Dardanelles and Bosphorus are, as might be inferred from the land which forms their channels, what are termed up and down, that is north-east and south-west; but sometimes the northerly squalls are troublesome. The weather, however, is mostly delightful, the heat being softened by that silvery mist which blends the features of landscape without concealing them; and in sailing up the Propontis towards Constantinople, prospects of singular and varying beauty open upon the eye of the navigator. But in truth there are very thick and damp fogs at times.

Black Sea terrors.

From the inexperience of the early navigators, and its then alarming distance from their homes, the Black Sea was thus named as expressive of the Cimmerian darkness of its fogs and tempests. But under the Ευφημισμος which flatters the evil genii, and still makes the utterance of the word death a rudeness, the Black Sea was soothingly dubbed the Euxine (*favourable to strangers*), although notoriously treacherous and unsafe—'Quem tenet Euxini mendax cognomine littus.' Modern commerce has changed all this; for though there are sometimes mists of a density sufficient to alarm a Greek sailor, hard storms are rare, and, when they do occur, seldom last more than twelve hours without considerable abatement. During the summer, north winds prevail, and south in the beginning of autumn and spring. Major-General Monteith told me, that at Kalla and Poli, on the east coast of the Black Sea, the hardest gales are almost invariably from the west, throwing up a rise of four feet in the waters along the shores of Mingrelia, and at the same time causing the rivers to overflow their banks on the low grounds of that neighbourhood. Shortly before his arrival there, a Russian transport was driven on shore in a black fog, and sixty lives lost.* Mists.

Dr. E. D. Clarke tells us that during violent east-winds in the Sea of Azof, the water retires in so remarkable a manner, that the people of Taganrók were able to effect a passage on dry land to the opposite coast; a distance of nearly fourteen miles. And he adds, 'but when the wind changes, which it sometimes does very suddenly, the waters return with such rapidity to their wonted bed, that many lives are lost. In this manner also, small vessels are stranded. We saw the wrecks of two which had cast anchor in good soundings near the coast, but were unexpectedly swamped on the sands. The east wind often sets in with great vehemence, and continues for several weeks. They have also frequent gales from the west; but very rarely a wind due north, and hardly ever an instance in which it blows from the south. This last circumstance has been attributed to the mountainous ridge of the Caucasus, which intercepts the wind from that quarter.' (*Travels, part* i. *chapter* xiv.) That accomplished traveller, in allusion to the 21st verse of the fourteenth chapter of Exodus, pronounces this to be a phenomenon 'which offers a very forcible proof of the veracity of the Sacred Scriptures.' But such comparisons may do more harm than

East wind at Taganrók.

Clarke's inference.

* For General Monteith's experiment at Kalla, to determine the height of the Black Sea, vide p. 153.

good: the dry land here is not occasioned by a miracle, for in this instance the sea was not a wall on the right and on the left, nor was there a pillar of fire by night and cloud by day. The doctor, therefore, must merely have intended it as an illustration, not as a proof of the words of Moses.

Variable weather in the Levant.

In the Levant, the temperature of the atmosphere is more variable than that of most other parts of the Mediterranean, as it alters considerably with each fluctuation of the wind: yet along these eastern shores, in common with the neighbouring regions, the *imbatto*, or regular land and sea breeze, prevails in the absence of stronger winds. But at some distance from land these periodical breezes are felt only within a small compass; and, as in the cases already mentioned, it is not uncommon for vessels to sail by each other in different atmospheric currents. Thus I once passed within hail of a ship on the opposite course, yet both of us with flowing sheets before the wind! This sea, as indeed in a degree are all others, is the grand means of softening the temperature of the air; whence every cold and raw gale becomes much milder by passing over it, and hot breezes are reduced to a refreshing temperature by the same process.

Weather at Cyprus.

The island of Cyprus affords an epitome of the usual Levantine weather, as the action of the breezes is confined to a comparatively circumscribed space. In the general progress of its seasons, the heats increase as the summer advances, and would be altogether insupportable were it not for the cooling imbatto, which begins to blow at 8 A.M. the first day of the season, increases as the sun advances till noon, when it gradually declines, and at 3 P.M. entirely ceases. Nothing is more easy to comprehend than the cause and course of this wind: between 8 and 10 A.M. the land is sufficiently heated to rarify the atmosphere over it greatly,—the cool air upon the sea consequently expands and forms a strong current to the land. Towards sunset, the sea being thus heated, something like an equilibrium takes place. About an hour after sunset, the imbatto generally dies away: an almost dead calm ensues, and at about 1 or 2 A.M. a light air springs up from the land, which continues for about an hour after sunrise. But before these winds terminate for the season, they become extremely violent. This imbatto is considered as a sea breeze on the north-west of Cyprus, and a land one on the south-east. The falling of the wind is usually succeeded by moisture, which renders the air somewhat heavy; but it is dissipated in the evening by a breeze springing up daily at that time. In summer this wind blows till four in the morning, in autumn and winter not till day-break, while in spring it does not continue longer than midnight. Those winds which arise in the beginning of summer, cease about the middle of September: and this is the period of the most intense heats, there being no breeze to attenuate them. Fortunately, however, they are not of long duration; and about the middle of October they sensibly decrease, as the atmosphere then begins to be freighted with watery clouds. The north winds, though possessed of some good characteristics, are disagreeable in summer, on account of the injury they inflict on the cotton plants, which are sometimes withered thereby to the very roots; and coming from the high mountains of Asia Minor, they are often very cold. But the principal cause of failure in the crops of Cyprus is drought, for the earth is often parched up—as it were—from the end of April till the middle of October.

Cyprus imbatto.

Other winds.

North winds.

The coast of Syria has, on the whole, a very fine climate, albeit there are a few drawbacks, for while the mountainous districts undergo a tempestuous and gloomy winter, the summer of the plain is oppressively hot. Throughout the year, the winds are considerably influenced at different seasons by the lofty summits of the Taurus and Lebanon, by which their intensity, direction, and force are varied. On the upper portion of this coast, along the flanks of Lebanon, and about the roadstead of Alexandretta, the

Climate of Syria.

sudden gusts of wind descending from the mountains, called *rageas* (ghaziyah), must be looked out for when the peaks are capped with clouds: some of these are exceedingly violent, though transient, and are but little felt at a wide offing where the true wind, which blows over those peaks, is found. The north winds are for the most part dry and salubrious, yet cold and often strong; while the south ones are mild and moist, accompanied by rain; those from the east are laden with mist; and the western, though often stormy, produce clear skies and exhilarating effects. These winds differ essentially according to the position of the ship's station, but they rarely blow very violently without a corresponding effect on the mercury. There is not much thunder, either in summer or winter, and when it does occur it is generally during the rainy season from November till March. The land winds, which in summer are very light, extend but to a short distance, commencing usually towards sunset, and continuing till sunrise; afterwards the sea-breeze commences, and subsides more or less about an hour before sunset, sometimes dropping altogether. But occasionally the sea-winds blow most furiously, and this harbourless coast then becomes a dead and perilous lee-shore. This was very seriously experienced by our squadron under Admiral Sir Robert Stopford, in December, 1840, after his attack on Acre, when the *Zebra* was stranded high and dry, the *Pique* cut away her masts, and various casualties were suffered. On this occasion the *Bellerophon*, a new ship of 80 guns, was obliged to cast some of her guns overboard; and it was only by the able management of Captain C. J. Austin (*see page* 274), and the surprising exertions of her officers and crew, that she was providentially preserved from being cast ashore upon an iron-bound coast, where not a soul could have been saved.

Rageas.

Effects of the winds.

Gale of 1840.

In treating of the meteorology in the Archipelago, certain inferences were confirmed by the statements of the ancient Greeks; but as regards this coast we can appeal to the higher authority of the Sacred Scriptures. Now an opinion has prevailed, that the north winds—which, blowing from the mountains in that direction, must be cold—are the bearers of wet: but this neither agrees with recorded observation, nor with what we read in the Bible. In the book of Proverbs (xxv. 23), Solomon says, most likely at Jerusalem, that 'the north wind driveth away rain:' such, at least, is the authorized version; but it must be admitted that others translated it—'the wind from the unknown land of the north is pregnant with rain.' Be that as it may, and admitting the difference of latitude, with the influence of Mount Lebanon, &c., the effect of this wind, as experienced by the late well-known Consul-General Barker, at Aleppo, is the same as mentioned by Job, near Damascus (xxxvii. 22), probably upwards of 1000 years before Solomon was born—'Fair weather (*gold-beaming clouds*) cometh out of the north.' Again, when Elijah's servant, on being sent the seventh time (1 *Kings*, xviii. 44) to the top of Mount Carmel to look out, reported that he saw a small cloud, 'like a man's hand,' rising from the sea—which, of course, was to the west of him—the prophet instantly predicted rain. A small dark cloud taking the nimbus form, with its rugged pendants resembling fingers, would be in keeping; for it is a natural and very common prognostic, which may be seen from the same spot to this day.

Allusions in the Scriptures.

Solomon.

Job.

Elijah's servant.

The climate of Lower Egypt is very hot in summer, though with cooler nights than could have been expected; with a mean annual temperature of 69°·3. On the coasts of the Delta, occasional rains commence with the fall of the year, and continue till March; during which time the west and south-west gales prevail; and as it then pours down for hours together, the Arabs designate those winds the *Fathers of Rain*. In March, the hot southerly wind called Khamsin (i. e. *Fifty*) commences, blowing two, three, or at most four days successively, and then subsiding only to

Lower Egypt.

begin again soon. Its presence induces disease, and loads the lurid atmosphere with warm vapours, while clouds of dust and small flies are wafted out to sea; but being a land wind, the water is generally smooth, even though it sometimes blows with hurricane force. It derives its name from its supposed limit between Easter and the æstival solstice. It is also called the Samúm (*in Turkish, Sámmyelí*), that is, the poisonous wind, from its suffocating heat. This in the central African deserts is often fatal; but in Egypt and Barbary—though oppressive and troublesome, from filling the air with columns of hot sand, they are not dangerous. I have, indeed, been inconvenienced by them, but never experienced any really ill effect. The heavy, hazy weather continues till the sultry east winds about the beginning of June may be said to usher in the summer, when there is sometimes hardly a breath of air stirring in the day-time, and not a cloud to be seen; but at night the *northers* set in, the surrounding air cools rapidly, and the dew falls densely. About St. John's day (24th of June), westerly and north-west winds refresh the air, and they continue more or less till September, with an atmosphere generally dry and clear. The north wind brings health and enjoyment; and by blowing the laden clouds into and beyond Abyssinia, insures a regular supply for the Nile. These, however, are rather bodies of Mediterranean vapour than clouds, collecting into masses as they advance over the valley and lower ranges of hills to the lofty mountains of Africa, where, being refrigerated and condensed, they fall in periodical rains, and are carried back to their native sea: thus confirming the preacher's *geological* inference —'unto the place from whence the rains come, thither do they return again.' (*Ecclesiastes*, i. 7.) But though the northern winds are welcomed as benefactors, since the Nile is then sluggish from the damming up of its waters, a learned but not scientific writer has presumed that they cause the unhealthy season.

The Samúm.

St. John's day.

Solomon.

Circumspection will generally gain a fore-knowledge of the harder gales of the Egyptian coast, though the oscillations of the mercury are confined to a very limited range. In March, 1822, I observed a slight fall in the barometer, at a moment when the atmosphere had a most suspicious aspect; and from taking advantage of this prognostic, we scarcely strained a rope-yarn, while at the same time the Turkish fleet lost two fine large frigates, three corvettes, and a brig, together with nearly 800 people. As this passed—so to speak—under the eye of the sagacious Mehemet Ali, he made numerous inquiries, in the course of which I was able to impress him with the use and importance of the marine barometer: little could I then anticipate the extraordinary fleet he was so soon to build and equip! **Barometer indications.**

Between the Delta of Egypt and the Lesser Syrtis, the sea winds from west, round by the north to east, are frequently violent and sudden, of which I have already recorded an instance respecting Lord Exmouth's squadron being caught, at page 90: but the weather in general is very fine, the summer heats being moderated by breezes from the offing along the coast, and the winters are remarkably mild. The nature and direction of the local winds may be tolerably well inferred, by an attentive meteorologist, in watching the form and colour of the clouds; those hot ones from the south often assuming the tint of the desert below them, as is especially seen at the back of Tripoli, from the offing. Their apparent changeableness has method and regularity, and even with the Mantuan's 'omnia ventorum concurre prælia vidi' in mind, however short their revolutions may be, we cannot but be struck with the constant periodical return of each cardinal wind, and its appropriation to certain seasons of the year, under solar influence. Thus, when the sun approaches the tropic of Cancer, the winds from the east change to the north, and become pretty constant to that direction through the summer: and towards the end of September, when that lumi- **Coast of Lybia.** **Regular alternations.**

nary repasses the line, the winds return to their eastern quarters. When the sun approaches the tropic of Capricorn, the winds become more variable and tempestuous, and frequently blow very hard from north-west and west: and when he returns towards the equator about the end of February and March, southerly winds may be expected. This periodical constancy and atmospheric circulation must have attracted notice from the earliest times; and the Son

Solomon. of David is borne out in saying—'The wind goeth toward the south, and turneth about unto the north; it whirleth about continually, and the wind returneth again according to his circuits.' (*Ecclesiastes*, i. 6.)

Greater Syrtis. The once-dreaded Lybian Gulf must not be passed without being noticed, as the seat of our earliest tales about whirlwinds, whirlpools, quicksands, vapours, and all possible marine perils: to say nothing of the monsters and spectral apparitions so learnedly discussed by Diodorus Siculus (*Lib.* iii. *cap.* 3). All these, however, save fogs and the surges occasioned by northern winds of long range blowing home on the coast, have disappeared: and to such natural

Saráb. phenomena must be added the Saráb, which Europeans now call *mirage*, a singular effect of unusual refraction so frequently seen in this and other arid shores of the Mediterranean, as well as elsewhere. This *deceptio visus* is the 'parched ground (*sultry vapour?*) which shall be-

Isaiah. come a pool' of Isaiah (xxxv. 7); the Saráb (*vapour of the desert*) which Mahomet says 'the thirsty traveller thinketh to be water, until, when he cometh thereto, he findeth it to be nothing!' (*Koran, chapter* xxiv.); and the deceitful sea of the desert of Sogdiana, described by Quintus Curtius (*lib.* vii. *cap.* 5); of which, probably, our loomers, flying Dutchmen, Capes Flyaway, and other deceptions from vertical or lateral refraction, are mere modifications of the action of the sun and earth on the different densities of the lower atmospheric strata. When the sun has heated the sandy plains, and by reverberation the air above them, the clear cerulean sky is inverted by the mirage into an extensive sheet of translucent water, in which the eminences and objects around are reflected, and of course reversed as they would be on the surface of a lake. On one occasion, near the west side of the Syrtis, the illusion which I witnessed was so perfect, that it was with difficulty I could persuade Mr. Edward Tyndale—whose extreme thirst made him long to reach the water—that the supposed lake was receding from us as we advanced, until our amused Arab companions pointed to another saráb formed in the space over which we had ridden. Another which I saw in Egypt was so distinct, and the desolate sands rendered so enticing and picturesque, as to make me for a moment doubt whether I was in my right senses. Instance of illusion.

Mirage is not confined to the arid wastes of north Africa: the temperature of the Mediterranean is of course modified and affected by the wind, while the refractive power of the atmosphere as naturally varies with its density, and its density with its temperature; but these again are strongly modified by the sun-burnt wastes adjacent. Hence the effect is carried to a certain height, and is productive of strong looming, so that places are sometimes seen which are otherwise generally concealed from the view of each other by the convexity of the globe. In my account of Sardinia (*page* 80), I mentioned the appearance of the mirage over the plain of Campidano; and I also saw it most distinctly in the neighbourhood of Manfredonia, as well as on the plain of the Bojana, in the Adriatic. But the most remarkable effect of irregular refraction recorded, is the celebrated aërial display in the Faro of Messina, which has for ages astonished the million, and perplexed philosophers. It is called *Fata Morgana*, from its being supposed to be a spectacle under the influence of a Fairy Queen, the 'Morgian la Fay' of popular legends. It is said to occur in sultry, calm weather, when the tides, or streamed-up waters, are at their highest, and when the sun shines from that point whence its inci- Other sites of mirage. Fata Morgana.

dent rays form an angle of about 45° on the water. At such times, they tell us, multiplied images of all the objects existing on the two lines of coast—as castles, arches, towers, houses, trees, animals, and mountains—are presented in the air with wonderful precision and magnificence.

Padre Minasi's story.

Padre Minasi assures us that, in addition to obvious appearances, numberless series of pilasters, superb palaces with balconies, armies of men on foot and horseback, and many other strange figures, are seen in their natural colours and proper action, as in a catoptric theatre; and there exist paintings and engravings of the wonderful phenomenon. Still, on the whole, I cannot but repeat the conviction to which inquiry led me, and which I published as far back as 1824 (*Sicily and its Islands, page* 109):—'I much doubt, however, the accuracy of the descriptions I have heard and read, as I cannot help thinking that the imagination strongly assists these dioptric appearances, having never met with a Sicilian who had actually seen anything more than the loom or mirage, consequent on a peculiar state of the atmosphere; but which, I must say, I have here observed many times to be unusually strong.'

Fogs.

Yet though the Gulf of Syrtis is now free from spectral illusions—fogs, mists, and sea-frets are still to be met with; and the accumulation of vapours is sometimes so great as to obscure the solar rays, a time when—as the poor Roman Campanians have it—*il sole si vede, e non si vede,* while the face of the luminary, in revealing itself, has the rusty-iron tint alluded to by Virgil.

Cum caput obscurâ nitidum ferrugine texit.

These fogs, in general, are unlike the damp mists of the north, being the dry, thick haze of which the air is full in the warm season, in most parts of the Mediterranean, occasioning little inconvenience or depression of spirits. Indeed, it would be more propitious to vegetation around, if the vapours so frequently seen dissipating in the lower regions of air, were more frequently condensed and precipitated in showers. There are meteorologists, however, who insist that the production of winds depends chiefly on the condensation of vapours; and that the direction of any wind is according to the situation of the condensing vapour; while its strength is as the velocity of such condensed vapour, and the quickness of its condensation. The course being thus indicated, M. Mariotte thinks the intensity may be brought under mechanical computation; for wind being only air in motion, and air a fluid subject to the laws of other fluids, an investigation of the ratio of specific gravities, times, and impulsions, will give the force. This is a conclusive Q. E. D. to some inquirers; but the process speaks more for the soundness of the formula than for the possibility of obtaining the data. When the leading phenomena respecting the distribution of heat, and the distribution and effects of so rare and expansible a body as vapour in the atmosphere, shall be better known, the inferences may become infallible. **Mariotte.**

It must not, however, be supposed that this sea is without thick humid fogs, as well as the dry ones here mentioned:* some of them have singular refractive powers where, from the nature of the country, sudden cold is induced by changes of wind; the specific gravity of the air being increased, and its ascent thereby retarded, it becomes a dense medium both to sight and sound. I once witnessed a curious effect of fog-looming at Scoglietti, on the south coast of Sicily. I was pulling on shore, where some of the inhabitants of that little port and Captain Henryson, R. E., were standing on the beach to wait for my landing. As we approached, the group appeared like a barrack, which gradually split into vertical portions as we advanced, and on approaching still nearer, separated more and more, until on our arrival they became palpably men. On another occasion, in May, 1812, when off Majorca—an island not at all **Humid fogs.** **Fog-looming.**

* Every navigator of the shores of Venice, where the chief land-marks are *campanili*, or steeples, must recollect how they are often vexatiously hidden by fogs.

Fog off Majorca.

subject to fogs—in a line-of-battle ship, and during an impervious haze, with the wind easterly, we all at once plainly heard human voices; this was partly owing to the power which fog has of transmitting and conducting sound, for the people proved to be further from us than we apprehended. At length we saw the mast-heads of several vessels, and shortly afterwards discovered their hulls, magnified by the medium into those of two-deckers. Aware that we had no such ships in this direction, we beat to quarters, cleared for action, and stood for the nearest. As the atmosphere suddenly cleared off, we found ourselves in the midst of an Algerine squadron of two frigates, two brigs, and two corvettes, under the command of Omar Bey, afterwards Dey of Algiers, when that city was attacked by Lord Exmouth.

Dew.

There is yet a point in Mediterranean meteorology which must be named, because greatly misunderstood: viz., Dew; another visible evidence of the aqueous vapour pervading the atmosphere. Those who talk of heavy dews 'falling,' and suppose that they may be deemed a kind of rain, think they might be allowed for as a shower; but would they recollect that an inch of water over an English acre is about 100 tons, and the dew mostly a humefaction, however copious the depositions may occasionally be, they would perceive the extreme difficulty of approaching such a question under such varied hygrometrical conditions and frigorific impressions. But in a more practical view, Dew is a standard weather-predictor. Entirely distinct from the evaporation which we have already treated, the next evaporating power will be as the difference between that force of vapour answering to the temperature at which dew would begin to act, and the temperature to which the evaporating substance is exposed; and this is called the *Dew-Point*. Now any sudden change in the dew-point is accompanied by a change of wind. Professor Daniell says, 'an increasing difference between the temperature of the air and the temperature of the point of condensation, ac-

Professor Daniell.

companied by a fall of the latter, is a sure prognostication of fine weather; while diminished heat and a rising dew-point infallibly portend a rainy season.' This is obviously correct, for it is from the latent caloric contained in vapour that the force of wind is derived; whence it follows, that when the dew-point is high, there is sufficient *steam-power* in the air to produce a violent gale, since then the quantity of vapour in the air is greatest. The hygrometer which I used was, as already stated, one of De Luc's construction; Lieut. Beechey employed Leslie's, but its rapid consumption of ether in that climate was a serious objection. Daniell's (*see page* 215) *wet-and-dry bulb* hygrometer had not yet made its appearance; but I cannot resist pointing out the passage in Pliny (*Nat. Hist., lib.* xviii., *cap.* 35), translated by Holland, which led to that ingenious invention:— The dew-point. Pliny.

And to conclude and make an end at once of this discourse, whensoever you see at any feast the dishes and platters wherein your meat is served up to the bourd, sweat or stand of a dew, and leaving that sweat which is resolved from them either upon dresser, cupbourd, or table, be assured that it is a token of terrible tempests approching.

I greatly regret that I was then unaware of another wonderful link in the chain of meteorological knowledge, which I could several times have contributed to unravel, by collecting specimens. In my account of Sicily and its Islands (*page* 6), I mentioned that on the 14th of March, 1814, on a warm hazy day, thermometer $63\frac{1}{4}°$, and barometer 29·43 inches, it rained in large muddy drops, which deposited a very minute sand, of a yellow red colour. Since this record was published, similar *dust-rain, blood-rain*, or *scirocco-dust* has attracted philosophical inquiry; and the crowning of the beautiful theory of atmospheric circulation only awaits the obtaining and examination of additional samples. By the zealous exertions of Professor Ehrenberg, the revealment of a truly wondrous and invisible working and vitality in myriads of infusoria pervading the atmosphere, has followed the microscopic scrutiny of this dust. Among the organisms, the Professor has recognised poly- Scirocco-dust. Ehrenberg. Organisms.

gastrica, phytolitharia, and many varieties of siliceous-shelled infusoria, which minimum types of life constitute, perhaps, so large a proportion as one-fifth of the whole quantity examined. What cyclical relation these creatures have in regard to different atmospheric strata, still remains for continued inquiry; but it is ascertained that they float in the air together with masses of fixed terrestrial matter, as flint-earths, chalk, and ferruginous oxides! It has also been found that the Mediterranean dust, and that of the Atlantic, possess a striking similarity of organic composition; and by a chemical analysis of the latter, recently made at New York, by Mr. W. Gibbs, it appears, that with 100 as unity, there were—

Analysis of the dust.

	Parts.
Water and organic matter	18·53
Flinty earths	37·13
Clayey earths	16·74
Iron oxide	7·65
Oxide of Manganese	3·44
Carbonic acid chalk earth	9·59
Talc earth	1·80
Alcali	2·97
Natron	1·90
Oxide of Copper	0·25
Total	100·00

A word to remind.

The main aim of these pages is to awaken the intelligent mariner's attention; but lest the general reader should be alarmed about the squalls, and fogs, and compounded atmosphere here necessarily enumerated, we may assure him—without a reminder that sudden transparency is ominous—that he will otherwise meet with brilliant and diaphanous skies. The atmosphere, for the greater part of the year, is so clear, that it gives brilliancy and life to everything in view; and the evening tints at such times are equally marvellous and delicious. Most of the Mediterranean shores, in the summer months, are subject to a whitish vapour in the sky, softening to a silvery haze, and forming a medium through which all objects present both delicate colours and picturesque appearances; and sometimes with the singular property of making headlands, edifices, and mountains seem more elevated than they really are; and this aërial translucence, when influencing highly rarefied moisture, is the reason why distant objects appear to be much nearer in fine weather just before the approach of rain. It was in weather of this description that, in September, 1822, while at anchor among the Tremiti Isles, in the Adriatic, we were enabled to see that singular effect of solar atmosphere, the zodiacal light, with striking distinctness, presenting a sloping, luminous pyramid upwards of 20° above the horizon, 8° or 10° wide at the base; and in such a state of the air, I have frequently enjoyed glorious views of finely-coloured double stars—as α Herculis, γ Andromedæ, and ε Boötis; and on one occasion the cluster in the sword-handle of Perseus was surpassingly gorgeous. A good index of atmospheric modification is found at Malta, which generally affords only a sea-horizon around; but in some states of the weather, the summit of Mount Etna becomes distinctly visible, although it is 110 miles distant, and once I must really have seen half of it. The 31st of January, 1822, was a wonderfully clear day, and that grand volcano so obtrusively perceptible to the naked eye, that I took its bearing by an azimuth compass from the tower of the palace, when the rhomb was exactly N. 27° 12′ E.; and it formed, from the same place, an angle of 110° 31′ with Civita Vecchia church.

Aërial effect.

Zodiacal light.

Mount Etna.

The climate of Tunis is one of the finest in the world, and its air is pure, serene, and wholesome; the thermometer ranging, in general seasons, from about 45° to 87°, with an average mean temperature of 68·5°; and all the revolutions of the weather, with rare exceptions, are between 29·10 and 30·30 inches. During the summer and early autumn, rain is unusual, but it is looked for towards the middle of October; and should it not fall till later in the year, a scanty following harvest is predicted. After the rains have com-

Tunisian climate.

menced, they continue with great violence for eight or ten days, when hunters for antiquities repair to the many neighbouring ruins in search of coins and other antiques, laid bare by the showers. From thence to the spring, a fine period for Europeans generally ensues, for the winter —perhaps improperly so called—can only include the months of December and January, during which fresh winds and heavy rain render the air chilly and raw. The spring is warm, but hot weather sets in towards the middle of June, and lasts in its fervour till September; the coasts, however, are attempered by a constant sea-breeze, which blows from about 9 A.M. to near sunset. Notwithstanding this corrective, the land-winds are almost insupportably sultry, and bring with them clouds of fine sand, which darken the air, and penetrate into every recess. During a scorching sirocco, in July, 1822—in which the thermometer rose to 93° in the afternoon, and fell only to 84° in the night—one of my seamen, a fine youth, employed on the Lake of Tunis, was overcome, and fell a corpse in the boat.* Navigators making landfalls hereabout in the winter should be sure of their reckonings in foggy weather, for there is not much sea-room. In July, 1797, the *Aigle*, a 36-gun frigate, commanded by the late Sir Charles Tyler, ran upon Zembra Island, and was totally wrecked. In February, 1808, the *Hirondelle*, a cutter of 14 guns, was lost on this coast, and only four men saved out of a crew of fifty; and on the 7th and 8th of March, 1821, a heavy gale, which ravaged most of the Mediterranean shores, was so vehement in Tunis Bay, that three frigates, three cor-

Tunisian seasons.

Scirocco at Tunis.

H.M.S. *Aigle*.

Hirondelle.

* Cloudless skies for weeks together, are wearisome enough, and the trite exclamation of the late Captain Fothergill, may be in point. This eccentric officer was returning from India, where he had served for years: coming on deck, when entering the English Channel in a foggy November morning, 'Hah,' said he to the lieutenant of the watch, 'this is what I call something like—none of your cursed eternal blue skies here—a fellow can see his own breath now!' The seasons of Tunis, as above stated, differ from those of Labrador; for according to a remark of the gallant Benbow, still preserved at the Admiralty, he tersely recorded—'There is a winter of nine months, and d——d bad weather the other three!'

vettes, two brigs, and a schooner of war, with about twenty sail of merchantmen, were wrecked, and more than 1800 men were drowned. (*See page* 92.) **Tunisian loss.**

Off the hills of the Ras Sebah Rus (*Seven Capes*), the headland so greatly dreaded by trading vessels, violent gusts are occasionally felt; but their approach may be inferred by the descent of light airs in fine weather, shown in the little playing eddies termed cat's paws. **Mountain gusts.**

Algeria has a fine climate and salubrious atmosphere, the winters being mild, and the summers, excepting for an occasional scorch from the Desert, far from insupportably hot; insomuch that Dr. Shaw, in the account of his residence there, has said that he found the thermometer contracted to the freezing point only twice in twelve years, and then under very unusual circumstances. And he adds, what is partly confirmed by my own experience, and partly by information which I collected, that the winds from the east are common from May to September; and that then the westerly winds take place, and become the most frequent. Sometimes, also, particularly about the equinoxes, they exert the force and impetuosity which the ancients have ascribed to the *Africus* or south-west wind, here called *Labbetch* (*Libeccio*). 'The winds from the west, the north-west, and north,' he continues, 'are attended with fair weather in summer, and with rain in winter. But the easterly winds (*Levanters*), no less than the southerly, are, for the most part, dry, though accompanied in most seasons with a thick and cloudy atmosphere. The barometer rises to $30\frac{2}{10}$ or $30\frac{3}{10}$ inches with a northerly wind, though it be attended with the greatest rains and tempests. But there is nothing constant or regular in easterly or westerly winds; though, for three or four months together, in the summer, whether the winds are from one or the other quarter, the mercury stands at about 30 inches without the least variation. With the hot southerly winds, it is rarely found higher than $29\frac{2}{10}$, which is, also, the **Seasons at Algiers.** **Africus.** **Levanters.**

Usual indications. ordinary height in stormy, wet weather from the west.' These remarks are substantially correct; but a little closer attention would perhaps have shown the observant Doctor that the barometer generally rises with winds from the north to the east, and falls with those from the contrary points; and that though the range is confined within a few lines, the indications are evident.

Winds. Although pregnant with salubrity to the coast inhabitants, the most troublesome winds to seamen are those from north-north-east, north, and north-north-west, which, however, are preceded by an on-shore swell two or three days beforehand. Yet, on the Christmas-eve of 1797, the H.M.S. *Hamadryad.* *Hamadryad* frigate was surprised by a norther, and literally blown on the beach of Algiers Bay, where she was totally wrecked. On the 15th of September, 1823, the H.M.S. *Adventure.* *Adventure* was at anchor off the lighthouse of that place, rolling prodigiously to the precursing swell, with an overcast sky, and a tremulous barometer. Knowing it was a time to expect hard weather, I asked for a berth within the mole; but finding the Dey disinclined to accede, the anchors were weighed, and we clawed off only just before a boisterous storm set in, when, as Mr. M'Donnel, our Consul-General, informed me, eight vessels were lost. Frequency of wrecks. Indeed, so frequently are fragments of wreck strewed about the several strands, that a gale from the northward is termed 'the Majorca Carpenter,' in allusion to the direction of that island from Algiers.

Charles V. While on this topic, it is advisable to notice the disaster which befel that proud emperor, Charles V., who here received a humiliation that must have taught him the vanity of human greatness: and as the principles of meteorology continue the same in all ages, the instance will still be an example in point. The successes of the Algerine corsairs, and their descents even on the coast of Italy, so alarmed and vexed Pope Paul III., that he earnestly solicited that potent monarch to gird up his loins against those audacious infidels. The appeal was not made in vain; for besides being elated on one side by his victories at Tunis, Charles was nettled on the other by the loss of his fortress before Algiers, by the indignities heaped upon his governor there, and by the many aggressions committed against his subjects. A tremendous armada was equipped, which he determined to command in person; and that nothing should be wanting to stimulate zeal, and render the enterprise both powerful and successful, the pope published a bull, promising a plenary absolution to all such persons as should embark, and a crown of martyrdom to all who should fall in the conflict. No fewer than 500 bottoms of all sorts, including 120 men-of-war, and 20 of the largest imperial galleys, were quickly fitted out, and, besides the numerous crews, 30,000 choice troops were put on board. In addition to the regular forces, numbers of the nobility, knights of Malta, and gentry flocked to the standard, among whom were some Englishmen, at their own expense; and so great was the general confidence, that many ladies also embarked. This mighty fleet, conducted by the famous Andrea Doria, cast anchor in the Bay of Algiers on the 26th of October, 1541, which was about three months too late in the year; for the general depth and exposure of the bight between the capes of Temedfús and Al-Kanátir, render it at all times liable to the rolling swells just mentioned, and in the winter season it is ever notoriously unsafe.

Spanish expedition.

Its strength.

A. Doria.

The arrival of such a force threw the corsairs into the utmost consternation, inasmuch as their best men were at that moment dispersed in the provinces to collect the annual tribute. In this dilemma, the Dey behaved with singular judgment and resolution; and being duly summoned by the emperor's herald to surrender, with a promise of many favours if he consented, replied with some humour, that 'he should take the man for a madman who would follow the advice of an enemy.' Meantime, Don Carlos had

Conduct of the invaded.

The landing. already experienced the inconvenience of the bay, in being obliged to disembark his troops through a heavy surf, the roughness of which compelled the men to wade ashore, and precluded the landing of tents and necessaries, while frequent falls of rain rendered their situation most comfortless. However, the general spirit was excellent; each individual did his best; the heights were gained, and the imperial pavillion was pitched on the eminence above the city, on the spot still called the Emperor's Castle. Here they maintained an encampment, though furiously assaulted

The sortie. by a sortie of the besieged, until their matches were extinguished and their powder damped by heavy rains. Now the weather must have presaged a storm from the time of the armada's bringing to; and had that simple monitor, the marine barometer, been then in use, the dreadful calamity which ensued might have been avoided, on a thirty or forty hours' notice. The observant Doria had, indeed, apprehended the mischief, from various natural indications, and warned his imperial master; yet, having no positive data to adduce, such as would have been afforded by the mercury, the emperor perhaps hoped—and the wish 'was father to the hope'—that the fresh breeze then experienced was already at its maximum. However, on the night of the 28th of October, after the repulse of a

The storm rages. sanguinary sortie had fatigued the whole camp, the gale increased to a furious hurricane from the north, accompanied by deluges of rain, which threw the unsheltered Christians into the greatest distress, and destroyed almost all their ammunition and provisions. As daylight advanced, a horrible scene opened upon their eyes. The ships in the bay, on which their safety and subsistence depended, were

Its ravages. most of them driven from their moorings and bilged; and both sea and coast were covered with broken wrecks, spars, goods, and drowned bodies. Thousands of Moors and Arabs of both sexes, beholding this destruction, rushed to the sea-side, stripped naked those who gained the shore, and then speared them without mercy. The number of square-rigged vessels alone which perished during that dismal night, was not fewer than 140; and many of those that rode till the morning, fearful of foundering at their anchors, as the storm still raged and the sea rolled home, slipped and ran aground on the sand between Temedfús and the Wad Haréj, thinking at least to save their lives: but, as soon as the wet and weary multitudes landed, they were inhumanly butchered, being unable to make any resistance Scarcely more than one-third of the armament escaped.* Loss of ships.

Morocco is necessarily very warm, but not so much as might be expected from its geographical situation; the interior being cooled by the mountain winds. The coast experiences the alternations of land and sea breezes, while the climate is at once mild and healthy. The seasons are divided into the dry and the wet, the latter generally being from November till March. From Algiers along the coast of Morocco, to the Strait of Gibraltar, the winds continue to follow, in great measure, the direction of the coast; being generally from west-south-west round by north to the east; the former being most prevalent in winter, and the latter in summer. Excepting as a land breeze near the shore, the south wind seldom blows steadily; though it is occasionally both hot and violent, raising the thermometer several degrees, and forming a marvellous contrast in its effect on the spirits between a souther and a north-west wind. Between Melîlah and Ceuta, vessels must not be caught in the bad seasons by a north-easter, which is apt to rise suddenly, and with a high sea. Breezes from the east often draw round to the south, and are sometimes—especially in the autumnal months—immediately followed by a west wind: the westerly winds, if light, are accompanied by fine clear weather, as before described; but when strong,

Climate of Morocco.

Leading winds.

Weather indications.

* In this gale, the sanguinary Hernando Cortez lost all the matchless jewels with which he hoped to have bought a return to the Emperor's favour. It is an ill wind that does no good!

they are cloudy, with a high sea; and if in winter they veer to the north, accompanied by a swell from that quarter, a brisk gale may be looked for. The weather is treacherous in the winter season, and should therefore be watched: in February, 1799, two vessels of war belonging to the Báshá of Tripoli were wrecked in Tetuan bay, when such was the driving sea, that only twenty-one men were saved of the two crews; and in November, 1801, the *Utile*, sloop-of-war of 14 guns, commanded by Captain Canes, foundered in a heavy storm, on her passage from Gibraltar to Malta, when all her crew and passengers were lost.

H.M.S. *Utile.*

---

§ 3. Damage by Lightning.

THUS far on Mediterranean weather, the various branches of which are all and severally exhibited therein under energy and effect. But of all the detriments to Britain's bulwarks and maritime life, none is more dreadful, when the sudden juncture breaks upon us, than lightning. By this term seamen do not mean those lambent displays of electricity which appear in sheets or balls, and are unattended with danger: they emphatically apply it to the full development of opposite electricities in commotion, with unbalanced fury flying at beast, man, tree, tower, and ship. Yet by the aid of experimental philosophy, this mighty and subtle agent is now all but reduced to the careful seaman's command; for though it would be desirable to avoid the contact of electric fluid under any circumstances, its powers can be regulated and restrained in their devious course by metallic conductors. This, indeed, forms one of the proudest mental and practical feats of comparatively our own times: and the Roman emperor who proclaimed a reward to the inventor of a new pleasure, should have been by the side of Franklin when he first enjoyed the gratification of drawing the lightning from the clouds, in order to witness the disparity between material and intellectual enjoyment. But

Electric discharges.

Franklin.

how would Franklin's triumph have been enhanced, could he have dreamt of the almost countless wonders to which he opened the way, as now performed by galvanism and all the various branches of amenable electricity—to measuring time, annihilating distance, and making lightning convey our very words through sea and over land from one end of the world to the other.

Electric telegraph.

Since my return to England, this question has attracted the strict attention of Government, and a Committee was appointed by the Admiralty to inquire into it, under parliamentary authority. On this occasion, I was applied to officially as to certain rumours they had heard. My reply was printed in the House of Commons' Report; but as seamen are not often in the habit of consulting those costly blue folios, I shall here insert it:—

A Committee appointed.

Bedford, 7th June, 1839.

Sir,—In answer to your letter of the 5th instant, I beg to state, that about the end of September, 1824, writing from memory, His Majesty's ships *Phaeton* and *Adventure* were moored inside the mole at Gibraltar, when a violent thunder storm took place. I was writing in my cabin in the evening, but was interrupted by a startling crash, followed by a cry of 'The *Phaeton's* on fire.' I instantly ran upon deck, turned up the hands, veered away upon the fasts, hove-in the bower-cables, and manned the boats; but the flames were quickly extinguished, principally, Captain Sturt told me, by the cool exertions of one of his men, who was therefore expressly recommended. Her foremast, I understood, was rent from the truck to the deck, some sails and rigging were set on fire, and several seamen struck down.

H.M. ships *Phaeton* and *Adventure.*

On this occasion, the *Adventure's* conductor was rigged, but the *Phaeton* was unprotected. The vessels were about a cable's length apart, and they were the only ships there. Many of my people felt a kind of electric shock more than once that night, but we did not sustain the slightest damage.

My own opinion of the conducting power of metallic wires, and therefore the vast utility of lightning conductors, indifferent as their construction and adaptation seemed to be, was very strong in their favour; and I have laboured hard to propagate this feeling, in opposition to the notion of their being dangerous, from attracting the lightning; an opinion which cannot but be deemed absurd, since it infers that the masts, and not the ship, form a point in the electrified surface. Indeed, it would be a comfort to the service, as well as an amazing saving in spars, canvas, and gear, were the laws and indications of meteorology more strictly attended to.

Conductors.

During many years passed at sea, I had known of several disasters occasioned by lightning, and also of various ships being struck, and escaping destruction as if by a miracle. This led me so to consider the subject, that, in my written orders, the officer of the watch was directed, whenever the

Precautions adopted.

weather appeared threatening, whether at sea or in port, to hoist the conductor, which was kept, not in a store-room, but in a box fixed to the stool of the after main-topmast backstay; and both officers and men were carefully instructed to place it so that the spindle should be well above the truck, and the chain carried into the water, clear of the cross-trees, top, and channels, by outriggers.*

Under these precautions I feel a confidence tantamount to conviction, that at least the spars of His Majesty's ship under my command were saved in several severe thunder-storms which she encountered in the Gulf of Lyons, the Adriatic and Ionian Seas, and in the Lesser Syrtis, the electric fluid having been seen to descend the chain, and pass overboard into the sea, without damage to the ship.

H.M.S. *Queen.*

I happened to be on board the *Queen*, of 74 guns, when an electric discharge shivered her main-topmast to chips, and fatally damaged her mainmast, in the harbour of Messina, in 1815. On this occasion I remarked to Sir Charles Penrose, who had his flag flying on board her, that the amount of injury now inflicted would supply all the ships on the station with lightning conductors. If I remember rightly, this ship carried the useless and dangerous appendage of a spindle upon her truck.—I have, &c.,

W. H. Smyth, Captain R.N.

To Waller Clifton, Esq.,
Secretary to the Lightning Committee.

Conflicting evidence.

In this inquiry, as in the case of very many others, it may be seen how difficult it is to get at the fact when no notes are taken at the time. From the conversation which took place between Sir Charles Penrose and myself, when I made the remark just related, it is obvious that there could have been no conductor up when the *Queen* was struck. But on the Committee's application to Captain Bird, who was a midshipman on board that ship, he stated that 'he was pretty sure it was up, and his shipmate, Mr. Bisson, thought the same.' Such contradictory evidence was baffling to inquiry; and the Committee then applied to the present Admiral Coode, who was captain of the *Queen* at the time, and he most distinctly replied that they were not up, because 'the Admiral, whose flag he had the honour to be under, had an objection to using the imperfect ones then supplied to the navy.'

Capt. Bird.

Admiral Coode.

---

* I ought here to have added, that I further directed, if the top-gallant masts were struck, that the back-stays should not be sheep-shanked, but stoppered down to their respective stools.

The labours of the Committee terminated in a full conviction of the utility of conductors, when handled properly, and of the great advantage which would result to the public from adopting those fitted on the plan of Mr. William Snow Harris, with whom I have been long in communication: this able electrician was knighted shortly afterwards, and pensioned for his skill. Many instances of damage by the electric fluid to our Mediterranean fleet are recorded in the preceding pages; yet the value of the adoption, both as to life and treasure, may be still more enforced by submitting the following list of the casualties which occurred during the time of my service on that most important of our fleet stations. Inference also points out that the electric fluid has destroyed various vessels to which the term 'missing' has been applied; and I well remember the Malta government-packet *Blucher* sailing for the Ionian Islands at the beginning of 1816, in a thunderstorm;—she was never heard of afterwards.

Sir W. S. Harris.

The *Blucher*.

Name.	*Guns.*	*Date.*	*Remarks.*
AJAX	74	*June*, 1811	Off Gorgona. Main-topmast shivered by the lightning, and mainmast disabled. This ship was again struck off Toulon, in 1813.
ALBION	74	*Dec.* 1818	At Malta. Mainmast struck, and one man killed. The mainyard was also wounded. (I was told of an alarming rumbling in the hold, which continued for several seconds.)
APOLLO	38	*Aug.* 1811	Mediterranean station. Spars wounded, but the particulars not correctly ascertained. The cabin bellwires were fused.
BARFLEUR	98	*Oct.* 1813	Off Toulon. Fore-topmast shivered, foremast damaged, light-room windows of fore-magazine shattered, and its door forced open. The danger was imminent.
BLAKE	74	*March*, 1812	Coast of France. Main-topgallant-mast shivered, the lower rigging set on fire, and two men hurt. (This made Captain Codrington so warm an advocate for conductors, that he told me he would never go to sea again without one.)
BUZZARD	10	*Sept.* 1812	Off Minorca. Lost main-topmast and topgallant-mast, mainmast wounded, and starboard pump split.
CHANTICLEER	10	*Oct.* 1822	At anchor before Corfu. The mainmast shivered from the truck to the deck, and the latter covered with chips and splinters.

Name.	Guns.	Date.	Remarks.
CUMBERLAND	74	*Aug.* 1810	Faro of Messina. Mainmast struck by lightning, the upper spars shivered, and the main-top set on fire.
CUMBERLAND	74	*Sept.* 1810	Faro of Messina. The ship again struck within a week of the above accident, and the spars disabled. Several of the men experienced a temporary blindness.
EAGLE	74	*Nov.* 1811	In the Adriatic. The lightning struck the foremast, and wounded one man. Some of the gear was damaged.
EAGLE	74	*Jan.* 1812	Off Ante-Paxo. Lightning struck the mainmast, burst off one of the hoops, and wounded ten men.
EAGLE	74	*Jan.* 1812	Off Corfu. Mainmast twice struck, and set on fire. The rigging and spars damaged, and Captain Rowley with many men were knocked down.
FREDERICKSTEEN	32	*March,* 1812	In the Piræus. Fore and mainmasts struck, other spars damaged, and two seamen stunned.
HIBERNIA	120	*Aug.* 1813	Gulf of Foz. Foremast and main-topmast damaged, two men wounded, and many experienced electrical shocks.
KENT	74	*July,* 1811	Off Toulon. Mainmast ruined, mizen-mast shattered, and the whole of the spars damaged. One man killed and several scorched. (On going on board this ship soon afterwards, I was told by Lieutenant Lord Napier—a warm advocate for conductors—that many men were slightly affected by the electric agency.)
LARNE	20	*Feb.* 1820	Off Corfu. Slight damage to the spars and gear. Several men knocked down, one killed on the spot, one died soon after.
LEVIATHAN	74	*Oct.* 1812	Gulf of Lyons. Main-topmast rent by the lightning, and the mainmast slightly damaged.
OCEAN	98	*Sept.* 1813	At anchor off the Rhone. Main-topmast split in pieces, and mainmast damaged. Obliged to return to Port Mahon, and thereby weaken the fleet under Sir Edward Pellew.
ORLANDO	36	*Jan.* 1813	At Smyrna. Main-topmast and topgallant-mast destroyed, and mainmast wounded. Several men hurt. Ship obliged to go to Malta to refit.
PHAETON	46	*Sept.* 1824	Gibraltar. Foremast shivered from the truck to the deck, and set on fire. Several men struck down. Other spars, and several sails, greatly injured. (See my letter to the Lightning Committee, page 308.)
PHŒNIX	36	*Feb.* 1816	Archipelago. Mainmast much damaged, and three men hurt. Wrecked shortly afterwards.
POMONE	44	*Nov.* 1811	Off Tavolaro. Fore and mainmasts struck; main-royal burned. One man killed and four wounded.

Name.	*Guns.*	*Date.*	*Remarks.*
Pompee	80	*Oct.* 1812	Gulf of Lyons. Fore and main-topmasts disabled, main-topgallant-mast splintered. One man killed, three wounded, and several stunned.
Queen	74	*March,* 1815	Messina harbour. Main-topmast destroyed, mainmast damaged, and main-deck beam injured. Obliged to return to Malta. (See my letter to the Committee, page 303.)
Redpole	10	*Oct.* 1822	Corfu. Main-topgallant-mast injured, and one man partially deprived of sight.
Repulse	74	*April,* 1810	Coast of Catalonia. Ship struck twice. Mainmast splintered from the truck to the deck. Seven seamen and a boy killed, three mortally wounded, and ten more or less hurt.
Resistance	44	*June,* 1811	Off Gorgona. Mainmast damaged and set on fire; main-topmast and topgallant mast destroyed. Two or three bulk-heads smashed.
Royal George	100	*Sept.* 1813	Off Toulon. Particulars not ascertained from the log-book; but many men knocked down, and others stunned.
San Josef	112	*Sept.* 1813	Mouth of the Rhone. Main-topmast and topgallant-mast shivered, and some gear injured. Deck covered with splinters and chips.
Scipion	74	*Aug.* 1813	Off Toulon. Main-topmast shivered, and mainmast damaged. Obliged to quit the fleet to refit.
Sultan	74	*Sept.* 1812	Off Tavolara, Sardinia. Mainmast, topmast, and topgallant-mast split in pieces, and some gear set on fire. She had been struck off Mahon before this, when seven men were killed and three wounded whilst furling the jib.
Swiftsure	74	*Sept.* 1813	At anchor off the Rhone. Main-topmast shivered to pieces, and several men much affected by the electric agency. (An officer assured me that 'a very little more might have created a panic.')
Union	98	*Sept.* 1813	Off Toulon. Main-topmast shivered in pieces, and much gear damaged. A marine had his sight injured, and several men were stunned.
Unite	36	*June,* 1811	Off Gorgona, in company with the *Ajax* and *Resistance* (*which see above*). Fore and main-masts ruined, upper spars shivered in splinters. Many men badly hurt, and one lost overboard.
Warrior	74	*Aug.* 1810	At Messina. Ship sharply struck, but particulars not ascertained. (Captain Spranger told me it was an alarming shock.)
Ville de Paris	120	*Oct.* 1811	Off Toulon. Mainmast shivered and ruined from the truck to the deck, the rigging damaged, five men hurt, and much other damage. (In the *Rodney,* we were close to this ship, when the accident occurred.)

These are the casualties of a comparatively short space of time, and which happened to ships with which I was acquainted; but had I had leisure for a further examination of the log-books at the Admiralty, I could probably have given more. Sir W. S. Harris, taking a larger range, has arrived at some very important results; and he kindly handed me the following details, deduced from sixty-five vessels struck by lightning in the Mediterranean :—

Ratio *per mensem.*

Months.	*Ships struck.*	*Months.*	*Ships struck.*
January	7	July	3
February	6	August	4
March	8	September	11
April	2	October	10
May	2	November	6
June	1	December	5

And the times of being struck, he thus tabulates, the hours in the following enumeration being inclusive :—

Hours.	*Vessels struck.*
12 A.M. to 12 P.M.	27
12 P.M. to 12 A.M.	45
6 A.M. to 6 P.M.	37
6 P.M. to 6 A.M.	33
12 A.M. to 6 P.M.	14
12 P.M. to 6 A.M.	21
6 A.M. to 12 A.M.	29
6 P.M. to 12 P.M.	15

Times of liability.

From these elaborated results, it appears that liability to lightning is greatest in the autumnal months; and that about three-tenths of the whole number of cases have occurred between midnight and sunrise. But it is also evident, that the chance of damage is greatest between sunrise and noon, upwards of four-tenths occurring in that quarter of the day; and least between mid-day and sunset. By a laborious investigation, Sir William also arrived at the following general deductions :—

The liability of lightning to strike on any given point appears to be as follows :—

In 2 out of 3 times	it strikes upon the	topgallant-mast or highest point.
1 in 5	,, ,,	topmast or next highest point.
1 in 7	,, ,,	lower mast or next highest point.
1 in 50	,, ,,	hull directly.

From this it may be inferred, that the electrical discharge is occasionally determined towards ships in directions more or less oblique to the masts and hull. Other deductions.

The liability of lightning to fall on one or more of the masts simultaneously, is as follows:—

In 2 out of 3 instances, a ship is struck by lightning on the mainmast.				
1 in 5 times	,,	,,	foremast.	
1 in 20 ,,	,,	,,	mizen-mast.	
1 in 200 ,,	,,	,,	jib-boom.	
1 in 6 instances the yards and sails are struck together with the masts.				

A ship may be struck by lightning on the fore and mainmasts about the same time, or on the main and mizen-masts at the same time, or even on all three masts simultaneously, but in no case on the fore and mizen-masts simultaneously, independent of the mainmast.

In such cases lightning has fallen on the fore and mainmasts together, in about once in 20 times; on the main and mizen-masts together, once in 40 times; on all the masts, once in about 200 times.

Permanent conductors.

During the progress of my inquiries and experiences, Harris's permanent lightning-conductors had not been invented, or I should have eagerly embraced his beautiful principle. But I cannot better close this section than by giving the opinion of Captain Robert Fitzroy thereupon. 'During the five years the *Beagle* was occupied in her voyage, she was frequently exposed to lightning, but never received the slightest damage, although supposed to have been struck by it on at least two occasions, when—at the instant of a vivid flash of lightning, accompanied by a crashing peal of thunder—a hissing sound was heard on the masts; and a strange, though very slightly tremulous motion in the ship indicated that something unusual had happened.'

PART IV.

OF THE SURVEYS AND GEOGRAPHICAL INVESTIGATIONS OF THE MEDITERRANEAN SEA.

---

§ 1. Early Ages.

Archaic notices.

AMONG the indefinite traces of the early origin of navigation, it is perceptible that maritime commercial intercourse of one nation with another on a considerable scale first took place on the shores of the Mediterranean; and to the spirit and enterprise of the Phœnicians, or Canaanites, must probably be assigned the merit of being the primæval traders, a consequence of their progressive steps in civilization. Unfortunately, this great people have not transmitted any writings to us, but their merchants are mentioned in Scripture as equal to princes, and it is clear that for many ages they had no rivals in navigation; whence they acquired a high degree of opulence while the rights and duties of community were still only dawning in Greece. It is probable that the Phœnicians supplied the Hebrews and Semitic people with foreign commodities, for there must have been a taste for inland traffic in Palestine. This is shown in the only authentic history of that very remote period which has descended to us; wherein the existence of early caravan-traffic is exemplified in the sale of Joseph by his brethren. But as to trading by sea—though express allusion to such

Jacob.

intercourse is made in the death-bed prophecy of Jacob to Zebulon (*Genesis*, xlix. 13), about 1700 years before our era—it was so long before the Hebrews became sailors, that there is no distinct indication of international sea-commerce before the time of Solomon, and even his fleets were navigated by, and perhaps hired from, the men of Tyre. Meantime, the Egyptians, from a superstitious aversion to venture afloat, took an utter dislike to all maritime expeditions.

Phœnician trade and wealth. The Phœnicians, though deprived of a part of their territory by Joshua, are seen in various fragments of ancient writers, not only to have traded with Cyprus, Rhodes, Greece, Sardinia, Gaul, and Spain, but also to have ventured beyond the Pillars of Hercules 1250 years before Christ; and the extent of their undertakings is well shown, in the enumeration of the goods and articles which constituted the riches of Tyre in Ezekiel's time (B.C. 500). **Ezekiel.** Thenceforward the spirit of commerce was lighted up in Carthage, Greece, and Rome: extending to their colonies, and the barbarian nations around the inner sea; where it flourished, though under many vicissitudes, through the classic and middle ages, and from them to the present times. But this torrent of commercial prosperity has subsided to a gentle stream: in other words, from having engrossed and monopolized the trade of all the ports of the then known world, it has spread over the whole globe, fostered by the progress of art, science, and civilization. **Change of commerce.** The discovery and colonization of the magnificent continents of America, and the opening of the ocean-route to India, produced an important change in the commercial intercourse between Europe and the East, as well as a great increase in its magnitude, by avoiding the enormous cost of conveying by land the commodities of India to the shores of a sea where neither periodical winds, nor available currents, offer facilities for expeditious navigation. The route by sea superseded the traffic by land, and revolutionized the intercommunication of the whole world; so that the important trade which had passed for nearly 3000 years through the Mediterranean, collapsed to nearly its present state.

Remark. It will hence be seen, that as the *mare internum* was so long and unceasingly traversed by triremes, galleys, argosies, and every description of shipping which war or commerce demanded, the wish for an accurate knowledge of its coasts and harbours would gather strength from necessity, so as to be continually more and more desirable; and accordingly, from the earliest dawn of nautical and geographical efforts, directions for the coasting navigation of the Mediterranean have been collected and evulgated. It will, therefore, be of interest to cast a glance over the successive steps by which an advance has been made from primitive efforts to our present approximate perfection; more especially as no other portion of the globe was examined through so many ages—insomuch that it may fairly be reiterated—

Nullum est sine nomine saxum.

Early surveys.

Charts, or delineations resembling them, however rude, were probably coeval with the earliest navigation of those shores, and the primitive essays of geographical delineation.

Moses.

Moses, so far back as 1500 years before our era, laid down with considerable precision, the boundaries, mountains, cities, and towns in the Holy Land: and after him, his

Joshua.

successor, Joshua, despatched some selected men especially appointed, to gather such information as to form an intelligible report of the principal features of the country. 'Go and walk through the land,' said the son of Nun, 'and describe it, and come again to me, that I may here cast lots for you before the Lord in Shiloh.' It may be assumed that the Hebrews had acquired this branch of knowledge during their Egyptian bondage, since it is known to have been cultivated immemorially in the valley of the Nile, but especially in Upper Egypt: and Apollonius Rhodius expressly states, that the Argonauts—upwards of 1200 years before our era—derived their hydrography from the same source. The geographical information of the Greeks

Homer.

in the time of Homer (about B.C. 900) may be inferred from his writings, by which we find that he knew something of Egypt and Lybia, and the Erembi, or Arabs; but his knowledge was only general, except among the Cyclades, and their immediate neighbourhood. In his description of the shield of Achilles, the earth was figured as a disc surrounded by a flowing ocean (*literally river-ocean*), like an egg in a vessel of water; or Job's 'thick darkness a swaddling band for it' (xxxviii. 9). Some imagine that these waters were only intended to represent the Mediterranean, because the stars described represent its situation in the northern hemisphere.

From Hesiod some geographical hints are obtained, **Hesiod.** by which a further trace of knowledge will be observed. In his time the centre of Greece was considered as the centre of the earth, and Sicily was so distant, as only to be just known as the land of wonders; while to the north all was fable beyond the Euxine; and such was the state of Mediterranean navigation, that none but pirates ventured, at the risk of their lives, to steer directly across from Crete to Lybia. Thucydides (I. 3 & 4) asserts that it was not till **Thucydides.** the use of the sea had opened free communication among them, that the Greeks ever acted in joint confederacy, and that the Trojan war was the first instance of such union: yet he adds that Minos was master of a navy, with which he cleared the Cyclades of pirates, for the more secure conveyance of his own tributes. Now, though these sea-rovers were wont to land and surprise unfortified places and scattered villages, there must also have been laden vessels to rob. However, from the siege of Troy, it seems that the Mediterranean was common and open to all men, till the time of the Emperor Justinian; whence it was, that the Roman laws granted an action against any person who should molest another in the free navigation and fishing therein.

The Greeks extended their practical geography by the **The Greeks.** system of colonization, and their maritime movements were aided by the sea-cards of the Phœnicians; they appear however, to have soon surpassed their teachers, by introducing regularity into the pursuit, and establishing it upon stable principles. Thales taught the sphericity of the earth; and his disciple, Anaximander, the Milesian, is considered by Agathemer as having compiled the first scheme of geographical tables, 550 years B.C. According to Greek reports, he also constructed the first map of the world; but had not the books of the Carthaginians been destroyed, we might have had a different account. Herodotus (*Terpsichore*, 49) particularly mentions a tablet of copper (χαλκεος πιναξ), which was shown by Aristagoras, the rebel prince of Miletus, to Cleomenes, king of Sparta (B.C. 495), upon which was engraved (ενετετμητο) the circuit of the whole earth, the sea, and all the rivers: and from an anecdote related by Ælian (*V. H.* iii. 28), we may gather that about a century after this time maps were used for public information at Athens; for when Socrates wished to humble the vanity of Alcibiades, he pointed to a table of the world, (πινακιον εχον γης περιοδον) which hung up, and bid him look for Attica, and then examine his own fields there. Herodotus (*Thalia*, § 135) also details the fitting out of a nautical exploratory expedition, by order of Darius, son of Hystaspes: it consisted of two triremes and a large transport, under the direction of fifteen Persians of approved reputation, or known ability, whose orders were to examine the sea-coasts and emporia of Greece most carefully. When they had reached it, inspected and delineated (απεγραφοντο) its most important places, they passed over to Tarentum in Italy, where the surveyors were seized as spies, and the rudders of their vessels unshipped. As Joshua's party executed the first cadastral map, and Hanno's may be deemed the earliest voyage of discovery, so this expedition may be considered as the earliest maritime survey on record.

Thales. Anaximander. Aristagoras. Socrates. First nautical survey.

Scylax.

A work which immediately followed the above-mentioned expedition, deserves especial mention, since it is the very prototype of sailing directories for the Mediterranean.

It was a periplus for the guidance of navigators, compiled by Scylax, the Carian geographer, somewhat irreverently dubbed Darius's pilot. This work, which has come down to us, though in a corrupted state, is a brief enumeration of the countries along the shores of the Palus Mæotis, the Euxine, the Archipelago, the Adriatic, and all the Mediterranean; it commences with the Strait of Gibraltar, and proceeding along the coasts of Iberia and Gaul, round by the Islands and Levant, returns to the same point, and then describes the western coast of Africa, along the Atlantic as far as Cerne—the last portion being evidently borrowed from the *Periplus* of Hanno.*

Herodotus. Considering the state of information in his day, Herodotus was himself one of the most valuable geographers of remote antiquity; and he boldly declared that the Mediterannean, the Atlantic, and the Erythræan Sea, were but parts of one ocean; but that the Caspian is a distinct sea, communicating with no other. Respecting the dimensions of this inland sheet of water, he says—'A swift-oared boat would measure its length in fifteen days, and its extreme breadth in eight,' (*Clio*, § 203.) These measures were rejected by his successors; and it was not till the eighteenth century that the Caspian re-assumed the form which the Father of History had given it. **Xenophon.** Both Xenophon and his great contemporary, Hippocrates, **Hippocrates.** made considerable additions to the physical and moral knowledge of geography; **Aristotle.** but Aristotle's was the master-mind which sought inference from all available materials. Thus, reasoning on the hypothesis of the earth's being a sphere, he concluded that Spain must be a place of departure for the Indies; an idea which—with all its imperfections of distance—must be pronounced the first suggestion of a voyage across the

* Timosthenes, Admiral of the Fleet to Ptolemy Philadelphus in the Red Sea, wrote an express treatise on sea-ports, of which Pliny has preserved a few fragments. Pausanias, one of the last of the ancient topographers, also throws light on the shores of Greece.

Eudoxus. Atlantic, notwithstanding the honour has been claimed for that clever adventurer, Eudoxus of Cyzicus. Thus, although as yet but few traces of mathematical accuracy crop out, we perceive that geography had made a considerable advance; for coasting charts, as well as itinerary maps, became indispensable to the leaders of naval and military expeditions. Insomuch that Alexander the Great, himself no mean geo-

Nearchus. grapher, dispatched his admirals, Nearchus and Onesicritus, for maritime and hydrographical purposes, and also employed Diognetus of Bæton to survey the countries through which he passed; from whose documents the writers of the following ages took many particulars. Seleucus, one of

Patroclus. Alexander's successors, sent Patroclus, the admiral of his fleet, on several maritime explorations.

Ancient chartography But though we infer that science was assuming a new face and form, it is difficult to trace the gradual approach to any tolerable success in the art of chartography. We know nothing of the style of the illustrations above mentioned; and the inscribed columns of Sesostris, as well as the depicted conquests of Ptolemy Evergetes, may have been rather relations and descriptions than maps or charts. We are told, indeed, that Theophrastus died possessed of certain

Dicæarchus. maps of the world; and that Dicæarchus of Messina, in Sicily, his contemporary, executed the drawings of some coast surveys which he had made in Greece, the plotting of which, as Agathemer observes, was bounding the land by a simple straight line (τομη ευθεια ακρατω), wherefore he must take rank as the first chart-maker that we know of. This surveyor was in great estimation for accuracy, and Cicero (*Ep. ad Atticus, l.* vi. *c.* 2) thus commends him:—'I have supposed that all the cities of the Peloponnesus were maritime, upon the strength of no obscure authority, but such a one as you approve of, I mean the geography

Egyptian surveys of Dicæarchus.' But though we are unable to ascertain the amount of their merit in mapping, the labours undertaken by the early geographers, and their approach to truth, may be estimated from various fragments of historic record: and it is not a little surprising that the ancient Egyptians —who travelled neither by land nor sea—should have made a trigonometrical survey of their country with such exactness, that we are at a loss to surmise the means by which they acquired so much precision. It is true that among their priests was one styled the Sacred Scribe, or Hiero-grammatist, whose qualifications are supposed to have included astronomy, cosmography, the chorography of Egypt, and everything concerning the Nile; yet we marvel how a man could become thus qualified unless there were both instruments and maps. The partial distances, by means of which the early writers, and among others, Herodotus, have given the complete length of Egypt, are taken in nearly a straight line; so that between Pelusium and Syene was a distance of 7° 37′ 7″ by the ancient observations, which differs only $\frac{1}{403}$rd part from that ascertained by the moderns, which amounts to 7° 38′ 15″.

Sacred Scribe.

Meantime maps and itineraries were objects of the greatest solicitude among the Romans, and that politic people exhibited painted representations of the conquered countries at their triumphal pageants. Polybius dwells on the care with which they plotted the countries through which it was likely that Hannibal would pass, at the beginning of the second Punic war; and no fact is better known, than that Julius Cæsar gave the idea which produced the *Antonine Itinerary*, in having ordered a survey of the whole Roman empire.* Arrian, named the second Xenophon, was employed by Hadrian—himself an experienced traveller—to examine the shores and trading places of the Black Sea, then considered by navigators to be a voyage of no small difficulty; and the Περιπλους Ευξεινου Ποντου—or

Roman surveys.

Julius Cæsar.

Arrian.

---

* From the familiar mention by Ovid (*Fasti* vi. 277) of the globe, 'Arte Syracosiâ suspensus in aere clauso,' it would appear that the Archimedean glass orb was not forgotten in Rome.

Arrian's Euxine. circumnavigation of the Euxine Sea—shows his fitness for that purpose. We are at a loss to know how he made the measurements by which his coast distances were ascertained; but in the storm which compelled him, after much distressful suffering, to bear up for Athenæ, one of his vessels being wrecked, he expressly tells the Emperor (*page* 117, *Ed. Blancard*), that everything on board was saved, not the crew and furniture only, but also 'the nautical instruments.' This—should τα σκευη τα ναυτικα mean anything beyond tackling—would imply that there were scientific tools in each of his ships, and, of course, trained men to use them. Indeed, there must have been a sort of engineer Vegetius. corps in the Roman forces, for Vegetius (*lib.* iii. *cap.* 6), in showing the importance of obtaining the exact topography of every seat of war, adds, 'We are told that the greatest generals have carried their precautions on this head so far, that not satisfied with the simple description of the country wherein they were engaged, they caused plans to be taken of it on the spot, that they might regulate their marches by the eye with greater safety;' a duty which would hardly be imposed on the uninstructed. The well-known line of Propertius indicates that men studied the material world from painted forms—

Cogor et e tabula pictos ediscere mundos.

One of the completest surveys of the Roman empire was begun and finished in the reign of Antoninus, and is well Antoninus Pius. known as his *Itinerary*, before alluded to; the maritime part of which shows the want of skill in the Mediterranean seamen of that epoch. All the ports which it was necessary to touch at in sailing from Achaia to Africa are enumerated, and how the mariners were to drag their course along the land to the west coast of Sicily, before they took their departure to the south. This itinerary was drawn up with all the labour and skill then procurable, and was esteemed a work of no common excellence; but it has probably suffered from errors of transcription.

The true science of geography, however, lingered through an infancy of unwonted tediousness, since it received but little improvement from the time of Thales to the establishment of the famous school of Alexandria. To be sure, Meton narrowly missed obtaining the latitude of Athens, by a solstitial observation in June, 432 B.C.; and Pytheas, the intrepid navigator of the North Sea, scientifically surveyed Lipara and Strongyle (*Vet. Scholiast, ad Apollon. Rod. l.* iv. *v.* 761), and actually determined the summer solstice at Massilia (*Marseilles*) by means of a gnomon 120 parts in height, the shadow being 42 parts all but one-fifth; that is, the two lengths were to each other as 600 to 209—proportions which gave 70° 47′ for the solar altitude. Although, by its doubt of the spherical form of the earth, the *science* of Herodotus is entitled to but small respect, still he must be allowed to have promoted it greatly, from the subject and number of his communications: these were drawn from his own observations in his various travels, and the accounts given by other travellers, in all of which—especially where he had the advantage of writing from what he saw himself—there is a strong vein of sound and sensible observation. But Eratosthenes—surnamed 'Surveyor of the Earth'—introduced a regular and solid system into geography, which, though deficient, was yet a great advance: he formed a consistent parallel of latitude, by tracing a line over certain places where the solstitial shadow was observed to be of the same length, as from Gibraltar through the island of Rhodes, through Taurus in Lycia, and over Syria to the Indies. From the central position of this line, with respect to the then known parts of the world, it became a standard of reference for the period; and the imperfect determination of places by ratio of climate, which had been widely adopted, was superseded by the method of observing the duration of their longest and shortest days. Erastosthenes also traced the first approach to a regular meridian, by means of an imaginary line passing through Rhodes and Alexandria, as far as Syene and Meroe:

State of science.

Meton.

Pytheas.

Herodotus.

Eratosthenes.

and it is not improbable that his suggestions were useful to Agatharcides, his contemporary and successor as President of the Alexandrian library, during the survey of the Erythræan Sea.

Agatharcides.

Hipparchus.

The next grand step was taken by Hipparchus, the ablest of ancient astronomers, who transferred the celestial latitudes and longitudes to the terrestrial globe, and introduced the stereographic projection. But the excellent conditions thus pointed out by science, were so little attended to until the days of Ptolemy, that Strabo, the prime geographer of the Augustan age, considered them as perplexing and unfit for ordinary use; while Vitruvius and Pliny—whose geography, nevertheless, is both full and curious—never give the least hint of their existence.

Ptolemy.

His merits.

Thus the true principles lay dormant, till the celebrated Ptolemy revived them 250 years afterwards, and applied both latitude and longitude to such itineraries, nautical surveys, and other materials as he could collect, drawn from observations which were of course less accurate than the principles upon which they were founded: in a word, this energetic geographer taught the projections on the plane of the meridian as the readiest method for arranging a map of the earth, in which the equator and parallels are arcs of circles, and the meridians arcs of ellipses, the eye hanging over the plane of that meridian which passes over the middle of the inhabited world. These were great strides towards the elevation of geography to a place among the exact sciences; and in the then state of practical mathematics, geometry, and astronomy, as well as the defective construction of the instruments in use, it was hardly to be expected that precision could possibly be attained. We may, therefore, rather feel regret than surprise that he fell into many and great errors; such as his flattening-in the north coast of Africa to the amount of $4\frac{1}{2}°$ to the south in the latitude of Carthage, while Byzantium was placed 2° to the north of its true position, thus increasing the breadth

His errors.

of that very sea where we should expect his greatest accuracy. Nor was this all: for the extreme length of the Mediterranean was carried to upwards of 20° beyond its real limits. Ptolemy trusted, it seems, chiefly to Marinus of Tyre, as able a geographer as ignorance of astronomy permitted, but still a broken reed to lean upon; and though he adopted the Tyrian's division of space into degrees and their parts, he had an erroneous system of projection and graduation. After his time there was a long period barren in discovery and commercial enterprise, and his geography was the standard guide until long after the revival of learning. As Ptolemy assigned 700 stadia for a degree of latitude, his errors were not so great in that ordinate as in his attempts at longitude: yet it should not be forgotten that his gravest mistake, in bringing China near to Europe, proved eventually, as was remarked by D'Anville, the efficient cause of the greatest discovery of the moderns, by leading Columbus to reckon upon sixty degrees less than the real distance from Spain to India. But was not Aristotle (*see page* 315) the true precursor?

Marinus of Tyre.

D'Anville's remark.

As the ancients possessed no means for critically measuring horizontal angles, and were unaided by the compass and chronometer, correctness in great distances was unattainable. On this account, while the eastern portion of the Mediterranean approached a tolerable degree of truthfulness, the relative positions and forms of the western shores are surprisingly erroneous. Strabo, a philosophical, rather than a scientific geographer, set himself, says the scholiast, to rectify the errors of Eratosthenes—'but Strabo made more mistakes than he:' and though he drew a much better contour of the Mediterranean, yet he distorted the western parts by placing Massilia 13½° to the south of Byzantium, instead of 2¼° to the north of that city. Strabo had a good education; still he seems to have possessed a very moderate share of astronomical knowledge, and was not so good a mathematician as his long residence at

Strabo.

Strabo's capacity.

Alexandria ought to have made him. In describing the countries which he himself had visited, he is generally very accurate, save where he relies upon Homer; and that a large portion of his work resulted from his own observation, is shown by a passage in the second book, where he says—

Strabo's travels.

I shall accordingly describe partly the lands and seas which I have travelled through myself, partly what I have found credible in those who have given me information orally or by writing. Westwards I have travelled from Armenia to the parts of Tyrrhenia adjacent to Sardinia; towards the south, from the Euxine (*near which he was born*) to the borders of Ethiopia. And perhaps there is not one among those who have written geographies, who has visited more places than I have between these limits.

Although Amurath III., about A.D. 1580, caused observations to be made which reduced the latitude of Byzantium to $41\frac{1}{2}°$, and the error in the position of Carthage was noticed in 1625, it may be said that Ptolemy's gross inaccuracies were continued upon maps till the middle of the seventeenth century, even Sanson's being 15° in excess in the length of the Mediterranean; though it was soon afterwards curtailed by the observations of M. de Chazelles.* Among the materials used by Ptolemy in composing his geographical system, it must be recollected that the itinerary measures of the surveyed Roman provinces usually exceeded the truth; since many of the documents were supplied by men of limited acquirements even in that day. In thus extracting matter from shallow computations and textual errors, besides giving the incongruous mass of details an approach to the solidity and unity of a mathematical basis, surely the labour of Ptolemy, with 'all its imperfections on its head,' was a welcome and valuable gift to his contemporaries, and also to posterity. The first

Ptolemy.

Value of his geography.

* In the ninth canto of the *Paradiso,* Dante describes this *Valley of Waters* as winding between the discordant shores (*tra discordanti liti*) of Europe and Africa, and assigns its length by astronomical tokens; it being, he says, noon in Palestine when the sun is rising in the Strait of Gibraltar. He also, in the twenty-sixth canto of the *Inferno,* mentions the Pillars of Hercules as boundaries not to be passed—

'Acciochè l' uom piu oltre non si metta.'

authentic maps, deserving the name, with which we are acquainted, are those found in the early MSS. of his geography, originally drawn by Agathodæmon, an Alexandrian map-maker, who lived in the fifth century; of these there is a splendid copy in the British Museum, of apparently about the year 1350, which formerly belonged to M. de Talleyrand. These appear to have been copied in the editions of 1462 and 1482, and thus was preserved the outline of the Mediterranean which had been received as accurate by geographers from A.D. 150 downwards. They were certainly indifferently drawn, but at all events their various relations are better expressed than in the Theodosian map, a valuable painted itinerary better known as the Peutingerian Table, now preserved in the imperial library at Vienna: it was intended, it seems, rather to show the high roads in the empire than the sea-shores. In it the Mediterranean is so reduced in breadth, that it resembles a long canal, and the site, form, and dimensions of its isles are displaced and disfigured; yet the artist must have flourished seventy or eighty years after Agathodæmon.

Agathodæmon.

Theodosian map.

These remarks may be illustrated by giving a direct view of the reductions of the ancient measurements, from three of the principal stations in ascertaining the length of the Mediterranean Sea; they being nearly in the same line east and west, and in the third climate. The latitudes were estimated in stadia reckoned from the equator, and are not so violently discordant as might be expected from such a method, Eratosthenes giving 25,450 stadia—Hipparchus, 25,600—Strabo, 25,400—Marinus of Tyre, 26,075—and Ptolemy, 26,833, as the length between the equinoctial line and Syracuse, or rather the place which they designated the Strait of Sicily. But the longitudes run rather wild; they are reckoned in stadia from the *Sacrum Promontorium*, or Cape St. Vincent; the numbers given by Eratosthenes being 11,800—by Hipparchus, 16,300—by Strabo, 14,000—by Marinus of Tyre, 18,583—and by

Ancient observations.

Longitudes deduced from them.

Ptolemy, 29,000, as the arc from thence to Syracuse. Other authorities for the dimensions of the Mediterranean might be cited, as Dicæarchus, Agrippa, Artemidorus, and above all, Polybius, the friend and adviser of Scipio Æmilianus, and a man equally distinguished as a soldier, historian, and geographer; but as their works are only known by fragments and quotations, besides there being some confusion *inter se*, I have not brought forward their figures. Pliny (*Nat. Hist. lib.* vi. *c.* 38, Brotier) praises their personal zeal in boldly braving fortune on such hazardous service.—'Hæc est mensura inermium, et pacata audacia fortunam provocantium hominum.' A curious illustration is, however, offered by the above-mentioned authorities: for this reduction of the numbers by assuming 700 stadia to a degree of latitude* for a plane projection in the 36° parallel, and 555 for the corresponding degree of longitude, is mostly from M. Gosselin's *Recherches sur la Géographie Systématique et Positive des Anciens*, and also his *Observations Préliminaires et Mesures Itinéraires* appended to the large French edition of Strabo. From the elaborate lists therein drawn up, for the different values of the stadium under different degrees of a great circle of the globe, the following tabular view is adjusted. To this, for the sake of immediate comparison, I have subjoined the determinations resulting from my own observations; though in order to meet the most probable stations of the elder geographers, I have referred to Europa Point, the centre of Syracuse, and Pompey's Pillar. The last, however, is very uncertain, since the differences in the table are large, recollecting that

Polybius.

Pliny.

Gosselin.

Alexandria.

---

* Major Rennel, and other modern geographers, suppose that the Greeks used several kinds of stadia, varying from 696 to 750 in a degree; but there are some who still hold that the stadium was an invariable standard at all times and in all countries, and that it was 203 English yards in length, which is nearly five feet more than the stade usually chosen. The value of Pliny's geography is vitiated by the fact proved by D'Anville—namely, that he indiscriminately reckons eight stadia to a mile, without reference to the difference between the Greek and Roman stadium.

Alexandria must have been one of the best determined of the ancient latitudes. Ptolemy, who had the advantage of being preceded by Timocharis, Eratosthenes, and Hipparchus, used 30° 58′ in the computations of his syntaxis, but adopted 31° in his geography: his place of observation, therefore, was probably south of the Serapeum. Probable sites.

Authorities.	Gibraltar.		Syracuse.		Alexandria.	
	Latitude.	*Longitude.*	*Latitude.*	*Longitude.*	*Latitude.*	*Longitude.*
Eratosthenes .	36° 21′ 25″	4° 17′ 9″	36° 21′ 25″	16° 51′ 26″	31° 0′ 0″	36° 8′ 34″
Hipparchus. .	36 20 0	4 17 9	36 34 17	23 17 9	31 8 34	38 48 30
Strabo . . .	36 17 8	2 51 25	36 17 8	20 0 0	31 8 34	32 8 34
Marinus of Tyre	36 0 0	3 34 17	37 15 0	26 32 50	31 0 0	41 25 43
Ptolemy . . .	36 0 0	3 34 17	38 20 0	26 32 49	31 0 0	41 25 42
Smyth . . .	36 6 30	5 20 40	37 3 30	15 16 50	31 10 45	29 53 59

§ 2. Middle Ages.

SUCH was the state of Mediterranean geography before the decline of learning, and such was its condition through the greater part of the middle ages, though their maps were occasionally improved by nautical experience and popular observation. Some of the learned of those days believed that the Mediterranean was so named because it passed through the midst of the earth, dividing it into two equal parts; but St. Asaph, who 'flourished' in the sixth century, is said to have written, in one of his mystical doctrinal illustrations, that it was also called the Meridian Sea, because it was to the south of the earth,—few of his era having heard anything of the actual south sea. Yet Cosmas, surnamed Indico-Pleustes, or Indian navigator, had written his *Christian Topography*, at Alexandria, in the preceding age, and must have been known to theological writers. Though accurate in his commercial particulars, this old navigator absurdly considered the whole ocean and seas

Mediæval opinions.

St. Asaph.

as forming a flat surface, bounded by walls supporting the firmament; and of this ocean he supposed the Mediterranean to be one of the four great navigable gulfs.

The Arabian geographers.

The Arabian geographers, as Ibn Yúnis, Abú-Ríhán, Abul Hasan, 'Abd al Atíf, and Abú-l Fedá, Sultan of Hamáh, and strictly speaking, the only *scientific* geographer the Arabs ever had—made various maps and plans of places; but Edrisi, of whom I have already spoken as a cosmogonist, on the whole, was perhaps the most eminent person of that school. His great undertaking—*Nuzhat, &c.*, (*see page* 115)—throws very considerable light upon the geographical system adopted by the Arabian writers. In this work he describes the earth as circular, with a circumference of 132 millions of cubits, or 33,000 miles, which he divides into 360 degrees. The Sea of Damascus, or the Mediterranean, he estimates as being 1136 parasangs in length, from its eastern extreme to its discharge into the Atlantic Ocean (*Mare tenebrosum*), which, on the low estimate of 30 stadia to each, amounts to 34,080 stadia, or rather more than 3901 miles, estimating each stadium at 604·4 English feet; but this apparently enormous aberration is extremely uncertain.* Edrisi's book was intended as a description of the geographical delineation of the world, which he made on a circular silver table (*dáyireh*)† for Roger II., King of Sicily; this was copied in the famous *Mappa Mondo*, by Frà Mauro (*cosmographus incomparabilis*), and by Martin Behaim of Nuremberg, so that it was, for upwards of three centuries, the pattern for all the maps of the earth. Though this remarkable table has been lost for ages, drawings of it are preserved in Arabian MSS., especially in one of the fifteenth century, in

The Sea of Damascus.

Frà Mauro.

* From a memorandum given me by the late Major Rennell, the Grecian παρασαγγα, whence the Persians got their *fursung*, was 5468·7 English yards. This would reduce the above length to 3529 miles and 140 yards.

† Dr. Pococke (*nomen venerandum*) renders it 'globe,' which would have required *kurrah* instead of *dáyireh*.

the Bodleian Library at Oxford, an engraving of which has been given in Dr. Vincent's *Periplus of the Erythræan Sea*, page 656. In the British Museum (*Add. MS.* 11,695), there is a coloured map of the world of about 1100, arranged according to the ideas of the Arabian geographers; in which the earth is represented of a quadrangular form, surrounded by the ocean—'like an egg in the water.' Here the Ægean Sea joins the Mediterranean, which is represented straight, at a right angle in the centre of the map. **Dr. Vincent.**

The Arabians did not follow the Greeks in their choice of a meridian, as they preferred the African coast to the Fortunate or far-west islands. But they afterwards substituted the *Khubbah Harinah*, or Cupola of Arina, the site of which is still only presumptive, although Abú-l Fedá's allusion has been ingeniously tested by the difference between the true and the inhabited horizons of the Alphonsine Tables. After all the discussions which have taken place, the Khubbah may have been an imaginary point; for Humboldt, after wading deeply into the inquiry, came to the conclusion, that—'the more passages were compared, the obscurer the subject became.' **Arabian meridian.** **Khubbah Harinah.**

Marino Sanuto, a Venetian noble surnamed Torsello, who published the *Liber Secretorum Fidelium Crucis*, about the year 1320, was a great voyager and traveller, and constructed a chart of the Mediterranean Sea, which has long been lost; but its outline may be observed in his planisphere, preserved in the *Gesta Dei per Francos*, published by the Bongars in 1611. Though this is certainly one of the earliest, the exact epoch of the first proper chart of this sea in the middle ages is not known. According to Señor Capmani, in his *Questiones Criticæ*, such was used by the Spanish navigators as early as 1286: and he also relates, as a certain fact, that the galleys of Arragon were officially furnished with nautical charts in the year 1359. Yet the very invention of the projection is ascribed, by others, to the celebrated Henry, son of John, King of Por- **Marino Sanuto.** **Capmani.**

tugal. An outline of the Mediterranean by the brothers Pizzigani, bears date in the early part of the fourteenth century, when the geographical treatises called *Imago Mundi* had crept into notice.

Pizzigani.

Shortly afterwards, various local surveys and descriptions of Mediterranean regions appear, and some not at all deficient in point of execution. In the British Museum, there is a valuable manuscript volume by Christopher Bondelmonte, in the commencement of the fifteenth century, in which the Cyclades and Ionian Isles are very fairly set forth.* At Oxford, I was shown a very neat illuminated manuscript of 66 pages of vellum, in oak boards, superscribed—'Chest le rapport que fait messire Guillebert de Lannoy, Chevalier, sur les visitations de plusieurs villes pors et rivieres par lui faittes—tante en Egipte comme en Surie. Lan de grẽ nr̃e signe^{r} mil · cccc · vingt et deux.' It appears that our Henry V., notwithstanding David Hume's sneer, actually did contemplate making a crusade to the Holy Land; and as a first step, he dispatched Sir Gilbert de Lannoy, evidently a well qualified officer, to survey the state and condition of the harbours and arsenals of Egypt and Syria. The result is the accurate and intelligent official report before us, in which the various anchorages, soundings, landing-places, fortifications, munitions of war, produce, and supplies of wood and water are diligently noticed, forming an authentic account of the hydro-geography of those countries 430 years ago. In 1478, on Moceniga's being elected Doge of Venice, a Captain Bartolommeo, who had made many voyages, and 'trod every rock in the Ægean,' published an account of the Archipelago, with wood-blocks

Bondelmonte.

G. de Lannoy.

Bartolommeo.

* There are various copies of this work abroad, both printed and manuscript; and some fac-similes of Bondelmonte's maps are given in the *Liber Insularum Archipelagi, a G. R. L. de Sinner. Lipsiæ*, 1824. The geography of the middle ages has attracted the researches of Cardinal Zurla, M. Jomard, and Joachim Lelewel; but the best collection of mediæval maps, perhaps, is that recently published at Paris by Viscount Santarem.

of the islands, a sonnet expressive of the features and peculiarities of each, the ports and produce, with the bearings and distance from one another. It commences with Cerigo and ends with Cyprus: and entering from the westward, shows the various islets, with the rock on which the *Nautilus* was so dismally lost in 1807, adding the following express advice to beware of them at night:— Poetical directory.

Sta inverso grieco il Poro e la Poresa
Fa che de note te guardi da esa.

At the time that Bartolommeo was making his levantine voyages, the Mediterranean was also under the examination of the great Columbus, whose knowledge of geometry, astronomy, and cosmography had been fostered by a commander of the same name and family under whom he served; and his brother was, moreover, a professional compiler of geographical charts. The effects of the exertions of such men are discernible on comparing the earlier maps and charts with those of the fifteenth and sixteenth centuries: and among those which have passed under my own examination, I cannot but enumerate the fine set of *portolani*, now preserved in the British Museum, as they will, in all likelihood, remain there for ages most authentic evidences. The kind aid of Sir Henry Ellis and Mr. J. Holmes,* together with the liberality of the trustees, enable me to give the following list:— Columbus. Portolani.

MS. Arundel, 93, *art.* 7.—'Christophori Bondelmontii, Liber insularum Cycladum atque aliarum in circuitu Sparsarum, cum earundem schematibus.' This is the collection which I have mentioned already; and there is also a second on the Archipelago in general. Bondelmonte.

Bondelmonte was a Florentine priest, who wrote about 1420—1422. His name and the date are found by the initial letters of the chapters of his work, which, taken in order of succession, form this sentence—'Cristoforus Bondelmonti e Florentia Presbiter nunc misit Cardinali Jordano de Ursinis, M.CCCC. Christi.'

* To Mr. Holmes I am the more particularly indebted, since it was owing to the admirable systematic order in which he had arranged the geographical department of his charge, that all my inquiries were as easy as satisfactory.

Add. MS. 11,547.—A portolano, or collection of sea-charts, drawn by
Benincasa. Grazioso Benincasa, of Ancona, in the year 1467. It contains five charts, drawn on vellum, of which No. 1 is the Black Sea, Asia Minor, and the eastern part of the Mediterranean; 2, the Adriatic Sea, the Archipelago, and the central portion of the Mediterranean; and 3, the part from Rome to the Straits of Gibraltar on the west. There is also another by the same hand (*Add. MS.* 6390), which is described by Tiraboschi; in contour and details it substantially resembles the former one, but the fourth chart is inscribed—'Gratiosus de Benincasa, Anchonitanus, magnifico viro Prospero Camulio, Medico Genuensi, fecit, 1468.'

Plut. clxiii.—In this press, which contains upwards of one hundred manuscripts of the Arundel collection, is another copy of the *Insularium* of Bondelmonte, written in 1485, with coloured maps, or rather bird's-eye views. The Museum possesses two other copies, but inferior, of the same work.

The Cornaro maps. *Egerton MS.* 73.—This is a fine portolano, containing thirty-five charts on vellum, executed by different Venetian artists, about the year 1489. It formerly belonged to the Cornaro family, and was afterwards in the library of St. Mark's, at Venice, where it was examined and described at great length by Cardinal Zurla. In this valuable atlas are contained no fewer than twenty-six charts of the Black Sea, Adriatic, and great divisions of the Mediterranean Sea, by Piero Roseli, Zuan di Napoli, Grocioxa Benincaxa (*sic*), Francesco Becaro, Nicolò Fiorin, Francesco Cexano, Zuan Soligo, Aloixe Cexano, Domenego Dezane, and Nicolò de Pasqualin. The book also contains tables of solar and lunar motions, moveable feasts, and planetary influences; and a sailing directory is appended, of which the part containing the ports of the Mediterranean closes with—'Qua compie tute le staree del mar mediterano,' &c.

Cademosta. *Old Royal Libr., MS.* 14, c. v.—A portolano containing seven charts on a plane scale, executed on vellum, at the commencement, apparently, of the sixteenth century; and which belonged to Lord Lumley, who died in 1609. Five of these charts are devoted to the Mediterranean Sea and its divisions. Contemporary with these, or in the year 1520, was published the earliest printed sea-directory which I have seen; it is intituled *Portulano del Mare*, and was printed at Venice without the author's name, but directly attributed to Cademosta, the noble navigator of that city.

Cardinal Rovere. *Add. MS.* 11,548.—A plane-scale chart, drawn on vellum at Ancona, in the year 1529; the name of the artist is obliterated. Under the date are the arms of Cardinal Giulio Feltri della Rovere, son of the Duke of Urbino. It contains the whole of Europe, with the Black Sea, the Mediterranean, and the coast of Morocco.

Spanish portolano. *Add. MS.* 9947.—A Spanish portolano, containing four charts on a plane scale, executed on vellum, three of which represent the Mediterranean Sea and its divisions.

Add. MS. 10,132.—A portolano containing five charts, executed in 1538, by a native of Ancona, whose name has been erased, the inscription running—'I.H.S. Conte . . . . Anconitano la facte nel año M°CCCCXXXVIII.' Three of the charts are devoted to the Mediterranean and Black Seas.

Old Royal Libr., MS. 20, *E.* ix.—A choice manuscript 'booke of Idrography made by me Johne Rotz,' for King Henry VIII., in 1542, contains a general chart of the Mediterranean Sea. J. Rotz.

Add. MS. 10,134.—A portolano containing three maps, executed by some Italian artist about the year 1550, which formerly belonged to Nicholas Canachi, a pilot, of Patmos, as we are told on the title-page; where, under a coloured drawing of the Virgin and Child, appear the two following inscriptions:— Canachi.

'E chesto llibro sta di Nicolo Canachi dell' isola di
Sa. Gioane di Pattino, pillotto di mare.'

Ε τουτον το χαρτην εναι του Νηκολου Κανακι του
Πατηνιο τη οπου στε κί στί να Λεγορνο.

The second and third of these maps relate to the Mediterranean and its large divisions.

Egerton MS. 767.—A portolano containing four charts, of which two relate to the Mediterranean Sea; it is rudely drawn and coloured on vellum, apparently by a Venetian artist, about the middle of the sixteenth century.

Add. MS. 5415 A.—A portolano consisting of nine large charts on vellum, drawn on a plane scale by Diego Homem, in 1558. As it is very highly finished, and ornamented in gold and colours, with the arms of the respective sovereigns emblazoned on the various countries, it is considered to have been executed for Philip the Second; but the arms of Spain, which were impaled with the coat of England, have been defaced. Of these charts, No. 6 represents the coasts of the Mediterranean from the Straits of Gibraltar to the Morea, with the Adriatic. Homem.

Add. MS. 9810.—A large chart of the coasts of Europe, with the Black Sea and the Mediterranean, richly ornamented with drawings of figures, tents, &c., and it is inscribed—'Jacobus Veschonte de Maiolo composuit hanc cartam in Janua, anno Domini 1562, die x. Octobus.' De Maiolo.

Harleian MS. 3450.—A portolano of eighteen charts on a plane scale, of which three specially relate to the Mediterranean, by Joan Martines, of Messina, in the year 1578; they are elegantly drawn on vellum, in colours and gilding. *Harleian MS.* 3489, is another collection of charts by Martines, very similar to the above, but larger; and there is a third (*Add. Sloane MS.* 5019), drawn in 1582. J. Martines.

Add. MS. 9811.—A chart of the Black Sea and Mediterranean, on a plane scale; it is inscribed—'Joanne Riczo alias Oliva, figlio de Mastro Dominico, in Napole, a di 7 de Novembre, anno 1587.' J. Oliva.

Bibl. Cotton. Julius, E. 11.—A volume containing neat pen-and-ink drawings of sixty-eight Mediterranean islands, intituled 'Isulario de Antonio Millo, nel quale si contiene tutte le isolle dil mar Mediteraneo, &c., A.D. 1587.' *Add. MS.* 10,365, is another copy of this work, written in 1591, in which Millo is styled Armiralgio di Candia. A. Millo.

Add. MS. 10,041.—The Black Sea, the Archipelago, the Adriatic Sea, and the rest of the Mediterranean, on a plane scale upon vellum, and of about the year 1600; it is inscribed—'Mayde by Thomas Lupo, in Shadwell, neere unto the mill.' T. Lupo.

Cautionary remark.

I have been more particular in citing these curious documents, because a lesson is thereby afforded us as to the mischief of an indiscriminate neglect of old surveys, and the danger of a blind reliance upon the newest compilations. For in all those charts of the fifteenth and sixteenth centuries, as well as in the rare manuscript portulano (*circa* 1450) presented to the Bodleian Library by the late Mr. Francis Douce, and likewise on the maps of Nicholas Vallard of Dieppe, Peter Plancius, and Paolo Gerardo, many shoals are sufficiently well-placed for a regardful seaman to have avoided them. Yet the most perilous of these afterwards disappeared from the charts, until I restored them, owing to the incredulity of certain navigators, and the carelessness of those smatterers employed by the ship-chandlers, who were long the purveyors of the seaman's scientific wants; among the most serious omissions, the cause of an awful sacrifice of life and treasure, we may instance the following:—

Cape de Gata.

The outer shoal off Cape de Gata, which was omitted by Tofiño, was noticed by the earlier hydrographers, and even appears to have been examined by Bartolommeo Crescentio, who, in 1585, surveyed the vicinity of Algiers, which city he brands even then as 'infamissimo albergo di corsari, et gravissimo danno et onta di Christiani.' In the directory which he afterwards published, the above rocks are accurately described; yet, from the subsequent omission of the outer one, we might have lost the *Belleisle*, of 80 guns, Captain J. Toup Nicolas, in 1840. On this being announced, I addressed the editor of a professional journal on the subject; and when the *Belleisle* returned to England, Captain Nicolas wrote me a letter from Plymouth, 20th January, 1841, in these terms:—'I have to thank you much for your late communication to the editor of the *Nautical Magazine*, in support of my statement relative to the rock off Cape de Gata that we discovered on our passage home. We passed it within half our ship's length. It was, in my opinion, not much larger than our launch, and possibly has three fathom or more upon it. It looked quite green. Had we not been going so fast—between eight and nine knots—I should certainly have shortened sail and examined it. . . . Our look-out men and every one were looking for the rock *within* us on our starboard bow, when to my astonishment the signal-man on the royal-yard called out, 'A rock *close to us* on the *port* bow!' which was instantly repeated to us by the look-out man at the jib-boom end. We all ran over to the port side of the poop, and I at the same moment ordered the helm to port, when the rock appeared close outside our lower-studding-sail boom. Every one saw it distinctly, and it appeared a miracle that we escaped it.'

Captain Nicolas.

The rock in Palamos Bay, on which a Spanish line-of-battle ship was lost in 1796, when nearly every soul perished. Palamos rock.

The Cassidaigne shoal, a dangerous reef on a wash, lying between Marseilles and Toulon, is accurately placed in old portulani, but was omitted by Mount and Page, and their followers. In 1807, a vessel running for Ciotal when chased by one of our cruisers, struck on La Cassidaigne, and was knocked to pieces in a few minutes. Cassidaigne shoal.

The reef off Cape St. Tropez is well placed in the earlier plans, but was omitted in many of the charts of about sixty years ago; and many ships have since struck upon it. From the circumstance of the *Rhadamanthus* steamer having got aground there since my examination of it, the reef now bears the name of that vessel on the Admiralty chart. Reef off St. Tropez.

The Vado shoal, called also the Mal di Vitro and Secca de la Barbiera, on the coast of Tuscany, was well marked 350 years ago; and both it and the Melora off 'Ligorne' appear distinctly in the Egerton MS. 73, of A.D. 1489. Yet in 1793 we lost the *Amphitrite* frigate upon it; and so lately as June, 1848, the fine English steamer *Ariel* was wrecked on its northern shelf. It was well surveyed by our boats, in 1818 and 1823. The Vado shoal.

The Aphrico rock off Monte Christo is well marked by Benincasa and others, but omitted by the first compilers of the general Quarter Waggoner. Several ships have struck here, and a Genoese brigantine was totally lost so late as the summer of 1815. Aphrico rock.

Rock Pomo, off Lissa, is shown on the early manuscript plans, and is given by Coronelli in the great Atlante Veneto; but it was not inserted in the charts issued by the ship-chandlers about 1790. Pomo.

The reef off Cape Bianco, Corfu. This appears in several of the portulani, and is very fairly figured by Coronelli; yet so little was it known of late, that two or three of our ships of war got aground on it while I was in the Mediterranean. Bianco reef.

The Gaio rock, between Paxo and the coast of Albania, was also lost from the charts; yet the Venetians had to record the loss of a valuable treasure-ship on it, and so late as 1817 two of our frigates struck there. Gaio.

The Patella shoal, off Prevesa, was well known to Gorgoglione and Mesfud, but disappeared since their time; it was again restored in the late war by one of our cruisers, the *Topaze* frigate, running upon it, where she lay several hours, but luckily the weather was very fine. Patella.

The shoal off Cape Chiarenza, named Montagu by us, on account of the line-of-battle ship of that name having run upon it in 1810, while sailing on an expedition against Santa Maura, is well marked in the old works. Montagu shoal.

Capra reef. Crescentio's portulano, page 49, mentions that in 1595 the Ragusan ship *Berniccia* lost her rudder on the shoal off Cape Capra, Cephalonia. The chart supplied by the Admiralty in 1810, had it not. Capra reef.

The rock off Cerigotto—on which the *Nautilus* sloop-of-war was lost in January, 1807, and 58 of her crew perished miserably—is marked in the portolani, and omitted in recent charts. Nautilus rock.

Skerki.

The Skerki rocks, between Sicily and Tunis, on which several ships have been lost in the last half-century; particularly H.M.S. *Athénien*, of 64 guns, in October, 1804, when 351 officers and men, including the captain, were killed or drowned, besides many passengers. This danger is also well placed in a curious vellum manuscript of 1547, which Sir Thomas Phillipps, Bart., at my request, exhibited to the Royal Geographical Society, in 1851 (*see my Address to that Society, Geographical Journal, vol.* xxi., *page* lxxi).

Sorelle.

The Sorelle reef, off Galita, on which, in the year 1820, a Tunisian cruiser was lost; and whereon the *Avenger* steam-frigate struck in December, 1847, when the captain, officers, and seamen, to the amount of 246, were drowned, only one lieutenant, the surgeon, the gunner, two petty officers, two seamen, and a boy, being saved. I had given the name of Sorelle to the two heads of this dangerous rock—which are nearly on a wash with the water—because they lie opposite to the high rocks on the coast of Barbary called Fratelli (*Neptuni aræ*); but, singularly enough, I have since found that the latter went by the designation *do Soror;* see the chart of B. P. Sina, 1488, and other middle-age hydrographers.

Fumosa.

Fumosa reef, in Baia Bay. This range of rocks was well-known to the Neapolitan and Maltese pilots, and was accurately surveyed by myself. Yet when the English and French fleets, under Parker and Baudin, were watching the disturbances of 1848, in working out from under Pozzuoli, Admiral Baudin's ship struck on the Fumosa: the chart in use was a reduction of the large four-sheet Italian survey, whereon—although I had furnished Visconti with my coast contour and soundings—it did not appear!

Further results of such neglect.

Although in this enumeration I have merely alluded to the early portolani, without forestalling my text, I may here mention that the more recent plans and drawings preserved in the British Museum also reveal the awful neglect of our modern chart-wrights, and it was high time that Government should take so important an affair out of irresponsible hands. Among many other matters, the examiner will find on charts drawn more than a century ago, with bearings and leading-marks, many of the rocks supposed to be recent discoveries.* The noted shoal off Al Bekur, on which

* Though not connected with the Mediterranean, I cannot but recall a remarkable fact in point. Agatharcides, in describing the coasts of the Red Sea, 170 years B.C., says, that at the entrance of the Elanitic Gulf, there are three islands covering several harbours on the Arabian shore. Yet these islands, lying so conspicuously on a dangerous lee-shore, do not appear to have been noticed in any European chart or description, till, after a lapse of twenty centuries, they were restored to hydrography by Mr. Eyles Irwin, of the East India Company's service.

the *Culloden* struck, an accident which might have occasioned the loss of the battle of the Nile, was tolerably well drawn on the homely plans of Lorenzo Mesfud and Antonio Borg; and it was even published in Bellin's *Mediterranean Atlas* so far back as 1771. The shoals near the Egyptian coast also, on which, in 1800 and afterwards, so many of our vessels struck, besides our actually losing the *Cormorant* of 24 guns—the *Fulminante*, 10—and the *Parthian* sloop-of-war on them, were well known long before. The Lefkimo shoal, Corfu, on which several of our ships have struck, is well placed on the older surveys; and so is the Gomenizze shoal, in the channel of that island—whereon the *Bacchante* frigate lay many hours, and was obliged to throw her guns overboard to lighten her—on which Borg marks one *brazzo*. The bank which tails off Augusta, in Sicily—where we lost the *Electra*, of 18 guns, in March, 1808—is well drawn by the pilots of the Maltese galleys; and the channels of Trapani, on the west side of the island, appear to have been very fairly examined by them, although they remained nearly unknown to our cruizers. At the close of October, 1803, the fleet under Nelson anchored at the Madalena islands, which had recently been examined by Captain Ryves. When they had watered, placing the fullest reliance on the chart furnished by that officer, the ships beat out without any accident. In the following year, however, a line-of-battle ship, the *Excellent*, struck on a rock just outside the very centre of the channel, and two other dangers were found in the vicinity of the spot where our fleet had been beating. Admiral Sir R. G. Keats told me that he congratulated Nelson on having escaped so well, adding—'It is evident, my lord, that Providence protects you.' These rocks were known to the Maltese pilots, yet might have occasioned a ruinous loss at the opening of an eventful war. Again, the extensive reef off Marsa Scirocco in Malta, on which the *Alexander*, 74, was greatly damaged in 1799, is shown on those old plans; as is also the shoal in Carbonara Bay,

H. M. ship *Culloden*.

Other ships.

H. M. ship *Bacchante*.

Nelson at Madalena.

H. M ship *Alexander*.

Sardinia, on which the French lost two valuable store-ships in their ill-fated expedition of 1793.

Shoal at Elba.

Among a few documents of the kind which I presented to the British Museum in 1848, is a plan of the north-east part of Elba, surveyed on the 4th of June, 1772, by Lorenzo Mesfud—'Primo piloto sulla capitana Galera della Sacra Religione Gerusolimitana di Malta;' though rudely drawn, its soundings are correct, and the marks for a dangerous shoal—since omitted—in the inner channel are admirably given; namely, the inner side of Topi islet in one with point Pera, and Cape Vita on with Torre di Giove. Again, respecting the rock off Cape Matafuz, forming the east point of the Bay of Algiers: on my visit to this part in 1816, in passing Matafuz at rather more than a mile distant, I perceived a breaking sea in the offing; yet the wind being fresh, could take no particular notice of it at the time. But some time afterwards, on looking over some nautical plans by the pains-taking Mesfud, I found one with a shoal marked near the spot on which we observed breakers. I therefore gave directions to Lieutenant Slater, who commanded my tender, the *Nimble*, to examine it in 1826; he soon found the rock, and sounded the whole vicinity. He could not, however, discover less water than four-and-a half fathoms; and this was precisely in the position from the extreme point of Matafuz that Mesfud had placed it nearly sixty years before.

Shoal off Matafuz.

§ 3. Modern Operations.

17th & 18th centuries.

DURING the seventeenth and eighteenth centuries, many hydrographical works on the Mediterranean shores were published; but they were generally by mere compilers, of whom those only who dabbled in nautical science, or were known as surveyors, will be mentioned. Among the earliest of this class was Bartolommeo Crescentio, a Papal engineer. I have stated above that he was surveying the coast of Algiers in 1585; and having completed that work, which was personally presented to Pope Sixtus V., he drew up his treatise, *Della Nautica Mediterranea*, and published it at Rome in 1607. He treats of the construction and fitting of galleys; the regulation of arsenals, as evidenced in those of 'Genova, Ligorno (*sic*), and Corfu;' and in the valuable sailing directory which he appends, he describes several of the above-mentioned shoals. In 1612, Francesco Basilicata executed a survey of the coast of Candia; and a series of fifty plans of the different ports thereof, in well-finished pen-and-ink drawings, is preserved in the Royal Collection (cxiii. 104), now added to the British Museum; and, on close comparison, there are reasons for supposing that both Jean Oliva of Marseilles, and Gio. Ant. Magini, had access to them when in hand. To this epoch, and these men, we owe our freedom from many of the errors and objections with which their predecessors were oppressed, and especially those which regarded the length of the Mediterranean Sea.

Bart. Crescentio. **Fran. Basilicata.** **Magini.**

Chartography was now on the advance, both in its general and special departments; but the documents then produced, compared with the works of the present day, display many faults of calculation, omissions, and traces of ignorance. Still in their endeavour to remove difficulties, by the correction of some long-established error, or the supply of some new information, our predecessors are entitled to gratitude; and they left hydrography richer than they found it. The use of plane-charts had continued till their mistakes were exposed, in 1556, by Martin Cortes, the celebrated author of the *Art of Navigation;* and to correct which, Gerard Mercator had also published a chart with his method of keeping the meridians and parallels of latitude in straight lines as before, but increasing each portion of the latitude with its distance from the equator as a compensation. All this, however, was done without

Chartography. **Cortes.**

any fixed rule how to divide the enlarged meridian; the discovery of a method for thus ensuring accuracy was reserved to Mr. Edward Wright, of Caius College, Cambridge, who, in 1599, published the first table of meridional parts for that purpose, in his work, intituled *Certain Errors in Navigation detected and corrected;* and as all charts prior to this application were erroneous in the increase of the degrees of latitude, the present excellent line projections should rather be designated Wright's than Mercator's, even although his received some improvements before it acquired geometrical exactness.* The supposed knowledge, however, of pilots, was an impediment to the advance of accurate hydrography; for most commanders of ships placed implicit reliance on their dicta, as if so uneducated a class of men were gifted with intuitive precision. 'Pilots, now-a-days,' said Pigafitta, the companion of Magellan, 'are satisfied with knowing the latitude, and are so presumptuous that they refuse to hear mention of longitude:' and Martin Cortes, in his epistle to Charles V., asks 'How much more shall the same seem difficult to Solomon, if at these days he should see that few or none of the pilots can scarcely read, and are scarcely of capacity to learne?' And it may be remembered to how recent a day the marine adage of the three L's obtained, meaning that the essence of navigation consisted in lead, latitude, and look-out!

Edward Wright.

Pigafitta.

About this time, many portions of the Adriatic Sea were examined by the officers, Giovanni Vitelli and Gerolimo Benaglio; and their surveys were protracted and drawn by Car. Cappi, in 1630. Between the same year and 1646, a work was in preparation which, from the known energies of the author, was anxiously looked for: at length, in the last-named year appeared, in two large folio volumes, that celebrated book, the *Arcano del Mare*, by Robert Dudley, a natural son of the court favourite, Robert,

Vitelli and Benaglio.

Robert Dudley.

* It is not improbable that Mercator worked from a globe, with rhomb-lines drawn upon it.

Earl of Leicester, and who, after having achieved some daring maritime enterprises, settled, as Earl of Warwick, at Florence, when by the emperor's creating him a duke of the Roman empire, he assumed Northumberland as his title. Dudley was one of the most remarkable men of his day, and greatly attached to science in general: he laid the foundation of many nautical and commercial improvements, and suggested the process for draining the maremme between Pisa and Leghorn, which last he was instrumental in making a free port. The *Arcano del Mare* is replete with skilful projects for the advancement of maritime knowledge, as well in building and fitting ships as in navigating them, and instructing their commanders. It is full of charts and plans, of which that of the Mediterranean Sea, though imperfect, was the precursor and model of the noted French *carte réduite:* and when we consider the kind of materials then available, and the state of practical mathematics, both reason and justice demand our respect for this able pioneer of hydrography.

Dudley's works.

From the works already mentioned were constructed the collection of charts by the two Cavallinis of Leghorn, in 1644, as well as those of Nicholas Comberford, of 'Redcliffe,' in 1657, corrected with a few local re-examinations; and that of the coast of Catalonia in 1650, of which the original, on five sheets, is in the British Museum (*Royal*, lxxviii. 31). At length the Directory of Francesco-Maria Levanto appeared, and instantly became a favourite leader among masters and pilots: it was intituled *Prima parte dello specchio del mare, nel quale si descrivono tutti li porti, spiaggie, baje, isole, scogli, e seccagne del Mediterraneo. Dato in luce* 1664. This folio volume was the text-book of our Wapping chart-sellers—Thornton, Hack, Gascoyne, Page, Mountain, and others—while the contemporaneous works published in Holland, as *L'Europe marine* of Ulas Bloem, and the *Monde Aquatique* of Peter Goos, were professedly copied from the Italian portolani.

Other publications.

In the year 1679, the attention of the French government was strongly directed towards a more intimate acquaintance with Mediterranean navigation; in consequence of which, Captains Cagolin and Chevalier, of their Royal Navy, with some intelligent engineers, were sent to the coasts of Spain and Italy, as well as into the Adriatic and Archipelago. But I am not aware that any beneficial result to hydrography followed these movements, unless it be true that R. Bougard, Maître de Navire, who published his *Petit Flambeau de la Mer, ou la Véritable Guide des Pilotes Côtiers*, in 1684, and the Maltese pilots, Olivier, Michelot, and Therin, who furnished the first engraved chart of the Mediterranean in 1689, gained access to the French documents. The Chevalier de Tourville addressed a letter to the Minister of Marine, dated 22nd December, 1685, on the necessity of constructing a better chart than any in use, for the navigation of the Mediterranean Sea; though his ideas on marine surveying, and the means for carrying out his views, were not very clearly expressed. Olivier's and Berthelot's were plane charts, and full of faults: 'Dans beaucoup d'endroits,' says M. de Chabert, 'ces défauts en latitude alloient à plus d'un demi degré: dans la plupart il n'y avait pas seulement une échelle de latitude, et que sans s'embarrasser de la situation des différentes terres par rapport au ciel, elles étoient placées à peu-près dans leurs distances grossièrement estimées, et dans leurs directions, suivant la boussole, dont la déclinaison étoit mal connue ou absolument ignorée.'

Cagolin and Chevalier.

Maltese pilots.

Chev. de Tourville.

Chabert.

In 1686, and the two following years, M. Mathieu de Chazelles, Hydrographe de Galères at Marseilles, corrected various points of the south coast of France; which so recommended him to the authorities, that he was commissioned to visit Greece, Turkey, and Egypt, on a scientific mission, in 1693. He made numerous surveys, with the intention of constructing a general Mediterranean Atlas upon thirty-two sheets, and on his return, his project was approved of by the Academy of Sciences; but the whole

Chazelles.

was frustrated by a lingering illness, which terminated in his death on the 16th of January, 1710. Meantime, Henry Michelot, *Pilote Hauturier sur les Galères du Roi,* after thirty years of experience, had published a compendious sailing directory for this sea, which, though owing largely to Crescentio and Gorgoglione, became very popular with the French and English seamen, and is still in request among the coasting craft of the former. It has run through several editions, of which I possessed copies of those of 1709 and 1806. The Père Baudrand, the rival of the Sansons, had published the Principality of Catalonia, and the County of Roussillon, on two sheets, in 1693.

Michelot.

Baudrand.

Nor had the Italians neglected the study of hydrography, so far as it was then understood and practised, as is shown by numerous documents and volumes found in libraries. From 1685 to 1718 the *Atlante Veneto—Morea—Isolario,* and other geographical publications of the laborious Venetian cosmographer, Padre Vincenzo Coronelli, were in circulation; and they really are depositaries of a great fund of substantial knowledge. Some of the plates combine both map and picture, so as to convey a clear idea of the object represented, though often rudely, or even incorrectly drawn; and the profusion of accessories, mostly pregnant with meaning and interest, shows that no expense was spared in their production. Indeed, the efforts of this very industrious compiler were so effectually aided by the encouragement of the doges and nobles of Venice, together with the association designated *Gli Argonauti,* that he was enabled to publish more than four hundred maps, with copious explanations of them. One of his contributors, Paolo Gerardo, published a volume treating of the passage along the east coast of the Adriatic, and thence across the Archipelago to the Holy Land. The rest of the coasts are more slightly touched, and in the Archipelago there is a mere enumeration of courses and distances from isle to isle. A copy of this is preserved in the Bodleian Library, at Oxford; it is intituled *Il Portu-*

Coronelli.

Paolo Gerardo.

lano del Mare; nel quale si dichiara minutamente del sito di tutti i porti quali sono da Venetia in levante, e in ponente. Venetia, MDCXCIX.; and it was then '*ristampato*.'

Le Père Feuillée.

Cassini.

In 1699, le Père Feuillée, the well-known and useful astronomer, was despatched by Louis XIV. on a scientific mission into the Levant, in company with Jacques Cassini, the future opponent of Newton as to the earth's shape, then only twenty-two years of age. They were directed to determine the exact position of various cities and ports, and to pursue every measure for the improvement of navigation; but though there appears to have been no want of theoretical talent, the results which oozed out disappointed the expectation of practical men, who suspected that much information was suppressed by authority. In the same year, M. d'Ablancourt published a chart of the strait, by order of the king of Portugal; this was said to be drawn up from careful observations made by the most experienced mariners and engineers to show plainly every anchorage. M. d'Ablancourt was of French extraction, claiming descent from the Perrot family; and he is said to have been an able master in hydrography.

M. d'Ablancourt.

Meanwhile, the English were not unmindful of Mediterranean hydrography, albeit their intercourse with that sea was then on a very restricted scale. Mr. Charles Wyld, whose large plan of the road and harbour of Cephalonia is preserved in the British Museum (*Sloane*, 2439, *fol.* 29, *b.*), made several minutes on the Ionian Islands and the coast of Albania, in 1673, some of which fell into the possession of the earnest explorer of those parts, the late Mr. John Hawkins, of Bignor Park, whose name was not forgotten in Greece when I was occupied there. Captain John Kempthorne, who commanded the *Dover*, of 48 guns, appears to have had an express mission to that sea; as there are in the Royal collection of the Museum, a series of plans and views of Cadiz, Tarifa, Gibraltar, Genoa, Leghorn, Naples, and parts of Sicily, Malta, and the Greek Islands, exe-

Charles Wyld.

Captain Kempthorne.

cuted by him between the years 1685 and 1688. A large folio volume written by Sir Nicholas Miller, shortly afterwards, containing not only sailing directions, but very numerous views of headlands, and outlines of harbours on the coasts and in the islands of the Mediterranean, is worthy of notice, though rudely executed. But the most industrious, and perhaps the best qualified, of the explorers of that day, was Mr. Edmund Dummer, who made many local surveys and views, which were evidently used by the compilers of our *Quarter Waggoner*;* and for which he was rewarded with the post of surveyor and commissioner of the navy. This gentleman was sent out in the *Woolwich*, of 54 guns, commanded by Captain William Houlden (*Houlding* according to the acrostic in Chaplain Teonge's amusing diary), himself an experienced Mediterranean cruiser. In this ship he visited the coasts of Spain, France, Italy, Greece, and their islands; and in the British Museum (*Royal MS.* 40) is a folio volume which bears testimony to his industry and observation: it is intituled *A Voyage into the Mediterranean Seas, containing by way of journal, the views and descriptions of such remarkable lands, cities, towns, and arsenals, their several planes and fortifications, with divers perspectives of particular buildings, which came within the compass of the said voyage: together with the description of twenty-four sorts of vessells of common use in those seas, designed in measurable parts, with an artificial show of their bodies, not before so accurately done: finished in the year* 1685, *by Edmund Dummer.*

Sir N. Miller

Dummer

The eighteenth century opened with a request from the

* This is the title of a ponderous volume which was long the *ne plus ultra* of naval hydrography; insomuch that the official certificates from captains and masters, apologizing for making no improvement in the charts, stated that they had met with nothing but what was already in the *General Quarter Waggoner.* In the seventeenth century, books of charts were colloquially termed Waggoners—perhaps a corruption from Lucas Jansz Wagenaer, author of the *Spieghel der Zeevaert*, or Mirror of Navigation, published at Leyden in 1585.

Emperor of Germany not remotely dissimilar from that which I received in 1817 from the same quarter; namely, for the aid of an English officer to superintend a survey of the ports in Istria and Dalmatia, with a view of selecting a safe and convenient harbour for shipping in the Austrian territories on the Adriatic. On this occasion Queen Anne selected Doctor Halley, who had already acquired the *brevet* rank of a captain in the royal navy, although there were such men in the service as Swanton, Fairfax, Trevanion, Haddock, Saunders, Wager, and Harlow, as well as that other *brevet*-captain, the meritorious Dampier, then unemployed. By the year 1702 everything was arranged, and Halley departed for the Mediterranean; where he executed his task so satisfactorily, that the emperor presented him with a valuable diamond ring, taken from his own finger, and he also wrote a letter to the queen, expressive of his gratification. When re-surveying these parts in H.M. ship *Aid*, I was naturally anxious to learn the opinion of so eminent a predecessor, and sought a sight of his manuscripts through my Austrian colleagues; but could only learn that they might possibly be found, in course of time, in the *gurgite vasto* of the Vienna archives.

Halley.

Halley's papers.

Jesuits.

In 1708 the Jesuits, under the auspices of their sagacious and powerful brother, the Père de la Chaise, for certain purposes of their own connected with Avignon, intrigued with the minister—Count Pontechartrain—for permission to make an extensive survey of the coast of Provence: their prayer was granted, and they set about the enterprise with great apparent energy, but the results never appeared. Simultaneously with this undertaking, Señor Diego Cuelbis was engaged in drawing up his *Thesoro Chorographico* of Spain and Portugal, of which a manuscript copy is preserved in the British Museum (*Harl.* 3822), illustrated by pen-and-ink sketches. Shortly afterwards, Sebastian Gorgoglione, a skilful Genoese pilot, published his well-known *Portolano del Mare Mediterraneo*, a work which quickly

Diego Cuelbis.

Gorgoglione.

became a popular standard among seamen, ran through various large editions, and is still very extensively in use with the cabotage vessels, or coasters. The first edition, which I have never been able to meet with, though General Visconti of Naples aided my search, was dubbed *la veritabile e luminosissima face del mare*, by no less an authority than Admiral Angelo Emo, the last naval hero of republican Venice. The best known editions are— Various editions.

I. NAPLES . . . 1717.	III. PISA . . . 1771.
II. NAPLES . . . 1726.	IV. LEGHORN 1799.
V. LEGHORN . . . 1815.	

The next publication which had a decided *run*, was a book written in 1732 by M. Ayrouaud, pilot of the French king's galleys: this was a volume of Mediterranean harbours, bays, and roadsteads, with views of the most remarkable headlands, and *reconnaissances des attérages;* which though a *réchauffé* of all others, with views so coarsely executed as only to merit the title of 'ugly likenesses,' was considered by the native pilots to be an excellent accompaniment to Gorgoglione. The public approbation of this work induced the Marquis d'Albert—who, although very young, then presided at the hydrographic dépôt—to endeavour in 1737 to revive the project of M. de Chazelles; but the result of his exertions was the production of so bad a chart, that the editors themselves felt it requisite to announce, that it was far from the perfection such a work ought to possess. M. Ayrouaud. Marquis d'Albert.

The next who bestirred himself with zealous activity in the cause, was the Marquis de Chabert, a very intelligent officer in the French navy. This gentleman made a representation to the Académie Royale des Sciences, on the state of Mediterranean hydrography; which able discussion is printed in their *Mémoires* for 1759, page 484. In this, after enumerating the defects, he makes a direct proposition, 'pour former pour la mer Mediterranée, une suite de Cartes Marquis de Chabert.

exactes, accompagnées d'un Portulan;' asserting that all the existing graphic representations of that sea were wretched productions—*qui ne méritoient pas le nom de Cartes.* M. Rouillé, the then Minister of Marine, struck with so strong an assertion, appointed M. Chabert to sail on a cruize for the improvement of navigation. He accordingly sailed, and visited numerous parts of the Levant; but I am not aware that any public good resulted. In 1764, he again returned to the charge, and under the ministry of the Duke of Choiseul, was entrusted with another voyage of rectification. But although this cruize was an expensive one, the results were not communicated, however they might have been indirectly used in correction: for A. Drury, in his maps of the King of Sardinia's dominions, on twelve sheets, in 1765, together with the Republic of Genoa, is reported to have had access to all the documents in the dépôt at Paris. Be this as it may, the Marquis undertook another *campagne* in 1771, to fix various positions in the Archipelago; on which occasion he carried with him Ferdinand Berthoud's chronometer, No. 3, which had done good service during M. l'Abbé Chappé's voyage to California. In 1775, he made yet another *campagne*, with two of Berthoud's watches, and is said to have done much good work; but the navigator was not informed how. 'Ce travail, très-étendu, n'a pas été publié,' is the remark in the *Histoire de la Mesure du Temps.* Many have called this expedition his last scientific Mediterranean voyage; but it happens that there is in the British Museum (Add. MS. 15, 326, 14) a plan of which the title is—*Plan du passage de l'isle Longue, ou Mavro Nisi, à la côte orientale de la Attique, où se trouve le mouillage de la Mandri, levé en* 1787, *par M. le Marquis de Chabert.*

Chabert's first cruize.

Second.

Third.

Fourth.

Fifth?

Coeval with the early exertions of this nobleman, the Sieur Bellin, Ingénieur de la Marine, et Censeur Royale, and a Fellow of our Royal Society, published his *Description du Golfe de Vénise, et de la Morée,* his Corsica, and

Bellin.

his elaborate general Atlas. The latter was certainly the best 'got up' compilation which had as yet appeared; and though the style of the engraving may appear rather coarse, still it is equal to the method of surveying at the time, and there are many reasons why it should always find a station in a hydrographical library. It will be recollected that it was to secure a copy of this atlas, as well as to purchase, at any price, plans of all the French Mediterranean ports, that Lord Camelford resolved upon going himself to Paris in the winter of 1798, while his ship—the *Charon*, 44—was fitting at Woolwich: but that strange step led to his being arrested at Dover, and superseded from his ship, on which he indignantly quitted the navy. Bellin's Atlas. Lord Camelford

Meantime, Cassini de Thury was throwing his grand chain of triangles over France: still, without detracting from the merits and exact particulars of his geometrical description, it must be confessed that his contour of the Mediterranean coast of that kingdom is both incorrect and ill-drawn, conditions quite unexpected at such hands. Between the years 1780 and 1785, several tolerable charts of the south coast of France were issued from the Parisian press; yet they also proved inferior to what was looked for from their rumoured command of official documents. Baron de Zach, in his various visits to that part, between the years 1787 and 1805, made many astronomical and geodesical observations at Marseilles, Toulon, and Hyères; though, as he remarks, 'English fleets and French suspicions' prevented his going to work on a large scale. In 1792, the French Dépôt de la Marine published a very useful chart of the space extending from the mouth of the Rhone to Villa Franca, on a scale about one-third that of Cassini's: this chart soon became well-known, but although it confessedly was a greatly improved specimen of hydrography, it still contained some serious errors. Cassini de Thury. Baron de Zach.

During these operations, various surveys had been made in the Archipelago, under the direction and influence of

Count de Gouffier.

the Count de Choiseul Gouffier, so justly celebrated for his antiquarian and artistic researches: and several anchorages in the Levant were amended by our naval officers—as Messrs. Clancy, Kirby, Atkinson, and Captain John Stewart, of the *Seahorse* frigate—though their improvement in chartography was but humble. Our travellers also rushed to those shores in shoals: yet among the numerous (I had almost said innumerable) volumes printed by these gentlemen, though undoubtedly we gather much general information, enjoy numbers of accurate pictures, and discuss many points of scholarship, still we find but little which bears upon exact science; and among the myriads of adventurers, scarcely anything is added to our hydrographic knowledge.

English tourists.

Contemporaneously with the Count de Choiseul Gouffier's operations in Greece, was the survey of Southern Italy, by a corps of engineers and draughtsmen, under the direction of G. A. Rizzi Zannoni, a clever man, though he spread rather more sail than his stability warranted; and who was greatly patronized by King Ferdinand. The work was conducted on a very extensive scale, and appeared to be worthy of the activity displayed; but the product was not of sufficient accuracy to warrant the great expense incurred. After years of boat and field work, Zannoni published a costly volume, comprising the coasts of Naples and Calabria, on twenty-three large and fairly engraven sheets; under the title, *Atlante Marittimo delle due Sicilie;* Sicily proper, however, was not treated of beyond the Faro of Messina. The interior space was also mapped by the same surveyors, and engraved by Nicholas de Guerra; and these two atlasses formed the main basis for the large publication which Bacler Dalbe brought out in 1802, as the *Carte Générale des Royaumes de Naples, Sicile, et Sardaigne.* In the year 1798, a chart of the Adriatic Sea, on nineteen sheets, was published at Venice as the combined work of Zannoni and Vincenzo di Luccio; the last being the doge's pilot, and one who had spent fourteen years in

Zannoni.

Bacler Dalbe.

examining the waters for hydrographical purposes. But though he is said to have delineated 413 shoals, and 410 islets more than had appeared in any former chart, and to have ascertained the direction and variation of all the currents throughout what he terms that 'dangerous navigation,' I found his work replete with the grossest errors. De Luccio's chart.

No maritime nation in Europe had, towards the close of the last and at the beginning of the present century, published so great a series of excellent charts as the Spaniards, whose Joachims, Luyandos, Malespinas, Ciscars, Bauzàs, Ferrars, Espinosas, and others, carried nautical science to all the littoral parts of the globe; and their labours have enjoyed the highest estimation among the officers of our own navy. In 1783, after due preparation, a survey of the entire line of the coast of Spain, both in the Atlantic and the Mediterranean, was ordered by the government of Spain. This was executed by the cadets of the naval academies under the direction of Don Vincente Tofiño de San Miguel; and the finely engraved charts of the coasts and harbours, being drawn from large scales, form an elegant collection, in two folio volumes called the *Atlas Maritimo de España*, accompanied by a description of the shores and directions for the navigation thereof, in two quarto volumes, intituled *Derrotero de las Costas de España; Madrid*, 1789. Spanish surveys. Tofiño.

The views of the Spanish hydrographers were then extended to other coasts, and their corrections were as valuable as various. In 1802 Don Dionysio Alcala Galiano, and Don Josef Maria de Salazar, obtained several chronometric differences between important stations; and they fixed numerous geographical positions at the Dardanelles, Constantinople, Smyrna, Candia, and the north coast of Africa. From these labours the best chart of the Mediterranean which had yet appeared was constructed and published in 1804, with the following title:—*Carta esférica que comprehende las costas de Italia, las del Mar Adriático, desde* Galiano and Salazar.

Galiano's chart.

Cabo Vénere hasta las islas Sapiencia en la Morea, y las correspondientes de Africa, parte de las islas de Corcega, y Cerdeña, con las demas que comprehende este mar, ec.

Spanish loss at Trafalgar.

But the battle of Trafalgar crushed the Spanish pursuit of maritime science; for it unfortunately happened that the only captains of ships killed in their fleet during the decisive combat on the 21st of October, were the three reputed the most accomplished officers they possessed; namely, Galiano, Alcedo, and Chirucco, who respectively commanded the *Bahama*, *Montenez*, and the *San-Juan Nepomuceno*, of 74 guns each.

Political events had increased our own acquaintance with the Levant, and the absence of accurate knowledge respecting some parts of it, was brought under the attention of the Lords Commissioners of the Admiralty; who, as it was of importance to ascertain more completely the nautical resources of Asia Minor, resolved to employ a frigate to make a detailed survey of it. On this occasion it was fortunate for the intellectual and professional character of the navy, that the *Fredericksteen*, of 32 guns, commanded by that excellent and well-informed seaman, Captain Francis Beaufort, happened to be then stationed in the Archipelago, and consequently was selected for that service; since an officer better qualified for settling the hydrography, and describing the venerable relics of antiquity of those interesting shores, did not exist. This survey, begun in July, 1811, was pursued with diligence, until it was unhappily terminated on the 20th of June, 1812, on Captain Beaufort's being desperately wounded by a party of assassins at Ayyás, a bay on the north side of the Gulf of Iskenderún. Despite of the interruption and delay attending this disaster, the fruits of the mission were a systematically digested atlas of charts and plans, a nautical memoir of them, and a descriptive volume upon Karamania, of such merit as truly to meet the words—*indocti discant et ament meminisse periti.*

Captain Beaufort.

He is wounded.

Having thus given a rapid sketch of the labours of those who preceded me, I am enabled to make a tolerable summary of the state of these matters at the opening of the nineteenth century: and it will be seen that, though hydrography sprung up, as it were, in the Mediterranean, and was strengthened by the local knowledge of intelligent navigators and pilots, yet our present enlarged and accurate familiarity with the waters and coasts of that sea, is the result of the improved state of observation chiefly in our own times. Witness, among other remarkable cases which might be cited, the astonishment and agreeable surprise of those engaged in our Egyptian expedition, in the winter of 1800; when beating about in a furious gale, they unexpectedly found themselves sheltered in one of the finest harbours in the world, the Bay of Marmerícheh, which had never been heard of by a single man in the whole armament. (*See page* 78). No such port can any longer remain hidden on our charts; and now, what with the great improvement in their execution under superior methods of observation, a completer course of points chronometrically determined, lighthouses more numerous and effective, with an increased number of better-placed buoys and beacons, daily advances are made in insuring the safety and seconding the skill of the navigator. Remark. Striking incident.

Nor is it difficult to perceive a cause for the state of mediocrity in which the hydrography of the Mediterranean so long remained, especially in those parts remote from the principal trading-places, in the constant and inveterate warfare between the Cross and the Crescent, in which defeat was ever attended with slavery and woe. Added to this, nautical science must have been considerably weakened by the division of the pilotage into two branches, the *navigazione di altura*, or open-sea work: and the *cabottaggio*, or coasting voyages; the second being held as secondary in rank compared with the first. By this rupture of what in reality should only constitute a whole, an impe- Probable causes.

Causes of neglect.

diment to progress was surely created. While a select few of the best seamen bestirred themselves in the cause, by giving their attention to the establishment of the true course and distance from one place to another, and thereby fixing their latitudes and longitudes, the majority were content with lead and look-out, and the monitory dicta of the printed directories. To the combined action of these two causes, together with political jealousies, and the operation of quarantine guards, it is to be imputed that a sea somewhat limited in extent, but of vast importance in a geographical and commercial point of view, and of exceeding interest in various respects and places, should have remained for so many ages, comparatively speaking, so imperfectly known.

Yet merit was displayed.

Still, much had been achieved, and much allowance should be made for the greater portion of what remained undone: and when we weigh the merits of our predecessors, we must consider them chronologically, for hydrography is peculiarly a progressive science, and as such has always been so incomplete as to demand a continual correction of its errors. And further, the contour of a coast, the depth of the waters which wash it, and the surface of the adjoining land, may undergo strange alterations within the period of even a single century, from volcanic and electric agencies, the combined action of winds, temperature, pressure, and the whole train of atmospheric affections. These, however, are only a part of the causation, for we must add the wearing of tides and currents, local anomalies, the works and efforts of man, extraordinary convulsions by land and sea, encroachments of the land on the sea, or the sea on the land, the lowering of mountains, the elevation of plains, and all the other influences in a constant state of activity. Nothing, therefore, can be more unjust or unsound, though nothing is more common, than for geographers to condemn most unsparingly the labours of their predecessors, without adverting to the circumstances, *pro et con*, in the history

Injustice of hyper-criticism.

of each case. An advance towards excellence will probably be made in every future age, though an absolutely correct and perfect chart can never be formed, as long as the powerful but invisible agents here enumerated continue to act—complete perfection can never be obtained by any work of man. Thus Cellarius, Riccioli, Merula, and Salmasius, forgot to whom geography owed its rise to the dignity of a science, and were very unduly severe in their censures on Ptolemy's mistakes, instead of ascribing them to the defective knowledge and imperfect instruments of his age; while their own labours are now criticised with as little candour by the writers of the present day; who, if they reflect, may form some notion of the estimation in which themselves will be held by the learned geographers of A.D. 2500! Improper censure.

§ 4. The Author's Surveys.

SUCH then was the actual state of our hydrographical knowledge of the Mediterranean Sea when I first went to Spain, in the *Milford*, of 74 guns, bearing the flag of Rear-Admiral Sir Richard Goodwin Keats. But of the surveys above-mentioned, a portion only were known to us generally; and the miserable charts of Heather, Norie, Blachford, and other ship-chandlers, were officially used in our ships of war. There was hardly a plan of any considerable harbour of our own execution, which had sufficient merit to illustrate a master's log-book, except some detached draughts by Admiral John Knight, and the Maddalena islands by Captain G. F. Ryves. My commencement.

Being appointed to the command of a large Spanish gun-boat, the *Mors aut Gloria*, on the 4th of September, 1810, one of the first acquisitions I made—and it was through the courtesy of the late excellent Admiral Valdes—

Tofiño's Atlas.

was the maritime survey of Don Vicente Tofiño, already mentioned. Gratified, however, as I was by the execution and elaborate details of that beautiful work, I could not but soon perceive that there were various omissions, and not a few errors of commission; and between the date just mentioned, and the close of 1812, I had many opportunities of examining the local minutiæ, and making additions. My respect for so really valuable a publication—at that day the very first work of its kind—prevented my pursuing this course so earnestly as I should otherwise have done: but as even that work was susceptible of improvement, I made various sketchy re-examinations of such portions as duty carried me to, and especially of Cadiz and its environs. And these corrections were continued, as well outside the Strait of Gibraltar as on the Mediterranean Coasts of France and Spain, and amid the Balearic Islands, in the *Milford*, the Spanish gun-boat, an armed transport, and the *Rodney*, 74.

Too implicitly relied upon.

Having returned to England, I was requested in May, 1813, by my friend and recent gun-boat commodore, Sir Robert Hall, to join him in the Anglo-Sicilian flotilla, then employed in defending Sicily against the French under Murat. Previous to going out, I consulted with Captain Hurd, the hydrographer, to whom my late contributions to his department had introduced me, respecting the state of the Mediterranean charts; and having provided myself with some superior instruments, I offered every exertion in behalf of his office, whenever my other duties would allow an opportunity of so doing. My object, however, in thus proposing to act as a nautical surveyor on the strength alone of my professional stock of practical knowledge of the subject, went merely to compass the construction of a chart which should answer the ends of navigation better than those which were in use;* for I had no hope of having

Appointment to the Sicilian flotilla.

Captain Hurd.

---

* Malte Brun said he was always in doubt when consulting a Mediterranean chart; Baron de Zach found the positions in the Indian Ocean were

time enough to lay down, with rigorous exactitude, the many leagues of coast in question. On this occasion the Admiralty archives were thrown open to my scrutiny, and every aid was given to my inquiries by the worthy captain, as well as by his excellent assistants, the late Mr. Walker, and his son, Mr. Michael Walker, who is still (1853) employed in the hydrographical office.

The Messrs. Walker.

From the quantity of documents placed before me on this investigation, of the respective merits of which there was no criterion, I became impressed with the opinion which Captain Hurd and the Messrs. Walker had already entertained —namely, that they were provided well enough with detached surveys, but that geographical points were wanted for adjusting them to. In the hydrographer's own words, he was 'in possession of sufficient documents to construct a chart of the Mediterranean Sea, but was greatly at a loss for latitudes and longitudes to dress it by:' those before him being so vague and conflicting, that he had not the position of a headland or a lighthouse that could be depended upon; insomuch that even the breadth of the entrance of the Adriatic was unknown! And previous to my sailing, he sent me the following memorandum, as a guide:—

The hydrographer's opinion.

Our knowledge of the coasts and neighbourhood of Sicily is extremely deficient; and although there are the three observatories of Palermo, Naples, and Malta, the exact position of any one of them is undetermined.* We are also unacquainted with the true place of the important land-fall Maritimo, which, we are assured by experienced officers, is placed in the charts

---

better settled than those of North Africa; and among the hydrographic remarks in Captain Beaver's logs for 1801, are found—'We are now working up between the Sporades and Asia, but can put no faith in the 'sea-cards,' as none of the islands are accurately placed, and many are entirely omitted.'—'The passage between Samos and the Formiche is disgracefully laid down.'—'The land we marked last evening for Cape Gallo must have been Matapan, but the charts are all so infamous that it is impossible to ascertain where one is, without running close in.'—'We are now off Toro, which is placed at least thirteen miles south of its proper latitude.'—(*See Life of Captain Philip Beaver, page* 154.)

* Meaning, no doubt, that they had not been officially communicated to his office; and that therefore he was unable to pronounce exactly upon them.

Captain Hurd's minute.

twenty miles too far to the westward; and Cape Bon, on the African shore, six or seven too much to the eastward. This, if true, constitutes a most serious error, as the Esquirques, Keith's Reef, and various other dangers, at present scarcely known, lie in the fair-way, and nearly mid-distance between the Sicilian and African shores.

All the charts of Sicily that I have examined are at variance with each other; and, from our having no good authority for either, we are at a loss which to select as the best: but there are many reasons for supposing a portion has been placed by a compass north, without allowing for variation, and the adjoining parts by a true one: nor have we any particular plans to be depended upon.

The Æolian group, Pantellariæ, Ustica, and Lampedusa, with several smaller islands, are not properly placed in any of the published charts, and very little is yet known as to their history, exact number, or relative positions with each other; and there are some shoals supposed to exist on the southern and western coasts of Sicily, which it would be praiseworthy to search for, as many unaccountable losses and disappearance of vessels have taken place, at various times, in those parts.

We have no official chart descriptive of the coasts of Malta and Goza; and as these islands now form a part of the British Empire, it will be necessary to have them hydrographically examined, and their dangers pointed out. The coasts of Sardinia and Corsica are but imperfectly known, though we have some fair detached plans of their ports; it would therefore be very desirable to obtain accurate observations on the most material points, and to examine the shores as far as they may be practicable.

Sir Robert Hall.

On my arrival in Sicily, Sir Robert Hall—who was not one of the class, so common at that time, that deemed the charts in use 'good enough'—kindly offered every aid by suiting my flotilla employment, as far as he could, to the proposed hydrographical researches. On local examination, it appeared that Rizzi Zannoni's large plan of the Faro of Messina, which Captain Hurd had handed to me as meeting all the wants of the seaman, was replete with errors, and unsatisfactory in its details: I therefore bestirred myself in making a new survey of that interesting Strait, although my first intention was only the chronometric measurement of arcs. My means, on the whole, were rather powerful, for a good vessel and crew were allotted to me, and the stores of the arsenal were at my requisition. Besides being armed with two excellent chronometers—one (*Earnshaw*, 825) belonging to myself, and the other (*Arnold*, 807) to the Admiralty, I had also been furnished by the hydrographer with a 5-inch theodolite, a micrometrical telescope,

Faro of Messina.

a sextant, and a station-pointer. My own stock of working tools consisted of a portable transit; a 10-inch reflecting circle, reading to 20″ of arc; a 9-inch quintant, divided by a vernier to 10″ of arc, with a stand and counterpoises, made expressly for me by Troughton; a dipping-needle; a variation dial; a finely-divided circular protractor, with spring points; a 3½-foot achromatic telescope, with an object-glass of 2¾-inches diameter; a Gregorian reflector, of 5 inches aperture; a Rochon prismatic telescope, and some minor instruments, including a well-poised marine barometer, three of Six's thermometers, and De Luc's hygrometer.

Instruments.

The political crisis and activity of that period occasioned my being ordered about between Sicily, Calabria, and Naples rather abruptly; by which, though my principal survey was interrupted, I was enabled to obtain some very satisfactory chronometric runs. These were carried to the Observatory at Palermo, where the able and amiable Abbate Piazzi always afforded me every assistance; and where I got drilled into a more regular system of astronomical observation than I had heretofore been able to learn. By these means, many of the adjacent capes and headlands were determined, and the sinuosities between them were reconnoitered as occasion served.

Abbate Piazzi.

At length, public affairs took a decisive turn, and, in the summer of 1814, the evacuation of Sicily by our army was resolved upon. It was a stirring moment; and Sir Robert Hall having been called to command on the lakes of Canada, I was left, with Colonel Robinson of the Marines, to deliver over the army-flotilla to the Sicilian government, after winding up its affairs, and paying off the greater part of the men. It now struck me that a favourable moment offered for effectually examining the coasts of the island and its dependencies; and not being under orders either to join Sir Edward Pellew's fleet or to return home, I determined upon remaining on the spot. Naselli, the minister

Political change.

of Marine, was friendly to my intention, as were also several of the principal functionaries, and I therefore experienced no difficulty in borrowing from the government one of their finest gun-boats, a large paranzello manned with thirty Sicilians; to which was supplied a capital luntra, or boat like a whaler's, but larger, being sharp at both ends, and double-banked for eight oars. Thus equipped, I prevailed on my friends, Captain Henryson of the Royal Engineers, and Lieut. Edward Thompson of the Royal Staff-Corps—who were likewise waiting for final orders—to accompany me round the island; and these gentlemen, as my guests, gave me the only personal assistance I received, aiding me greatly in sketching the topography and fortifications during the time occupied by my nautical and astronomical operations, and assisting in the reduction of the various observations.

The gun-boat.

Captain Henryson. Lieut. Thompson.

A general peace now took place; the fleet was ordered to England, and Rear-Admiral Penrose came out, with a reduced squadron, to take charge of the Mediterranean station. This worthy and accomplished officer, after making himself duly acquainted with all the bearings of the case, warmly approved of the step I had taken; and as he considered the object of too public an interest to be carried out on individual means, he communicated his views to the Admiralty. Shortly afterwards, on my submitting a portion of the survey to his examination and care, he wrote the following official letter to the Board:—

The new commander-in-chief.

H.M.S. Queen, at Sea, 4th April, 1815.

The letter.

SIR,—Lieutenant Smyth having delivered to my charge some finished plans of ports in Sicily, requesting me to forward them for the inspection of the Lords Commissioners of the Admiralty, I have promised to do so by the first safe opportunity.

I feel it my duty to add, that the celebrated Piazzi, as well as the officers of engineers, and all other judges, give ample testimony to the extreme accuracy of the observations and calculations of Lieutenant Smyth, and I have had opportunities of comparing some on the spot, which fully corroborate it. His written remarks, both in a nautical and military point of view, are very valuable; and he has the advantage of uniting great celerity of operation with extreme exactitude.

The respectable light in which he is held by all the Sicilian ministers and authorities will enable him to act with much greater effect than any other person.

I venture to press the merits of Mr. Smyth with more confidence because he was entirely unknown to me, till I saw the utility of his professional labours in Sicily.

The very great errors detected in former charts, exhibit the value of the present survey in a strong light.

I have the honour to be, &c.,

(Signed) C. V. PENROSE, *Rear-Admiral.*

To J. W. Croker, Esq., Admiralty.

I was promoted to the rank of Commander in September, 1815; but notwithstanding the admiral's friendly exertions, and they were often renewed, I remained shifting entirely on my own means, and without any official instructions, till the spring of 1817. By this time I had made tolerably-detailed surveys of the coast of Sicily, Malta, and the neighbouring islands; besides having ascertained various geographical points on the shores of Italy and North Africa. Moreover, I attended Lord Exmouth in his first expedition to the Barbary States, with my paranzello; and when at Tripoli, prevailed on him to obtain from the Báshá of that Regency the permission by which I afterwards made the excavations at Leptis Magna (*see the Appendix*), and examined the surrounding country. These several services having been performed under the eye of the Rear-Admiral, he again and earnestly urged the Admiralty to supply me with proper assistance; and in this he was warmly seconded by Lieutenant-General Sir Thomas Maitland, the energetic governor of Malta.

Lord Exmouth.

Leptis Magna.

Sir Thomas Maitland.

In the meantime, a circumstance happened which considerably strengthened the case. Early in June, 1816, the French corvette-gabarre *La Chevrette*, commanded by Captain Gauttier du Parc, arrived at Valetta on a duty similar to that which I proposed to execute—namely, to make chronometric runs, for the purpose of adjusting the various detached surveys already made. A most friendly intercourse was maintained between us, and I assisted in placing his circle on the very spot which had lately been occupied

Capt. Gauttier.

by my own. A comparison, of course, took place; and both the Admiral and the General expressed themselves highly gratified on finding, that the mean result of the French operations gave precisely the same position for the Palace tower, as that which I had already given in. Such a conclusion was gratifying, although its full agreement was accidental; still a representation of the fact was officially forwarded by Sir Charles Penrose to the Admiralty, on the 18th of June. From Captain Gauttier, from Lieutenants de Lloffre, Gay, and Matthieu, from MM. Benoist, Allegre, Richard, Jacquinot, and Berard, and indeed from every officer of that ship, I received the most marked kindness and respect; and our intercourse was under an open and unreserved communication of instruments, methods, and documents. But in spite of all this being well known to the local naval, military, and civil authorities, and it was indeed matter of publicity, there was a spirit of detraction abroad which endeavoured, though fruitlessly, to sow discord between us. Some comments on my work, anything but complimentary, were made in one or two of the Sicilian papers, which drew an indignant reply from the Abbate Piazzi; and Captain Gauttier wrote the following letter to the editor of *L'Osservatore Peloritano*, a copy of which was forwarded to the Admiralty, and another to me, through the intelligent M. Angrand, the French consul at Malta. The editor says—

Agreement of observations.

Officers of the *Chevrette*.

A bad spirit at work.

Piazzi and Gauttier.

Essendo corso un' involontario equivoco nell' articolo, riguardante le osservazioni idrografiche fatte dal Sig. Gauttier, inscritto nell' ultimo numero del nostro Giornale, ci affrettiamo a rettificarlo col publicare la seguente lettera, tradotta dal francese, a quest' oggetto dirizzataci dal perlodato Sig. Gauttier.

A bordo della Gabarra, la Chevrettè, *nella rada di Messina, li* 19 *Settembre,* 1817.

SIGNORE,—Nel vostro foglio del 17 corrente avete fatto qualche cenno sulla missione idrografica di cui sono incaricato. Io non so d' onde avete potuto procurarvi questi dettagli, ma sono stato assai sorpreso quando ho letto l' articolo che parla di osservazioni fatte dal Sig. W. Smyth, che io ho rettificato, come voi dite. Vi priego, Signore, di smentire questo articolo per essere del tutto erroneo.

All' epoca del mio passaggio per Malta, io ed il Sig. Smyth ci abbiamo reciprocamente communicate le osservazioni da noi fatte sulle coste della Sicilia, ed abbiamo avuto la soddisfazione di trovarle perfettamente d' accordo.*

Ho l' onore di salutarvi.

(Firmato) *Il Capitano di Fregata, Cav. di S. Luigi, e della Legion d' Onore,*

P. GAUTTIER.

At length the Admiralty gave me an official appointment to proceed with my *reconnaissance* of the Mediterranean shores: and on the 7th of May, 1817, the *Aid* sloop-of-war arrived at Malta, to bear my pendant. Maritime surveying, however, was not then well understood even at head-quarters, or a faster vessel, with at least one tender, would have been equipped. But instead of this, the *Aid*—on being inspected by the Admiral, the Commissioner, and myself, was found to be in want of several material requisites for the service she was about to proceed upon, and even to require substantial repairs. A partial remedying of these defects detained her till the 27th of June, when I was happy to find myself in a more efficient position than I had hitherto been. During the time the ship was in the dockyard, I concerted measures with the commander-in-chief for carrying my own plan of operations into execution; this plan was, in the first place, to go to the Channel between Sicily and Malta, and there complete my examination, while waiting for the expected ship which was to embark the architectural relics I had collected at Leptis Magna for the Prince Regent. After those remains should have been

Arrival of the *Aid*

Her state.

Plan of proceeding.

* In November, 1829, I had the pleasure of receiving a letter from Sig. N. Cacciatore, the Astronomer Royal of Palermo, stating that in the previous August he had measured a base-line, and laid a series of triangles, from Trapani to Maretimo, and thence to Cape San Vito, the whole of which, down to every rock and shoal, he found to agree so perfectly with my survey, that he could not but publish the results. His words are—'Io nel mese passato ho mesurato una base, ed ho fissato una serie di triangoli sulla costa, e nelle isole di Trapani, Favignana, Levanzo, Maretimo, e Capo S. Vito. Debbo dirle, che ho trovato tutt' i punti della costa, tutt' i scogli, e tutte le innumerabili secche di quei paraggi, notati col massimo rigore ed esatezza nella sua carta idrografica. Io sto descrivendo questo lavoro che pubblicherò; e con piacere annunzierò che le di lei osservazioni, e descrizioni, le ho trovate tutte rigorosamente esatte.'

stowed in the vessel's hold, under my inspection, I proposed next to repair to the Ionian Islands: to which, from our then recent occupation of them, our ships were often sent; and on which, from the utter worthlessness of the government charts, they as often ran on shore. It was also evident, that besides furnishing a series of geographical points, as stipulated, the features of the coast would absolutely demand a re-examination; and that though the Admiralty might possess tolerable plans of some of the principal ports, still we could fill up many a gap. And lest untoward events should interrupt such a course, or draw me away during its execution, I adopted a *festina-lente* method, entirely—so far as I know—my own; namely, to take the ship—as much as possible under easy weather for the chronometer-rates—to the various ports, as normal stations for a principle of mensuration obviously simple and accurate, thence to cast angles wherever they could be thrown or continued, and to fill in the less-broken shores between, by the boats and patent-log runs. By such means I hoped to effect the primary object of fixing latitudes and longitudes, to develope many tracts which were all but unknown to our charts, and to rectify others which had been imperfectly surveyed. On discussing these several points with Sir Charles Penrose, I had the satisfaction of receiving his hearty acquiescence.

Ionian Islands.

Chronometer runs.

Shortly after I had commenced carrying out these views, a correspondence was opened with me through the medium of Sir Thomas Maitland, by Baron Potier des Echelles, a major of the Austrian staff, respecting the Adriatic Sea; the large chart of which then in use, though vaunted as an actual survey of Vincenzio di Luccio, was absolutely deemed a disgrace to hydrography. It seems that, after their military occupation of Italy in 1799, the French had been actively examining the shores and lagoons of the Venetian territory, and completing the observations already made thereon by their countryman, Pierre A.

Baron Potier.

Forfait. In 1808 and 1809 they ordered some detached surveys to be made on its eastern shores, under the celebrated Beautems Beaupré, whose works approach nearer to perfection than any that hitherto have been made in that quarter. From the examinations then obtained, together with other occasional observations and corrections, a very tolerable *Piloto Pratico*, or coast directory, from Trieste to the mouth of the Tronto, was compiled by the geographical-engineer, Ignazio Prina, and published in 1816. In the meantime, the Austrians—who had previously employed a party of staff-officers under Marshal von Zach, a brother of the well-known astronomer, in making a special survey of the Venetian states—when they re-entered Italy in 1816, recommenced their geographical labours on the shores of the Adriatic Sea. They had made a considerable advance along their own coasts when they heard of my mission; whereupon I was formally applied to for the purpose of giving their operations a maritime completion, as well as to carry on a continuation of the survey along the Turkish shores as far as Parga, where respect could then be commanded only by the British flag. In consequence of this proposition, and fully empowered by the Admiral, I repaired to Naples early in 1818, and there entered into a convention with Marshal Koller, Count Nugent, Colonel Visconti, and Baron Potier, by which I engaged to blend all the detached operations of the several parties into one maritime work, and to complete the eastern shores to Parga, Corfu, and Paxo. For this purpose it was agreed that I should embark with me four Austrian staff-officers, namely, Baron Potier, Baron Gränzenstein—Marshal Koller's brother-in-law, Baron Jetzer, and Lieutenant Lapie; with two Neapolitan engineers, Captain Soldan and Lieutenant Giordano. Moreover, an Austrian sloop-of-war, the *Velox*, of 20 guns, commanded by Captain Pöltl, was placed under my orders; and I was to have the occasional assistance of the gun-

Beautems Beaupré.

Ign. Prina.

Marshal von Zach.

Marshal Koller. Count Nugent.

Austrian and Neapolitan officers.

boats stationed in the principal ports. To this force, Colonel Visconti afterwards added two more officers of the Neapolitan staff, Captain Chiandi and Lieutenant Bardet.

Untoward reserve.

All this was highly satisfactory. But I ought here to say, that notwithstanding these shores had been thus under examination, the results thereof were not before the public: and though a slight outline *lucido*, or transparent tracing of a general chart, was furnished by my collaborateurs, I was not made aware of the nature and extent of what had been actually achieved by them, and deposited in the archives at Milan. Nor did I ever see any of Beaupré's work till Captain Gauttier showed me several manuscript surveys of harbours, with which he had been entrusted by the Dépôt de la Marine at Paris. Indeed, such was the general ignorance of Adriatic hydrography, that Captain Hurd, in a representation which he made to the Admiralty, said—

Captain Hurd's official note.

All the charts of the Adriatic that I have seen are erroneous in regard to its eastern shores, except a small Venetian one, which is, however, published on so small a scale as to be of little use to the navigator. There are grounds for supposing that the coasts of Albania and the Morea have had but little scientific attention paid to them. I would therefore recommend a survey to be undertaken of all that part comprehended between Ragusa and the island of Cerigo, including the Seven Islands; or, if this should not be judged practicable, that correcting observations at all the principal points of land should be made, to enable us in some measure to correct these errors, as far as the same can be done by such means. A judicious and skilful observer employed on this duty, would very soon make us acquainted with everything necessary to enable us to form a chart thereof, sufficiently correct for all general purposes.

Course adopted.

Not being able clearly to ascertain what my co-operators had accomplished, until visiting the Geographical Institute at Milan towards the close of our labours, I adopted Captain Hurd's statement as a guide. Accordingly, with the excellent means then at my disposal, a chronometric chain was run over the whole sea, and extended into detail by copious triangulations; copies of which, together with particular plans of the harbours we resorted to, were promptly sent both to Milan and Naples. But throughout the whole of these proceedings there was certainly much less of reserve and mystery at the latter place than had crept into the former; my direct application for information was overlooked, and it consequently happened that much time was expended in doing work twice over. However, the object which was steadily kept in view being accomplished by the close of the year 1819, I discharged the Austrian sloop-of-war and the foreign officers; an arrangement which brought forward mutual expressions of regret, for those gentlemen and ourselves had always been on the best of terms, and for nearly two years they had ever obeyed my directions with alacrity and good-will. 'Au moment,' wrote Baron Potier, the senior officer embarked, 'que je regrette de me séparer de vous, je me rappelle doublement toute l'attention et amitié avec laquelle vous m'avez comblé, et que je me ferai toujours un devoir de vous réitérer vivement à chaque occasion.'*

The Austrian and Neapolitan officers depart.

Potier's adieu.

Having thus secured a chart of the Adriatic Sea, similar operations were continued through the southern Ionian Islands, with the opposite shores of Albania and the Morea. This being completed by June, 1820, my *reconnaissance* of the west coast of Italy was resumed, and we were busily employed on the Riviera of Genoa, when I was suddenly recalled overland to England in the winter of that year: the ship followed soon afterwards, and was paid off at Deptford on the 22nd of January, 1821.

Return to Italy,

and England.

During this time, Captain Gauttier had continued his chronometric runs, and he had annually obtained the permission of his government to meet me. In the numerous

* From what is about to follow, I should mention that the Emperor of Austria afterwards presented me with a valuable gold box set with diamonds; and two of my best stations—Budua and Pola—were marked each by a small stone pyramid on the spot where the instrument stood, with this inscription—

OBSERVATIO ASTRONOMIÆ
AB W. H. SMYTH
ANGLORUM NAVIS AID PRÆFECTO,
REGNANTE IMPERATORE
FRANCISCO I°, M DCCC XVIII.

and unreserved communications which occurred with this highly-efficient officer, I found his methods and practice so truly good, as to call for the utmost reliance on the results. On making a comparison of our respective works, we always found a fair agreement wherever we observed on shore; but that our secondary points sometimes differed: thus, writing to me in March, 1819, he says: 'Vous trouverez ci-joint la position géographique de tous les principaux points de l'Adriatique que j'ai fixé l'année dernière ; vous y verrez que les points qui se trouvent communs dans votre travail et le mien s'accordent en longitudes. Nos latitudes different bien d'avantage, mais j'ai peu d'observations E et O.' As this letter followed one in which he showed me his intention of triangulating the whole of the Archipelago, and its boundary coasts, it struck me that by an easy arrangement we could mutually benefit each other, and the correction of the chart of the whole Mediterranean Sea be speedily effected; therefore, as the document alluded to was the basis of my succeeding operations, it should here appear in full:—

Gauttier's observations.

My views thereon.

Ministère de la Marine et des Colonies,
Paris, le 5 Fevrier, 1819.

MONSIEUR ET AMI,—Je n'ai pas l'avantage de recevoir de vos nouvelles depuis l'époque où nous nous sommes séparé à Corfou; je crois cependant que vous devez être actuellement en Angleterre; c'est pourquoi je vous adresse ma lettre dans ce pays, le Général Brisbane ayant la bonté de se charger de vous la faire parvenir.

Gauttier's excellent process.

J'ai fait cette année quatre stations dans l'Archipel, sur les sommets de Milo, Zéa, Paros, et Naxie, la base qui va me servir à determiner tous les sommets des îles de l'Archipel a été mésurée au moyen d'un grand nombre de séries de hauteurs de la Polaire prises avec le cercle astronomique, qui ont déterminé la latitude de chacun de ces points à moins de deux secondes, et comme le gisement de cette base, d'après les azimuts observés à Milo, est le Nord 1° 15′ 48″ Ouest: je suis sûr de sa longueur, que j'ai trouvée exactement de 57 milles, à moins de 2″.*

---

* In a subsequent letter M. Gauttier entered into minute details respecting this base-line, and the several series of Pole-star altitudes taken with his excellent repeating-circle by Le Noir. It seems that the final results of these series differed only four sexagesimal seconds among themselves; and there is reason to believe that the mean result may be found in the very close limits of which these observations are susceptible.

Je vous envois la position géographique des sommets de toutes les îles des Cyclades déterminés au moyen de ma base: ce sont des triangles sphériques qui ont servi à déterminer ces points en supposant la terre ronde. On a calculé pour chacun l'angle au Pole, qui donne leur différence en longitude, et puis la distance polaire, ou le complément de leur latitude.

Gauttier's labours.

On compte terminer cette année, au dépôt, la construction de la carte de la Méditerranée en deux feuilles; comme la mer Adriatique entre dans la première de ces feuilles, vous m'obligeriez beaucoup de m'envoyer ce que vous m'avez promis sur cette mer, que je vous prie d'adresser au dépôt de la marine, en cas que je ne sois plus à Paris. Vous voyez que la carte que nous allons publier est une carte routière. Dans quelques années d'ici, quand j'aurai eu le loisir de construire tous les petits détails dont j'ai les matériaux, on publiera alors des cartes particulières de l'Adriatique, de l'Archipel, et de la partie la plus orientale de la Méditerranée. J'espère finir la campagne prochaine tout l'Archipel; il ne me restera plus à faire que la mer Noire pour les campagnes suivantes.

Je désirerais bien avoir encore le plaisir de vous rencontrer cette année, mais je compte me rendre directement à Milo. Si vous étiez cependant à Zante à cette époque, et que je fus contrarié à l'entrée de l'Archipel, j'y relâcherais pour avoir le plaisir de vous voir, ainsi que Madame Smyth, à laquelle je vous prie faire agréer mes hommages respectueux.

J'ai l'honneur d'être, avec les sentimens d'un parfait attachement,

Monsieur,

Votre très humble et obéissant serviteur,

P. Gauttier.

My opinion thereon adopted by Lord Melville.

When consulted therefore by Lord Melville, then First Lord of the Admiralty, on my arrival in England, upon the state and prospects of Mediterranean hydrography, as time was a far greater element in such considerations then than it is now, it became a duty to represent my conviction of the inutility of Gauttier and myself going over the same ground, with objects so nearly the same. I then informed his lordship of the French operations, and assured him that, after careful examination no hesitation could remain as to their accuracy. I also showed, that if they saved the necessity of my working in the Archipelago, it would enable both of us to give a better completion to our respective labours; for there were various points of my own in the Western Basin which required additional attention; and that most of the space between Algiers and Alexandria, on the north coast of Africa, was hydrographically a blank. His lordship was pleased so to approve of my remarks as to authorize me to proceed to

Paris, in December, 1820, empowered to enter into an arrangement with Admiral de Rossel, the hydrographic director, for an official interchange of the projected labours. On arriving in that city, the proposition was most favourably entertained by Admirals Count Rosily-Mesros and De Rossel, and the members of the Board of Longitude, MM. Delambre, Arago, and Beautems Beaupré; and I received a copy of all the French results in the Archipelago, Levant, and Black Sea, which were then reduced. To my great disappointment, Captain Gauttier was in quarantine at Toulon; but he was made acquainted with every step taken, and wrote to me—'Soyez bien persuadé de ma reconnaissance pour la manière franche et loyale de vos communications, et j'espère que vous pensez de la même manière à mon égard.'

Admiral de Rossel.

My proposition accepted.

This account is the more detailed, because it will prove no slight was thrown on the labours of my contemporaries; while my conduct in respect to recommending the French operations for adoption will also show the high value which I placed upon time. Now, it appeared that certain remarks on the want of confidence shown me by the Milan authorities had given umbrage, and that a letter of mine to Baron de Zach, which he printed in his *Correspondance Astronomique*, gave pain to an individual to whom nothing personal was intended. It was the system, and not the person, I meant to impugn; for, in consequence either of delay or neglect, or the tiresome and tedious forms of office, the grounds were not communicated upon which the Imperial Geographical Institute at Milan was putting our joint work together for publication; and my friends in the Topographical Office at Naples were equally in the dark. Colonel Campana, the Director of the Imperial Institute, at length forwarded to Baron de Zach a prospectus of an Adriatic Atlas for insertion in his widely-circulated periodical; on which the Baron directly inquired of me what degree of precision he might safely assign to the geodetical points thus handed

Coolness with Col. Campana.

Reasons for my conduct.

to him.* Not being in actual possession of the means they had adopted, I returned the following reply, dated 15th March, 1826:—

Vous me demandez, Monsieur le Baron, jusqu'à quel degré de précision on peut compter sur les positions géographiques des lieux dans le golfe de Vénise, gravées sur la carte directrice de cette mer, publiée au dépôt des cartes à Milan, et que vous avez rapportées dans le viiie volume, cahier v., p. 490, de votre correspondance astronomique.

Je vous dirai donc que tous ces points ont été déterminés en premier lieu, géodésiquement par un canevas de triangles, qui a été conduit le long des côtes par le Colonel Ferdinand Visconti. Tous ces points ont été réduits au méridien et à la perpendiculaire du clocher de St. François de Ripatranzone, d'où enfin on a tiré les longitudes et les latitudes. En second lieu, plusieurs de ces endroits ont été déterminés par moi astronomiquement, c'est à dire, les longitudes par des chronomètres, les latitudes par des hauteurs méridiennes des astres. Pour vous donner une preuve dans quelles limites les longitudes ont été déterminées, afin que vous puissiez en juger par vous même, je vous rapporterai ici quelques exemples, qui vous feront voir l'accord qui règne dans ces déterminations faites selon les différentes méthodes, ce qui a servi de contrôle, et pour ainsi dire, de pierre-de-touche à tout ce travail. Vous savez aussi, Monsieur le Baron, que le Capitaine Gauttier a de même parcouru la mer Adriatique; cet habile officier de la Marine Royale Française y a également fait plusieurs bonnes déterminations; or, voici l'échantillon d'un tableau qui fera voir cet accord:—

Visconti's triangle.

Captain Gauttier.

Long. d'Otranto	selon le Capt. Smyth	16°09′50″	de Paris.
,, ,,	selon le Capt. Gauttier	16 09 00	,,
,, ,,	selon les triangles du Col. Visconti	16 09 30·1	,,
Long. de Brindisi	selon le Capt. Smyth	15 38 17	,,
,, ,,	selon le Capt. Gauttier	15 36 40	,,
,, ,,	selon les triangles du Col. Visconti	15 37 59·9	,,
Long. de Bari	selon le Capt. Smyth	14 32 40	,,
,, ,,	selon les triangles du Col. Visconti	14 32 04·1	,,

* This opinion was asked because the Baron was aware of some of the particulars; and a notice which he published after visiting the *Adventure* in 1823 (*see his Correspondance Astronomique, vol.* iv. *page* 143), will show that reserve formed no part of my character:—'Le 12 du mois d'Août, M. le Capitaine Smyth est venu relâcher avec son observatoire flottant dans le port de Gênes. J'ai eu la seconde fois le plaisir et l'avantage de revoir, et de m'entretenir avec ce marin distingué sous tant de rapports. Cet habile officier a eu la bonté de me communiquer, et de me faire voir avec sa franchise ordinaire, tous les travaux qu'il a fait depuis que nous nous sommes vus la dernière fois. Il m'a montré tous ses journaux, observations, plans, cartes, soit gravées soit dessinées, il n'avait rien de caché ni pour moi, ni pour personne. Il ne craint pas les communications; sûr de son fait, ses travaux peuvent supporter l'œil du scrutateur. Il ne fait aucun mystère de ses observations, car les Anglais ne pensent pas que des longitudes, des latitudes, des azimuts, des bases, et des triangles peuvent être des secrets d'état.

Long. de Corfou selon le Capt. Smyth	17°	35′	23″	de Paris.
„ „ selon le Capt. Gauttier	17	35	50	„
„ „ selon les triangles du Col. Visconti	17	35	41·4	„
„ „ par l'éclipse d'Aldebaran* . . .	17	34	41	„

Remark. To this letter I added a full list of the reduction of Visconti's triangles, in French metres, which is appended in the fourteenth volume of the *Correspondance;* but as there was no mention of the points adopted by the Milan Institute, nor any names of the Austrian staff given—which, as already shown, were in these particulars unknown to me—it gave rise to a little warmth. Colonel Campana wrote a statement for publication, which he forwarded to Baron de Zach, and which that gentleman, after consulting me, printed in his fifteenth volume, page 51. To preserve consistency, I here reprint it:—

Campana's letter. Je m'étais flatté, Monsieur le Baron, que d'après les détails explicatifs de la manière dont on a déterminé les différentes positions géographiques gravées sur la carte directrice de l'Atlas de la mer Adriatique publiée par cet I. R. Institut géographique Militaire, et contenu dans l'annonce de l'Atlas que j'ai eu l'honneur de vous envoyer, et que vous avez eu la bonté d'insérer en partie dans votre Correspondance Astronomique, on n'aurait pas revoqué en doute le degré de précision de ces positions.

Mais comme la lettre de M. le Capitaine Smyth publiée dans le ive cahier du volume xiv. de la même correspondance fait présumer qu'on n'est pas tout à fait tranquille là-dessus, vous me permettrez, M. le Baron, d'ajouter ce qui suit, savoir: que tous les triangles qui s'étendent sans interruption le long des côtes depuis Budua (Dalmatie) jusqu'à Sta. Maria di Leuca (Royaume de Naples) ont été mesurés, soit par les officiers de l'état major général Autrichien, soit par ceux de l'état major Napolitain, avec beaucoup de soin; c'est pourquoi les latitudes et les longitudes des différents points qui en ont été deduites doivent mériter la préference, sans vouloir contester pour cela le mérite de celles des savans marins, qui dans la suite ont déterminé quelques unes de ces mêmes positions.

Du reste, l'accord assez satisfaisant qui se trouve entre les longitudes déterminées géodésiquement, et celles qui l'ont été par les méthodes pratiquées par les marins rapportées par M. le Capitaine Smyth dans la lettre ci-dessus citée, peut servir à faire juger du degré de précision de l'Atlas en question depuis Budua jusqu'à Parga (Albanie Turque), où les positions n'ont été fixées que par les méthodes des marins. C'est pour compléter l'échantillon

* This occultation obtains a place here, because it seems to have been admirably observed, under the most favouring circumstances, by Inghirami at Florence, Oriani and Carlini at Milan, and Captain Chiandi at Corfu. It occurred on a beautiful evening, [illegible]th March, 1813.

donné par M. le Capitaine Smyth que je prends la liberté de vous présenter ici la comparaison des longitudes de quelques autres points de l'Atlas:— Comparison of points.

Noms des Lieux.	Provenance. (Longitudes comptées de Paris.)		
	Triangles de Dalmatie.	Cap. Smyth.	Cap. Gauttier.
Galiola Ec. dans le Quarnero . .	11° 50′ 17″	11° 49′ 55″	11° 49′ 30″
Selve, Eglise	12 21 38	12 20 28	——
Sansego, isle, sommet	11 57 33	11 57 20	——
Arbe, Eglise	12 25 29	12 25 00	——
Pomo, Ecueil	13 07 25	13 07 10	——
Cazza, isle, sommet	14 10 39	14 10 57	14 10 30
Lagosta, signal Trigonom. . .	14 31 30	14 31 08	14 31 10
Ecueil S. Niccolò di Budua . .	16 31 08	16 30 32	16 30 30
Sta. Maria di Leuca	16 02 40	16 02 57	——

This letter occasioned a reply from me to Lieut.-General the Baron Prochaska, Chef d'Etat Major at Vienna, because I thereby accidentally learned, in July, 1826, what ought to have been officially communicated to me in the spring of 1818. The want of a more liberal unanimity was, however, injurious to the Austrian publication; for on its appearance I received a complaint from Visconti, respecting the conduct of our co-operators, in arrogating the whole work to themselves; and adding that he would forthwith construct and publish a more correct chart of the Adriatic, with a strong reclamation. But this should, as well as the preceding citations, be recorded in his own words:— Baron Prochaska. Visconti's anger.

Queste nozioni mi sarebbero utilissime, poichè mi sono proposto, appena sarà in Milano publicato il rimanente di dare io una carta più semplice, più adattata all' uso de' Marini, più economica, e soprattutto più esatta, mentre ho veduto che nella suddetta 1ma parte fatta a Milano vi sono degli errori sulle latitudini e longitudini d'Otranto, Fano, S. M^{a} di Leuca, Sasseno, Linguetta, Corfù, etcetera, e sulla distanza d'Otranto a Capo Linguetta. E siccome la detta carta a Milano è stata publicata come fatta tutta dagli uffiziali dello Stato Maggiore Austriaco, senza far memoria nè di voi, che trovavasi inciso ne' fogli terminati al 1814; così mi propongo ancora di rivendicare la proprietà di ognuno, facendo conoscere al pubblico a chi si deve il rilievo d'una costa, o d' un porto, a chi lo scandaglio, a chi la latitudine o la longitudine osservata ecc^{a} e l' epoca del lavoro d' ognuno, e si vedrà che agli uffiziali dello Stato Maggiore Austriaco non si devono che poche vedute, e qualche altra piccola cosa.

To resume the narrative. Early in July, 1821, I again left England, in the *Adventure* sloop-of-war, with instructions conformable to the agreement already made in Paris. These orders principally directed me to re-examine the doubtful parts of my selected portion of the Mediterranean Sea, and to make a running coast-survey of the deficient parts of the detail; all to be completed, if possible, within three years. Thus instructed, and furnished with more efficient means than before—though still in want of a tender—I laboured to act up to the spirit of the Admiralty instructions, in the manner which will be presently noticed. Having accomplished the points and coast-line required, the same was duly announced to the Board, and we prepared to return home in the autumn of 1824. But their lordships having strengthened my force in the spring of that year, by the addition of a fine 10-gun cutter, the *Nimble*, I took upon myself to leave her under the command of Lieutenant Slater, who had won his commission by zeal and attention to the work in hand, to execute several secondary though important details during the time that my charts and plans would be under completion. For this purpose I furnished the necessary instructions to that officer, directing him to examine the mouth of the Magra river, between Genoa and Tuscany; to add more soundings to specified portions of those coasts; to sound round the shoal of Capo Vita, on the north-east of Elba; to re-examine our coast line of Algiers; and especially to search round the rock off Cape Matafuz, and between it and the Cape, for a second danger reported to me by a Sardinian.

H.M. ship *Adventure*.

The *Nimble* tender.

Orders to Lt. Slater.

These arrangements having been made, I returned to England, and paid off the *Adventure* in November, 1824. The Board of Admiralty was now furnished with a series of latitudes and longitudes fixed by Captains Gauttier, Beaufort, and myself, from the Straits of Gibraltar to the Sea of Azof; and by the accomplishment of this mission, hydro-

Return to England.

graphy was presented, for the first time, with the true figure of the Mediterranean Sea. My documents being both numerous and various, necessarily demanded much time to prepare them for publication: to meet, therefore, the immediate wants of navigation, a large general chart of the whole was drawn up under my inspection by Captain Graves—then a midshipman—and issued for public use in 1825. It was with this purpose in view that I tolerated a plate forty-seven inches by twenty-eight inches, to fit the large sheets of paper called antiquarian; but for the others adopted half the double-elephant sheet, as economical, easy to work, handy to use singly, and measurable without crease if bound in an atlas. Practical utility being my object, and as but one chart or plan is used at a time, I studied the nature and importance of a coast or port for cruising upon, and afterwards arranged the scale of each according to the portion contained on the half-sheet. In order to methodize in other respects, the north was always placed vertically upwards, as a meridian line, and especial care was taken not to crowd the details, leaving all the dangers, and objects to be avoided, as obvious as possible to those who have to consult charts in fresh weather, and with indifferent light.* Under such treatment the subject of scales—so interesting to chamber theory—became a very secondary concern; for no navigator pricking off his ship's place, with a chart before him and compasses in hand, ever cared a straw about what scale the next sheet might be pro-

Work by Mr. Graves.

My views of a chart,

And consequent method.

* On this account, I cannot but look upon the method introduced by Beautems Beaupré, of contour-lines round a port, with dots for depths, as more adapted for the engineer than the seaman. To preserve perspicuity, I inserted no more soundings on the charts than were deemed necessary to a proper comprehension of the included space, and unnecessary rhumb-lines were omitted; the coasts being understood to be entirely clear of danger, unless otherwise expressed. Under this system, the lead is trustworthy; and on most coasts it will be found that inshore the bottom is generally sand, and in the offing, blue clay.

tracted upon. Each was therefore severally treated as an individual case, rendered amenable by its size to its contents. Those who desire to form a correct idea of the relative proportions of different countries, can always refer to the general chart. In library atlases, the number of arbitrary scales might be reduced with advantage; but the *study* and the *ship* require very different treatment.

Nature of the materials. The materials for filling these sheets must now be mentioned, since, in pronouncing an opinion upon any work, the intention of its projector ought not to be overlooked. I therefore repeat, that my whole wish was to secure as much information as possible in a given time; owing to the constant apprehension of being broken off. Consequently, My views. in executing this arduous duty, my aim was rather to obtain substantial good, than minute or absolute accuracy; for which latter another kind of establishment would have been demanded. The chief quality, therefore, to which I lay claim, in the conduct of the survey, is unwearied diligence; and by resorting to the practical rather than to the theoretical application of mathematics, and exerting my stock of professional knowledge in a given and decided direction, I was able to enter into competition with some Assistance received. of my superiors in acquirements. When the *Aid* first came out to the Mediterranean, there was not an officer or other person on board who had any practical acquaintance with maritime surveying: but there was no lack of zeal and inclination, and when I first instructed and afterwards personally directed them, they all became useful, and some of them extremely so. Even the mere youngsters were handy in the boats; and they who are vulgarly termed *idlers* employed themselves in making statistical and other inquiries for me. It is, therefore, with great pleasure that the names of those who served with me in the *Aid* and *Adventure* are here given; merely remarking that my chief obligations were to Messrs. Slater, Graves, and Elson. The following lists contain the officers' names, &c., nearly in the order in which they joined:—

Names and particulars.

Gun-Room Officers.

	In 1853.
Lieut. John Hose	*Dead.*
,, C. R. Malden	*Reserved H. P.*
,, Josiah Oake	*Captain.*
,, F. W. Beechey	*Captain.*
,, Hen. E. Coffin, volunteer	*Captain.*
Master, Thomas Atwell	*Dead.*
,, Thomas Elson	*Dead.**
Surgeon, James M. Madden	*Retired.*
,, Abraham Courtney	*Retired.*
,, John Stephenson	*Drowned.*
Purser, Robert Young	*Dead.*
Assist.-Surgeon, William Clarke	*Died a Surgeon.*
,, ,, John Campbell	*Surgeon.*
,, ,, William Beg	*Dead.*
,, ,, Thaddeus Porter	*Dead.*

Mates and Midshipmen.

	In 1853.		In 1853.
J. F. Dessiou	*Drowned.*	Henry Bush	*Dead.*
M. A. Slater	*Died Commander.*	Thomas Dutton	*Dead.*
Ed. Holland	*Commander.*	Wm. Robinson	*Dead.*
Wm. Skyring	*Killed, Commander.*	W. Sidney Smith	*Captain.*
Ed. Tyndale	*Died a Lieut.*	Robert Sholl	*Dead.*
James Cooling	*Died a Lieut.*	Henry Raper	*Lieut.*
Thomas Graves	*Captain.*	Mayne Lyons	*Killed, Lieut.*
Nelson Elliot	*Died a Master.*	Alfred Miles	*Commander, dead*
James Wolfe	*Died a Commander.*	James B. West	*Commander.*†

Remarks.

These gentlemen, as already said, were all and severally desirous of rendering their exertions effective; and I endeavoured to appoint each to the kind and degree of work for which he appeared best qualified. Upon those who

* I took Mr. Elson out of the *Weymouth* store-ship, on observing his activity in embarking the marbles at Leptis Magna; and he afterwards served with me seven years as Master. He died in 1848, Master-Attendant of Woolwich yard, having written me a cheerful letter in the morning of the day on which he expired.

† Officers were occasionally sent from the flag-ship, as the present Lord Adolphus Fitzclarence, Captain Charles Howe Fremantle, his brother Henry —who died on board—Mr. J. J. Smith (*retired*), and two or three others; but as they merely joined for their own instruction, they are not entered upon the above list.

more immediately bore a share in the survey, I inculcated the necessity of appropriating their labour to its intended purpose; and in order to conduce to the consistent economy of time and means which the tenour of my mission demanded, distinctly described the proper degree of accuracy expected at their hands. I also pointed out the readiest method of attaining the desired end in the respective data with which they were to furnish me; while an uniformity in the method and manner of drawing and reducing was established. And here a few more words may be necessary to elucidate our proceedings, even at the risk of repetition in some details.

Viscount Melville.

On returning to the Mediterranean in 1821, I had arranged with Lord Melville to carry out a present from our government to the Báshá of Tripoli, in acknowledgment of the assistance he had formerly afforded me; and to obtain permission from him for the completion of our survey of the Greater Syrtis. To effect this fully, I showed his lordship the benefit there would be, if, while the ship should be employed on the hydrographic details, a land party were simultaneously to proceed along the shores, the whole site being replete with objects of antiquarian and geographical interest. For this purpose there was a volunteer well versed in the Arabic language and customs in my former messmate, the late Captain G. F. Lyon, who had recently returned from Murzúk (*see my letter to Lord Melville, in the Appendix*); and so fair did the opportunity of exploring the Cyrenaica seem, that the celebrated Sig. Belzoni, then lately returned from Egypt, offered to accompany me. From a change of circumstances, however, Lyon went with Captain Parry on the Arctic expedition, and Lieutenant Beechey, who had been on Parry's memorable first Polar voyage, was appointed to the *Adventure*, to supply Lyon's place; and instead of Belzoni, we embarked Mr. Henry Beechey, the Lieutenant's brother, an old acquaintance of my own, well known as an Egyptian

Capt. Lyon.

Messrs. Beechey.

traveller. To these gentlemen I added Mr. Edward Tyndale, a midshipman, who had travelled with me in Africa; Mr. Campbell, assistant-surgeon; and a volunteer, Lieutenant Henry Edward Coffin, whose uncle, Admiral Sir Isaac Coffin, was going out in the ship as my guest. I had myself already examined and fixed the coast of this unfrequented gulf as far as below Isa on the west side, and from Kharkarah to the northwards on the east side. The land-party had therefore to proceed round the bottom of the Syrtis, and from thence to the examination of the ruins in the Pentapolis, and the whole country round Cyrene. Lieutenant Beechey's force.

This section of the survey, comprehending the exposed space between Isa and Kharkarah, is the only portion which I did not personally see: and am therefore bound to state that, on a complete knowledge of the means and method employed to carry out my instructions, I was fully satisfied with the Lieutenant's results. Considering it unsafe to carry the ship farther into a gulf of which we knew nothing, Mr. Elson was despatched in the ship's launch, expressly fitted for the occasion, to make a surveying cruize of the intervening shores. Lieutenant Beechey's charge, indeed, comprehended the topography of the space between Tripoli and Derna, the mensuration of which was to be adjusted to some of my determinations, as detailed in his interesting volume. But I sorely regretted that untoward circumstances utterly prevented the party which he led, from proceeding through Derna to Alexandria; as that region, then all but unknown, would have been accurately examined before political changes had increased the obstacles to a full investigation. His work. My regret on his being interrupted.

After quitting the coast of Africa, I returned to complete my charts of the Italian islands in the western basin of the Mediterranean, including Corsica, Sardinia, and the channels near Elba; the whole being connected by triangles with the Tuscan and Roman shores. On the 12th of November, 1823, while thus employed, the *Lloiret,* a French brig Return to Italy.

M. Allegre. of war, commanded by Mons. Allégre, sought refuge from an easterly gale in Port San Pietro, Sardinia, where the *Adventure* was then riding. As this gentleman was formerly one of Gauttier's officers, and therefore acquainted with me, he related that he had since served three years with Captain Hell on a detailed survey of the Island of Corsica. Having kindly brought his documents on board, and made a comparison with our operations and results, we found a general agreement in the points and such portions of the coast as were completed by both: but, moreover, the details of Corsica were so elaborately laid down by the French officers, and bore such internal evidence of extreme accuracy, that, as I told Captain Allégre, there was not the slightest occasion for my returning thither, to survey between the points which I had already established there.

French survey of Corsica.

Such being the outline of my re-examination, it may still be necessary to dwell a moment upon the system we pursued, before giving a table of the geographical positions which form the framework of the general chart.

The course followed.

In the course of our operations, we determined the latitudes and longitudes of a certain number of principal places in our best manner, and then used them as consecutive points on a series of bases; using triangulation by the theodolite between them where obtainable, and filling up others by the most eligible means afforded by the ship and boats. Every port was thus considered a station; and where the hostility of the natives, or the quarantine regulations, were inimical to landing, islets or rocks off such coasts were always resorted to. With the exception of those just mentioned at the bottom of the greater Syrtis, the whole of the latitudes and longitudes were entirely under my own computation; the keeping of the chronometer-rates, only, having been latterly consigned to the care of Messrs. Slater and Graves. The principal harbours were surveyed by myself, with occasional assistance, and the more open bays by the other officers; while some of the minor places were

Chronometer rates.

Surveys.

sketched in from measures obtained by means of painted lengths on poles, and the application of Rochon's micrometer thereto. We examined the most remarkable banks, with unsparing diligence. Rochon's micrometer.

In early days I had seen a cylindrical cartridge-box, with its bottom cut out and the orifice glazed, used for examining the state of a ship's bottom by immersion, as a cure for the reflection and refraction of the rays of light at the surface of the water which impede distinct vision: and in order to facilitate subaqueous inspections, I had a tube constructed in the shape of an overgrown speaking-trumpet, well glazed, and steadied in the water by a large grommet of lead. Upwards of twenty years after this had been publicly used, a similar instrument was advertised as a new American invention, under the style and title of the 'Water Telescope.' A water-gazing tube.

The lines of coast between the ports were mostly sketched on a patent-log basis by Mr. Elson, the master, whose activity and seamanship were in constant demand; and the skill with which he managed our schooner-rigged launch, and accommodated her to circumstances, partly compensated our being without a tender. These running surveys, constituting properly a maritime *reconnoissance*, were laid down on a very large scale, and afterwards reduced and subjected to the points previously established.* Use of the patent-log.

As the chronometric bases on which these longitudes depend, and by which they are connected with the Palermo Observatory, are of first-rate import, a brief sketch of our working routine may be acceptable in a general sense, how- Chronometric bases.

* In my instructions to the officers, the amount of labour was apportioned according to the local circumstances; for before geology had attained its present rank, I had observed that the depths of the sea follow the nature of the shores—the slopes of the one varying with the nature of the other. Thus, high and rocky cliffs have deep water, and are pierced with harbours; while low coasts are generally shallow, and destitute of ports. It may therefore, from this and other peculiarities, be held, that a low shore is a growing one, and a high shore is a wasting one. Wherever the *Zostera marina*, or riband-like grass-wrack, is found, shoals may be expected, for it detains silt, mud, and sand, till a bank is gradually formed.

ever trite such details may be to the practised nautical astronomer. On this second trip we followed precisely the same plan and method as was before stated, but with the increased power and confidence which experience gives.

Additional instruments.

To the instruments already mentioned, the Admiralty had added four more chronometers—namely, No. 12 of Pennington, and Nos. 320, 547, and 553, of Arnold; another 7-inch theodolite; and a very beautiful 15-inch altitude-and-azimuth circle, with good levels, and a capital telescope. And I should mention that, when in Paris, Mons. Arago took me to the house of M. Breguet the elder, to show me

Transit-telescope.

a newly-contrived transit-telescope, fitted with a chronometer No. 2741 to the eye end, which he qualified as a '*Compteur des secondes, des dixièmes de seconde, et des centièmes par approximation:*' this instrument, on examination, appeared to possess such advantages as a portable transit, that I purchased it for myself, and the Lords of our Treasury granted it a free passage through the Custom-house. Breguet on this occasion presented me with one of his exquisitely-sensible metallic thermometers.

In pursuing our progress, I made as short passages as were consistent between the principal ports, in order the

Chronometric rates.

better to obtain their chronometric differences of meridian; the 'sights' for watching the truth and permanence of the rates being taken regularly on arrival, by the method of equal altitudes, and corrected for the true refraction in the existing state of the atmosphere. The harbour-rates assumed—determined by the observed daily rates in the port before sailing—were always estimated for the time elapsed between the run from one place to another; and though the march was watched by daily observation whenever the weather permitted, the results of the extreme series, only, were employed in determining the longitude of departure

Harbour sights.

and arrival. The harbour sights were always taken on shore, in a quicksilver artificial horizon, with reflecting instruments well adjusted, and the index-error ascertained at the time of taking them; and they were invariably the mean of three altitudes of the upper, and three of the lower limb of the sun; while the corresponding times were carefully registered from my job-watch (*Earnshaw's pocket-chronometer, No.* 825), which was forthwith carried off and compared with the stationary time-keepers on board. The 9-inch quintant was the favourite instrument for this purpose; but, when the celestial altitude was very great, Troughton's reflecting circle was substituted. And occasionally when, owing to clouds or other causes, altitudes could only be procured on one side of the meridian, more than usual care was bestowed on determining the exact corrections necessary for instrumental and object errors, refraction, &c. Mode of observing.

The standard chronometers were placed on hair cushions wedged with cork, where the temperature, as far as we could contrive, was so uniform and constant as to be no task on the compensation; for the cabin never had a fire in it while I commanded, and was little liable to sudden transitions, causes which might otherwise disturb chronometric action. Their several rates were therefore easily scrutinized; and the discrepancy of each individual time-piece was valued by its allotted weight, in the summation of the products. Standard chronometers.

Many of the principal latitudes were taken on shore with the 9-inch quintant and artificial horizon, and with the reflecting circle and sextants; but some of the first class were obtained with the fine 15-inch altitude-and-azimuth circle, by a mean of sets with the face of the instrument alternately turned to the east and the west. Thus we hoped to clear our results from probable errors of division and ex-centricity; and it was always steadily mounted on a cask filled with sand, and most carefully adjusted, so as to serve as well for time as for altitudes. Here both sun and stars were employed for latitude, and were always observed at the instant of meridian passage, except Observed latitudes.

in a few cases when that could not be exactly attained, and then the horary angle, either east or west, was duly observed, and the reduction to the meridian computed. The moon. The moon was not used for this object, on account of the liabilities of irradiation, diameter, and tabular errors. The sea-horizon was never resorted to in these processes, except in a few rare instances where, from moral or physical impediment, landing to secure a latitude was impracticable; and then the object was carefully brought on the true east or west line from us, in order to do away with the arbitrary reduction which is consequent upon a general compass bearing. The dip-sector. Captain Gauttier had lent me a dip-sector—then a new introduction—made by Lenoir, after one of Wollaston's, to obviate some of the objections to the natural horizon; but I found it so troublesome to use, and showing such discordance between the results and theory, that I soon abandoned it. This would not have been satisfactory to the inventor: if the principle be true, it ought to be a requisite instrument, no matter how difficult to use, because the horizon may be out by almost any quantity, so that results may be egregiously bad in spite of the goodness of sextant, and skill of observer. But the dip-sector appears to be dependent on the principle of opposite points of the horizon being equally affected by any abnormal state of the refraction; whereas, though this may be the case in ordinary states of the atmosphere, it is not likely to be so in the extraordinary cases, where its correction would be most in demand, and where the effect would probably be confined to a very limited azimuthal range.

Nature of observation. Lunar distances, eclipses, and sidereal occultations were at first diligently observed, and recorded among the determinations: but finding some of these in fair accordance with the chronometric measurements, and others, equally well taken, widely differing; and also seeing that besides the known deficiencies of the tables, they were influenced by the existing state of the atmosphere, the tone of the eye, and the power of the instrument, and consequently must be inadequate for precision,* I discontinued these nice and most delicate observations, save for practice; thus abandoning the method of getting *absolute* longitudes by astronomy, for that of *differences* by time. The great accuracy and extreme simplicity of mensuration by chronometers, left nothing to be desired on this head, especially where the lengths of the runs were so arranged, that a comparatively speedy return to the starting point, allowed an estimate of the probable error of a determination.

Relative longitudes.

The geodetical angles were generally taken and reduced by myself, except in the northern portion of the Adriatic Sea, where I was largely assisted in this arduous duty by Captain Soldan, of the Neapolitan staff, and Lieutenant C. R. Malden. In particular places a base was generally measured on a selected spot of ground, with a tested Gunter's chain; and the line was lengthened at pleasure, by means of boarding-pikes stuck in the ground at a chain's distance from each other. The angles at each end were then taken with a truly-adjusted theodolite, an instrument which—as it gives the horizontal arc intercepted between the verticals of the two stations without reduction—unites celerity with precision. But where the sides were long, and the large circle was used, the difference that exists between the three angles of a theoretical plane triangle, and the three observed by reason of the sensibly spherical form of the earth, became appreciable, even when such sides were not more than a dozen miles in length. The excess thus occasioned may be corrected by treating the angles of the chords, and the chords themselves, as the angles and sides

Geodesy.

Capt Soldan and Lt. Malden.

Value of the theodolite.

* I should, however, remind the reader that the above is stated without any intention of undervaluing that beautiful method. The quantity of motion to be measured in lunars is only $\frac{1}{30}$th of that which is employed in getting the time by chronometers (it being the diminutive amount of the proper motion of the moon as opposed to the great amount of its diurnal progress); this is the circumstance that makes any error in the observation tell so largely on its result, and renders the conclusions so rough.

of a plane triangle; but the theorem of Legendre more simply shows, that if one third of the spherical excess be deducted from each angle, the opposite sides become proportioned to the sines of the corrected angles, and their magnitude may, therefore, be calculated by the rules of plane trigonometry.

Spherical correction.

In re magnetism. When I commenced surveying, the instruments for getting the variation of the compass were very inferior to those now in use; and the method of swinging a ship to ascertain the effects of its mass on the needle in different azimuths had not yet obtained—a method which has rendered the determinations in a ship so trustworthy for absolute magnetic quantities, in the variation at sea. The essay of Captain Flinders, however, concerning the differences in the magnetic needle arising from an alteration in the direction of the ship's head, printed in the *Philosophical Transactions* for 1805, and his experiments a few years afterwards, had revealed to me the tendencies and power of local attraction; which had only been puzzles to Dampier, Cook, and Löwenhorn, and remained uninvestigated by Downie. Therefore, in determining the corrections, and watching the affections of the compass, but little was ultimately used which was performed on board; although, in pursuing the usual methods for sea-practice, every precaution was taken to guard against accidental derangements and alterations. The variation of the needle was very readily established wherever we formed a meridian line for observatory purposes on shore; and in the secondary requirements we resorted to a large and well-graduated universal solar-dial, furnished with spirit-levels, and carefully adjusted to the sun's altitude at culmination. But whenever necessity compelled us to resort to amplitudes, they were taken at the instant when the altitude of the inferior limb of the sun above the visible horizon was equal to the difference of the semi-diameter and of the horizontal refraction increased by the dip—in other words, when the

Variation of the compass.

Captain Flinders.

Cook, &c.

Methods adopted.

centre of the sun was in the true horizon—a process less troubled with unequal refractions in the Mediterranean than in our latitudes, and therefore susceptible of considerable accuracy. Azimuths at sea, when the sun was not obtainable at rising or setting, were determined from the observed difference and compass bearing of altitude in a measured interval of time; though sometimes they were taken by the sun's angular distance from a terrestrial object. These, however, were for practice rather than use, since, as has been already said, we depended chiefly on shore work, whereby the non-verticality of the sight-vanes, and the effect of local attractions on the oscillations and vibratory movements of the needle, were more easily remedied. The inclination, and estimated magnetic intensity, were carefully observed with a well-made dipping-needle at our principal stations; on which occasions we usually shifted the poles of the magnet, and took readings with the face of the instrument alternately to the east and to the west, in the plane of the magnetic meridian.

Azimuths.

Magnetic dip.

The mariner's compass is too important an instrument in all the purposes of navigation, not to merit distinct mention; and what has been here advanced as to the great improvements since my survey, will receive illustration from a passage which I wrote in 1848, on receiving a book from the late Captain E. J. Johnson, Superintendent of the Navy Compasses:—

Captain Johnson

In order to carry out his (*Johnson's*) representations, a memorandum for the proper treatment of compasses on ship-board has been issued by the Admiralty. In this document it is directed to remove all iron to the distance of seven feet from the binnacle, as of old; that mixed metal or copper be used in place thereof, for bolts, keys, and dowels; and that iron tillers are not to range within seven feet of the compasses. The binnacles are to be at least four feet and a half apart, and they are no longer to be fitted with doors, so as to render them dirt-lockers, or depositories of improper materials. But a most important improvement is this: that in every ship a closet is to be constructed for the reception and keeping of the compasses, under the express charge of the Master, who is to see that the cards are never packed with poles of the same name nearest to each other—that is, that the north end of one needle shall be placed next the south end of its neighbour.

Admiralty mem.

This is assuredly a real reform in the steering department, and one which

Compass reform.

must awaken recollections of a grave tenor in those who are able to look back to the day when the compasses were turned-in higgledy-piggledy with the hooks, thimbles, marline-spikes, and iron implements of all sorts in the boatswain's store-room. The case is wonderfully altered; instead of the ship-chandlers' contract concerns—with inefficient suspension, indifferent needles, bad pivots and caps, and contemptible gimbals—the Service is now provided with machines of a first-rate description and trusty character. The needles are made of the best clock-spring steel, and perform their oscillations in truly-balanced copper bowls; the pivots supporting the card are pointed with a material harder than steel to work into the ruby cap: and, instead of leaving it to its fate, as of old, it is directed that the card should be raised whenever the compass is to be moved, or the guns fired. Some of the straw-pickers have branded the present standard compasses as costly, seeing their price consists of as many sovereigns as those they displaced did of shillings: but what is this difference when we weigh a trustworthy instrument against one which is all but worthless! What is the sum of twenty-five or thirty sovereigns for the most important machine in a ship, and one to which the beautiful chronometer is only secondary! The same critics, to be sure, are ready to remind us that the ocean has been passed in safety before Captain Johnson was born; even so—but how far the expenses of bad reckoning have been carried, or the absolute ruin of numerous fine ships may have been owing to a similar cause, is buried in a dense fog of oblivion. In a similar train of thought our author observes, 'While the tides and currents of the ocean—imperfect logs—inaccurate charts—unsteady steerage—inattention to the lead—stress of weather—defective ships—defective equipment, or defective management, may be the cause of loss, it would be fallacious to assume that the greater number of wrecks are caused by errors of the compass; but that many have occurred in consequence of these, there can be no doubt whatever.' Be it also remembered that the annual average number of British shipwrecks is stated to be about 547, or, as we formerly observed, a ship and a half a day.

The new compasses.

Absurd objection.

Reply.

Surveys of extreme accuracy are, of course, invaluable, and can only be obtained by much time, uninterrupted labour, and heavy expense; but for a reconnaissance of some thousands of miles, with every probability of interruption, immediate utility alone could be aimed at. After fixing the co-ordinates of latitude, longitude, and height, my whole effort was speedily to compile and correct a chart which should meet every want of the navigator. The boat-runs between the ports to which I have alluded, consisted, therefore, merely in sailing along on a patent-log base, or any other feasible scale; and the delineations thus made were afterwards reduced to my points and positions for adjustment. The whole being laid down for a certain end, the rough drawing was then reduced to the dimensions

Coast surveys.

Patent log.

which I considered adapted to the nature and extent of the examination, and the maritime importance of the place. The boat-bearings were generally taken with Kater's hand azimuth-compass; but in cases of moment, magnetic rhumbs were seldom resorted to when astronomical ones could be obtained; and all the shelves, shoals, and leading soundings, were fixed by sextant angles to assumed stations on shore.

Kater's compass.

Shoals fixed.

Among other desiderata, I determined to sound all that part of the sea, which was thus under my charge, to very unusual depths, to confirm the existence or non-existence of reported banks and dangers. This is a point upon which very particular attention was bestowed, since, although convinced that many shoals are reported which do not exist, yet I am also satisfied, that because an alleged rock or shelf is no longer to be found, it cannot be thence positively concluded that it never existed. Some such might have been thrown up by submarine volcanoes, and afterwards submerged, as, among recent and well-known instances, the Sabrina and Graham Islands; and many a *vigia* may have originated where mistakes of vision or mere imagination misled the judgment: drifting trees were taken for wrecks, while fish-scules, spawn-patches, meeting of currents, and local discolorations, passed for imminent perils. Thus, although some dangers have been noted without sufficient grounds, there can be no reasonable doubt that others are in like manner questioned without sufficient reason. Still, nothing in hydrography demands more circumspection than the act of erasing doubtful dangers from a chart; and as precaution is an acknowledged source of security, mariners should 'open their eyes' when they approach the sites of those previously reported. The class of would-be *savans* who absurdly hold that rocks grow under water, are easily furnished with a cause when an unexpected danger is announced; but their opinion is not the

Soundings.

On rocks and shoals.

Paradox.

less absurd because it was once pretty widely entertained,* and still lingers. Nor can I quite quadrate with the more modern doctrine of the disintegration and dispersion of such rocks; for when beneath the surface of the sea, rocks cannot be subject to such decomposing conditions as are produced by the active influence of the atmosphere. Oxygen, in the necessary quantity for the production of sufficient oxidation to weaken rocks at great depths, would hardly be afforded by the air contained in waters; and I have already shown the length of time during which many of the Mediterranean shoals have been recorded by navigators.

Rock geology.

Remark.

Holding it, therefore, on these grounds, to be highly improper to expunge dangers from the charts, however sceptical we may be as to the original authority that placed them there, it has always been my opinion that, though troublesome, every reported rock ought to be strictly searched for; and the pains I took in quest of supposed dangers—such as the Thisbe shoal, the Fox rock, l'Entreprenante reef, shoals north of Minorca, &c.—can only be known to those who sailed with me. Besides this, my general practice was to catch a very deep cast of the lead on every favourable occasion, for the chance of picking up a bank: so that, what with soundings, experiments for temperature, and drawing up water from great depths, calm weather was not an idle time upon our decks. On the assumption that no danger could be without a bank, my trials in the vicinity of *vigiæ* were made with from 150 to 600, and even 800 fathoms of line; and though doubts may remain relative to several reported dangers, still as it cannot be presumed that they do not exist, I have marked their reputed places with a note of interrogation. Indeed, when so many visible effects of the expansive and explosive gases

* Here, of course, I do not include the case of corallines; nor the gradual accretion of sedimentary strata assisted by pressure from the superincumbent water, and natural calcareous and ferruginous cements.

of submarine volcanoes are noted, it is quite clear that shoals may be up-heaved and again submerged, as in the instances already described.

Until my mission, all the charts in use exhibited a long bank between Cape Creux and Toulon, with from 40 to 70 fathoms of water upon it. It was called the Roches Molles; and it was very material that no ship should get far into the Gulf of Lyons in the night, or in thick weather, lest then gaining soundings she might imagine herself on the Roches Molles with plenty of sea-room. I therefore determined to examine this bank very particularly, but could not find it, although the whole vicinity was searched under casts of from 500 to 800 fathoms without striking bottom. This, coupled with a want of corroboration of its existence among the seafarers of Provence and Languedoc, induced me to drop it from my survey. There was another bank equally notorious, and with shallower water marked, shown for a century on paper, between Minorca and Asinara, under the name of Caccia. In the Admiralty chart supplied to our fleet, it bore 13 fathoms water; but in Mount and Page's edition of the *General Quarter Waggoner*, 1717, it shows the alarming notice—"sometimes 2 fathoms." Now had this shoal existed, we might have had a disagreeable acquaintance with it in the *Rodney*, 74, during a gale of wind to which she was exposed in January, 1812; but neither then, nor afterwards, could we substantiate the fact. In addition to my own exertions, finding that the Sardinian coral-fishers were unacquainted with such a spot, I made no scruple of omitting it also, thereby clearing a bugbear from an important navigation.

Although I thus expunged some supposed banks from the charts, in two cases only was my usual caution departed from, where actual dangers were marked on presumed authority. The first was between Capri and Cape Campanella, on the south entrance of the Bay of Naples: here a shoal was marked nearly in mid-channel, on Zannoni's

Supposed shoals expunged.

and other charts, and it had caused many ships to be carried round the western side of Capri, in preference to the shorter route. This shoal we could not find, nor even hear of among the fishermen: and having mentioned the matter to my friend Visconti, he employed some gun-boats, and swept the whole ground so completely, that we were quite satisfied there could be no danger there. The second

Thisbe rock.

instance was in rejecting the rock at the entrance of the Strait of Gibraltar, on which it had been asserted that H.M.S. *Thisbe* struck, at about 3h. 30m. A.M. on the 12th of August, 1804. For this I searched in vain; and on afterwards becoming acquainted with Mr. Corner, who was first lieutenant of that ship at the time, he assured me he did not know how any bearings could have been taken, as it was quite dark even after she had forged off. Therefore no doubt could be entertained of her having run upon the

Sir George Cockburn.

Cabezos. But in 1825, Sir George Cockburn put a memorandum in my hand, reporting that a merchantman was nearly lost upon the Thisbe; on which I remarked that some men thought it safer to state that to their owners, than to acknowledge the vessel had got upon well-placed and well-known rocks; and gave both him and Sir Edward Parry, the then hydrographer, my reasons for disbelieving the story. As the declaration, however, came with Lloyd's official strength, H.M.S. *Mastiff*, Commander Copeland, was ordered to the spot, and several weeks of expensive labour were wasted in confirming my impression.*

* In a similar manner, in the summer of 1849, a vessel reported that she had struck upon the Entreprenante rock, ninety miles east of Malta; whereupon H.M. steamers *Rosamond*, *Oberon*, and *Spitfire*, were sent to the ground which I had so often passed over. The mate of the vessel owned afterwards that the whole was a falsehood, advanced to cover their having struck upon a point of Malta. Since then another search has been instituted. Between the 17th and 23rd of April of this year (1853), H.M. ships *Retribution*, *Modeste*, *Niger*, and *Spitfire*, sounded for twenty miles around the site, with from 500 to 2570 fathoms of line out, and no bottom. A costly matter this, merely to remove a mare's-nest!

From the efforts we made, together with my constantly gaining all the local knowledge of the pilots and fishermen of the various ports I anchored in, it is very improbable that there is an unknown hidden danger within the limits of my chart. Still, to render navigation secure, to the chart thus furnished, should be added the prudence and skill of the intelligent seaman. Opinion.

In conducting these examinations, the soundings between twenty and sixty fathoms were usually taken with Massey's sounding-machine; but in greater depths we used solid leads, there being an apprehension of a collapse of the hollow cylinder forming the air-tube of the wings or vanes, at depths exceeding 100 or 150 fathoms.* Even with this defect there were great advantages in the use of this admirable instrument, for in moderate soundings the true vertical depth was easily ascertained, without the trouble of heaving-to; and we repeatedly reached the ground in upwards of forty fathoms while going six or seven knots, only rounding-to for catching the angles of objects in view. But our men became practically expert; while the machines were always kept in complete order, and duly tested as to accuracy. We were also furnished with Birt's buoy-and-nipper, but found it more ingenious in theory than satisfactory in practice. In very great depths we therefore resorted to the older method, of which experience somewhat lightened the labour, and helped to overcome the resistance and friction which the line had to encounter. Sounding machine. Birt's buoy.

The temperature of the sea at various depths was frequently taken, and registered with that of the atmosphere at the time of observation. The instruments used were Six's thermometers, which were compared before placing Sea temperature. Method practised.

---

* On informing Dean Buckland that one of these air-tubes was crushed quite flat under a pressure of about 300 fathoms, he suggested that the cylinder should be fortified by the introduction of transverse plates, acting on the principle of the chambered portion of the shells of Nautili and Ammonites. (*See his Bridgewater Treatise, vol.* i., *pp.* 345 *and* 349.)

them in the cylindrical copper cases attached to a white line by which they were immersed, and the index-floats duly noted. By these experiments we established various local peculiarities, which have been mentioned in the preceding pages; but could advance nothing in favour of Colonel Williams' theory of 'Thermometrical Navigation.'

Heights. Both in our shore-surveys and in sailing along the coasts, we always noticed the heights of the mountains; but though our observations were corrected for refraction of light and the curvature of the earth, they lay no claim to great precision, being taken merely for the direction of the navigator, by thus affording him pretty fair means of knowing his distance from the shore. Some were settled from the offing by their angular heights, with a reflecting instrument and a patent-log base; others on land by the barometer, the boiling-water point, their zenith distances, and some by observing their depressions to the horizon: still the results were all entered only as available approximations. And it should be mentioned that they are estimated above what is termed the 'level of the sea,' which, however, is not so uncertain in the Mediterranean as in those waters which experience the elevations and depressions of greater tidal power.

Meteorology. The meteorological phenomena were constantly noted and duly recorded with those instruments which we had at command—namely, the barometers, sympiesometer, thermometers, and hygrometer, which were subjected to as few disturbing influences as possible: in order to avoid interference with our other various duties, the regular time of observing those instruments was fixed for 8 A.M., when the chronometers were wound up and compared; and the observations required for correcting the refraction, we took when the astronomical operations which called for the correction were in hand. These registers aided very materially the conclusions recorded in Part III. of this work, and I am satisfied that the study of these matters is equally important to seamanship, agriculture, personal comfort, and medical science.

Such was the adopted routine, and though greater expertness might be found, it is hoped that it would have proved difficult to exhibit more zeal and perseverance. In so extensive a range of operations, and such a mass of troublesome arithmetical calculations, errors are unavoidable, though none, it is trusted, have crept in which can materially affect the accepted results; for these a liberal allowance will be made by those who have learned from experience the complex difficulties of such an undertaking. Other surveyors, with more time, better means, and a longer practice of the art, will improve the details from time to time; but I believe the coasts are now so approximately thrown into their proper form, that courses may be confidently shaped by them, which, it will be recollected, was very far from being the case when this survey was commenced. For instance, in the most important channel which divides the Mediterranean Basins, the island of Pantellaria bore S.E. of Maretimo on the charts supplied by the Admiralty until 1820, whereas it is actually S. by W.: and again, on entering the Adriatic, and shaping a course by the isle of Fano, its very portal, to Cape Linguetta, in Albania, that course would be N. 17° E. by the Admiralty chart, but it is really N. 5° W.; so that a ship trusting to the official documents at night, and not exerting the restless vigilance which characterizes true seamanship, would run on shore under the Acroceraunian cliffs.

Remark.

Former errors.

An attempt was also made to delineate the general topographical features of the coasts and harbours, and it is hoped that the endeavour was not unattended with success: still it is not possible, under what may be termed rather a revision than a survey, to meet the local knowledge of every critical observer on the spot. Public works demand, it is true, strict examination, but hyper-criticism only warps the judgment, and gives attention a wrong direction; indeed, in

Drawing of the charts.

Harsh critics.

most cases it is just as unfair to catch up an accidental and unimportant omission, as was the supercilious conduct of that Greek who found fault with a map of Greece as incorrect and useless, because his father's house in Athens was not noted on it! Such carpers must be reminded, in the

Liebig.

words of *Liebig*, that our duty really is, not to point at the supposed blemishes of others, but to labour onwards in the cause of accuracy ourselves: 'It is startling,' says he, 'when we observe that all the time and energy of genius, talent, and knowledge are expended in endeavours to demonstrate each other's errors.'

Remark.

My charts have long been in the hands of the service, and have been used by the fleets of all nations; they therefore are open to the most stringent criticism of professional men. The foregoing strictures should, however, be kept in view the while, in order that the nature of my intentions may be understood; and to this may be added, that, as far as concerns myself, I only regretted, on quitting the Mediterranean at the close of 1824, with the enlarged means and experience we then possessed, that I could not begin the whole of my work again. Regrets, however, at not having attained a higher degree of perfection, are now unavailing; this part must, therefore, be closed with a catalogue of the sheets of surveys which I handed in to the Admiralty, and which have, with a few exceptions, been long engraved and published in a form not very different from the following enumeration of the manuscripts:—

I.

General Chart.

A chart of the western division of the Mediterranean Sea, on a scale of $31\frac{1}{2}$ geographical miles to an inch, or about $\frac{1}{2300000}$ to nature (*see page* 373). Under the title the following note, somewhat bearing upon the whole survey, is appended:—'The basis of this chart is grounded on an entire new series of determinations by Captain Smyth, from astronomical, chronometrical, and geodetical operations. The details of the coasts of France and Spain, with their dependant islands, are in great part from the charts of Tofiño, Cassini, and Hell; and other most authentic documents, examined and corrected on the spot. The west coast of Italy and its islands are new surveys, in the execution of which much assistance was rendered, in the vicinity of Naples, by Colonel Visconti. The Adriatic Sea is constructed from the united operations of the Austrian, Neapolitan, and English officers, employed under Colonels Campana and Visconti, and Captain Smyth. The coast of Africa is laid down under such examinations as circumstances permitted, by Captain Smyth and his officers; and the whole intervening sea has been examined and sounded with so much attention, as to leave little probability of any unmarked danger existing. The reported shoals marked with a note of interrogation have not yet been found, though frequently sought for; but are still under search by the tender which Captain Smyth left in the Mediterranean for that purpose.'

II.

Plan of the bay, harbour, and environs of Cadiz, with a view of the Alameda, on a scale of $\frac{1}{58000}$, or $\frac{4}{5}$ths of a mile to an inch. SPAIN.

III.

The Strait of Gibraltar, on a scale of $\frac{1}{111000}$, or $1\frac{1}{2}$ miles to the inch; with views of the land from the Cabezos shoal, to the west of Tarifa.

IV.

The Rock of Gibraltar ($\frac{1}{12000}$, or $\frac{1}{6}$th of a mile to an inch); with a view of the new mole-head, Ape's Hill and Ceuta in the distance.

V.

General chart of the coast of Spain, from Gibraltar to Alicant, and the opposite shores of Barbary ($\frac{1}{910000}$, or $12\frac{1}{2}$ miles to one inch); with plans of Malaga, Almeria, Port Genovés, San Pedro, Carbonera Bay, Port Aguilas, Melillah, and Alboran; the general scale of the first six being $\frac{1}{49000}$, or $\frac{2}{3}$rds of a mile to one inch.

VI.

The harbour of Cartagena ($\frac{1}{26000}$, or $\frac{2}{7}$ths of a mile to an inch), with two views; and the Columbretes Rocks ($\frac{1}{26000}$, or about $\frac{1}{3}$rd of a mile to one inch), with two views of them.

VII.

A sheet of Spanish ports—namely, Turilla Bay, Peniscola, Calpe, Altea Bay, Isle Grosa, and Alicant; the last on a scale of $\frac{1}{49000}$, or $\frac{2}{3}$rds of a mile to an inch.

VIII.

General chart of the coast of Spain, from Alicant to Palamos and the Baleares ($\frac{1}{880000}$, or about 12 miles to an inch); with plans of Barcelona, Tarragona, Grao of Valencia, Port Iviza ($\frac{1}{75000}$), Palma Bay ($\frac{1}{24000}$), and Port Cabrera ($\frac{1}{12250}$).

IX.

Plan of the harbour and environs of Port Mahon, in Minorca, on a scale of $\frac{1}{11500}$, or about $\frac{1}{6}$th of a geographical mile to one inch; with a view of Lazzaretto isle.

X.

General chart of the south coast of France, and part of Catalonia, on a scale of $\frac{1}{700000}$, or about $9\frac{3}{5}$ miles to an inch; with plans of Callioure, Vendres, Cadaques, Tusa, Blanes, and Palamos. FRANCE.

XI.

FRANCE. The coast of France, from the mouth of the Rhone to Riou isle, containing the gulfs of Foz and Marseilles ($\frac{1}{83000}$, or $1\frac{1}{4}$ mile to the inch); with a view of Planier lighthouse.

XII.

The port and roads of Marseilles ($\frac{1}{15000}$, or $\frac{1}{5}$th of a mile to an inch); and the Cassidaigne Rock ($\frac{1}{80000}$, or about $\frac{4}{5}$ths of a mile to the inch), with views of Cassis and the Bec de l'Aigle.

XIII.

The harbour and road of Toulon, with the adjacent coast, on a scale of $\frac{1}{70000}$, or nearly one mile to an inch.

XIV.

Chart of the coast of France, from the Peninsula of Giens to Cape Roux, or from Hieres Bay to the Gulf of Frejus ($\frac{1}{130000}$, or $1\frac{1}{5}$ miles to an inch); with a view of the fort on Port Cross island.

XV.

Chart of the coast of France and Italy, from Cape Roux to Monaco ($\frac{1}{78000}$, or one mile to an inch); with a view of the Lerins isles from Cannes, and a plan of Monaco ($\frac{1}{25000}$).

XVI.

ITALY. The harbour of Villa Franca and its vicinity, on a scale of $\frac{1}{13750}$, or $\frac{1}{5}$th of a mile to one inch; with views of the town and castle of Villa Franca and the city of Nice.

XVII.

Chart of the coast of Italy, from Ventimiglia to Piombino, or the Gulf of Genoa ($\frac{1}{440000}$, or 6 miles to the inch); with plans of Savona ($\frac{1}{24500}$), Gallinara ($\frac{1}{38500}$), Gorgona ($\frac{1}{37000}$), and Finale ($\frac{1}{75000}$).

XVIII.

A sheet of plans, containing the road and vicinity of Vado, the Bay of Noli, and Porto Maurizio, each on a scale of $\frac{1}{24500}$, or about $\frac{1}{3}$rd of a mile to one inch.

XIX.

A sheet of plans, containing Genoa harbour ($\frac{1}{14500}$), with a view of the lighthouse; with Porto Fino and Sestri a Levante, on a scale of $\frac{1}{24500}$, or $\frac{1}{3}$rd of a mile to the inch.

XX.

The Gulf of Spezia ($\frac{1}{28500}$), and a plan of the road, town, and environs of Via Reggio ($\frac{1}{84000}$, or $1\frac{1}{3}$ mile to one inch).

XXI.

A sheet of plans, containing Capraja island ($\frac{1}{50000}$), the Mouth of the Arno ($\frac{1}{37000}$), and the town and road of Leghorn on a scale of $\frac{1}{35000}$, or about $\frac{1}{2}$ a mile to one inch.

XXII.

General chart of the west coast of Italy, from Piombino to Civita Vecchia, and the Tuscan Islands ($\frac{1}{280000}$, or about $3\frac{4}{5}$ miles to the inch); with plans of Pianosa isle ($\frac{1}{35000}$), Port Campo ($\frac{1}{48500}$), Piombino ($\frac{1}{50000}$), Formiche of Grosseto ($\frac{1}{59000}$), and Gianuti isle ($\frac{1}{50000}$).

XXIII.

A sheet of plans, containing Giglio island, on a scale of $\frac{1}{84000}$, or about $\frac{6}{7}$ths of a mile to the inch; Palmajola Channel ($\frac{1}{47000}$), Porto Longone ($\frac{1}{24500}$), Porto Ferrajo ($\frac{1}{10500}$), and Orbitello. ITALY.

XXIV.

General chart of the west coast of Italy, from Civita Vecchia to the Gulf of Naples ($\frac{1}{365000}$, or about $5\frac{1}{4}$ miles to an inch); with plans of Terracina ($\frac{1}{25500}$), the Mouth of the Tiber ($\frac{1}{64000}$), Mount Circello ($\frac{1}{80000}$), Gaeta ($\frac{1}{31500}$), Porto d'Anzo ($\frac{1}{10750}$), and Civita Vecchia ($\frac{1}{13000}$).

XXV.

The Ponza Islands, on a scale of $\frac{1}{85000}$, or $1\frac{1}{6}$ geographical miles to the inch; with views of Zannone and Capo di Guardia, and a plan of Port Madonna ($\frac{1}{10230}$).

XXVI.

The Crater or Gulf of Naples, and its islands ($\frac{1}{95500}$, or $1\frac{1}{3}$ geographical miles to one inch); with views of Ischia and Capri.

XXVII.

General chart of the west coast of Italy from Naples to Cape Vaticano (scale $\frac{1}{530000}$, or about $8\frac{3}{5}$ miles to an inch); with plans of i Galli Rocks ($\frac{1}{23300}$), Pæstum in Agropoli Bay ($\frac{1}{90000}$), and Dino Isle and Bay ($\frac{1}{36500}$).

XXVIII.

The island of Corsica, with the Tuscan islands ($\frac{1}{500000}$, or $6\frac{4}{5}$ miles to the inch); with plans of San Fiorenzo, Isola Rossa, Calvi, Porto Vecchio, and Bastia (each on the scale $\frac{1}{42500}$). CORSICA.

XXIX.

A sheet with a plan of the Gulf of Ajaccio, on a scale of $\frac{1}{18000}$; and the Road of Capo Corso ($\frac{1}{15500}$), with two views of the land.

XXX.

The Strait of Bonifaccio, between Corsica and Sardinia ($\frac{1}{59500}$, or nearly a mile to an inch); with plans of Lavezzi and its rock, the harbour of Bonifaccio, and the isle of Cavallo.

XXXI.

A general chart of the island of Sardinia, on a scale of $\frac{1}{510000}$, or 7 miles to an inch; with small plans of Port Longo Sardo and the Bay of Tortoli. SARDINIA.

XXXII.

The Gulf of Asinara, on the north-west coast of Sardinia ($\frac{1}{130000}$, or $1\frac{2}{3}$ miles to the inch); with a plan of the Road of Porto Torres, and a view of Castel Sardo.

XXXIII.

The north-east coast of Sardinia and its adjacent islands ($\frac{1}{95000}$, or about $1\frac{1}{3}$ miles to one inch); with plans of Maddalena and Porto Cervo, and a view of Capo dell' Urso.

XXXIV.

SARDINIA. A sheet of Sardinian ports—namely, Ports Conte and Alghero ($\frac{1}{82000}$), with a view of Capo della Caccia; the Channel of San Pietro ($\frac{1}{82000}$), with a view of Point Colonne; and Cagliari Bay ($\frac{1}{27000}$), with a view of the city from the anchorage.

XXXV.

The south coast of Sardinia, on a scale of $\frac{1}{275000}$, or about $3\frac{2}{3}$ miles to the inch; with views of San Pietro, and the Gallo Rock, off the west point of San Pietro.

XXXVI.

SICILY. A general chart of Sicily, Malta, the adjacent islands, and parts of Italy, Sardinia, and Africa. Scale $\frac{1}{830000}$, or about $11\frac{1}{3}$ miles to one inch.

XXXVII.

A map of Sicily, on a scale of $\frac{1}{513000}$, or 7 miles to the inch; reduced and corrected from Baron Schmettau's large manuscript map on thirty sheets, lent me by the Sicilian government.

XXXVIII.

Chart of the west coast of Sicily, and the Ægadean islands ($\frac{1}{188000}$, or $2\frac{3}{5}$ geographical miles to the inch); with part of the gulf of Castell'a mare.

XXXIX.

A sheet of coast views:—1. From the shoal off Cape San Vito. 2. From the shoal off Emilia point. 3. Trapani from the Asinello rock. 4. Marsala from the outer shoal. 5. The town of Mazzara from the roads.

XL.

The anchorages and vicinity of Trapani, on a scale of $\frac{1}{88400}$, or about $1\frac{1}{5}$ miles to an inch; with views of Maretimo Castle and the Saracenic tower on Mount St. Julian, and an ancient coin of Eryx for identity of site.

XLI.

Chart of the north coast of Sicily and the adjacent islands ($\frac{1}{470000}$, or $6\frac{1}{3}$ miles to the inch); with views of the rock of Scylla and the Faro point.

XLII.

Plan of the island of Ustica ($\frac{1}{28800}$); with a view of the bay and town of Santa Maria, in the same island, and an ancient coin ascribed to it.

XLIII.

Plan of the environs and gulf of Palermo ($\frac{1}{71000}$, or about one mile to an inch); with views of Cape Di Gallo and Cape Zaffarano, and an ancient coin of Soluntum.

XLIV.

Plan of the bay and city of Palermo, on a scale of $\frac{1}{15000}$, or about $\frac{1}{5}$th of a mile to an inch; with a view of the Ponte dell'Ammiraglio over the Oretus, and an ancient coin of Panormus for identity of site.

XLV.

A sheet of coast views:—1. Bay of Palermo. 2. Cefalu bearing east by south five miles distant. 3. The channel between Sicily and the Æolian Islands. 4. Distant views of the entrance into the Faro Channel. SICILY

XLVI.

Plan of the Lipari group, or Æolian Islands, on the north coast of Sicily; to a scale of $\frac{1}{145000}$, or about two miles to the inch.

XLVII.

Plan of the bay of Lipari ($\frac{1}{20000}$), with an ancient coin of the island, and a view of the city of Lipari. Plan of Olivieri Bay ($\frac{1}{17300}$), with an ancient coin of Tyndaris, and a view of Cape Tindaro.

XLVIII.

A sheet of coast views:—1. The channel between the islands of Lipari and Vulcano. 2. Panaria, Basiluzzo, &c., and Stromboli, from Exmouth Bank. 3. The Æolian Islands as seen from the Penrose Rocks. 4. The Strait of Messina, from the anchorage on the shoal off the Faro point.

XLIX.

The city, bay, and promontory of Milazza, on a natural scale of $\frac{1}{11300}$, or about $\frac{1}{6}$th of a mile to one inch; with a view of the town and castle, from near the Tonnara.

L.

A general chart of the east coast of Sicily and the south part of Calabria. Scale $\frac{1}{300000}$, or four miles to the inch.

LI.

Plan of the Faro or Strait of Messina ($\frac{1}{28600}$, or $\frac{2}{5}$ths of a geographical mile to an inch); with a view of Scilla Castle.

LII.

Plan of the city and harbour of Messina, on a scale of $\frac{1}{7700}$, or about $\frac{1}{10}$th of a mile to the inch. An ancient coin of Messina for identity of site and symbol.

LIII.

A sheet of coast views:—1. The city and harbour of Messina. 2. Mount Ætna, as seen from off Schisò point. 3. View of the Cyclop islets. 4. The city and port of Catania.

LIV.

Plan of the bay and environs of Taormina ($\frac{1}{15000}$); with views of the city of Taormina and Schisò point, and ancient coins of Tauromenium and Naxos.

LV.

Plan of the town and harbour of Augusta ($\frac{1}{20500}$, or $\frac{1}{4}$th of a mile to an inch); with a view of the town and Torre d'Avola lighthouse, and a coin of the ancient Megara.

LVI.

SICILY. The city, harbour, and environs of Syracuse, on a natural scale of $\frac{1}{13400}$, or about $\frac{1}{6}$th of a mile to the inch; with a view of the port from the temple of Jupiter Olympius, and two ancient coins for identity of site.

LVII.

A sheet of coast views:—1. The port and castle of La Bruca. 2. The city and port of Syracuse. 3. Cape Passaro bearing south-south-east about six miles. 4. The town and road of Alicata.

LVIII.

General chart of the south coast of Sicily ($\frac{1}{380000}$, or $5\frac{1}{6}$th miles to one inch); with a plan of Alicata, and views of Cape Passaro from off the tonnara, and from the south.

LIX.

Plan of the city, environs, and anchorage of Girgenti ($\frac{1}{33700}$, or about $\frac{1}{2}$ a mile to an inch); with a view of the city from the temple of Æsculapius, a coin of Agrigentum for identity, and another of the tyrant Phintias.

LX.

A sheet of coast views: — 1. The mole of Girgenti, as seen from the temple of the Virgins. 2. Appearance of the south-west point of Sicily. 3. The island of Pantellaria. 4. Cape Dimitri, off Goza. 5. Appearance of Malta and Goza when passing Comino.

LXI.

The town and port of Pantellaria ($\frac{4}{5500}$), with a view of the town, and an ancient coin of Cossyra; also a plan of the harbour of Lampedusa on a natural scale of $\frac{1}{7500}$.

LXII.

Plan of the island of Linosa ($\frac{1}{32000}$, or $\frac{1}{2}$ a mile to an inch), with a view of its south coast; and a plan of Lampedusa and Lampion on a scale of $\frac{1}{68000}$, or nearly a mile to the inch.

LXIII.

MALTESE ISLANDS. Hydro-geographic map of the Maltese islands and rocks, on a natural scale of $\frac{1}{94000}$, or one $\frac{1}{2}$ mile to one inch.

LXIV.

Plan of St. Paul's Bay ($\frac{1}{8500}$, or 0·11 mile to an inch); with a view of the tower and battery on Koura point, and a view of the Salmona palace.

LXV.

The city, towns, fortifications, and harbours of Valetta, on a natural scale of $\frac{1}{8850}$, or 0·12 mile to an inch; with a view of the castle and lighthouse of Sant' Elmo, one of the castle of Sant' Angelo, and a third of Valetta from a distance.

LXVI.

Plan o Marsa Scirocco bay ($\frac{1}{9300}$, or 0·13 mile to the inch); with a view of its commanding fortress, St. Lucian's tower-redoubt.

LXVII.

General chart of the south-east coast of Italy, from Cape Spartivento round Cape Santa Maria di Leuca, and into the Adriatic to Polignano ($\frac{1}{570000}$, or nearly 8 miles to an inch); with plans of Cotrone ($\frac{1}{8750}$), Taranto ($\frac{1}{82000}$), and Gallipoli ($\frac{1}{30500}$). **CALABRIA.**

LXVIII.

A sheet containing plans of the harbour of Brindisi ($\frac{1}{31750}$, or 0·44 mile to the inch), the port of Otranto ($\frac{1}{38000}$), and the Tremiti or Diomedeæ isles ($\frac{1}{30500}$, or 0·41 mile to an inch). **NAPLES EAST.**

LXIX.

General chart of the east coast of Italy, from Monopoli and Polignano to Fossaceca ($\frac{1}{400000}$), or 5½ miles to one inch; with plans of Barletta ($\frac{1}{32000}$), Viesti ($\frac{1}{24750}$), Manfredonia ($\frac{1}{26500}$), and Pianosa Rock ($\frac{1}{30500}$).

LXX.

General chart of the east coast of Italy, from Fossaceca to Rimino, on a natural scale of $\frac{1}{400000}$, or nearly 5½ miles to the inch; with plans of Ortona ($\frac{1}{17000}$), Fano ($\frac{1}{22200}$), Rimino ($\frac{1}{22200}$), Pesaro ($\frac{1}{30000}$), Sinigaglia ($\frac{1}{22500}$), and Porto Nuovo ($\frac{1}{48000}$). **PAPAL STATES.**

LXXI.

Plan of the city, fortifications, and port of Ancona, on a natural scale of $\frac{1}{12400}$, or 0·17 mile to the inch; with a general view of the citadel and mole as seen from off Mount Conero.

LXXII.

General chart of the coasts of Italy and Istria, from Rimino to Cape Promontore ($\frac{1}{350000}$, or 4·8 miles to one inch). This comprehends the north part of the Adriatic Sea, and is locally termed the Gulf of Venice. **VENICE.**

LXXIII.

A sheet containing a particular plan of Venice and its anchorages ($\frac{1}{53500}$); Porto di Chioggia on the same scale; and the free-port of Trieste ($\frac{1}{12000}$). With a view of the city and the Porporello, from the anchorage off Malamocco.

LXXIV.

A sheet of Istrian ports—namely, Pirano ($\frac{1}{43500}$), Omago ($\frac{1}{13800}$), ports Quieto and Cittanova ($\frac{1}{27300}$), Parenzo ($\frac{1}{11250}$), Orsera and the Lemo Canale to Rovigno ($\frac{1}{57500}$), and Port Veruda ($\frac{1}{26000}$). **ISTRIA.**

LXXV.

The harbours of Fasana and Pola, with the Brioni islands, on a scale of $\frac{1}{20000}$ to nature, or 0·27 mile to an inch; and a view of the amphitheatre and watering-place.

LXXVI.

A general chart of the coasts of Croatia and Dalmatia, from Cape Promontore to Slozella, comprehending the Quarnero, Quarnerolo, Morlacca, Maltempo, and Zara channels, on a scale of $\frac{1}{350000}$, or 4·8 miles to the inch; with plans of Kerso ($\frac{1}{41000}$), Porto Re ($\frac{1}{19250}$), San Pietro di Nembo ($\frac{1}{57600}$), and Unie Bay ($\frac{1}{34000}$). **CROATIA.**

LXXVII.

DALMATIA. A sheet containing ports of Croatia and Dalmatia—namely, Port Augusto, in Lossin Piccolo ($\frac{1}{20500}$) Port Beguglia ($\frac{1}{57600}$), with a view of Bianche lighthouse, Zara and its harbour ($\frac{1}{13900}$), the strait of Pasman ($\frac{1}{39500}$), Morter Canale ($\frac{1}{39000}$), and Port Tajer ($\frac{1}{45700}$).

LXXVIII.

Plan of Port Sebenico, with the outer channels and Vodizze road ($\frac{1}{54000}$), the port of Ragosnitza ($\frac{1}{57000}$); and the bay of Spalatro ($\frac{1}{11000}$).

LXXIX.

General chart of the coast of Dalmatia, from Zara Vecchia to Ragusa Vecchia ($\frac{1}{422500}$, or $5\frac{4}{5}$ miles to the inch); with plans of Pelagosa rocks ($\frac{1}{30500}$), and ports Lago and Rosso on Lagosto island ($\frac{1}{55500}$).

LXXX.

A sheet of Dalmatian ports, containing Lessina and its Canale ($\frac{1}{50000}$); Port S. Giorgio in Lissa ($\frac{1}{20500}$); Valle grande of Curzola ($\frac{1}{52200}$); the Canale di Curzola ($\frac{1}{117500}$); Port Milna of Brazza, in the canale di Spalatro ($\frac{1}{39000}$); and Porto Palazzo in Meleda ($\frac{1}{32000}$).

LXXXI.

Ragusa and the Kalamota channels, with the rocks and bay of Ragusa Vecchia. On a natural scale of $\frac{1}{77500}$, or about one mile to a linear inch.

LXXXII.

ALBANIA. General chart of the coast of Albania, from Ragusa Vecchia to Port Palermo, with a part of the opposite coast of Italy ($\frac{1}{487000}$, or $6\frac{3}{5}$ miles to the inch), and an enlarged sketch of the coves under Kimara.

LXXXIII.

Plan of the gulf of Cattaro (*Bocche di Cattaro*), on a natural scale of $\frac{1}{76500}$, or about one geographical mile to an inch, with a plan of Porto di Budua ($\frac{1}{31500}$), and the isle of S. Niccolò.

LXXXIV.

A sheet of Albanian ports—namely, Antivari bay ($\frac{1}{59000}$), Dulcigno road ($\frac{1}{42500}$), Durazzo bay ($\frac{1}{82500}$), Aulona or Valona bay ($\frac{1}{135000}$), Port Palermo ($\frac{1}{36000}$), and Parga ($\frac{1}{19500}$).

LXXXV.

IONIAN ISLANDS. General chart of the channels of Corfu, with the adjacent coast of Albania ($\frac{1}{238000}$, or about $3\frac{1}{4}$ miles to an inch); with plans of Alipa and San Niccolò in Yliapades bay ($\frac{1}{20000}$), Port Gayo in Paxo ($\frac{1}{10000}$), and Port Laka in the same island ($\frac{1}{21400}$).

LXXXVI.

The town and road of Corfu, with the environs from Ulysses rock to Porto Govino, on a scale of $\frac{1}{28000}$, or $\frac{1}{2}$ of a mile to a linear inch; with views of the town and citadel.

LXXXVII.

Ionian Islands.

General chart of the central Ionian Islands, with the opposite coast of Greece from Parga to the mouth of the Alpheius, with the gulfs of Arta and Patras ($\frac{1}{350000}$), or $4\frac{4}{5}$ miles to an inch); with four views from particular points.

LXXXVIII.

A sheet of Ionian ports, containing Santa Maura and its vicinity ($\frac{1}{70600}$), Port Vliko in Leucadia ($\frac{1}{52500}$), Dragamesti and the Echinades ($\frac{1}{76600}$), Port Vathi in Ithaca ($\frac{1}{15000}$), Port Argostoli in Cephalonia, with two views ($\frac{1}{111000}$), and Zante Bay ($\frac{1}{32800}$), with a view of the bay from the lazzaretto.

LXXXIX.

Morea.

A general chart of the west coast of the Morea, from Gastouni river to the Gulf of Koron ($\frac{1}{257000}$), or $3\frac{1}{3}$ miles to the inch); with a plan and view of the Stamfane or Strivali rocks ($\frac{1}{65000}$), and Mothoni or Modon, Port Longona of Sapienza, and the road of Koron, each on a scale of $\frac{1}{32000}$.

XC.

Plan of the town and harbour of Navarin, or Neo Kastro, with the Paleó Kastro or ancient Pylus, on a natural scale of $\frac{1}{37000}$, or half a geographical mile to the inch.

XCI.

General chart of the coast of the Morea, from Venetico Island to Kyparisi, with the Cervi and Cerigo channels into the Archipelago ($\frac{1}{250000}$, or $3\frac{2}{5}$ miles to the inch). Also plans of Port Nikolo in Cerigo ($\frac{1}{36000}$), Kapsali bay in Cerigo ($\frac{1}{25000}$), and Port Potamo in Cerigotto, also $\frac{1}{25000}$, or $\frac{1}{3}$ of a mile to an inch.

XCII.

Egypt.

Chart of the coast of Egypt, from Al Awaïd to the Rosetta mouth of the Nile ($\frac{1}{220000}$, or about 3 miles to the inch); with views of Abukeer Castle, and the Arab's Tower.

XCIII.

Plan of the city, environs, and harbours of Alexandria, on a scale of $\frac{1}{26000}$, or $\frac{1}{3}$ of a mile to the inch; with the Pharos enlarged ($\frac{1}{3125}$), and a view of it. Also a view of Alexandria from the anchorage.

XCIV.

General chart of the north coast of Africa, from Alexandria to Ras al Halal ($\frac{1}{1276000}$, or $17\frac{1}{2}$ miles to an inch); with plans of Ras al Halal ($\frac{1}{195000}$), Ras et Tyn ($\frac{1}{78000}$), Marsa Tebruk ($\frac{1}{80000}$), Dernah ($\frac{1}{72000}$); and Ishaïlah rocks, Marsa Labeit, and Marsa Mahadda, each $\frac{1}{200000}$, or $2\frac{4}{5}$ inches to a mile.

XCV.

Plan of the Gulf of Bombah and the adjacent isles, on a natural scale of $\frac{1}{55500}$, or $\frac{5}{7}$ths of a mile to a linear inch; with a view of Bhurdah Isle from the north-east.

XCVI.

TRIPOLI.

General chart of the coast of Barbary from Marsa Susah to Misratah, forming the Gulf of Sidra or Greater Syrtis ($\frac{1}{1350000}$, or about $18\frac{1}{2}$ miles to an inch). On this sheet are also plans of Marsa Bureigah ($\frac{1}{27000}$), Gharah Rocks ($\frac{1}{104500}$), Benghazi ($\frac{1}{14400}$), Marsa Susah ($\frac{1}{9300}$), Tolmeïtah ($\frac{1}{20500}$), and Marsa Zafran ($\frac{1}{14400}$).

XCVII.

A general chart of the coast of Barbary, from Melhafah in the Syrtis to Karkarish on the west of Tripoli ($\frac{1}{400000}$, or $5\frac{1}{2}$ miles to the inch); with a plan and view of the ruins of Leptis Magna ($\frac{1}{84000}$), and Marsa Ugrah ($\frac{1}{30000}$).

XCVIII.

The harbour and environs of Tripoli, on a scale of $\frac{1}{16800}$, or $\frac{1}{5}$ of a mile to an inch; with a view of the fortifications from the outer roads, and another from the middle of the harbour.

XCIX.

TUNIS.

A general chart of the coast of Barbary, from Ras al Amrah in Tripoli to Tabulbah in Tunis, including the Gulf of Khabs or Lesser Syrtis ($\frac{1}{700000}$, or about $10\frac{1}{2}$ miles to the inch); with plans of the Bukal channel of Jerbah ($\frac{1}{288000}$), and Tripoli Vecchio ($\frac{1}{53000}$).

C.

General chart of the coast of Barbary, from Cape Africa in Tunis to the Fratelli rocks ($\frac{1}{500000}$, or $6\frac{4}{5}$ miles to the inch); with a plan of Mehediah or Africa city ($\frac{1}{95500}$), and another of the Fratelli rocks ($\frac{1}{55500}$), or about $\frac{6}{7}$ths of a mile to an inch.

CI.

A sheet of Tunisian ports, containing the Bay of Bizertah, the details of Cape Bon, Monastir Bay and the Kuriah Isles, and the lake and environs of Tunis, with the vestiges of Carthage; the last on a natural scale of $\frac{1}{94000}$, or $1\frac{1}{3}$ of a geographical mile to one inch.

CII.

ALGERIA.

A sheet of Barbary plans, namely, the port and isle of Tabarkah, $\frac{1}{3}$ of a mile to one inch; the bay and beaches of Ustorah, and the Galita Islands, the last on a scale of $\frac{1}{24250}$, or $\frac{1}{3}$ of a mile to an inch; with a view of Galita and Galitona.

CIII.

General chart of the coast of Barbary, from the Fratelli rocks of Tunis to the Pisan rocks of Algeria ($\frac{1}{650500}$, or nearly 9 miles to an inch); with plans of Bujeyah ($\frac{1}{157000}$), the Pisan rocks ($\frac{1}{32000}$), Port Jigeli ($\frac{1}{24000}$), Kolah ($\frac{1}{72000}$), Al Kal'ah cove ($\frac{1}{17000}$) and Bonah bay ($\frac{1}{193000}$). Also views of Cape Carbon and the Pisan rocks, and the town and castle of Bonah.

CIV.

A general chart of the coast of Barbary, from Bujeyah in Algeria to the Zaphran Isles on the coast of Morocco ($\frac{1}{1105000}$); with plans of the cove at

Sidi Ferej, and the port of Waharan ($\frac{1}{10700}$). The remainder of the coast of Barbary, from the Zaphran Isles to the Atlantic Ocean, is on plate No. V. MOROCCO.

CV.

A sheet with the city, bay, and environs of Algiers ($\frac{1}{53700}$), and the Zaphran Isles, the scale of the latter being $\frac{1}{10000}$, or about 0·15 mile to one inch; with two views, one of the city of Algiers, the other of the Ja'ferei, or greater Zaphran.

Such being the results of my surveys and re-examinations, I am prepared to show—more in confidence than presumption—that, however much these charts may fall short of that fulness in detail and delicacy of finish which more time and strength would have enabled us to give them, they are quite equal to every reasonable requirement of the navigator; and generally also to the engineer and the inquiring traveller. But that is not all: the school thus formed has flourished, and my survey may be said to have been continued into the East. When Captain Copeland was despatched to the Levant, two of my officers—Cooling and Wolfe—were placed with him; while Messrs. Elson and West were making use of such opportunities as offered, on the same station. At length my zealous *élève*, Captain Graves, after returning from an arduous voyage to Magellan's Strait under Captain P. P. King, in our old ship the *Adventure*, assumed the surveying tiller in the Levant, and most successfully guided an enlarged and efficient establishment for many years. The effect of unanimity and talent has been truly gratifying; insomuch that there results a mass of Archipelagan charts and plans of so high a quality in detail, accuracy, and finish, that any naval officer may be proud on scrutinizing them. Altogether, whatever improvement in the art of marine surveying may yet arise, it can safely be asserted that Mediterranean chartography can never again incur such reproaches as those recorded on pages 354 to 356. Forty years have, indeed, worked wonders in meeting the scientific wants of the seaman.

On the charts.

My survey extended.

Captain Graves.

Prediction.

PART V.

ON THE ORTHOGRAPHY AND NOMENCLATURE ADOPTED; THE GEOGRAPHICAL POINTS — OR CO-ORDINATES OF LATITUDE, LONGITUDE, AND HEIGHT—OF THE MEDITERRANEAN SHORES; WITH THE VARIATION OF THE MAGNETIC NEEDLE, AND OTHER NOTANDA.

---

§ 1. On the Orthography and Nomenclature adopted.

Geographical points.

WE have now arrived at the fundamental end and aim of all the before-mentioned operations—namely, the register of the Geographical Points by which all the former Mediterranean charts are reformed; and as this tabular exhibition contains certain symbols for reference, they necessarily require some explanation in order to obviate needless subsequent repetition.

Prefatory matter.

In the first place, the attempt at reconciling the discordant orthography and even the nomenclature of islands, towns, ports, and headlands, ought to be expressly stated, in order to prevent misunderstanding where an apparent discrepancy occurs; and in the identifying of ancient and modern sites now offered, the reader must accept of my responsibility, instead of being troubled, through a long series, with fathoms of discussion and contending authorities. Thinking, with Cervantes, that annotations such as those he alludes to—'Goliah, Golias, or Goliat the Philistine'—rather retard than illustrate, I have endeavoured to enrol only what is demanded by the object in view; and throughout the survey I have sought to preserve the local name in actual use, the spelling of which has been as much as possible fashioned by that of the inhabitants. Thus, in the coasts of Spain, France, and Italy, the orthography of the natives is carefully followed—except where the spelling has been so Anglicized by custom, and apocope, as to have become vernacularly adopted into our language; and a departure from such adaptations might be stigmatized as an affectation by correct writers, however they may have become familiarized by our continual intercourse with the Continent, and by the example of some whose education teaches them French rather than English grammar. In Greece and Barbary, on the other hand, where the alphabets totally differ from our own, I have declined receiving the names through any intervening channel, and have transferred, as well as obvious imperfections will admit, from the common pronunciation directly into English. Our borrowing Eastern names through the filter of other tongues of western Europe, has been both absurd and mischievous; and it is curious what singular errors and misconceptions have originated from a cause apparently so insignificant.

Usage.

Differing alphabets.

Yet although along the coasts of France and Italy I have almost invariably followed the French and Italian nomenclature, on that of Spain a slight *discretion* has been used where the name is not truly Spanish. That language is one of the most pure in Europe, but its guttural enunciation has hampered their writing of foreign terms—as Guadalquiver (*Wad el Kebír*), Alfaques (*El Fakkah*), Oran (*Wahrán*), Mazalquiver (*Marsa 'l Kebír*), Algeçiras (*Al Jezeïrat**). These are certainly merely naturalized words, for, as Don Quixote told Sancho, 'the name of Albogues is Moorish, as are all those in our language be-

Remark.

* This common word for island is usually spelt with *t*; but the final *h* is only sounded as *t* before a following vowel—Jezeïrat-u-l-khadhrá, the Green Island, is the complete name. Jezeïrat should be pronounced so as to rhyme with 'fire at.'

Spanish spelling.

ginning with *al*;'* but even in the old Spanish the orthography is by no means uniform, and certain letters, which in that language have the same sound to a Spaniard's ear, are often used for each other, as *b* and *v*—*c*, *z*, *s*, and *ç*—and *g*, *j*, *x*. Some confusion has arisen, also, from our rendering their word *Monte* by mountain, whereas it often merely marks a copse or thicket; and others of their geographical terms are incurably difficult for the mouth of a foreigner, as Jaraicijo, the name of a Spanish town whence the Duke of Wellington wrote his pithy letter to General Eguia, in 1809, and which no one but a native ever pronounced properly. If spelt *'Haraïcého*, it might be better aspirated by an Englishman at sight.

Names with meaning.

Variations in orthography are not confined to Spain; and though I have *almost* followed the French and Italians on their own coasts, it has been with that degree of caution that custom has not been violated without substantial reason. On this account no alteration in the spelling has taken place which would injure the sense, since so many names have meanings—as the puntals, Olla, Cabezos, Palos, and Palomas, of the Spaniards; the Sèche, Fourmigues, Gabinière, of the French; the Bonaria, Capraja, Maremme, of the Italians; and the Cranæ, Styli, Zancle, Gaïderonesos, Drepanum, Myconus, Hydrussa, and Strongyle, of the Greeks: not a few being derived from parts of the human frame, as *brow* and *foot* of a mountain; an *arm* of the sea,

* 'Yeste nombre Albogues es Morisco, comolo son todos aquellos, que en nuestra lingua Castellana comiençan en AL.' While on this topic, a glance may be allowed at the recent rendering of Cape Trafalgar into *Head of Laurels*, and the consequent unnecessary compliments to Nelson. Taraf-al-ghúr literally means Cape Cave, or Cavern Point: *taraf* when rapidly uttered is either *tarf* or *traf*, meaning extremity, angle, side, direction, &c. *Al ghár* may signify bay-tree, the ancient laurus; but it is very unlikely that an Arab would call a point of land covered with laurels, *taraf al ghár;* and certainly, from well knowing the spot and its elemental visitations, I should have strong doubts of a bay-tree's ever having grown there. That both wind and sea have unceasingly attacked it for many ages, is attested by its aspect and the adjacent shoals.

and its *sleeve;* a *tongue* of land; a *ness*, or *nose* of land; a *vein* of ore; a *head*-land, &c. The old geographers of Italy, before the complete change had taken place which resolved the Latin into an Italian language, were nearer to our mode of writing than they are now. In the fine Portolano of Canachi already mentioned, *Legorno, Florentia,* and *Neapolis*, appear for Livorno, Firenze, and Napoli of the present day; and assuredly the classic enunciation of the two last cities assimilates more with the English names —also our adjective, *Neapolitan*—than with the modern Italian. Leghorn he has even written in Greek characters, Λεγορνο (see page 331); and others of the same epoch (*circa* 1550) term it Legorne, Ligorna, and Ligorno, which last was adopted by Crescentio, in 1607.*

Italian spelling.

Leghorn.

Greek names.

The general rule for spelling Greek names, in the Latin mode, is hardly applicable to modern Greek, wherein accent and emphasis are now one and the same thing, whereas anciently accent was—as its name signifies—intonation: and where the consonants and vowels have very different sounds from those given to them by the ancient Greeks and Romans. Well-educated Hellenians have latterly been anxious, almost to affectation and pedantry, in their attempts to restore or to appropriate the ancient geographical names to places

* I have been the more particular in this statement, because some of our superficial linguists—following each other—have stigmatized these names as being exclusively British violations of lingual purity. A floundering wit asserts that though the English profess to abhor assassination, they have made no scruple of murdering the names of both the Tuscan capital and its seaport: 'And who,' he asks, 'would ever have thought that such a word as Naples could have possibly been Anglicized from the sweet-sounding Napoli?' Purdy (*Mediterranean Directory*, 1826, *page* 91) has it 'LIVORNO, the chief port of Tuscany, commonly, by the French, called *Livourne;* by the English, *Leghorn:* a barbarism sanctioned by custom.' 'We will now transport the reader,' says Conder (*Italy, vol.* iii. *page* 51) 'to the bustling commercial city of Livorno, which John Bull only knows by the uncouth name of Leghorn.' Even the *Penny Cyclopædia*, less courteous than Mr. Purdy to established right, refuses to describe the port under 'Leghorn,' saying, 'see Livorno.' On referring to the *modern Italian* word, the parentage of the former is thus made over to us:—'Livorno, called, by corruption, Leghorn by the English.'

celebrated in the annals of their classical ages. But even the renowned cities of Athens, and Thebes, and Eleusis, with the islands of Delos, and Lemnos, and Cos, and many other time-sanctified places, of which the names have *never* been changed nor altered since the early days of Greece, have assumed such trivial appellations in maps and charts as to leave few traces of their original to the eye, or at least to the ear of the experienced inquirer. These changes are owing to several causes, exclusive of the operation of time and the influence of a general decay in the Greek language; we may enumerate four of the most obvious:

Causes of change.

From religious motives.

I. *By the zeal and piety of the Christian emperors,* who were ever making dedications to the Virgin Mary, or to angels, saints, and martyrs of the Greek Church. An instance of this may be seen at Ephesus. When the worship of Diana was abolished in that renowned city, its temple, and the place itself, were dedicated to and named after St. John the Evangelist, its first bishop. Leucadia has in a similar way become Ἅγια Μαῦρα, or Santa Maura. Many other places have been christened *Panagìa, San Giorgio, San Michele, San Demetrio:* and in like manner the peninsula of Athos assumed the name of Ἅγιον Ὄρος, or Monte Santo. Stavros, the Cross, has become a common and respected designation.

II. *By Latin or Frank Conquerors of the Eastern Empire.*—An example of an ancient Hellenic name changed by the modern Greeks, and abruptly corrupted by the Frank conquerors, occurs with respect to the island of Eubœa. The modern Greeks seem to have discontinued the use of that name, and to have called the place Εὔριπος, *Evripo,* from the celebrated channel which divides it from Bœotia. But when the Franks took possession of it, they seem to have mistaken εἰς τὸν Ἔγριπον for ν Ἔγριπον, calling it first, Egripo, and then Negropon. The bridge which crosses this channel, uniting Eubœa to the main land, may have suggested the addition of a final syllable: and thus

Euripus.

becoming Negroponte, the Black bridge, has usurped the honour of Euripus.

A similar wish to assign a significant name to the principal port or harbour of Athens, induced the Venetians to change the name of Piræus to Porto Dracone, and afterwards to Porto Leone, in allusion to the sculptured lions at the extremities of the piers of the artificial harbour. So, also, they degraded Mount Hymettus into Monte Matto, whence the Trelo vouni, or mad mountain of the Turks. The promontory of Sunium, in Attica, was called Capo Colonne by the Venetians, as Phigalìa is called Styli (*the columns*) by the modern Greeks, on account of the marble colonnades of the temples still remaining there. So Port Prasiæ, in Attica, gained the designation of Porto Raphti, from a statue on an islet there which resembled a raphtis, or sempstress in attitude.

Porto Leone.

Styli.

Raphti.

The Cyclades were called by the modern Greeks Dodeka-nesi (twelve islands); but on the successes of the flag of St. Mark they became Duca-nesi, in honour of the Dukes or Doges of Venice; and as the Isthmus of Corinth is about six miles across, its name was changed to Hexamili, an appellation which has almost become generic for an isthmus, such as that of the Thracian Chersonesus, and others.

Duca-nesi.

Hexamili.

III. *By the domination of the Turks.*—The changes superinduced by the Turkish conquerors on these corruptions by modern Greeks and Franks, have still more disfigured the names of celebrated places. Thus Ephesus, after having been changed into Ἅγιος Θεολόγος by the Greek Christians, has been corrupted by the Turks into Ayasolook, to avoid sounding the γ and the θ. For the latter reason Thessalonica was made into Saloniki. Among other contractions and deprivations, they are accused of having carried the abuse of the preposition εις, and the accusative, in the formation of names, to its present puzzling condition, by which a whole sentence is mistaken for a

Turkish changes.

Ephesus.

Constantinople. proper name. Thus the lengthy word Constantinople was reduced to ἡ πόλις, the city, by way of special eminence over all other cities. Besides which, the 'going to town,' expressed in Greek by 'στηνπόλιν, or εἰς τὴν πόλιν, pronounced 'Stambolin, has given rise, first to the names of Istambol and 'Stambòl, and then to that of Islambol, which in Turkish means the City of Islamism, or of the true Mahometan faith.*

Lepanto. Naupactus. But perhaps the most violent change of this nature was the complete substitution of the Gulf of Lepanto for the Sinus Corinthiacus, which has properly been lately restored to Gulf of Corinth on our charts: Naupactus, says d'Anville, became Lepanto 'by a strange depravation of the name Enebect, formed by the Greeks from that of Naupact'—a place for building ships: so the Genoese called the Palus Mæotis by the name *Mare delle Zabacche.* A friend suggests that Lepanto may only be a vulgarized corruption of Levante, from its being the eastern opening, which would in some degree continue the ancient name if we derive that from ναῦς, a ship, and πακτὰ, doors, a ship channel.

Errors by travellers, and by Greek pilots. IV. *By the more recent corruptions of travellers, &c.*—These consist of the mistakes made by such Frank sojourners in the Levant as are ignorant of the orthography, pronunciation, etymology, or grammar, of the Greek language, and yet are in sufficient numbers and station to secure the assumption of error. Thus also the modern Greek pilots and mariners have adopted the names of places in their own seas, that are most familiar to the various foreigners who frequent the shores of Greece and Asia Minor; though these names are a strange mixture and corruption of Hellenic, Romaïc, Latin, Frank, and Turkish. From such

---

* See the excellent Sir George Wheler's *Journey to Greece* (*fol.*, *Lond.* 1681, *p.* 178); a valuable authority for its date, however faulty in illustrations.

acquiescence many of the misnomers retain their places in our charts and maps.

It is thus, even to the time of Dr. Chandler's visit, that Athens seems to have been called and spelt Setines, or Setenes, by the Franks, whereas it is merely a bad pronunciation of the ancient ἐς 'Αθήνας, or 'ς 'Αθήνας, 'to Athens,' by foreigners who were not aware that it is the accusative case, and who, being unable to pronounce *th*, substituted the *t* for that sound. By a similar process Θῆβαι, or 'ς Θήβας, has become Stevas, or Steves. So Eleusis, or 'ς 'Ελευσίνα, has become Slefsina and Lefsina; Leuce is Lefke; Lemnos, or 'ςτὸν Λῆμνον (sub. νῆσον), is now called Stalimene; Cos, or 'ςτὸνΚῶν, is Stanchio; Delos is Standili and Solili; Ithaca becomes Teaki and even Val di Compare, and its port, Vathì. In this last it should be remembered that B is always pronounced V by the modern Greeks, the sound of B being represented by MΠ in the Romaïc, so that they spell Bonaparte, Μπωναπαρτε. Thus, their employing NT for our D, their Δ as well as Θ being *dh* or *th*, the χ a strongly aspirated H, besides the Latins having substituted the C, which may be either hard or hissing, as *k* or *s*, for their unmistakeable K, and other differences of pronunciation and spelling, show the difficulty of representing the names of one language by the alphabet of another. Now without violent reform, it seems that our charts would be more intelligible to Levantine pilots, were Kephallonía—Vostitsa—Avlona—Tserígo—Kenkhries—and the like, to be thus written, instead of Cefalonia—Vostizza—Valona—Cerigo—and Cenchri.

Athens.

Other changes.

Greek spelling.

Such are some of the causes of confusion in geographical orthography and parlance, and since they followed commerce and intercourse, some of them were inevitable. But the anomaly here complained of—namely, that of writing the proper names of a foreign language which has a different alphabet—was wilful. My own difficulties in that line in-

Remark.

creased greatly along the shores of North Africa; for though I comprehended Greek sufficiently to wade through some of the grosser corruptions, my knowledge of Arabic, and its Moorish dialect, was small indeed. The principle to be followed, therefore, was, as far as sound could guide me, to write the word on the spot, using the vowels as in Italian, because of the simplicity and invariability of their pronunciation and orthography—which usage also prevails throughout the south-east and centre of Europe; and the consonants as in English, each with one unchangeable sound. Having thus written them to the best of my ability, I generally, where a native capable of so doing was to be found, got them also written in Arabic, with their significations, as a means of future correction; and these, fortunately, have been most carefully scrutinized by my friend, the Rev. George Cecil Renouard, whose recognised knowledge of the oriental languages is a guarantee for the system adopted. Hence many *bizarre* words were erased from the charts, and in those substituted, the English reading will sound so as to be intelligible to the native. Indeed, in this respect our enunciation favours the change, as may be instanced in the Gallicized word marabou, for marábut, a saint or devotee, and thence a small chapel built over his grave—*the whited sepulchre of St. Matthew?*—to be seen on most of their points and headlands, and which, instead of the full word it makes in the mouth of a native, became the marraboo of our sailing directions. There are many of these blunders, but one may be cited ere it is quite cleared off the charts. At Smyrna, in a former day, no foreign ship was allowed to anchor before her boat had reported her name and nation to the officer of a post on a projecting headland, where the Turkish sanják, or banner, was hoisted. This cape, therefore called Sanják Búrnú, became of consequence in the charts and directories; and the French *hydrographes*, adapting it to their own euphony, dubbed it Pointe St. Jacques, the which our *savans* duly translated Point St.

Arabic names.

My mode.

Marábut.

Risible blunder.

James, the name under which it appeared, till very lately, in our Admiralty charts.*

Another cause of trouble has been the difficulty, in many instances, of assigning the exact sites of various ancient places of considerable historic importance; a necessity involved by the classical and memorable facts and events of those interesting shores. The delta of the Rhône has altered surprisingly since Strabo wrote, as is readily traceable. Notre Dame des Ports, a harbour in A.D. 898, is now a league from the shore; and Aigues Mortes—as already stated—the sea-port whence St. Louis embarked for Palestine only in 1248, has retired inland to the distance of five miles (*page* 13). Ravenna is at present in the midst of gardens and meadows; and Ostia is surrounded by fields. The isle of Lada, where the Athenian fleet took up a station in the days of Thucydides, is lost in the alluvion formed by the Mæander; and that of Minoa, the out-post of Megara, is not now traceable, through its having become part of the coast of that vicinity,—and the actual spot, as reconcilable to its relations with Nicæa, is involved in the greatest uncertainty. The strong town Œniadæ, stated to have been at the mouth of the Achelous in the days of Thucydides, is nowhere to be found in that vicinity; and there is much confusion in reconciling history and geography in respect to the depositions of that river, as affecting the islets of Oxiæ and Echinades. The identity of Sphacteria and Pylus with the vicinity of Navarino in our charts, would seem to be sufficiently obvious; but scrutiny has shown the existence of various puzzling inconsistencies, for which the reader is referred to the able dissertation of my late friend Dr. Arnold, appended to the second volume of his edition of Thucydides, pp. 399—407,

Changes in places.

Delta of the Rhône.

Ravenna.

Lada.

Nicæa.

Oxiæ.

Pylus.

---

* This will remind the reader how the authorities were puzzled to ascertain the identity of Peter Gower, who was such an authority among Freemasons in Locke's time. The name proved to be a corruption of the French *Pytagore*, or Pythagoras (Πυθαγόρας)!

which was written with my survey, on a very large scale, under his eye.

Remark. Such conditions involved much difficulty; but by local examination I was able to confirm or reject the views of Cluverius, Cellarius, d'Anville, and other geographers, who laboured with fewer advantages than myself.

---

§ 2. Respecting the Tabulated Points.

Table of positions. THESE being the remarks with which it was considered necessary to precurse the Table of Positions, we can now proceed to the principal register of my undertaking; trusting that, although—as in everything human—there will be yet a more delicate precision obtained, my determinations may — combined with those of Captain Gauttier —substantially constitute the Mediterranean landmarks of the nineteenth century. And indeed, it is not a little satisfactory to find that they have now been in constant use among navigators of all nations, for upwards of a quarter of a century, without any material alteration or correction having been made or even suggested. A proof is therefore afforded, that however short they may be of absolute perfection, they are relatively correct; a condition which meets the true demand of navigation.

Arrangement. The Register, or table of these maritime points, has been diligently arranged, and the information which it contains is pretty fully expressed, although in as condensed a form as was deemed desirable. It may be observed in explanation, that where the positions are assumed from numerous observations, those results were rejected whose mean was very discordant with the extreme terms; earnestly hoping, that whatever acme of perfection practical astronomy may yet introduce into geographical details, the points tabulated in the following pages will be found tolerably chained together, within reasonable limits.

But I must here advert to an apparent discrepancy which has occasioned me a little disquietude on this head, since it would involve the exactness of my assumed zero, the very starting-point of the chronometric comparisons: and this arose from a doubt as to the longitude of the Royal Observatory on the palace of Palermo, with which all the positions are mediately or remotely connected. On commencing operations in Sicily, the Abbate Piazzi gave me a note stating the longitude of the pillar on which his great circle stood, as 13° 20′ 15″ east of Greenwich; and I had some conversation with him upon the observations from which that result was deduced. It was mainly founded on the mean given by an occultation of λ Virginis on the 12th of June, 1791, and by the solar eclipse of the 5th of September, 1793; for which he obtained trustworthy corresponding observations from Dr. Maskelyne himself. By these he placed Palermo to the east of Greenwich, as follows:—

On the normal point.

Piazzi's note.

Observations.

Phenomena.	*Time.*		*Arc.*
	H. M. S.		° ′ ″
By the immersion of λ Virginis . .	0 53 23	...	13 20 45
Beginning of the solar eclipse . . .	0 53 17·7	...	13 19 25·5
Termination of ditto	0 53 22	...	13 20 30
Mean . . .	0 53 20·9	...	13 20 13·5

As we had pitched a marquee on the mole-head for erecting a portable transit-instrument during my stay, Piazzi also furnished me with the distance from the meridian and perpendicular, in Sicilian palmi, from the palace to the lighthouse adjoining the spot on which my transit was mounted; and which, when protracted, agreed nearly with an azimuth bearing then taken from the lighthouse to the palace observatory=S. 28° 10′ W. true. So eligible a nautical station was, of course, assumed as a standard; it gave a longitude of 13° 21′ 56″ east for the normal point of my departures, and from it my several arcs were measured with all the precision I could attain.

Palermo light.

Some years afterwards, Baron de Zach was desirous of

Baron de Zach.

inserting a specimen of my method of treating chronometric runs, in his *Correspondance Astronomique;* and as his object was to call the attention of geographers to the subject, I supplied him with a detailed *sample,* showing the manner of connecting some places on the coast of Barbary with the palace at Malta, which he printed in 1822. Among the remarks which were then drawn up in illustration, I mentioned the position of the Palermo observatory which Piazzi had given me; and this so strongly excited the notice of my colleague, General Visconti, that he appealed to me under the idea either that there was a misprint in the Baron's pages, or that there must have been some misunderstanding between Piazzi and myself. He had, he said, toiled through all the Sicilian observations, and was inclined to believe that the observatory was at least 1′ more to the east than I had represented it to be: 'but,' added he, 'in that case what will become of the longitudes of Tripoli, Malta, Alexandria, &c., all of which you have connected with that off Palermo? and how happens it that your longitude o Corfu, which of course is connected with Malta, and therefore with Palermo, should accord so well with that derived from the occultation of Aldebaran, and with that by me adopted at Naples?' To this the only reply was, that I had reason to think, even from his own queries, that the position of the lighthouse—which gave $8^m\ 51^s$ between Palermo and Messina, and $4^m\ 15^s{\cdot}4$ between Messina and Malta—would eventually be found very near the truth. He then begged of me to remeasure the arcs between Malta, Palermo, and Naples—"così i vostri cronometri vi darebbero un' esatta differenza di longitudine tra Napoli e Palermo, e meglio che qualunque osservazione d'eclisse, o d'occultazione. La nostra longitudine è ancora incerta per vergogna nostra, e sarebbe pur bella cosa che questa determinazione importante la dovessimo a voi per il primo." This, for the attainment of precision as nearly absolute as possible with the means at our disposal, assuredly would have been done, but that

General Visconti.

My reply.

Visconti's request did not reach me till after my arrival in England in 1824. I therefore sent him a few documents on the subject in question, begged him to refer the whole to Piazzi for his re-consideration: and further mentioned that I was satisfied that we could not be much in error in the position adopted. But, recollecting the liability of chronometers to sudden and often inexplicable changes of rates, under causes acting so differently that the irregularities are sometimes opposed, my dependance on the results would not have made me confident to a mile. Remark.

Here the matter rested for some years, for neither of my zealous correspondents would have been satisfied without a definitive measurement under full means and practised hands. But it was taken up at Paris by M. Daussy, *Ingénieur Hydrographe*, whose elaborate calculations of no fewer than ten occultations, extracted from Piazzi's observations, are published in the *Connaissance des Tems* for 1835. By this severe labour he has arrived at the result, that the Palermo Observatory is $44^m\ 4^s$, that is, 11° 1′ east of Paris, or 13° 21′ 15″ east of Greenwich. Still although this conclusion may yet influence the position of the observatory, I must, for cogent reasons, retain the situation in which I place the lighthouse, since from thence I settled Malta; a station whence all my runs since the year 1816 were carried and valued, as the chronometers were always rated at the Anchor Wharf, in the dock-yard, and transported from thence by triangulation to the Palace in Valetta. This, therefore, is a matter upon which we must dwell a moment longer; and first, for the remarkable agreement between Gauttier and myself, I am enabled by the kindness of the present Admiralty authorities to cite a letter from the Commander-in-chief to their Lordships. This is dated from Malta, 18th June, 1816:— M. Daussy. Sir C. Penrose.

I found, in conversation with Captain Gauthier of the *Chevrette*, French king's ship, that he had several scientific associates; but that their plan was not to go in search of and ascertain the various dangers in these seas (a point, however, of the utmost importance), but to fix, by means of five excellent watches, 110 different points, and borrow materials for the details.

It is highly to the honour of Captain Smyth that the mean of the observations of those eight French astronomers (*for Valetta*) came within a second, and in other instances a small fraction, of those made by him unassisted.

Remark.

It should be remarked, that however gratifying this incident was, so close an accordance in the result can only be viewed as accidental. But we had another opportunity for finding that our *modus operandi* did not differ materially; since a total eclipse of the moon occurring on the 9th of June, while the *Chevrette* was in the harbour, the following were the comparisons thereby afforded:—

Lunar eclipse.

Phenomena.	*Gauttier.*				*Smyth.*		
	H.	M.	S.				
Disappearance of the white light . .	1	37	57·3	...	(*not noticed.*)		
Commencement of the true shadow, or disappearance of sun's light . . .	1	51	40·5	...	1	51	39·7
First appearance of the penumbra, or partial reflected light	2	35	40·5	...	(*not observed.*)		
End of the penumbra	2	49	42·1	...	2	49	41·5
End of the eclipse	3	58	50·4	...	3	58	48·6

Having reason to be satisfied with the position of Valetta, it became, as I have said, my standard Mediterranean point: but attention had been so strongly called to Piazzi's probable oversight, that it was desirable to remeasure the arc between Palermo and Malta. This was accomplished, in 1846, both by Captain Graves, and the well-known Russian Admiral Lütke, with very powerful means; and an arc nearly 3^s shorter than mine was carried to the *Observatory*. But my published stations were—Palermo *lighthouse*, 13° 21′ 56″, and Valetta palace, 14° 30′ 50″; a matter to which Captain Graves paid great attention, and produced an arc = 4^m 34^s·365 from the light to Spencer's monument in the harbour of Valetta. This spot is 1750 yards distant from the palace flag-staff, on a bearing N. 27° 35′ east, by compass; which, corrected for variation (13° 53′ west, in 1846), places his station 427 yards (in time, 1^s·056) to the west of mine. The comparison, therefore, stands thus:—

Graves and Lütke.

The new arc.

					M.	S.
1816.	Smyth	... ...	3 chronometers	... ...	4	35·6
1846.	Graves	... ...	10 chronometers	... ...	4	35·421

Now the arcs for these important maritime positions being, as it were, identical, the question follows, shall the longitude of the two places be shifted a mile or more to the east, in order to meet a quantity not given to me by the excellent astronomer of Palermo? Besides the remarkable accordance of almost all my standard positions, the longitude of Alexandria, by various persons and methods, may afford a further reason for not disturbing them. Having briefly alluded (*page* 416) to the sample of my operations with which Baron de Zach was furnished, it will be eligible to show the use made of that communication by Mons. M. P. Daussy, at present *chef* of a useful class as yet unknown to our navy—*Ingénieurs Hydrographes.* The extract* which follows is from the additions to the *Connaissance des Tems* for 1832, pages 60-3; the essay being under the title of *Déterminations des positions géographiques du Caire, d'Alexandrie, et de quelques autres points de la Mediterranée.* Remark. M. Daussy's examination.

Les observations du Capitaine Smyth donnent encore un moyen de déterminer la longitude d'Alexandrie. Comme les différences qu'il a obtenues par ses chronomètres sont rapportées dans la *Correspondance Astronomique* de M. de Zach, nous pourrons les comparer à ce qu'a trouvé M. Gauttier. En général, on ne saurait trop donner de détails lorsque on opère au moyen des chronomètres; car quoique ces instrumens précieux donnent entre les mains de personnes habiles et soigneuses des resultats très exacts, comme ils ne donnent que des différences, que les erreurs par conséquent peuvent s'accumuler, et qu'une nouvelle détermination change nécessairement tous les points environnans, il est essentiel de connaître les points que l'on peut regarder comme servant de départ, et la manière dont ils sont rattachés les uns aux autres. Extract from his essay.

Il serait donc important de rapporter toujours dans ces sortes d'opérations, non seulement la longitude moyenne et la marche des montres, mais encore l'état de chacune d'elles sur les tems moyen du lieu, ce qui mettrait à même de vérifier les longitudes calculées et d'adopter le résultat qui paraîtrait le plus probable. C'est ce qu'a fait en partie M. Smyth dans la *Correspondance Astronomique,* et ce qui nous permet de combiner les différences qu'il a obtenues avec notre détermination de la longitude de Malte.

En 1816, M. Smyth détermina la différence de longitude entre Malte

---

* The passage in Baron de Zach's *Correspondance Astronomique,* thus cited by Mons. Daussy, is but an abridgment of the paper which I forwarded; and the Baron says (*vol.* vii., *page* 548), 'Le Capit. Smyth donne encore ici tout le type de calcul.'

M. Daussy's essay.

(à l'obsérvatoire du grand-maître) et Tripoli (maison du consul Anglais); il obtint par une moyenne 5′ 20″·33, ce qui, rapporté au château du Pacha par une petite opération trigonométrique, lui a donné pour ce point 5′ 19″·50.

En Septembre, 1821, il alla encore de Malte à Tripoli. Il observa cette fois sur un rocher de la rade, 45″ de degré à l est du château; ses chronomètres lui donnèrent pour différence de longitude entre Malte et le rocher:—

5′	16″·27	moyenne, 5′ 16″·05.
5	15 ·87	
5	12 ·47	
5	13 ·87	
5	18 ·27	

Enfin, la même année il détermina encore la différence entre Malte (le lazaret) et le même rocher de Tripoli; il obtint par ses chronomètres,

5′	8″·9	moyenne, 5′ 14″·30.
5	13 ·5	
5	19 ·7	
5	17 ·4	
5	12 ·0	

Mais le lazaret de Malte est de 1″·73 de tems à l'ouest de l'observatoire du grand-maître; les observations rapportées à ce point donneraient donc pour différence avec le rocher de Tripoli 5′ 16″·03. Les observations de 1816, rapportées au même point, donneraient pour sa différence avec Malte 5′ 16″·50: la moyenne entre ces trois résultats, qui diffèrent très peu les uns des autres, est 5′ 16″·19.

En 1816, M. Gauttier avait aussi obtenu de son côté la différence entre Malte et Tripoli (maison du consul de France); ses chronomètres lui avaient donnée au bout de 27 jours,

Les n°s 23	...	4′	59″·37	moyenne, 5′ 14″·30.
80	...	5	11 ·24	
94	...	5	20 ·31	
2741	...	5	26 ·28	

ou, en rapportant au même rocher, 5′ 10″·5; ce qui diffère de 5″·69 d'avec ce que M. Smyth a trouvé; mais il suffirait de négliger le résultat donné par le numéro 23, qui était une montre de poche, pour s'en rapprocher beaucoup. En effet, le milieu des trois autres serait 5′ 19″·28, ce qui, rapporté au rocher, donnerait 5′ 15″·48 presque la même chose que ce que trouve M. Smyth. Comme il s'agit seulement ici des déterminations de ce capitaine, nous adopterons sa différence de longitude.

Etant ensuite allé de Tripoli à Bomba, M. Smyth trouva entre ces deux points les différences suivantes:—

39′	52″·9	moyenne, 39′ 49″·32.
	47 ·9	
	42 ·9	
	54 ·2	
	48 ·7	

Le lieu des observations, à Bomba, était au fond du port.

Enfin, en 1822, il trouva entre Bomba et Alexandrie (pointe Eunoste).

26′	42″·27	moyenne, 26′ 42″·50.
	42 ·43	
	42 ·56	
	42 ·76	

Des opérations trigonométriques lui firent connaître que la pointe Eunoste était 1′ 30″·0 de degré ou 6″·00 de tems à l'ouest du phare; la différence entre ce dernier point et Bomba serait donc 26′ 48″·50. M. Daussy's essay.

Si nous réunissons ces différences, nous aurons—

De Malte à Tripoli	— 5′ 16″·19
De Tripoli à Bomba	+ 39 49 ·32
De Bomba à Alexandrie	+ 26 48 ·50
Donc de Malte à Alexandrie	1h·1 21 ·63
Longitude de Malte	48 44 ·40
Donc longitude d'Alexandrie	1·50 6 ·03

En 1822, M. Smyth avait encore déterminé la longitude de la pointe Eunoste de 1h 59′ 27″·84 à l'est de Greenwich, ou 1h 50′ 6″·24 de Paris, ce qui donne pour le phare 1h 50′ 12″·24. Il ne dit pas quel est son point de départ; mais nous croyons que c'était Malte, dont la longitude diffère peu de celle que nous avons adoptée; nous emploierons donc aussi cette détermination.

Réunissant ces différens résultats, nous aurons pour la longitude du phare d'Alexandrie:—

Par l'occultation d'Antarès	1h	50m	16s·4
Par celle de γ ♎ observée à Larnaca et rapportée à Alexandrie	1	50	5 ·2
Par les chronomètres de M. Gauttier, en allant	1	50	22 ·35
Par les mêmes, au retour	1	49	59 ·68
Par ceux de M. Smyth, par Tripoli	1	50	6 ·03
Par les mêmes, en 1822, par Malte	1	50	12 ·24
Moyenne . .	1	50	10 ·33
Ou	27°	32′	35″·0

La *Connaissance des Tems* donne 27° 35′ 0″, et Nouet dans son Mémoire, 27° 35′ 30″; c'est donc à peu près 3′ que nous trouvons à retrancher: nous avons eu 4′ 31″ à retrancher de celle du Caire; on voit que la position rélative de ces deux points éprouve par là peu de changement; en effet, la différence que nous trouvons entre eux est de 5′ 25″·6. Nouet avait eu par ses chronomètres 5′ 31″·7; mais on peut douter si ses observations n'étaient point susceptibles d'une erreur de six secondes.

It is now time to turn to the register of geographical points, as contained in the following table; which has been drawn up with careful attention, in order at once to economize space, and secure perspicuity. They are, of course, classed according to their general boundaries; but the headings of each series are placed rather with a view to broad geographical feature, than political distinction—a method adopted as offering a readier reference, than defining the smaller states would have allowed. For example, all the divisions and subdivisions between Nice and Spezzia (*olim* On the table of points. Divisions.

Liguria), are quartered under the authorized designation PIEDMONT—the domains of the sovranacci of Massa, Lucca

Divisions. the recent Etruria, and Piombino, are embodied under the name TUSCANY—the Papal States under that of ROME—and an abundance of principalities are merged together under ALBANIA, GREECE, TRIPOLI, and ALGERIA. The places thus ranged follow each other in chorographical order along the coasts, passing where convenient or necessary to islands and rocks, after which the coast is again continued.

Order. In the course round the Mediterranean, the same order is observed as that of the preceding chapters of this work—entering at Gibraltar, proceeding by the shores of Spain, France, and Italy to Greece; returning westward from Alexandria along the coast of North Africa, to the Strait of Gibraltar. But the Archipelago, Black Sea, and Levant, being Captain Gauttier's contribution, are appended as a

Latitudes. separate series. All the latitudes, it will be observed, are

Longitudes. north; and the longitudes, except where marked with a W on a portion of Spain, and the opposite coast of Morocco,

Heights. are always east of Greenwich. The heights of mountains and buildings are in English feet, and from the sea-level

Variation and dip. (*page* 391). The magnetic variation is constantly west, and the dip of the needle to south, or nadir. In the first column will be found the names of the places—and the particular spots of those places—where the observations were taken, or carried to by means of angles or bearings; and these are

Type of the names. printed in small capitals, Roman type, or italics. The first show the stations where our opportunities were of the best description, and the results entitled to the highest value on the list: the second indicate that though mostly taken on shore, or connected geodetically, the operations were more hasty or less advantageous than the first; while the third resulted from bearings, secondary angles, patent log-runs, and intersections, of a more hurried and less exact tenour. In other words, the three styles of printing the names, exhibit the classification according to my conviction of the relative degree of weight to be assigned to the points. With a labour more onerous than pleasing, these positions have been entirely re-examined since the publication of the charts, some of which—to meet the demand—were hurried out, during which process various small corrections occurred, and several local errors were detected—principally as to identity from sea-ward, where landing was then impracticable. On the whole, though we would advise chronometer-ratings to be made only at the first class of these points, as standards, it is hoped they will all be found in such accordance, relatively, as to enable ships to shape a proper course,—thereby meeting all the ends and purposes of practical navigation. Remark.

As the magnetic variation has been already spoken of (*pages* 384 & 5), it only remains to add, that these observations are rather of relative than absolute value; and though they cannot be forced into any very strict accordance with a set of isogonic curves which I drew up, they are the results of a long series of experiments with the means at my command, to show the needle deviations at given epochs. Whenever, therefore, we shall be able to ascertain, by direct measurement, the total magnetical force of the earth, then even these unpretending operations—however influenced by instrumental error and local affections—may become a chronological reference for its effects in the Mediterranean Sea.* I have also alluded (*page* 391) to the heights of mountains which we took as being merely approximative, although it is trusted that the majority of instances will be found very near the mark: but I had great distrust of those taken by angle of depression to the horizon, although perhaps they were near enough for the purpose of my register. Still, as the horizon is a very difficult line in other cases where ultimate accuracy is demanded (*see page* 382), Magnetic variation Heights. The sea-horizon.

* It must not be forgotten, that terrestrial magnetism is subject to secular variations.

I got Captain Graves to call at the island Al-Boran, which for such a purpose is an oceanic pivot, and there successively take the dip at morning, noon, and evening, on the cardinal points of the compass. This was accordingly done from its highest point, an elevation of about 68 feet, with a carefully adjusted 5-inch theodolite; and the experiment being new, the results are here given at full:—

Dip at Al-Boran.

Experiment.

Bearing.	Telescope. Direct.	Telescope. Reversed.	Mean.	
N.	6′ 00″	11′ 00″	8′ 30″	Observations taken at 7 A.M. *Thermometer*, 78° 0. *Barometer*, 30·12 in.
N.E.	5 30	11 00	8 15	
E.	4 30	11 00	7 45	
S.E.	4 00	11 00	7 30	
S.	4 30	11 00	7 45	
S.W.	5 00	11 30	8 15	
W.	5 30	11 30	8 30	
N.W.	6 00	11 30	8 45	
N.	4 30	10 30	7 30	NOON. *Thermometer*, 82°. *Barometer*, 30·13 in.
N.E.	5 00	11 00	8 00	
E.	4 30	11 30	8 00	
S.E.	4 00	12 00	8 00	
S.	4 00	12 30	8 45	
S.W.	5 00	12 45	8 52·5	
W.	5 00	13 00	9 00	
N.W.	4 30	12 30	8 30	
N.	5 00	13 30	9 15	SUNSET. *Thermometer*, 78°. *Barometer*, 30·13 in.
N.E.	5 00	14 00	9 30	
E.	5 30	12 30	9 00	
S.E.	6 00	13 00	9 30	
S.	*5 00	*10 00	7 30	
S.W.	*3 00	*10 00	6 30	
W.	4 00	17 00	10 30	
N.W.	4 00	13 00	8 30	

* Hazy.

On the Notanda.

Besides these geographical and physical conditions, hydrography requires information as to the nature of the various coasts, while navigation also demands a notice of the approach to those coasts, and the degree of capacity of the several ports and anchorages. Having partially attempted to supply these desiderata for general purposes, in the following register, without impinging on the advice and duties of a detailed directory, a word is necessary as to the form I have adopted. This cannot be better expressed, than by giving a passage from my address to the Royal

Geographical Society (*Journal*, Vol. xx), on the 27th of May, 1850:—

But among the many publications of the year I must select one which, though only a new edition, is entitled to a high place in your regard, because, on its being first launched, you discerned its merit, and awarded the Gold Medal as a mark of your approbation. I allude, Gentlemen, to the third edition of that truly useful work, Lieut. Raper's *Practice of Navigation and Nautical Astronomy;* a work in which the capacity, systematic method, and intelligence of the author are so strikingly evident. The book is greatly augmented in matter since its original appearance, but, from the excellence of its printing, it has not grown much in bulk; and the additions are such as to increase its utility. The most operose and remarkable feature of this edition, however, is the 'Table of Geographical Positions,' discussed and methodized upon a chronometric system, now consisting of no fewer than 8800 points, instead of the 2300 it first placed before us. From its bearing not only, as usual, the latitudes and longitudes of places, but also the dimensions of islands, state of anchorages, peculiarities of lights and lighthouses, depths of shoals, and other necessary details, I may fearlessly pronounce it to be the most accurate and comprehensive representation of the present state of maritime geography extant. To accomplish this, the author has devised a series of very significant symbols, and applied them to the expression of many important matters; indicating by their means watering-places, dangers, the character of the natives as friendly or hostile; the presence or absence of trees or bushes—whether as a means of identification, or as marking places where firewood is to be found—and distinguishing more especially the *cocos nucifera*, which, on account of its conspicuous form, and its affording both food and beverage, is an object of peculiar interest to the tropical navigator. By such symbols this table is made to contain, with scarcely any increase of size, a vast quantity of varied information: while the signs themselves, being founded on obvious or natural considerations, are easily acquired and retained. The author, in justifying the introduction of a scheme which a few years ago might have been considered a rash, if not a dangerous innovation, concludes his remarks by saying:—'The employment of symbols, therefore, on a more extensive scale than we have yet been used to, and that at no distant period, may be considered inevitable; and the present system, which has occupied my attention for several years, is proposed as so far deserving consideration, that it is constructed with rigid adherence to principles.'

Use of symbols.

Raper's Navigation.

This is important to the ends of tangible geography, as well in the construction and arrangement of tables, as in every description of cartographic composition. In a work of my own, which may one day be brought to light, I shall assuredly adopt Lieut. Raper's symbols in tabulating the results of observations; and I notice that Lieut. Maury, of the United States Navy, has greatly extended the use of such signs in his important Wind-and-Current Chart of the Atlantic Ocean. The imperative task in the question is, so to conventionalize the matter, that, as with music, the forms may be read and understood by people of all nations.

The presei book.

In the present instance symbols only are spoken of, for I am not yet prepared to advocate the abbreviations of

On abbrei ations.

words beyond the usual practice,* as each language will necessarily use its own method; to the injury of the general application so greatly desired. Herein, perhaps, antiquarian tendencies may influence me, having had occasion to recollect how the army of sufferers in the cause of truth was recruited by the uncial BM. of ancient tombs being rendered *Beatus Martyr*, instead of *Bonæ Memoriæ.* Hadrian struck a large brass medal on the 874th *birthday* (natali urbis) of Rome, in the legend of which a P has been disputed as meaning *populus, plebii, publici, primus,* or *parilia;* and even the well-known counter-mark on the early Emperors' medals—N.C.A.P.R., has been rendered by

On abbreviations.

Æneas Vico	Nobis concessa à Populo Romano.
Jobert	Nota cusa à Populo Romano.
Others	Nummus concessus à Populo Romano.

While a satirist insists that it should be read—*Non concessa à Populo Romano.* In the same spirit Coke, the noted manager of the Raleigh tragedy, in contempt of the continental travellers of his day, said "S.P.Q.R. was sometimes taken for these words, *Senatus Populus Que Romanus*—the senate and people of Rome; but now they may be truly expressed thus, *Stultus Populus Quærit Romam,* a foolish people that runneth to Rome."

Sir Edward Coke.

To some readers it may seem travelling out of the record to cite numismatics in a work like the present; but they may be assured that on the shores we have been treating of, an acquaintance with coins, medals, and marbles, is very important. By the unequivocal aid of these handy monitors, often more trustworthy than written records, I have obtained a satisfactory clue both to dates and places; insomuch that I even proved, respecting the age of Rome (*Descriptive Catalogue of Roman Large-brass Medals,* page 267), the preference of the vulgar computation over

Medals and marbles.

---

* Thus, in the confined space allowed in tabulating matter, I place *Var.* the usual abridgement of *magnetic variation*, though I would rather use a symbol: so, also, *Int.* is placed for magnetic intensity.

the chronology of Sir Isaac Newton. The ancients were well aware of the advantage of a systematic terminology in conveying accuracy of conception, and influencing the formation of ideas; although their knowledge was not sufficiently advanced, to particularize minutely and distinctly between positive and relative positions. Their adoption of symbols was on a far more extensive scale of operation than that which is here advocated; for they not only used them to cover moral mysteries, but also as types, emblems, enigmas, and hieroglyphics. These, it must be confessed, are sometimes not a little paradoxical and perplexing, since those elders not only represented moral things by natural things, but even natural by natural. Winged horses, sphynxes, human-headed oxen, and 'chimeras dire' were certainly fanciful enough; but the thunder-bolt for power—eagle and globe for sovereignty—laurel for victory—palm-tree for Judæa—wheat-ear for Metapontum—crab for Agrigentum—hare for Messana—cray-fish for Catana—bull for Tauromenium—horse for Carthage—goat for Thrace—silphium for Cyrene—labyrinth for Cnossus—tortoise for Egina, &c. &c., are sufficiently simple and obvious.* In treating upon this subject elsewhere, I stated that the epithets of symbol and device are often used indifferently, although the former strictly signifies a practical or figured metaphor, and the latter an allegory: the one simple, the other complex. Hence it is obvious, that figures on coins and medals are arbitrary devices, and hieroglyphics are absolute symbols, or significative.

Ancient use of symbols.

Symbols, however, which are unsusceptible of equivocation, and capable of being universally understood, should be encouraged by those who desire accurate perception with

Simple symbols.

* The reader may here be reminded, that *Admiral* Neptune, as Newton styled him, presided over the Mediterranean, as Oceanus did over the circumambient sea, or waters supposed to surround the whole earth: he was symbolized with the trident as a sceptre, a dolphin or aplustre on his hand, and his foot on the prow of a ship. (*See the vignette on the title-page.*)

and but one key. economy of time: and in hydrography it were a truly desirable consummation, that they should require but one key for the common use of seamen of all nations. True to this spirit in arranging the following table, though wishing also to have introduced a mark for the compass-variation, another for the dip of the needle, and others to denote places which fish frequent, and where fresh water, provisions, and refreshments are obtainable, I would not introduce one except through the hands of Lieutenant Raper; so that this department must await some future edition of his standard work. The following are those selected for my register, from the *Practice of Navigation*, and it is hoped they will be found plain, easy, and efficient:—

Lieutenant Raper.

Symbols adopted.

⚓ (*Anchor*) Anchorage for large vessels; ⚓′ good ditto; ⚓, bad ditto; $⚓_{\circ}$ no ditto.

⚓ Anchorage for smaller vessels; ⚓′ good ditto; ⚓, bad ditto.

⊞ Harbour for large vessels, or having always three fathoms water.

⊡ Harbour for smaller vessels, or having at times less than three fathoms.

⌒ (*Birds*) As birds frequent some places in preference to others, they may afford a means of indication.

[(*Boat-hook*) Landing; $[_{\circ}$ no landing.

β Break or breakers; $\beta\beta_{\circ}$ breakers at times.

! (*Note of admiration, surprise*) Denotes caution, or calls attention.

|| Channel, or passage. Mouth of a river.

δ Danger, dangerous; $\delta_{\circ}$ no danger, safe.

† (*Palm-tree*) Here the date-palm.

⟋ Rising gradually.

⌂ Rising in the middle.

⌣ Saddle-shaped. A valley.

⟍ Sloping downwards.

⌐ Sloping bottom, or change of soundings gradual, may be approached with safety by attention to the lead.

⊥ Steep or precipitous. *Note.*—This is quite independent of high. A headland may be low, yet precipitous.

⊤ Steep to, or bold to.

Ψ (*Tree*) Trees; Ψ well wooded.

✲ (*A tree without a trunk*) Brushwood.

Place.	Latitude.	Longitude.	Notanda.
SPAIN.	° ′ ″	° ′ ″	
San Lucar, tower...........	36 43 10	6 24 2w	‖ River Guadalquiver. ⊡. ⊤.
Rota, mole-head	36 36 40	6 20 10w	Var. 22° 30′ in 1810. Approach δ. $\beta\beta_o$.
CADIZ, S. Sebastian's light	36 31 51	6 18 10w	Height 151. Rise of tide 9½ feet. ⊞.
Chiclana, S. Aña church...	36 25 10	6 9 50w	510 feet. Hill ⌓. ‡.
Sancti Petri castle	36 22 45	6 13 0w	‖. High water 1^h 35^m. δ. β. ⌒.
Cape Trafalgar, turret ...	36 10 10	6 0 56w	δ at approach, but the cape ⊥.
Tarifa, lighthouse............	36 0 15	5 36 39w	⊤. South point of Europe.
Palomas rock, centre	36 3 35	5 25 24w	! for Pearl reef. δ. ⌊.
Algeciras, the pier	36 8 5	5 26 16w	‖. ⊡. ‡ in road. Mostly δ. !.
Gibraltar, signal-station ...	36 7 46	5 20 19w	Height 1255. O'Hara's tower, 1408.
GIBRALTAR, arsenal mole...	36 7 17	5 20 49w	High water 0^h 40^m. Rise & fall 4½ ft. ⊞.
Gibraltar, Europa point...	36 6 16	5 20 9w	Var. 21° 37′; Dip 61° 8′; Int. 232 (1824).
Al Koreïn rock..............	36 19 12	5 13 16w	Close to the shore, and small. ⌒.
Estepona, Marmoles point	36 25 17	5 7 25w	⦣ to Sierra de Bermeja. ⊥′. ‡.
Frangerola, castle	36 32 51	4 37 1w	On a small hill. ⊡. ‡. ✵.
Malaga, lighthouse	36 42 48	4 26 12w	Height 125. Var. 21° 5′ (1811). ⊡. ‡.
Velez Malaga, Torre del Mar	36 46 44	4 7 25w	‖. ‡, except in sea-winds. ✵.
Castel de Ferro, sanidad...	36 42 19	3 21 31w	‡, but !. ⦣ to Ugija range, 2700.
Torre Belerma	36 42 40	2 54 0w	‡, but ⊤, and !.
Almeria, torre del Tiro ...	36 50 50	2 31 7w	In the cove ⊥; in the bay ‡.
Capo de Gata, fort	36 42 59	2 11 56w	⊥. Var. 20° 42′ in 1813. ⌒.
Port Genovés, fort	36 44 15	2 7 18w	‡, but ! currents and flaws.
Cresta de Gallo..............	36 47 0	5 3 0w	S.E. of Ronda, 5950. Summit ⦣.
San Pedro, castle.	36 53 21	1 59 58w	⊡, the best of these coves.
Sierra de Gador	36 56 0	2 56 0w	7100; S.E. end of Apuljarras range.
Carbonera islet..............	36 58 22	1 55 24w	‡. Var. 20° 43′ in 1813. ✵. ⌒.
Cerro de Mula-hacen	37 8 0	3 28 0w	11,370. Peak of Sierra Nevada.
Mount Filabres..............	37 12 0	2 23 0w	1800 feet. A mass of white marble.
Aguilas, Fort S. Juan......	37 23 33	1 36 49w	⊥′ and ‡. Var. 20° 43′ in 1813.
Cartagena, fort Gateras ...	37 35 28	0 58 17w	Height 655. ⊤. δ_o. ✵.
CARTAGENA, mole-light ...	37 35 58	0 57 42w	⊞. Mag. var. 20° 44′ in 1812.
Mount Roldan	37 6 0	1 2 0w	1851 feet. This is of Murcia.
Cape Palos, turret	37 37 18	0 37 40w	$\beta\beta_o$ off Hormigas. Var. 20° 34′ (1813).
Cape Cervera, turret	38 0 1	0 37 58w	‡, open to sea-winds.
Lugar nueva, fort	38 12 7	0 31 54w	‖ Elche, ‡. ‡ in Tamarit road.
Plana I. Tabarca bastion...	38 10 15	0 27 31w	Low and flat, but ⊥. ⌒.
Alicant, mole-light	38 20 30	0 27 18w	95 ft. ‡′. M. Tosal 230. Castle-hill 550.
Benidorme isle	38 30 5	0 5 11w	Var. 20° 30′ (1813). ‡′, but exposed. ⌒.
M. Roldan, the gap	38 36 0	0 11 0w	The 'cuchillada,' or gash.
Point Ifac, or Calpe	38 37 18	0 4 10	‡′ over Altea bay, but !. ⦣.
Cape S. Antonio, hermitage	38 48 31	0 10 58	High, level, and ⊥. Off Xavia, ‡.
Denia, castle................	38 50 50	0 7 34	⊡ between the banks. Outside ‡.
Cullera, sanidad	39 10 48	0 14 38w	‖ Iucar. Admits barks over the bar.

Place.	Latitude.	Longitude.	Notanda.
SPAIN.	° ′ ″	° ′ ″	
Cape Cullera, tower	39 12 5	0 13 0w	174 feet. ⸦. ⊻. ⤳.
Valencia, cathedral	39 28 35	0 22 10w	Var. 20° 35′; Dip 63° 36′; Int. 236 (1813).
Valencia, Grao light	39 28 47	0 18 58w	45 feet. ‖ Turia. ⸦. Outside ⚓, but !.
Mount Espadan	39 54 0	0 24 0w	Marli for the coast.
Cape Oropesa, outer tower	40 6 12	0 9 55	⸦. ⊻, shoal under it.
Peñiscola, fort	40 22 53	0 25 0	⊥. ⸦. except with sea-winds, ⚓′.
Mount Peña de Bel	40 36 0	0 2 0	4000. Mark for approach from S.E.
Vinaroz steeple	40 29 10	0 27 57	⚓ in off-shore winds.
Monsia, east summit	40 37 0	0 31 0	880, ⸦ inland to 2500.
Alfaques, San Carlos mole	40 37 46	0 35 30	⊞. Var. 21° 0′ in 1813. ✻. ⤳.
Buda I, or Cape Tortosa...	40 43 10	0 54 7	‖ Ebro. !. Station, Gola del N. ⌐.
Port Fangal, point Fango	40 47 40	0 47 47	⊡. In the bay ⚓. ✻. ⤳.
Tortosa, castle	40 49 0	0 32 50	Ebro ⊡ for vessels of 50 tons.
Salou, mole-head	41 5 28	1 7 12	⚓′. Var. 20° 37′ in 1813.
Tarragona, cathedral	41 6 57	1 16 10	410 feet. ⊡. In the roads, ⚓′.
Montazut summit	41 24 0	1 25 0	3200. Mark for the coast.
Castel Fells, sanidad.........	41 16 30	1 57 58	⸦ 230. ⚓′ with !.
Torre del Rio	41 19 12	2 9 45	107 feet. ‖ river Llobregat. !. ⤳.
Barcelona, Monjui fortress	41 21 35	2 10 13	680. Var. 20° 45′ in 1813.
BARCELONA, mole-light ...	41 22 30	2 11 20	80 feet. ⊞. Outside ⚓′.
Mon-serrat, centre	41 36 0	1 48 0	4200. Excellent sea-mark.
Mataro, eastern fort	41 32 47	2 27 33	210 feet. Inside reef ⸦′, outside ⚓.
Point Toldera, or Tordara	41 37 45	2 48 0	‖ the Toldera. Var. 20° 40′ (1812).
Blanes, fort Santa Aña ...	41 40 0	2 49 47	⚓ in land winds, here and Lloret. ⊻.
Tosa, church at cove.........	41 42 36	2 57 35	⚓′, with ! δ_o. ⊻. Inland ‹.
S. Feliu de Guixols	41 46 13	3 1 0	⚓′, except in S.E. winds. ✻.
Pálamos, mole-head	41 50 57	3 6 25	⊡. ⚓, with !. Var. 20° 37′ (1813).
Cape S. Sebastian............	41 53 10	3 12 14	⊥, but ! the Hormigas. ⌐.
Medas isles, fort	42 3 40	3 13 15	‖ river Ter. Var. 20° 40′ (1813). δ_o.
Ampurias (Emporiæ)	42 9 0	3 3 20	Old castle of the Ampurdan.
Rosas, fort Trinidad	42 16 12	3 10 25	315 feet. ⊤ & ⊥. ⚓′. δ_o with !.
Cadaqués, church............	42 17 10	3 16 48	⊞. ⸦ to M. (*Fingers*) of Cadaqués.
Cape Creux, Masa de Oro I.	42 19 12	3 20 45	⊥. δ_o with !. ⌐. ⤳.
Santa Cruz della Selva......	42 18 0	3 12 25	⊤ & ⊥. ⊡. ⚓. Var. 20° 35′ (1813).
Cape Cervera, Carox tower	42 26 16	3 10 45	⊤ & ⊥. Boundary of Spain & France.
SPANISH ISLANDS.			
Columbretes I. M. Colibre	39 53 58	0 44 27	⊡. ✻. Var. 17° 41′ in 1823. ✻. ⤳.
Columbretes I. Ship rock...	39 51 20	0 43 32	⊤ & ⊥. δ_o with !. ⤳.
Formentera, I. P. Codolar	38 38 30	1 36 10	⊥. δ_o but ⌐. ‹. ⤳.
Formentera, I. P. Aguila	38 38 15	1 23 0	The pitch. ⊤ & ⊥. ‹.
Espardel isle, cove	38 48 5	1 29 25	⚓. With !, δ_o.
Iviza, Cape Falcone.........	38 50 20	1 24 0	⊤. ⊻. ⸦. ✻. ⌐.

Place.	Latitude.	Longitude.	Notanda.
Spanish Islands.	° ′ ″	° ′ ″	
Iviza, the citadel	38 54 0	1 27 8	⊞. Var. 20° 15′ in 1813.
Iviza, S. Eulalia rock	38 59 10	1 36 0	⊥, but ! the reef. ⌊.
Iviza, Togomago I.	39 2 56	1 38 40	⊤ & ⊥. δ_o with !. ⌊.
Iviza, Point Denserra	39 8 5	1 31 55	⊥, moderate height. ⚓$_o$. ⚲.
Iviza, Port S. Antonio	38 59 0	1 20 10	Watering place. ⊞. ⚲.
Iviza, Bedra islet	38 52 24	1 12 18	100 feet. Var. 20° 30′ in 1813.
Cabrera, P. Anciola	39 5 0	2 53 15	Pitch. ⊤ & ⊥. ⌊$_o$. ✻. ⌒.
Cabrera, castle	39 6 57	2 54 36	⊞. ⚲$_o$. Var. 19° 58′ in 1813.
Foradada islet	39 10 0	2 57 28	The pierced rock. δ_o with !. ⌊.
Majorca, Cape Salinas	39 13 30	3 3 25	The pitch. ∠ to Torre Gorta. ⚲.
Majorca, Port Colon	39 22 0	3 15 58	⚓′. ∠ to M. Salvador.
Majorca, Cape Pera	39 40 36	3 30 0	The tower. ⊤ & ⊥. δ_o.
Majorca, Alcudia church	39 49 40	3 8 57	In the bay ⚓′. ⊤.
Majorca, Cape Pinar	39 51 30	3 15 0	The pitch. ⊥. ⚲. ⌊$_o$.
Majorca, Pollenza castle	39 53 10	3 8 28	⊞ but ! winds. Var. 19° 50′ (1813).
Majorca, C. Formenton	39 56 48	3 15 10	Long hummocky tongue. ⊥. δ_o.
Majorca, M. Torellas	39 48 0	2 48 50	The Silla or Saddle ⌣∠. 5200 feet.
Majorca, Port Soller	39 47 58	2 43 0	The landing place. ⚓,.
Majorca, Dragonera isle	39 36 10	2 17 35	Upper tower light, 1180. ⊤ & ⊥.
Majorca, cape Llamp	39 32 0	2 23 20	⊤ & ⊥. ∠ to M. Galatzo. ⚲. ⌒.
Majorca, Palma mole	39 34 5	2 38 45	Light 37 ft. Var. 19° 54′ (1812). ⊡. ⚓′.
Majorca, cape Blanco	39 20 32	2 47 0	⊤, ⊥, and ∠ inland. ⚓$_o$.
Minorca, cape Dartuch	39 54 50	3 51 22	Low and flat, but ⊥. ⌊.
Minorca, Ciudadela	39 59 15	3 52 30	Octangular spire. ⊡ and ⚓.
Minorca, cape Bajoli	40 0 25	3 48 12	⊤ & ⊥. ∠ to Torre del Raam.
Minorca, Cala Caldera	40 3 5	4 2 0	⚓,. ∠ to S. Agata. Var. 19° 38′ (1811).
Minorca, C. Cabaleria	40 5 0	4 6 58	⊥. ∠ inland. ⚓$_o$.
Minorca, Port Fornelles	40 3 25	4 8 51	The fort. ⊞. δ_o with !. ⚲$_o$.
Minorca, Mount Toro	39 58 36	4 8 30	(El Tor) convent 1220. ⌒. Sea-mark.
Minorca, Colon isle	39 58 7	4 18 20	⊥ from seaward. Var. 19° 35′ (1812). ⌒.
Minorca, cape Mola	39 52 45	4 21 36	Atalaia. ⊤, ⊥. δ_o. ⌊$_o$. ⚲$_o$.
Minorca, Port Mahon	39 52 57	4 19 53	Quarantine isle. Var. 19° 30′ (1811). ⊞.
Minorca, Port Mahon	39 53 30	4 18 32	Arsenal sheers. Var. 19° 36′ (1813). ⊞.
Minorca, Ayre isle	39 48 30	4 18 25	⊥ outside. δ_o with ! in ‖. ⌒.
Minorca, Alaior tower	39 52 10	4 9 42	⊤ & ⊥. ⚓′ in north winds. ⚲.
France.			
Cape Béarn, light	42 30 48	3 8 48	752 feet. ∠ to the Pyrenees. ⊥.
Port Vendres, light	42 31 2	3 7 32	98 ft. ⊡. ∠ to fort S. Elmo, on ⌒.
Callioure, islet St. Vincent	42 31 20	3 5 40	⚓ but ! east winds. ∠ to fort S. Elmo.
Mount Canigou	42 30 0	2 30 0	9300 feet. ⌒. A sea-mark.
Canet, S. Marie tower	42 42 4	3 0 50	‖ of Tet. ⚓ in land winds. ⊤ with !. $\beta\beta_o$.
Perpignan, steeple	42 41 45	2 54 0	Rise to town 67 ft., to M. Forceral 1650.

Place.	Latitude.	Longitude.	Notanda.
FRANCE.	° ′ ″	° ′ ″	
Leucate, fort les Mattes ...	42 53 58	3 2 46	⊥. ⊤. In Franqui ⚓′. ⦣ 125 feet.
Grau de Sigean, light	43 0 55	3 3 55	33 feet. ‖ port Nouvelle of Narbonne.
Sérignan, douane...........	43 15 30	3 17 18	‖ of Orbe, ⊡. ⦣ to Beziers, 380 ft. ⌒.
Fort Brescou, light	43 15 44	3 29 0	30ft. ⊤ with !. $\beta\beta_o$. Var. 20°10′ (1813).
Mount Agde, light	43 17 56	3 28 10	415 feet. Useful sea-mark.
Cette, Mount S. Clair	43 24 0	3 40 0	620 ft. Mark for the Etang de Thau.
Cette, mole-light	43 23 55	3 41 35	88 feet. ‖ of the Etang. ⊡. ⚓ by ⊤.
Frontignan, steeple	43 27 0	3 44 30	Seen over etangs and drowned lands.
Montpellier, steeple	43 37 10	3 52 0	120 ft. Mark for Grau de Maguelonne.
Aigues-mortes, light	43 32 27	4 8 10	70 feet. ‖ Grau du Roi. ⊡. ⚓ ⊤.
Grau d' Organ, fort	43 26 30	4 24 22	‖ Petit Rhone. ⊤. ⚓.
Rhone Delta, la Camargue	43 20 35	4 41 0	130 f. ‖ Vieux Rhone, in drowned lds. ⌒.
Rhone Delta, S. Louis tower	43 22 55	4 47 48	‖ Rhone. Var. 19° 51′ in 1812. ⌒.
Rhone Delta, Tanpan douane	43 21 37	4 50 27	⚓′ in Gulf of Foz. M. Opica 1630, seen to N.
Port de Bouc, light	43 24 0	4 58 56	98 feet. ‖ Etang de Berre. ⊡. ⚓′.
Cape Couronne, pitch	43 19 20	5 3 30	⊤ & ⊥, but ! for Regas and Muet. ⚓.
Carré, l'Estes rock	43 19 30	5 9 33	δ with ! ⚓. ⚓. ⦣ to M. Tabouret, 490.
P. Mourrepiane, fort	43 21 10	5 19 54	⦣ Moulin du Diable 680, Pilon du Roi 2360.
MARSEILLES, fanal S. Jean	43 17 42	5 21 40	32f. ⊞. Road ⚓. Var. 19°30′, Dip 63°10′ (1820).
Marseilles, upper castle ...	43 16 58	5 22 7	Notre Dame de la Garde, 526 feet.
Daume I., fort Tourville ...	43 16 10	5 20 45	! shoals in d'If ‖. ⚓ Daume road.
Chateau d'If, tower	43 16 45	5 19 50	150 feet. ⊤ & ⊥. ! the above shoals.
Ratoneau I., castle	43 17 0	5 18 57	293 feet. ⊥. δ_o with !.
Pomegue I., tour S. Jean...	43 16 25	5 18 40	282 feet. ⊥. δ_o.
Planier isle, light............	43 11 55	5 14 17	131 ft. ! $\beta\beta_o$. Var. 19° 46′ (1820). ⌒.
Tiboulen isle, centre........	43 12 53	5 19 50	⊤ & ⊥. δ_o in Cape Croisette ‖.
M. S. Michel, semaphore...	43 13 28	5 22 25	(Collet du Rose) 1340f. Gardalaban 2270.
Riou isle, tower...............	43 10 34	5 22 55	⊤ & ⊥ to seaward. 510 feet.
La Cassidaigne rock.........	43 8 37	5 32 26	A wash at times, but ⊥ around.
Cassis, lighthouse............	43 12 49	5 31 55	⊡. Var. 19° 20′ in 1815.
Bec de l'Aigle, cape.........	43 9 55	5 36 36	! for δ in isle Verte ‖. ⦣ 410. ⚓$_o$.
Ciotat, Bureau de Pratique	43 10 20	5 36 40	⊡. ⚓. Mole light 40 ft. The new 394.
P. Grenier, or Carboniere	43 9 40	5 41 10	⊥, but ! the tunny nets.
M. Pilon de Beaume	43 20 0	5 46 30	3200 feet. A sea-mark.
Bandol isle, centre	43 7 36	5 45 10	⊥ outside, in ‖ δ. Inside ⚓.
I. des Embiez, fort	43 4 34	5 46 45	In the ‖ s δ. ⚓$_o$. ⌒.
Cape Sicie, semaphore......	43 3 12	5 50 55	1200 feet. ⊤ & ⊥, with δ. ⌊.
Cape Sepet, pyramid	43 4 32	5 56 57	⊤ & ⊥, ! for Rascas rock. ⌒.
TOULON, la Grosse-tour ...	43 6 6	5 56 8	⊞. Var. 20° (1815). M. Faron 1700. Coudon 2150.
Chateau de Giens...........	43 2 18	6 7 30	‖ of La Petite Passe ⊥, δ_o.
Hyères bay, fort Gapeau...	43 6 28	6 11 39	‖ the Gapeau. ⚓′ by ⊤.
Hyères bay, fort Bregançon	43 5 25	6 19 13	⊤ & ⊥. ‖ for boats. Var. 19° 45′ (1812).
Porquerolles I., light	42 59 15	6 11 55	262 feet. ⊤ & ⊥. δ_o. ⌊$_o$.

Place.	Latitude.	Longitude.	Notanda.
FRANCE.	° ′ ″	° ′ ″	
Port-Cros I., fort Man ...	43 0 32	6 24 58	⊡. ⊤ & ⊥, but ! in Titan ‖. ∠ 650.
Pt.-Cros I., La Gabinière rk.	42 59 0	6 23 42	⊥. ‖ for boats. ⌒.
Levant I., Phare du Titan	43 2 46	6 30 35	246 feet. ⊥, δ_o with !. ⚘.
Esquillade rock, turret......	43 2 18	6 32 5	⊥. δ_o. Var. 19° 50′ in 1812. ⌒.
Cape Cavalaire, redoubt ...	43 9 42	6 32 12	⊤ & ⊥. In road ⚓′. ∠ 1100.
Cape Taillat, islet	43 9 53	6 39 50	⊥ except close in.
Cape Lardier, Camarat light	43 12 14	6 41 30	427 feet. At base $\beta\beta_o$. ⊥.
Cape S. Tropez, islet	43 16 32	6 43 16	δ. ! among the shoals.
S. Tropez, maison de Pratique	43 16 38	6 39 0	⊡. In Grimaud bay, ⚓′.
Frejus, ancient amphitheatre	43 26 5	6 44 15	98 feet above the sea.
Frejus, S. Rapheau	43 25 15	6 46 22	⊡. ⚓′. Var. 19° 43′ in 1815.
Agay, the castle	43 25 45	6 52 25	⚓, but ! shoals of Isle O, & la Boute.
Cape Roux, summit	43 27 30	6 54 30	1500. ⚘ (Forêt d'Esterelles).
Napoule, maison de Pratique	43 31 20	6 56 30	⚓, but with !.
Cannes, fort S. Pierre......	43 32 48	7 0 32	⊡. ⚓ with !. Var. 19° 20′ (1823).
Lerins Is., Ste. Marguerite	43 31 20	7 2 35	Fort Monterey. ‖ for small vessels.
Lerins Is., St. Honorat ...	43 30 19	7 2 40	The abbaie. δ to south. !. ⚘. ⌒.
Gourjeau, or Iouan	43 33 56	7 4 30	Maison de Pratique. ⚓′, but enter !.
Garouppe lighthouse.........	43 33 50	7 7 54	340 feet. ⊥ under, but !.
ANTIBES, mole light	43 35 5	7 7 38	50 feet. ⊡. Var. 19° 30′ (1823).
Ville-neuve, castle...........	43 39 35	7 8 0	A mark on the hills.
S. Laurent du Var	43 40 40	7 11 15	Boundary of France and Piedmont.
PIEDMONT.			
Grenaglia point	43 39 30	7 12 12	‖ river Var. ⊤. ⚘. ⌒.
Nice, port Limpia sanitá...	43 41 16	7 17 12	⊡. ∠ to M. Mignons, mark from S.W.
VILLA FRANCA, arsenal flag	43 41 25	7 18 35	Arsenal ⊡. Harbour ⊞.
Villa Franca, fort Montalban	43 41 38	7 18 13	Var. 19°; Dip 64° 10′; Int. 245 (1823).
Villa Franca, lighthouse...	43 40 2	7 19 27	On point Mala, 225 feet.
Belluogo (Beaulieu), mole...	43 41 44	7 20 5	⚓ in shore winds. ∠ M. Leuza, 1690.
Point St. Laurent	43 42 24	7 23 25	⊤, but ! a shoal. ∠ to Eza, 1840. $⚘_o$.
Turbia, ancient trophy......	43 43 49	7 25 41	Sea-mark, 1650. $⚘_o$.
Monaco, castle flag	43 42 50	7 26 55	⊤ & ⊥. ⚓. ∠ to M. Nagel (table-hill).
Cape S. Martin, battery ...	43 43 20	7 32 10	⊥. ∠ to Col de Braus, 3800.
Ventimiglia, dogana	43 45 42	7 38 0	‖ the Roya. ⚓,. ∠ Col de Tende 5900.
San Remo, mole	43 49 10	7 50 28	⚓ in ld. winds. ⚘. Var. 19° 19′, Dip 64° 14′ (1824).
Monte Grande, summit ...	43 51 0	7 37 0	3100 feet. With M. Cougarde, a mark.
Cape dell' Armi	43 49 52	7 53 0	‖ of Taglia. ⊤. δ_o. ∠ to the Cornice.
Port Maurizio, convent ...	43 53 22	7 58 46	⊥,. Road ⚓. Var. 19° 40′ in 1820.
Cape delle Mele	43 57 58	8 11 0	⊥. δ_o. ∠ to Maritime Alps.
Gallinara islet, tower	44 2 6	8 13 5	⊤. & ⊥. δ_o. in ‖. Var. 19° 30′ (1820).
Monte Calvo, summit	44 10 0	8 9 30	2900. M. Melogno 3400. Marks for coast.
Finale, battery	44 9 56	8 19 3	⚓ with land winds. Var. 19° 20′ (1820).

Place.	Latitude.	Longitude.	Notanda.
PIEDMONT.	° ′ ″	° ′ ″	
Noli, convent S. Francis...	44 11 54	8 22 42	⚓ with land winds. Var. 18° 40′ (1824).
Bersezzi islet, ruins	44 14 0	8 24 50	‖ for boats. Cape ∠ to Mt. Invincibile.
Vado, fort S. Lorenzo......	44 15 27	8 24 32	⚓′. Mag. Var. 18° 32′ in 1824.
Savona, citadel...............	44 18 25	8 27 52	⊠. Var. 19° 15′ (1820). ∠ Col d'Altare 1600.
Cape Arenzano, pitch	44 24 12	8 39 15	⊤. & ⊥. δ_o. Inside ‖ of the Pizzo.
Rock Polla, centre	44 25 0	8 46 14	δ_o in ‖ to Castellazza.
GENOA, lighthouse	44 24 36	8 53 5	370. ⊞. Var. 19°15′, Dip 63° 40′ (1818).
Genoa, fort Diamante	44 28 48	8 55 12	∠ to Bocchetto pass, 2700.
Nervi, the palace	44 23 39	9 2 18	δ_o in approach. ⚓$_o$.
Porto Fino, the fort.........	44 18 15	9 14 4	⊞. Var. 18° 54′, Dip 64° 7′ (1820).
Sestri à Levante, fort	44 16 23	9 25 26	⚓′ in land winds. Manara pnt. ⊤ & ⊥. δ_o.
Levanto, landing place ...	44 10 55	9 38 17	Mag. Var. 15° 45′ (?) in 1820.
Mount Castellana............	44 4 0	9 50 30	The summit, 1610 feet.
Porto Venere, S. Pietro ...	44 3 10	9 51 50	Eastern side ⊞. Ϯ. ⚹.
Tino islet, lighthouse.........	44 1 58	9 52 31	384 feet. ⊤ & ⊥. δ_o.
La Scola, fort	44 3 20	9 52 56	⊥. δ_o in ‖ to Palmaria.
SPEZIA, fort Pezzino	44 4 37	9 52 12	High water 1ʰ 38ᵐ. Rise & fall 1½ ft. ⊞.
Spezia, city castle	44 6 25	9 51 23	Var. 18° 10′. Dip 63° 35′. Int. 237 (1820). ⊞.
Spezia, Lerici castle	44 4 32	9 54 42	⚓′. Var. 17° 59′ in 1823. ⚹.
Santa Croce, sanitá	44 2 54	9 58 10	‖ the Magra. Porto di Luni ⊠.
TUSCANY.			
La Marinella, ruins of Luni	44 3 16	9 59 18	Confines of Piedmont and Tuscany. ∠
L'Avenza, landing place ...	44 2 15	10 3 30	⚓′ with !. ‖ Carrara river. ∠ M. Sagro.
San Giuseppe, tower	44 0 38	10 7 8	⊤. ‖ river to Massa. Ϯ.
Fort Cinquale	43 58 35	10 9 4	⊤. ∠ to Monte Altissimo, 5200.
Motrona, church	43 55 30	10 13 27	⚓′ in easterly winds. Ϯ. Dist. M. Cimone 6400.
Viareggio, sanitá	43 51 51	10 15 19	⚓′. Var. 18° 30′; Dip 63° 5′, in 1823. Ϯ.
Serchio tower..................	43 46 48	10 16 30	‖ river Serchio. In road ⚓. Ϯ. ∠ M. Pisano.
Pisa, campanile	43 43 30	10 24 0	'Leaning Tower,' with grd. 255f. A coast mark.
Arno fort, flagstaff	43 40 50	10 16 40	‖ river Arno. ⊠. β on bar.
Leghorn, Marzocco tower...	43 34 15	10 18 7	Mag. Var. 18° 37′ in 1823.
LEGHORN, lighthouse	43 32 50	10 17 45	154 feet. Var. 17° 58′ in 1820.
Leghorn, Melora beacon ...	43 32 56	10 13 32	On the head of Melora bank.
Calafuria tower	43 28 50	10 20 0	⊥. ∠ to Monte Negro. Ϯ.
Castiglioncello tower.........	43 24 30	10 24 5	δ_o in ⊤. Cove for boats. ⚹.
Vada tower	43 21 7	10 27 20	⚓ in ‖ to the reef. Ϯ.
Mal di Vetro reef...........	43 20 0	10 20 48	⊤. ββ$_o$. !. Submerged ruins (?).
Cecina, palace	43 18 6	10 29 30	‖ river Cecina. Ϯ.
Castagneto fort...............	43 10 15	10 32 30	⚓ in land winds. ∠. Ϯ.
Torre S. Vincenzo...........	43 6 0	10 32 18	∠ hills of Calvi and Campiglia.
Port Baratto, Populonia...	42 59 30	10 30 10	⚓′, except in N.W. winds.
Piombino, palace	42 55 32	10 31 40	390 feet. Var. 18° 0′ in 1823.

Place.	Latitude.	Longitude.	Notanda.
TUSCANY.	° ′ ″	° ′ ″	
Fullonica, dogana	42 55 20	10 44 50	⚓′ off Portiglione. ⚘. ∠ to 550.
Troja islet, tower	42 48 3	10 43 10	⊤ & ⊥. ∥ requires !. ∠ to M. Maus, 980.
Castiglione della Pescaja...	42 45 57	10 52 52	∥ to Lake, & river Bruna. Var. 18° 18′ (1823). ⚘
La Trappola fort...........	42 41 0	11 2 20	∥ river Ombrone. ⚘.
Cala di Forno	42 36 57	11 5 24	⚓ in the cove. ✻.
Port Talamone, sanita......	42 33 20	11 8 19	⊡. ⚓. Var. 16° 57′ in 1823.
Talamonaccio tower	42 33 14	11 9 56	∥ river Osa. ⊤. ✻.
Orbitello, landing-place ...	42 26 38	11 12 38	In middle of a lake. ⌒.
Santa Liberata tower	42 26 30	11 8 59	Above ruins of Domitian's port.
Port Santo Stefano, mole	42 25 56	11 7 57	⊡. ⚓′. Var. 17° 28′ in 1823.
Mount Argentaro, telegraph	42 23 45	11 10 28	1750 feet. δ_o on coast below.
Port Ercole, fort Stella ...	42 23 34	11 12 0	⊡. ⚓ but !. Var. 16° 55′ in 1823.
Tagliata tower	42 24 50	11 17 36	Ruins of Ansedonia. ⊤. ⌒.
Formica di Burano.........	42 23 0	11 19 10	⊥ around. δ_o in ∥. ⌒.
Lake Burano, E. Graticciaja	42 22 57	11 27 30	∥ Chiarone. Boundary of Tuscany.
TUSCAN ISLANDS.			
Gorgona isle, Torre Vecchia	43 25 45	9 53 0	⊤ & ⊥. ⚓ in cove. Summit 1200 ft.
Capraja isle, castle	43 2 36	9 50 49	⊤ & ⊥. Var. 18° 53′ in 1818.
ELBA I. Port Ferrajo	42 49 5	10 20 30	Fort Stella light 192 feet. ⊞.
Elba I. Cape Vita	42 52 40	10 24 58	Var. 19°, Dip 62° 40′, in 1823.
Elba, Longona citadel flag	42 46 12	10 24 22	Focardo light 105. ⊞. Var. 19° 5′ (1823).
Elba, M. Calamita (Loadstn.)	42 44 0	10 24 30	1195 ft. Var. 19° 30′, Dip 64° 10′ (1823).
Elba, Campo tower	42 44 55	10 14 35	⊡. Further out ⚓. ✻.
Elba, M. Capanne	42 46 30	10 10 0	2700 feet. Mark for the ∥s.
Palmajola I. fort light......	42 52 2	10 28 40	344 feet. Var. 19° 10′ in 1823.
Cerboli I., ruin	42 51 44	10 33 0	⊥ on all sides. δ_o.
Pianosa I., Turco rock ...	42 32 40	10 9 14	⊥. ⚓ off the boat coves. ✻.
Africa rock, centre	42 21 40	10 8 5	6 ft. Var. 19° 20′ (1823). ! shoal 2¼′ N.
Monte Christo, summit ruins	42 19 14	10 20 0	1900 feet. ⊤ & ⊥. δ_o. ✻.
Formiche di Grosseto, N. rck.	42 34 45	10 53 5	N. rock 32 feet, S. one 13. δ_o with !.
Giglio, town spire	42 22 3	10 55 10	∠ M. Pagana, 1317. Port ⊡. Campese bay ⚓.
Giannuti, Spalmatoja bay	42 14 56	11 7 0	⊤ & ⊥. ⚓. Var. 18° 5′ in 1823. ✻.
CORSICA.			
GIRAGLIA I., redoubt	43 1 45	9 24 10	250 ft. M. Campana (opposite) 576. ⚓.
Cape Minervio, pitch	42 54 5	9 19 0	⊤ & ⊥. $⚓_o$. ! land squalls. ⚘.
San Fiorenzo, citadel	42 41 10	9 17 56	∥ of Lomio. ⚓. Var. 18° 21′ (1815).
Punta Peralto, pitch	42 44 15	9 13 12	⊥. ∠ to peak of Sierra Lortella.
Isola Rossa, islet battery ...	42 38 43	8 55 55	⚓ with !. Var. 18° 30′ (1815).
Calvi, the citadel	42 34 0	8 45 0	⊡. ⚓. ∠ to Paglia-orba, 8700 feet.
Calvi, Rivellata light	42 35 0	8 43 12	290 feet. ∠ inland to Capo Tondo.
Gargana (Gargalo) I. turret	42 22 6	8 32 0	⊥. Boat ∥ to Gardiolo. ⌒.

Place.	Latitude.	Longitude.	Notanda.
CORSICA.	° ′ ″	° ′ ″	
Cape Rosso, pitch	42 14 12	8 32 30	⊥ to Sbiro rock. ✻.
M. Rotondo, Admiral's nose	42 12 0	9 4 0	8900 feet. Excellent sea-mark.
Sagona, white tower	42 6 50	8 41 0	⊡ & ⚓. ⦣ inland to Monte Ricco. Ŧ.
Sanguinario I. light tower	41 52 45	8 35 35	320. Mostly ⊥, but ! Tabernacolo rock.
Ajaccio, citadel flag.........	41 55 10	8 44 30	⊞. Var. 18° 25′, Dip 62° 5′ in 1815.
Cape Mulo (*Muro*)	41 45 0	8 39 20	Pitch of the Sierra Kutefa. ✻.
Campo Moro, redoubt......	41 38 12	8 48 15	⊞. ⚓′ by ⊤ in gulf of Valinco.
Point Senetoza, tower	41 33 50	8 47 23	⊤ & ⊥. ⦣ inland.
Monachi rocks, highest......	41 27 10	8 54 30	40 feet. δ in the ‖s. β. !. ↝.
Porto Figari, landing place	41 28 30	9 4 0	‖ river Canale. ⦣ inland. Ŧ.
Cape Fieno, or Feno	41 23 40	9 6 20	The pitch of Monte Trinitá. ⊤.
BONIFACIO, middle tower	41 23 14	9 10 0	⊞. Var. 18° 5′, Dip 61° 39′ in 1815.
Cape Pertusato, light	41 22 10	9 11 30	325 feet. ⊤ & ⊥.
Lavezzi I. Arrini cove	41 20 30	9 15 44	⚓. ⊥, but ! for reef 1½′ south.
Cavallo I. Levant cove ...	41 22 5	9 16 32	‖ to Piana I. and main, δ. β. !. ↝.
Perduto islet, centre.........	41 22 18	9 19 0	⊥, except reef to S.E., with β.
Porraja rock, summit	41 23 40	9 16 22	δ in ‖, but practicable. $\beta\beta_o$.
Santa Manza, Capicciolo T.	41 25 5	9 15 50	⊞, but ! for north-easters.
Toro rocks, the highest......	41 30 20	9 23 0	⊥, except on the east. δ_o in ‖.
Porto Vecchio, church	41 35 30	9 18 0	⊞. Unwholesome air. Var. 17°58′(1815).
Porto Vecchio, Chiappa light	41 35 55	9 22 10	220 ft. ⊤ & ⊥, but ! a rock near base.
Pinarello bay, torre de' Corsi	41 40 0	9 23 20	⚓ with shore winds. ⦣ M. Cava, 5000. Ŧ
Fium-Orbo, Casa Fiesci ...	41 59 40	9 26 30	‖. ⦣ to M. Cappella, 6750. Ŧ. ↝.
Alleria fort	42 7 0	9 31 15	‖ of the Tavignano. ⊤. Ŧ. ↝.
Fiorentina tower	42 17 10	9 34 0	‖ Alezani. Ŧ. E. extreme of Corsica.
Punta d'Arco, tower	42 33 30	9 31 50	‖ Biguglia lake & Golo R. ⚓′ in road. ↝.
Bastia, mole-light	42 42 0	9 27 0	52 feet. ⊡. ⚓. Var. 18° 30′ in 1815.
Monte Stella, summit	42 48 0	9 25 0	4500 feet. Mark in Elba & Capraja ‖.
Finocchiarolo I. tower	42 59 15	9 28 0	⚓ both in S. Maria & Figarona bays.
SARDINIA.			
Point della Marmorata ...	41 15 50	9 13 45	With Falcone, N. point of Sardinia.
Longo Sardo, redoubt......	41 14 59	9 11 20	⊡. N. winds rake the port. ↝.
Capo della Testa, light......	41 14 28	9 8 15	On torre Santa Reparata, 220ft. ⚓. $Ŧ_o$.
Vignola tower	41 8 5	9 3 10	⚓. ⦣ to M. Giuncara, 1700 feet.
Monfronara tower	41 1 10	8 52 35	Isola Rossa & Cala falsa ⊥. ⦣ to M. Cucuru.
S. Pietro di Mare, chapel	40 55 45	8 48 46	‖ river Coguinas. ⦣ to Castel Doria. Ŧ.
Castel Sardo, high steeple	40 55 7	8 42 36	⊤ & ⊥. ⚓ with !. ⚓ in the coves.
Sardo rock, four fathoms	41 0 50	8 43 36	⊥. M. Spina (2650) in Gallura, S. 22°, E. 13¾′
Porto Torres, tower light...	40 50 31	8 22 51	49 feet. ⊡. ⚓′. Var. 18° 50′ in 1824.
Sassari, cathedral...........	40 43 40	8 33 20	Above Porto Torres 710. Osilo pk. 2200.
Asinara I. punta Caprara	41 8 0	8 18 25	(Lo Scorno). ⊤ & ⊥. $⚓_o$. ✻. ↝.
Asinara I. Trabucato tower	41 4 4	8 18 53	⊞. ⦣ Scomunica peak 1458. ✻.

Place.	Latitude.	Longitude.	Notanda.
SARDINIA.	° ′ ″	° ′ ″	
Capo del Falcone, tower ...	40 58 . 5	8 10 17	⌓ 610 feet. ⚓$_o$ under. ∟$_o$.
Cape Argentiera, pitch ...	40 44 10	8 5 26	⊤ & ⊥. ∟$_o$. ∠ to Rotondo peak, 1390. ✱.
Torre della Pegna	40 35 43	8 7 40	913 feet. ⊤ & ⊥. ⚓$_o$. ∟$_o$. ⌒.
Porto Conte, Capo Caccia	40 33 24	8 8 29	Summit 575 feet. ⊥.
Porto Conte, torre Nuova	40 35 40	8 10 38	⊞. Var. 19° in 1824. ✱.
Monte d'Oglia	40 37 0	8 14 0	Summit 1398. A mark for Alghero.
Alghero, torre Sperone......	40 32 47	8 16 49	⚓′. Var. 18° 55′ in 1824.
Cape Marargiu, pitch	40 19 52	8 20 46	⊥ to the rock. ∠ to 2550 feet.
Isola Rossa, Bosa tower ...	40 16 40	8 25 31	∥ river Temo. ⊡. ⚲.
Cuglieri, castle on peak ...	40 12 0	8 32 0	1300 feet. ∠ M. Ferru 2796. ⚲.
Cape Mannu, torre Mora...	40 1 44	8 20 35	⊥. In land winds ⚓ in cove.
Mal di Ventre Rock.........	39 58 58	8 16 0	δ_o in ∥ to the main. ⌒.
Coscia di donna (Catalano)	39 52 40	8 14 10	⊥ except on the N.N.E. δ_o with !.
Oristano, torre Grande ...	39 53 55	8 28 40	⊡. ⚓. Var. 18° 36′ in 1824. ⌒.
Oristano, the belfry	39 53 47	8 33 20	Deadly air in summer. ⌒.
M. Aci, Trebina peak	39 46 0	8 43 30	Triple-peak hill of pilots.
M. Arcuentu, the mark ...	39 35 35	8 32 0	The finger of Oristano, 2315.
Cape Pecora, pitch	39 27 0	8 21 22	⚓ ∥ Flumini-maggiore. ∠ to M. Linas, 4000.
Pan di Zucchero	39 19 44	8 23 0	⊥. ⚓$_o$. δ_o. A conical rock.
S. Pietro I. Gallo rock......	39 9 0	8 11 12	⊥ on all sides. δ_o in ∥. ⌒.
S. Pietro I. north summit...	39 9 40	8 16 28	Guardia dei Mori, 680 feet. ✱.
S. PIETRO I. Torre Vittoria	39 8 28	8 17 28	Below Carloforte. ⊞. Var. 19° (1823).
S. Antioco, Casteddu Crastu	39 4 20	8 26 0	⊡. In Palmas bay, ⚓′. ✱. ⌒.
S. Antioco, cape Sperone...	38 57 20	8 23 10	⊥ & ⊤. ∠ to M. Arbus, 780 feet.
La Vacca rock, summit ...	38 56 10	8 25 20	550. ∥ for a boat between it & il Vitello.
Toro rock, summit	38 51 58	8 22 44	⊤ & ⊥. 693 feet. ∟ but ⚓$_o$. ✱. ⌒.
Cape Teulada, pitch.........	38 51 48	8 37 12	⊥. Summit 780. S. point of Sardinia.
Cape Malfatano, tower ...	38 53 15	8 47 18	440 feet. ⊞. Var. 18° 28′ in 1824.
Cape Spartivento, extreme	38 52 28	8 50 47	⊥. δ_o with !. ⚓ in land winds to the E.
Pula. S. Macario I.	39 0 7	9 1 50	The tower, 310 feet. ⊡. ⚓′.
CAGLIARI, Arsenal mole ...	39 12 13	9 6 44	⊡. In bay ⚓′. V. 18°23′, D. 59° 13′ (1823). ⌒.
Cagliari, S. Elias' light ...	39 10 48	9 7 58	Over Laida rock, 248, ∠ to 340 feet.
Cape Carbonara, Cavoli I.	39 4 50	9 31 41	Ficaria turret, 80 or 90 feet. ⊥.
Serpentara I. torre Luigi...	39 8 30	9 37 0	⊥. δ_o with !. Var. 18° in 1824.
Capo Ferrato, pitch.........	39 17 58	9 39 16	80 ft. ⊥. ✱. ∠ to torre di Monte Ferru.
M. Budui, Sette Fratelli...	39 18 30	9 26 45	The Seven Peaks. Summit 3800, Station 2300.
Chirra islet, centre	39 31 48	9 41 30	⊥. δ_o. Mag. Var. 18° 20′ in 1824.
Cape Sferra Cavallo	39 43 10	9 42 5	Pitch of M. Cuadazzoni, 3342 ft. ⊤. ⊥.
Cape Bellavista, S. Gemiliano	39 56 20	9 44 15	300 ft. ∠ to Gen-Argentu peak, 5276. ⚓.
Ogliastra isle, summit	39 59 30	9 43 55	⊤. ⚓ all over Tortoli bay. ⌒.
M. Gennargentu	40 0 0	9 19 0	Sciuscia peak 6200 feet. ⚲.
Cape Monte Santo	40 5 58	9 44 40	Summit 2425 feet. ⊤ & ⊥. ⚓$_o$.
Orosei, S. Maria di mare	40 22 59	9 44 5	⊡. δ_o in bay. Var. 18° in 1824.

Place.	Latitude.			Longitude.			Notanda.
SARDINIA.	°	′	″	°	′	″	
Cape Comino, rock Rossa...	40	31	35	9	49	58	⊥. ∠ inland. E. point of Sardinia.
Monte Albo, Cupetti peak	40	35	0	9	40	0	2317. Mark for Posada & Siniscola. ⌒.
Petrosa point, Santa Anna	40	44	0	9	42	57	! in approach. ∠ to M. Mazzori 3200. ⚘.
Mount Limbara, Tempio	40	51	0	9	11	0	Balestreri peak, 4500 feet. ⚘.
Molarotto, or Tauladetto...	40	52	12	9	47	0	70 or 80 ft. and conical. δ_0 in ‖s. ⊥.
Molara isle, middle	40	51	20	9	43	18	The ‖s safe with !. ✱. ⌒.
Tavolara I. Cala di fuori	40	54	54	9	43	36	⊥. ⊤ & ⊥. δ_0. 1500 feet nearly.
Tavolara I. Spalma.di terra	40	53	10	9	40	40	⚓. ‖s δ, but practicable with !.
Terra nova, ruins of Olbia	40	55	25	9	29	15	⊞ but ! the bar. Var. 18° 5′ in 1824.
Cape Figari, extreme cliff	40	59	20	9	39	30	⊤, ⊥, & ⊥₀. Fine ⚓ inside l'Aranci.
Mortorio isle, east cove ...	41	4	10	9	36	0	!. M. Cogaora (2150) marks Port Congianus.
Porto Cervo, landing place	41	7	56	9	31	27	⊡. Pedestal on M. Mola, S.W.
Cape dell' Urso, the bear...	41	10	17	9	24	26	Mark for ⚓ in Arsachena sound.
Mezzo Schiffo, il Parau ...	41	11	7	9	22	30	⊞. The Agincourt sound of Nelson.
Peninsula delle Vacche ...	41	13	20	9	17	27	North point. ⊞ on both sides.
Caprera I. Tejalone peak...	41	12	40	9	28	18	750 feet. ⊥ to east. ‖ with !.
MADALENA I. Old Guardia	41	13	27	9	23	42	Var. 17° 36′, Dip 61° 28′ (1824). ⊞.
Spargi I. summit	41	14	32	9	20	46	⊤ & ⊥. δ_0 with !. ✱.
Razzoli I. lighthouse	41	18	16	9	20	31	270 feet. ⊥. ⚓ in Cala Longa.
ROME.							
Clementino palace...........	42	14	22	11	42	30	‖ of Marta. ∠ to Corneto & hills, 1250.
CIVITA VECCHIA, pier light	42	5	40	11	43	55	82 f. ⊞. Var. 17° 30′, Dip 61° 15′ (1823).
Cape Linaro, Chiaruccia T.	42	1	55	11	48	58	! reef off point. ∠ to Tolfa peak, 1500.
Torre San Severo	41	58	30	11	59	0	⚓ here to S. Marinella with land winds.
Palo, beach magazine	41	54	27	12	5	28	⚓. ∠ to Cervitari. ✱.
Rome, St. Peter's cross......	41	54	0	12	27	0	A sea-mark. Ground & edifice 650 ft.
Tiber, Fiumicino light......	41	45	49	12	11	39	⊡ in ‖. ⚓′ in the offing. ⚘. ⌒.
Tiber, Bocca di Fuimara	41	43	58	12	12	56	Ostia ‖. Torre Santo Vito. ⚘. ⌒.
Ardea, steeple	41	37	0	12	33	0	δ_0 on coast. ⚘. ∠ Albano 998, M. Cavo, 3150.
Porto d'Anzo, mole-light...	41	26	54	12	42	9	⊡. ⚓ with !. (*Ceno Portus and Antium.*)
Astura, rock tower	41	24	10	12	48	15	Ruins of Cicero's villa. ⚘.
Fogliano, beach tower	41	21	20	12	56	54	‖ into the lakes. ⚘. ⊤. ⌒.
M. Circello, S. Felice church	41	12	40	13	5	18	⚓ by ⊤. Summit ruins, 1730 feet.
Terracina, ancient mole ...	41	15	51	13	15	9	⚓′ in land winds. ⚘ around.
NAPLES.							
Torre Vetere of Fondi......	41	16	0	13	20	0	‖ lake Fondi. Frontier of Naples. ⚘.
Gaeta, Orlando's tower ...	41	12	20	13	34	16	⊡. ⚓′ in bay. Var. 17° 39′ in 1823.
Mola, la Sanitá	41	15	10	13	35	29	⚓′ with off-shore winds. ∠ M. Castellone.
Monte Massico, or Falerno	41	9	30	13	54	0	Mark for coast. δ_0 with ⊤.
Castel-Volturno, beach tower	41	1	40	13	56	25	⚓ in east winds. Bad air. ⚘.
Torre di Patria	40	55	55	14	0	40	⚓ with land winds. ⚘. ⌒.

Place.	Latitude.	Longitude.	Notanda.
NAPLES.	° ′ ″	° ′ ″	
Cape Miseno, pitch	40 46 30	14 5 10	280 feet. ⊤ & ⊥. In the port ⊞.
Baia castle, flagstaff.........	40 48 35	14 4 44	⊞. Mag. Var. 17° 21′ in 1820.
Monte Nuovo, crater	40 50 0	14 5 10	480. Over Lucrine & Averno lakes.
Pozzuoli, Caligula's bridge	40 49 15	14 7 10	The inner end. ⊥ in approach.
M. Camaldoli, convent......	40 51 26	14 11 30	1490. Excellent mark from the bay.
Naples, Castel S. Elmo......	40 50 39	14 14 28	Var. 17° 32′; Dip. 60° 37′; Int. 241, in 1817.
NAPLES, mole-light	40 50 18	14 15 36	161. ⊞. In bay ‡′ with ! for foul ground.
Mount Vesuvius	40 49 15	14 25 30	Crater of 1820, 3880 feet.
Pompeii, temple of Isis ...	40 45 0	14 29 0	Overwhelmed A.D. 79.
Castellammare (di Stabia)	40 41 34	14 28 12	Mole-head. ‡′. ∠ M. S. Angelo, 4700.
Sorrento, the dogana	40 37 39	14 22 30	⊥ & δ_o, but ‡, from depth.
Point Campanella, light ...	40 34 10	14 19 32	⊤ & ⊥. δ_o. ∠ to M. Costanzo, 1600.
Amalfi, madre-chiesa	40 38 0	14 37 0	⚓, but $‡_o$. Open to S. and S.E.
Salerno, the mole...........	40 39 35	14 45 0	‡′, but exposed from S.S.E. to S.W.
Torre di Pesto	40 23 0	14 59 35	Malaria around Pestum. ⚘. ⌒.
Cape Licosa, tower	40 13 45	14 53 0	(Leucosia) ∠ inland. ⊤. $\beta\beta_o$.
Port Palinuro, torre Prodese	39 59 40	15 14 45	Round point Spartimento, ‡.
Point degl' Infreschi	39 57 0	15 26 0	⊤ & ⊥. ∠ M. Bolgaria, 3950 feet.
Policastro, dogana	40 1 38	15 32 35	(Buxentum). ‡, but exposed. V. 17° 10′ (1815).
Castro-Cucco, torre Caja...	39 53 0	15 45 30	⚓, and ‡ in land winds. On a hill.
NEAPOLITAN ISLANDS.			
Palmarola I., cala Forcina	40 56 18	12 52 58	⊤. δ_o with !. ∠ 427 feet. ⌒.
Ponza I., signal-station ...	40 53 5	12 57 38	757 f. Except the Formiche, δ_o below.
PONZA, lighthouse	40 53 35	12 58 26	⊞. Mag. Var. 17° 23′ in 1815.
Gava islet, or la Gabbia ...	40 55 42	13 0 40	δ_o in Scoglietelle ‖, with !.
Zannone I., pt. Galatella...	40 57 42	13 3 15	⊥, δ_o but the Varo in la Gabbia ‖. ✲.
Botte rock, summit	40 50 10	13 6 0	68 feet. ⊥ around. ⌊.
Vandotena I., port S. Nicolo	40 47 38	13 25 42	(Pandataria). ⚓. ⊥. δ_o with !. ∠ 803 f.
Santo Stefano I., redoubt...	40 47 15	13 26 53	183 feet. ⊤ & ⊥ around. ✲. ⌒.
Ischia I., Forio sanità ...	40 44 10	13 51 10	‡ in east winds. ∠ M. Epomeo 2570. ⚘.
Ischia I., Ischia castle ...	40 43 54	13 57 42	‡′, ‖ clear to Vivara shoal.
Procida I., Chiupetto light	40 46 12	14 0 57	74 feet. ‖ practicable, but !.
Nisita I., tower redoubt ...	40 47 45	14 9 37	⊞. ⚘. ‡ in Bagnoli bay.
Capri I., Palace of Tiberius	40 32 46	14 15 19	860. ⊥ ⊤. $‡_o$. $⌊_o$. δ_o in ‖. S. of Capo di Monte.
Capri I., Carena point ...	40 31 58	14 11 53	⊥ to nearly close under. ∠ M. Solaro, 1900.
Galli rocks, Lungo tower...	40 34 40	14 25 50	⊤ & ⊥. δ_o in ‖ to Vivara rock.
CALABRIA.			
Dino islet, turret	39 48 5	15 48 40	‡ on north or south with !.
Cirella I., tower	39 37 0	15 50 0	⚓ on the N.E. ✲. ⌒.
Fuscaldo, torre San Giorgio	39 24 53	15 59 20	‡′ with off-shore winds. ∠.
Monte Cocozzo, summit.. ...	39 16 0	16 6 30	A mark for Amantea, Belmonte, &c.

Place.	Latitude.	Longitude.	Notanda.
CALABRIA.	° ′ ″	° ′ ″	
Cape Suvero, tower	39 2 53	16 8 47	⊥. ‡′, but exposed, before S. Eufemia.
Mezza Praja tower	38 53 50	16 17 0	! bad air as far as Maida. Ŧ. ⌐.
Pizzo, Murat's prison	38 47 0	16 12 45	(*Napigia*). ‡′, but ! the north-westers.
Monte Leone, castle	38 42 0	16 10 0	(*Vibo Valentia*). On a hill. Ŧ.
Tropea, madre chiesa	38 39 45	15 55 12	⊤. ɹ. Var. 16° 50′ in 1815.
Cape Vaticano, tower	38 36 58	15 51 48	Good sea-mark. ⊥ to pretty close in.
Gioja, middle of the town...	38 24 49	15 56 0	Deep water close in. Ŧ.
Bagnara, the church	38 16 57	15 49 40	Var. 17° 10′ in 1815. ɹ. Ŧ.
Scilla (Scylla), castle	38 15 4	15 44 36	‡ in the bay, but ! currents.
REGGIO, marina fountain...	38 5 42	15 39 47	⊥. Var. 16° 25′ in 1815. ‡′ off Arco.
Cape dell' Armi, torre Molaro	37 57 25	15 42 0	⊥ and δ_o. ⟋ inland. Ŧ.
M. Pentedattilo, gli unci...	37 57 30	15 46 30	Sea-mark. ⟋ M. Aspromonte 4400. Ŧ.
Cape Spartivento, tower ...	37 55 50	16 3 0	S.E. point of Calabria.
Point Bruzzano, tower......	38 2 23	16 9 15	ɹ. ✲. ⟋ to Bruzzano town.
Ruins of Locris	38 15 0	16 14 40	⊥, approach to the beach very deep.
Point Stilo, torre Verdera	38 29 0	16 35 20	⊥ to beach. Var. 16° 20′ in 1816.
Squillace, campanile.........	38 48 48	16 28 0	δ_o in gulf, yet Virgil's 'Navi fragum.'
Cape Rizzuto, torre Vecchia	38 57 50	17 0 46	⊥. ‡ with !. Ŧ.
Cape Nao, or Colonne......	39 5 22	17 13 28	! ruins forming a shoal off.
Cotrone, castle light.........	39 7 35	17 9 30	98 feet. ⊡. Var. 16° 40′ in 1816.
Point Alice, tower	39 24 0	17 9 0	⊥. δ_o in Gulf of Taranto with !.
P. Trionto, Bufalaria tower	39 35 0	16 47 18	‖ river Trionta. Ŧ. ⌐.
Capo Spulico, tower	39 57 10	16 35 30	⊥. ‖ of the Ferno. ‡ off Roseto. Ŧ.
SICILY.			
Cape San Vito, church......	38 12 26	12 45 50	⊤ but ! the reef of the point. $\beta\beta_o$.
Castell' a Mare, fortress ...	38 1 51	12 52 43	ɹ in the coves beneath.
Cape Uomo-morto, tower ...	38 12 40	13 6 10	δ_o in approach, with !.
Femina isle, tower	38 14 10	13 12 50	Outwards ⊥. Inside ‖ for boats.
Cape di Gallo, pitch.........	38 14 53	13 18 20	1692 f. M. Pellegrino, 1955. M. Cuccio, 3300.
PALERMO, mole-light	38 8 15	13 21 56	⊞. Var. 18° 45′, Dip 59° 12′ (1814).
Palermo Observatory	38 6 44	13 20 15	As given me by Abbate Piazzi.
Mount Catalfano............	38 5 40	13 32 0	1095. Pts. Gerbino & Zaffarana, ⊤ & ⊥.
Termini, castle	37 57 28	13 42 0	⊤ & ⊥. ‡′ with off-shore winds. ✲.
Cefalù, cathedral............	38 0 0	14 3 57	ɹ. ‡ in summer. Var. 18° 40′ (1814).
Sant' Agata, tower	38 1 30	14 36 32	⊥. ‡ with shore winds. ⟋ to Caronia. Ŧ, 2000.
Cape Orlando, castle gate	38 7 46	14 44 30	⊤, but ! shoals to the west.
Cape Calava, pitch	38 10 0	14 54 15	⊤ & ⊥. ‡′ in the bay to the east.
Port Madonna, convent ...	38 6 45	15 2 20	651 feet. ⊡. Var. 18° 10′ in 1814.
Milazzo, promontory light	38 15 58	15 14 10	262 feet. ⊤ & ⊥. Var. 18° 38′ (1814).
Milazzo, castle	38 14 6	15 14 17	320 feet. In the bay, ‡′.
Spadafora, palace	38 14 0	15 22 10	⊥. ‡. Fine beach for watering. ✲.
Cape Rasaculmo, telegraph	38 17 56	15 31 57	⊤, but β close under. ✲.

Place.	Latitude.	Longitude.	Notanda.
SICILY.	° ′ ″	° ′ ″	
Faro point, the light	38 15 50	15 40 40	70 feet. ⚓ on the spit, but ! currents.
Grotta point, rotonda	38 14 20	15 35 30	Betw. it & P. Pezzo, no bottom with 200 fms.
MESSINA, lighthouse	38 11 30	15 34 40	74f. ⊞. Var. 18°33′; Dip 58°56′; In. 270 (1815).
Mount Dinnamare	38 8 30	15 27 30	3112 feet. A good mark in the Faro.
Scaletta, castle	38 1 45	15 27 45	⊤. Temporary ⚓ south of the point.
Point S. Alessio, barbican	37 52 30	15 21 10	Betw. it & C. dell'Armi, no bot. with 750 fms.
Taormina, telegraph	37 48 15	15 17 40	890. ∠ Moorish castle 1305, & Mola 1519.
Mount Ætna, summit	37 43 31	15 0 0	⌒ 10,874 f. Radius of vision, 150 mil.
Riposto, prison tower	37 40 10	15 12 50	⊥. ⚓′ in off-shore winds.
Trizza, high Cyclop rock ...	37 32 0	15 10 5	⊡. ⊥ outside, but ! inner ‖s.
Catania, mole-head	37 28 20	15 5 15	⊡. ⊥ in approach. Var. 18° 5′ (1814).
La Bruca, the castle	37 16 20	15 11 35	⊤ & ⊥. ⊡. ✱. ⌐.
Agosta or Augusta, light	37 12 50	15 13 15	⊞. Var. 17° 40′ in 1814.
Magnisi tower	37 9 25	15 13 45	⚓ in Panagia bay.
SYRACUSE, lighthouse	37 2 58	15 16 50	⊞. Var. 17° 45′, Dip 58° 3′ in 1814.
Cape Morro di Porco	37 0 0	15 18 58	⊤ and ⊥. δ_o.
Lognina tower	36 58 15	15 15 0	⊥. ⚓ in the cove. ✱.
Avola, the tonnara	36 55 10	15 8 5	⚓′ in summer, and land winds. ⚲.
Vindicari tower	36 49 12	15 5 20	⊡. Mag. var. 16° 40′ in 1814.
Marzamemi tower	36 45 30	15 6 45	⚓′. ∠ to Pachino. ✱. ⌐.
Passaro isle, redoubt	36 41 30	15 8 56	⊤ & ⊥. ⚓. ‖ for boats. Var. 16° 24′ (1814).
Current isle, summit	36 38 10	15 3 5	⊤ with !. To the west δ. ✱. ⌐.
Pozzallo, fort	36 44 40	14 50 48	⚓′ in off-shore winds.
Cape Scalambra, tower ...	36 46 13	14 30 15	⚓. Approach with !. ✱. ⌐.
Scoglietti, chapel	36 52 34	14 27 25	⚓,. Near marsh of Camarina.
Terra-nova, Doric column	37 2 54	14 15 0	⊥ beach. ⚓′ in land winds.
Alicata, the castle	37 4 3	13 55 54	⚓. Mag. var. 16° 58′ in 1815.
Palma, marina	37 8 47	13 43 11	Summer ⚓. ⚲.
GIRGENTI, mole-light	37 15 39	13 31 40	⊡ & ⚓′ outside. Var. 17° 0′ in 1817.
Girgenti, temple of Juno ...	37 16 38	13 35 40	In Agrigentum. Dip 58° 5′ in 1814.
Girgenti, cathedral	37 17 44	13 34 6	A mark in taking anchorage.
Seculiana, the church	37 19 50	13 25 28	⚓. In off-shore winds ⚓′. ✱.
Cape Bianco, turret	37 22 25	13 16 27	!. Distant Calata-bellota peak, 3800.
Sciacca, castel Peralta......	37 29 50	13 4 46	M. Calogero 1035. Var. 17° 30′ (1814).
Cape San Marco, tower ...	37 29 15	13 0 20	⊤. δ_o with !.
Selinuntum, ruins of	37 36 14	12 46 32	Temple of Neptune. Beach below ⚓. ✱.
Cape Granitola, point	37 33 57	12 36 39	Approach with !. At night δ.
Mazzara, the citadel	37 39 56	12 33 59	⊡. ‖ the Salemi. Var. 17° 37′ (1814).
Marsala, cape Boeo chapel	37 48 10	12 25 10	⊡. Mole-light 55 feet. In road ⚓′.
San Pantaleo islet	37 52 54	12 28 14	Gate of ancient Motya. ⌐.
Torre Teodoro	37 55 45	12 27 50	‖ to Borrone & Favilla salterns. ⌐.
TRAPANI, Colombara light	38 1 53	12 30 18	⊞. Var. 17° 10′, Dip 58° 55′ (1815).

Place.	Latitude.	Longitude.	Notanda.
SICILY.	° ′ ″	° ′ ″	
M. S. Julian, Saracenic T.	38 2 58	12 37 5	2184 feet. Mark in the ∥s.
Cape Cofano, summit	38 7 21	12 42 48	⊤ & ⊥. ‡ off Messa tonnara.
SICILIAN ISLANDS.			
Stromboluzzo, summit	38 49 16	15 14 0	Also Strombolino. ⊤ & ⊥. $\lfloor_0$ 240 ft. ↼.
Stromboli, S. Bartolo church	38 48 12	15 13 10	Schicciola crater station 2171. Sum. 2800. $‡_0$.
Basiluzza, the ruin	38 39 50	15 7 54	⊤ & ⊥. δ_0 in ∥ to Panaria.
Ann's reef, three fathoms...	38 35 0	15 8 0	δ_0 in ∥ between it and Bottaro. $\beta\beta_0$.
Panaria, port Castello ...	38 37 40	15 2 55	⊞. The isle ⊥ around.
Penrose rocks, four fathoms	38 38 20	14 54 40	δ, being in mid ∥ Panaria and Salina.
Salina, Amalfi church......	38 35 40	14 47 35	⊥. δ_0. ∠ Mts. Salvatore & Vergine.
Bentinck shoal, 2½ fathoms	38 28 52	14 49 20	δ. ⊥. Safe ∥ to Scoglio del Bagno. $\beta\beta_0$.
LIPARI, the castle	38 27 56	14 57 50	⊞. Var 18° 50′ (1815). M. S. Angelo 990.
Pietra lunga, summit	38 25 40	14 54 40	⊤. $\lfloor_0$. Like a ship, ↼. ∥ to Vulcano δ_0.
Vulcano, sulphur works ...	38 23 19	14 55 56	⚓ in cove formed by Vulcanello. M. Aria 2400.
Felicudi, the church.........	38 34 5	14 29 37	⚓. Station on M. Permera 1950.
Canna rock, summit.........	38 35 2	14 25 42	⊤, ⊥, $\lfloor_0$. Like a ship, 286 feet. ↼.
Alicudi, the church	38 32 41	14 16 30	⊤ & ⊥ all round. $‡_0$. ✱.
Ustica, fort Falconara......	38 43 17	13 11 10	⊤ & ⊥. ⚓ in Santa Maria cove.
Ustica, Walker's rocks......	38 44 40	13 10 30	!. ⊥ all round, but two fathoms on.
Maretimo, the castle	38 1 10	12 3 55	⊤ & ⊥. ⚓′ ∠ to 2300 feet.
Levanzo, guard-house	38 1 38	12 20 29	⊤ & ⊥ around. δ_0. ✱. ↼.
Porcelli rocks, a-wash	38 4 30	12 26 45	⊥ and δ. !. $\beta\beta_0$.
Formiche, tonnara pier ...	38 0 37	12 25 53	⊞. Mag. var. 17° 15′ in 1815.
Favignana, fort Leonardo	37 57 40	12 18 30	⊞, and ‡′ for a fleet in the road.
Favignana, S. Catarina...	37 56 36	12 17 45	1249 feet. Excellent sea-mark. ↼.
Favignana, S. Catarina sh.	37 53 40	12 17 10	δ_0 in ∥ to Point Sottile. $\beta\beta_0$.
Skerki shoals..................	37 44 53	10 45 15	For this, & others around, see coast of Tunis.
Pantellaria, prison fort ...	36 51 15	11 54 29	⊞. Mag. var. 16° 15′ in 1817. ♆. ↼.
Pantellaria, Sataria point	36 45 40	12 4 20	∠ to 2213 ft. ∥ for boats betw. point & rock.
Linosa, landing cove	35 51 50	12 52 9	⊥. $‡_0$. Highest crater, 522 ft. ✱. ↼.
Lampedusa, Capo Ponente	35 31 0	12 29 57	⊤ & ⊥. 378 feet. ♆ ✱. ↼.
LAMPEDUSA, castle	35 29 19	12 35 10	⊞. Mag. var. 16° 23′ in 1822.
Lampion rock, ruin.........	35 32 47	12 19 50	140 ft. ⊤ & ⊥. Var. 16° 30′ in 1822.
MALTESE ISLANDS.			
Gozo I., Cape S. Demitri...	36 3 20	14 9 0	⊤ & ⊥. δ_0 in rounding the cliffs.
Gozo I., the castle	36 1 30	14 14 35	Summit 570 ft. Var. 16° 36′ (1816).
Gozo I., fort Chambray ...	35 59 37	14 16 55	⊞. δ_0 in ∥ to Comino.
Comino I., tower redoubt	35 59 6	14 19 48	⚓′ in the coves. ∥s quite clear.
Malta I., Torre Rossa......	35 57 31	14 20 54	Commands Melheha bay & Comino ∥. ✱.
Malta I., St. Paul's tower	35 56 26	14 25 25	⊞. ‡′. Traditional site of S. Paul's wreck.
Malta I., Civita Vecchia...	35 51 57	14 25 0	Cathedral, in the Rabatto, or suburb.

Place.	Latitude.	Longitude.	Notanda.
MALTESE ISLANDS.	° ′ ″	° ′ ″	
MALTA I., Valetta palace	35 53 55	14 30 50	240 feet. Var. 17° 21′ in 1816. ⊞.
Malta I., St. Elmo light ...	35 54 12	14 31 20	167 feet. Between the two ports.
Malta I., Dockyard sheers	35 53 0	14 31 10	⊞. Var. 17°; Dip 57° 42′; Int. 443 (1822).
Malta I., S. Thomas castle	35 52 15	14 33 45	⚓ in Marsa Scala. ! Mansciar reef.
Malta I., P. dell' Mare ...	35 49 47	14 33 10	⊥. δ_o in rounding, with !.
Malta I., Marsa Scirocco	35 50 15	14 32 30	St. Lucian's castle. Var. 17° 20′ (1816).
Malta I., Benhisa tower ...	38 48 56	14 32 20	∥ between the point and reef, but !.
Malta I., Bocca di Vento	35 52 40	14 21 40	$⚓_{,}$. ⦣ to Benjemma heights, 500. ✵.
Filfola rock, summit	35 47 12	14 27 0	⊤ & ⊥. δ_o in ∥. Var. 16° 25′ (1823).
NAPLES, *continued.*			
Torre Mattoni	40 22 36	16 50 30	∥ of the Bradano. δ_o. ɤ. ⚲. ⌒.
TARANTO, citadel	40 27 19	17 14 5	Var. 16° 0′, Dip 59° 55′ (1816). ⊞.
Cape Santo Vito	40 23 40	17 12 30	Light 23 feet. ɤ with !. ✵.
Port Cesareo, tower.........	40 13 0	17 55 50	Approach with !. ⚓. ✵. ⌒.
Gallipoli, castle	40 1 51	17 58 0	⊞. Before the city ‡, with !.
Ugento shoal	39 50 0	18 10 20	Giurlitto reef δ, β. Town 498.
Cape S. Maria di Leuca ...	39 47 53	18 23 12	476. Convent column. ⊤ & ⊥. ⌊ but $‡_o$.
Gagliano, cove	39 50 43	18 23 40	⦣ to town 495. ⊥. ✵. ⌒.
Cape Otranto, telegraph ...	40 7 20	18 30 16	East extreme of Italy. ⌒.
Otranto, castle	40 9 5	18 28 45	Mag. var. 15° 15′ in 1816. ⚓.
Lecce, cathedral	40 21 0	18 10 28	Capital of Terra di Otranto.
Torre di Cavallo	40 38 0	18 4 58	On the north δ. !.
BRINDISI, castello di mare	40 39 21	18 0 27	Var. 15° 6′, Dip 59° 42′ (1816). ⊞.
Torre di Penna	40 41 0	17 59 25	Cape Gallo ⊥. ɤ. ⌒.
Guaceto islet	40 42 45	17 50 30	⚓ in the cove. Outside ‡.
Monopoli, point Paradi ...	40 57 10	17 20 55	ɤ. With off-shore winds, ‡′.
Polignano, Paolo rock......	40 59 47	17 16 20	⊤. ⦣ to Mount Bagiolara. $⚓_{,}$.
Mola, castle	41 3 50	17 7 38	$⚓_{,}$. ✵. In land winds, $‡_{,}$.
Bari, the pier-head	41 7 56	16 54 29	$⚓_{,}$. Var. 16° 15′ in 1816.
Giovinazzo, turret...........	41 12 0	16 42 40	In Spiriticchio cove, $⚓_{,}$.
Molfetta, mole	41 12 44	16 37 13	Between the light and rock, ⚓.
Bisceglia, pier-head	41 14 25	16 31 14	Inside ⚓′. Outside δ_o. ɤ.
Trani, dogana	41 17 52	16 26 45	⚓′ in the port. In the road ‡.
Barletta, light	41 20 25	16 19 27	⊞. In the road ‡′. ɤ.
Torre di Rivoli...............	41 29 5	15 57 0	∥ river Carapella, and Lake Salpi.
MANFREDONIA, mole	41 37 40	15 55 58	Var. 14° 55′ in 1819. ⊞. ‡′ by ɤ.
Mt. S. Angelo, hermitage...	41 42 30	15 57 0	S. Angelo summit, 2400, ⌂.
Monte Calvo, the station ...	41 43 50	15 47 0	Highest peak of the Gargano, 3500.
Viesti, S. Croce rock	41 52 35	16 11 23	⦣ and ⚲. ⚓. Outside ‡, ɤ. ⌒.
Peschici, landing-place......	41 56 48	16 1 20	‡ with land winds. ⚲.
Varano, west tower	41 55 20	15 48 30	∥ into the fishery. ⚲.
Tremiti isles, telegraph ...	42 7 15	15 29 50	S. Nicola castle, 260. ⊤ & ⊥. ‡. ⌒.

Place.	Latitude.	Longitude.	Notanda.
NAPLES—*continued.*	° ′ ″	° ′ ″	
Pianosa islet	42 12 38	15 45 30	48 ft. Var. 15° 26′ (1819). ⊥. δ_o. $Ψ_o$. ~.
Mileto point, telegraph	41 55 44	15 38 10	On Cala-roscia tower. Ψ.
Termoli, telegraph	42 0 26	15 0 11	150. ⊥ with γ. ⚓. Ψ.
Monte Majella	42 5 0	14 6 0	8500 feet. On the flanks Ψ.
Vasto, campanile	42 6 36	14 43 20	600. On the hill Aimone.
Punta di Penne, turret	42 10 5	14 43 57	⊥ in approach, but γ.
Ortonammare, mole	42 20 29	14 26 27	Var. 16° 0′ in 1819. ⚓,. ⚓ in offing.
Chieti, steeple	42 21 15	14 11 0	1250. Mark for Ortona.
Pescara, madre chiesa	42 27 0	14 14 5	‖ river Pescara, from M. Majella.
Monte Corno, summit	42 28 0	13 35 0	Gran Sasso d'Italia, 9570.
Atri, cathedral	42 35 0	13 59 0	1590. Over Galbano, or Calvano, its port.
Vomano tower	42 39 0	14 3 15	‖ riv. Vomano. ∠ to M. Pagano, 1020.
Colonnella, steeple	42 52 32	13 52 30	Frontier post of Naples, 1080.
PAPAL STATES.			
Torre d'Ascoli	42 54 30	13 56 16	‖ river Tronto. ⚓. γ.
Grottamare, Lama fort	42 59 50	13 52 38	∠ to the town, 450.
Ripatransone, steeple	43 0 0	13 46 41	Standard point, 1750 feet.
Fermo, marina	43 10 10	13 48 30	‖ river Lete. ∠ to the city, 1200.
Recanati, port	43 25 48	13 39 50	∠ to the city, 1400.
Loreto, cathedral	43 26 42	13 36 50	On a height, 565 feet.
Monte Conero, chapel	43 33 14	13 36 5	Summit 1900 feet.
Porto Nuovo, Trave	43 34 48	13 35 0	⊡. ∠. γ. ✱ but $Ψ_o$. ~.
ANCONA, mole-light	43 37 40	13 30 3	130 feet. Var. 16° 26′ in 1819.
Sinigaglia, mole-head	43 43 20	13 13 9	⚓′ for boats. Outside ⚓′ with !.
Fano, lighthouse	43 50 57	13 1 0	γ. With off-shore winds ⚓′.
Pesaro, mole-light	43 55 31	12 53 58	Open, but γ. ✱.
Monte Luro, spire	43 54 47	12 46 10	Sea-mark for the coast, 980.
San Marino, steeple	43 56 30	12 27 0	A republic. Sea-mark 2470.
Rimino, mole-head	44 4 18	12 34 20	‖ Marecchia. Var. 16° 50′ in 1819.
Cesenatico, pier	44 12 46	12 24 20	In land winds ⚓′. Ψ.
Cervia, town tower	44 15 50	12 21 10	107. On with mole-light ⚓′. Ψ.
Ravenna, rotonda	44 24 55	12 12 40	Now far inland. Ψ.
Porto Primaro, battery	44 35 18	12 17 45	‖ for boats. γ. ⚓.
Comacchio, steeple	44 41 2	12 10 49	135. ‖ from Port Magnavacca.
Volano, telegraph	44 48 15	12 15 25	‖ of the Po di Volano. Ψ. ~.
Goro, Gorino battery	44 48 55	12 22 21	⊡. In Sacca dell' Abbate ⚓′.
VENICE.			
Porto della Maestra	44 59 11	12 27 40	Main ‖ of the Po. ⊡. ~.
Adria, belfry	45 3 25	12 4 0	Between it and the sea Ψ.
Port Brondolo	45 10 10	12 19 46	‖ of Brenta Nuova. ⚓′. γ.
Chioggia, Castel Felice	45 13 48	12 18 50	150. ⊡. Var. 17° 28′ in 1819.

Place.	Latitude.	Longitude.	Notanda.
VENICE.	° ′ ″	° ′ ″	
Fort S. Pietro	45 20 10	12 20 43	‖ of Malamocco. ⊞. In roads ⚓′.
VENICE, S. Mark's belfry...	45 25 48	12 21 40	315. Var. 17° 10′; Dip 65° 8′; Int. 248 (1819).
Fort S. Andrea	45 26 28	12 24 31	‖ of Lido ⊞. In the roads ⚓.
Cortellazzo, battery	45 32 7	12 45 20	‖ of river Piave. ⚓,. ⌒.
Caorle, steeple	45 35 39	12 54 37	155. ‖ of Livenza. Var. 17° 40′ (1819).
Port Tagliamento...........	45 38 30	13 6 5	‖ of the river. ⚶. ⚘. ⌒.
Port Lignano	45 41 20	13 9 57	⊡. Outside ⚓. ⚘. ⌒.
Grado, campanile	45 40 44	13 22 57	160. A mark for the coast. ⊤.
Aquilea, campanile	45 46 0	13 22 30	250. Seen over drowned lands. ⚘.
Point Sdobba, telegraph ...	45 43 40	13 33 0	‖ of the Isonzo. ⊡.
ISTRIA.			
Monfalcone, centre	45 48 20	13 32 14	La Rocca, 300 feet. ⚘.
Duino castle, flagstaff......	45 46 14	13 35 58	In Sacco di Panzano. ⊡. ⚓′.
TRIESTE, Sta. Teresa mole	45 38 49	13 46 15	Light 106 f. ⊡. ⚓. ∠ to the Karst, 1590.
Trieste, castle flagstaff	45 38 25	13 46 47	310. Var. 16° 54′, Dip 65° 13′ (1819).
Capo d'Istria, Sanità	45 32 32	13 44 12	46 feet. On an insulated rock. ⚘.
Isola, campanile	45 31 58	13 40 0	185. ⚶. In the road ⚓.
Pirano, San Giorgio belfry	45 31 18	13 33 54	240 feet. Var. 16° 5′ (1819). ⊡. ⊞′.
Cape Salvore, lighthouse ...	45 28 57	13 29 47	117. Cala Mosca of Bassania.
Omago, steeple	45 23 50	13 31 30	110 ft. ⊡. Outside ⚓. ∠ to Buje, 890.
Cittanova, battery	45 18 36	13 32 55	Here and Port Quieto ⊡, and ⚓.
Parenzo, islet convent......	45 13 34	13 35 10	⊡. ⚓′. Var. 16° 21′ in 1819.
Orsera, church	45 8 30	13 36 12	⊡, but ! in approach. ⚘.
Rovigno, S. Eufemia spire	45 4 36	13 37 39	330 feet. ⊥. ⊞. ⌒.
Dignano, church	44 57 25	13 51 14	Mark for Canale di Fasana.
Fasana, the mole...........	44 55 16	13 48 10	Between it and Brionis ⊞. ⚘. ⌒.
POLA, Olive islet	44 52 18	13 50 10	⊞. ⚘. Var. 15°, Dip 64° 38′ (1819).
Pola, Cape Brancorso	44 51 42	13 48 37	Summit, 149 feet. ⚘. ⊥.
Port Veruda, isle convent...	44 49 28	13 50 24	118 feet. ⊥. ⊡. ⚘. ⌒.
Cape Promontore, Porer rock	44 45 27	13 53 54	Lighthouse 111 feet. δ. !.
Port Bado, landing-place...	44 53 46	13 59 57	⊥. ⚶. ⚘ but ⚘$_o$. ! for boras.
Punta Nera, tower	44 57 55	14 8 30	⊤. ⊥. ∠ to M. Ostrina, 1760.
Albona, church..............	45 4 46	14 7 40	1160 feet. Shores beneath ⊥.
Fianona, steeple	45 7 51	14 11 0	640 feet. ⚶ but !. ⚘.
Mount Caldero, or Maggiore	45 16 32	14 11 57	4530 feet. Sea-mark around. ⚘.
CROATIA.			
Kastua, black castle.........	45 23 12	14 20 10	On a hill inland. ⚘.
Fiume, landing-place	45 19 5	14 25 43	‖ river Reka. δ$_o$ but wind. ⚓′
Porto Re, arsenal...........	45 16 0	14 33 36	⚓′, but torn by boras.
S. Marco islet	45 14 55	14 33 0	‖ Maltempo. ⊥ but !.
Kernovitza, chapel	45 6 30	14 50 0	A clean cove, but !.

Place.	Latitude.	Longitude.	Notanda.
CROATIA.	° ′ ″	° ′ ″	
Segna, mole-head	44 59 40	14 54 10	⊤, ⊥. Ravaged by boras.
Jablanaz, chapel	44 42 30	14 53 30	∠ to M. Velebich. ⚓.
Karlopago, mole	44 31 40	15 3 56	⊥, but ! boras. Var. 17° 10′ (1819).
Lukovo, landing-place	44 26 8	15 11 0	⊤ and ⊥, but !.
Castel Venier	44 15 0	15 27 40	‖ of Novigradi lake. !.
Novigradi, fortress	44 10 10	15 32 9	When inside ‖, ⊞. ⌐.
Karin, convent	44 7 0	15 36 10	‖ of Karisniza ⊡. ⚓.
CROATIAN ISLANDS.			
Puntadura, station	44 18 10	15 3 50	355. Dinara peak in Julian Alps, 8500.
Pago, fort Glubatz	44 19 22	15 15 10	Commands ‖ into Morlacca ‖.
PAGO I., landing-place ...	44 27 2	15 2 50	Land-locked, but ravaged by boras.
Pago, point Loni	44 42 10	14 43 30	⊤ & ⊥ with !. ⚓. ⌐.
Pago, M. San Vito	44 28 30	14 59 40	1150 feet. Theodolite station.
Maon Isle, chapel...........	44 26 20	14 54 0	⊤ & ⊥ in Pago ‖, but !.
Arbe I., steeple	44 45 7	14 44 45	⊥ but !. Var. 17° 0′ in 1819.
Gaglian rock	44 56 42	14 40 30	‖ to Besca-vecchia ⊤. ⌐.
Veglia I., madre chiesa ...	45 1 40	14 33 58	⚓ but !. ⚓.
Veglia, val Dobrigno	45 8 20	14 35 40	⊡ but ! for boras.
Kerso, Farasina convent ...	45 7 49	14 16 56	⊤. ⊥. ∠ to Mount Sys, 1680.
Kerso I., Sanità	44 57 36	14 23 50	⊡. In bay δ_o, but for winds !. ⚓$_o$
Kerso, Osero church.........	44 41 5	14 22 51	‖s ⊥, but !.
Galiola rock, centre	44 43 12	14 10 10	In the Quarnero ‖. ⌊. ⌐.
Unie I., Porto-lungo	44 38 35	14 15 38	⊥ but !. ⚓. ⌐.
Sansego I., mount Garbi ...	44 31 4	14 17 45	350 feet. $\beta\beta_o$ to the N.W. ⚓$_o$.
Lossini, mount Osero	44 40 16	14 21 38	1900. Isle also named Lossin Piccolo.
LOSSINI I., port Augusto	44 32 6	14 27 21	Arsenal. Var. 16° 58′ in 1819. ⚓.
S. Pietro di Nembo I.	44 28 0	14 32 10	Ilovatz chapel. ⊡ in ‖. ⌐.
Grivitsa rock.................	44 24 30	14 33 30	‖ of the Quarnerolo. ⌊.
Selve I., town church......	44 22 39	14 40 43	‖ to S. Pietro δ, !. ⌐.
Ulbo I., town	44 22 15	14 46 15	Mag. var. 17° 6′ in 1819.
DALMATIA.			
Nona, steeple	44 14 30	15 10 15	In the basin, ⊡.
ZARA, bastion S. Francesco	44 6 39	15 12 49	Var. 14° 13′, Dip 64° 20′ (1819). ⊞.
Mount Vratsavo	44 2 0	15 24 0	710 feet. Inland ∠ to 4900.
Zara Vecchia, steeple	43 56 27	15 26 40	In ‖, ⚓′ with !. ⌐.
Monte Nero, station	43 54 0	15 39 0	970. Mark for Lake Vrana.
Slozella, landing-place......	43 49 0	15 40 0	With ! ⚓′. ⚓ but ⚓$_o$. ⌐.
Sebenico, Castel-vecchio ...	43 44 15	15 52 45	⊤. ⊥. Var. 15° 8′ in 1819. ⊞.
Capo Cesto, tower...........	43 34 52	15 54 45	⊥ but !. ⊤. ⊡. ⌐.
Ragosnitsa, mole	43 31 17	15 57 48	⊤ & ⊥. ⊞. Var. 14° 30′ in 1819.
Port Manera, landing-place	43 29 36	16 0 28	⊥. ⚓. ∠ to M. Movar, 400.

Place.	Latitude.	Longitude.	Notanda.
DALMATIA.	° ′ ″	° ′ ″	
Trau, S. Mark's tower......	43 30 46	16 14 58	In ‖ Trau, Bua, and Salona ⊞.
SPALATRO, cathedral	43 30 11	16 26 10	Diocletian's palace. Var. 15° 0′ (1819).
Spalatro, fort Botticella ...	43 29 19	16 25 45	⊡. In bay ⚓,. M. Maglian, 550.
Almissa, convent	43 26 20	16 42 10	In the ‖, ⊤ and ⊥. ⚓.
Monte Borak	43 26 0	16 44 0	2800 feet. ⚘ and ✳.
Macarska, chapel...........	43 16 59	17 1 16	Var. 14° 45′ (1819). M. Sustvid, 3800, ∠ 5900.
Fort Opus, flagstaff	43 1 45	17 35 0	‖s of river Narenta. ↼.
Fort Smerdan	42 56 50	17 33 20	∠ to M. Ulico, 1800. Turkish confine.
Sabbioncello, point Ossit ...	42 59 50	16 59 30	‖ Curzola, ⚓′. ∠ to M. Vipere, 3160.
Sabbioncello, Val di Briesta	42 54 0	17 31 10	Access !, ✳. M. Sukino, 2050.
Monto-rogo, summit	42 46 0	17 56 0	2810. Slano between it & M. Tmor, 2965.
Isola Rudda, station	42 42 37	17 55 10	‖ of Kalamota, all ⚓′ and ⊞. ↼.
Ragusa, mole battery	42 38 16	18 6 39	∠ to fort Imperial, 1350. ⊡. ⚓. Inland, 4500.
RAGUSA, fort S. Marco ...	42 37 40	18 6 54	310. On Lakroma isle. Var. 16° 0′ (1819). ⚘$_o$.
Ragusa Vecchia, chapel ...	42 35 0	18 12 0	⊤ ⊥. ⊤. In Prahlivaz, ⊞.
Molonto, port Piccolo	42 27 5	18 25 0	⊥. ⊡, Mark, S. Elia, 1850. On the coast ⚓$_o$, ⊥$_o$.
DALMATIAN ISLANDS.			
Premuda, summit	44 20 20	14 36 30	⊥, except on the N.W. ↼.
Isto, magazine	44 16 15	14 44 50	∠ to Monte Guardia, 560.
Melada, Banastra point ...	44 12 18	14 48 58	In port Beguglia, ⊞.
Klib rock, or Diboskik......	44 13 35	14 54 0	⊤. ⊥. ⊥, but difficult. ↼.
Grossa, point Bianche light	44 9 10	14 48 40	‖ of the Sette Bocche. ⊥. !. ⚘$_o$.
Grossa, M. Vela Stratza ...	43 59 0	15 2 30	1100. Landfall for Grossa or Lunga.
Grossa, mount Krepassia..	43 54 24	15 6 50	Port Tajer, ⊤. ⊥. ⊞.
Incoronata, M. Opat	43 43 38	15 26 25	Summit of the isle, 760.
Curbabella, east peak	43 41 15	15 30 55	380 feet. ✳ but ⚘$_o$.
Sestrugn, summit	44 9 55	14 59 20	⊥ but !. ↼.
Eso, the port	44 1 52	15 5 40	⊥. ⊤. Var. 15° 50′ in 1819.
Ugliano, castle	44 4 39	15 8 42	879. ‖ of Zara, δ_o.
Pasman, church	43 57 20	15 22 58	δ_o in ‖ with !. Summit 893.
Vergada, summit	43 51 10	15 30 15	370 feet. ! in the ‖s.
Zut, summit	43 51 50	15 18 46	Station on Velikivak. ↼.
Morter, Gessera chapel ...	43 47 58	15 38 14	δ_o in ‖. ∠ to Broskitza, 359.
Zlarina, port	43 41 40	15 49 58	∠ to M. Batokio, 540 feet.
Smajan, summit	43 42 5	15 44 12	457 feet. ⊥ in ‖s, but !.
Zuri, mount Bohl	43 39 0	15 37 50	380. δ_o in ‖s with !. ↼.
Suilan, Aid rock	43 32 30	15 51 0	6 fathoms, ‖s around deep.
Zirona, port Grande	43 26 48	16 8 30	‖s δ_o with ! ⊡. ↼.
Solta, port Sordo	43 23 0	16 18 25	⊥. ⊡. ✳. ∠ to M. Stratsa, 695.
Bratsa, Milna church	43 19 23	16 27 16	δ_o in ‖. ⊞. ⚘. ↼.
Bratsa, Stjépanska church	43 20 36	16 39 10	‖ to main land ⊤, ⊥, & δ_o. ↼.
Bratsa, Bol cove	43 15 10	16 39 42	⊤ & ⊥. ∠ to M. San Vito, 2560.

Place.	Latitude.	Longitude.	Notanda.
DALMATIAN ISLANDS.	° ′ ″	° ′ ″	
Lesina, S. Giorgio tower ...	43 7 20	17 11 10	δ_0 in ‖s. ∠ M. Glavalikova, 1390.
LESINA, cathedral	43 9 10	16 26 29	⊞. Var. 14° 5′, Dip 62° 42′ (1819).
Lesina, M. S. Nicolo	43 8 30	16 30 0	2100. Mark in outer ‖.
Torcola, pt. Maslinizα......	43 5 28	16 42 35	⊥ on all sides. ✻. ⟵.
Bacili rocks, largest	43 4 57	16 34 30	δ_0, ⊤ & ⊥ around. ⟵.
LISSA, S. Francis steeple...	43 3 22	16 10 12	Var. 14° 0′, Dip 62° 51′, Int. 240 (1819). ⊞.
Lissa, Stupisca point	43 0 26	16 4 0	‖s δ_0. ∠ to M. Hulm, 1940.
Busi isle, station	42 58 10	16 1 27	790 feet. ⊤ & ⊥. ⚓′. ✻.
S. Andrea in Pelago	43 1 25	15 45 20	Ruins ∠ to summit, 1020. ⚓. ⟵.
Pomo rock, summit	43 5 35	15 27 25	1¼′ to W. by N. δ, otherws. ⊥. 100. ⌊₀. ⟵.
Pelagosa, M. Crocella	42 23 49	16 16 20	150 feet. δ_0 with !. ⟵.
Katsa, summit	42 45 56	16 31 12	830. ⊤ & ⊥. ✻. ⟵.
Katsiola, summit	42 44 50	16 42 30	✻. $\beta\beta_0$ on shoal W. by S. ⟵.
Lagosta, S. Rafael	42 45 39	16 49 0	⊤ & ⊥. ⊞. ∠ to M. S. Giorgio, 1390.
Lagostini rocks, Glovat ...	42 45 10	17 8 27	⊥ in approach, but ⌊₀. ⟵.
Curzola, Blatta mole	42 57 32	16 43 11	⊞. ⚓. Var. 15° 10′ in 1819.
Curzola, port Raciskie	42 58 0	17 0 56	⊞. ∠ to M. Dobravasca, 1880. ⚓.
Curzola, fort S. Biagio ...	42 57 30	17 7 18	⊞. Var. 14° 55′ (1819). M. Vipere, 3100.
MELEDA, port Palazzo ...	42 46 50	17 21 52	Ruined palace. Var. 15° 0′ (1819). ⊞. ⟵.
Meleda, port Suvra	42 44 50	17 35 0	⊥. ⚓. ∠ M. Grado, 1670 (Mezza Meleda).
Melada, M. Plagnak	42 42 10	17 42 58	1190. Mark for Val Sablonava.
S. Andrea, di Ragusa	42 38 10	17 57 20	Donzella chapel, 185. ⊤ & ⊥.
Marcano isle, station	42 34 37	18 10 51	⊤ & ⊥. δ_0 with !. ⟵.
ALBANIA.			
Cattaro, point d'Ostro	42 23 22	18 31 12	‖ of two 'Bocche,' ∠ 220, δ_0. ⚓₀.
Cattaro, point Morak	42 28 28	18 40 0	⊤. ⊥. ∠ M. Desviglie, 2541. ⟵.
Cattaro, city mole	42 25 25	18 45 47	⊞. ∠ M. Sella, 3240. Var. 14° 25′ (1818).
Cattaro, porto Rosa	42 25 22	18 31 40	⊞. ∠ M. Lustitsa, 1900 feet. ✻.
Monte Vetergnak	42 19 0	18 53 0	3960. Above Stagnevich convent.
Budua, M. S. Salvatore ...	42 17 45	18 49 25	Above the town, 1250. ⊞. ⚓₀.
BUDUA, S. Nicolo isle	42 15 45	18 50 47	Observation stone, 365 feet.
Antivari, old dogana	42 2 11	19 7 21	⚓. Var. 14° 57′ (1818), ∠ to 4500.
Dulcigno, la Cala............	41 53 58	19 11 49	⊞. In the road ⚓. ⚓.
Peregrino rock	41 51 47	19 15 40	⊥. Near ‖ river Bojana. ⌊. ⟵.
San Giovanni di Medua ...	41 48 20	19 29 0	⊞. ‖ Drino. ⚓. Var. 14° 0′ (1818).
Cape Rodoni, station	41 36 35	19 28 10	⊥ with !. ∠ 400 feet. ⚓.
Cape Pali, summit	41 23 5	19 24 14	In the bay ⚓′. ⚓. ⟵.
Durazzo, the mole	41 18 15	19 26 54	⚓′ with !. Var. 13° 50′ (1818).
Cape Laghi, tower	41 10 10	19 25 40	340. From base to Kavaja, $\beta\beta_0$.
Point Samana, centre	40 48 55	19 17 37	‖ river Tuberathi. δ at night. ✻. ⟵.
Mount Pegola	40 54 30	20 7 0	7760 feet. Peak over Berat.
Talao rocks, centre	40 38 0	19 18 30	‖ river Vojutza, or Poro. !. ⟵.

Place.	Latitude.	Longitude.	Notanda.
ALBANIA.	° ′ ″	° ′ ″	
AVLONA, dogana	40 27 15	19 26 20	⊞. Var. 14° 0′, Dip 60° 38′, Int. 231 (1818).
Avlona, fort Kanina	40 26 41	19 27 30	1360 feet. ⚓.
Sasseno isle, station	40 29 10	19 14 12	Summit 988. ⊤ & ⊥. δ_o. ⤙.
Cape Linguetta, extreme ...	40 25 37	19 15 0	⊥. ∠ to 2990 feet. ⚓.
Valle dell' Orso	40 19 12	19 20 35	⚓. ∠ 1546 feet, and 4300.
Monte Cica	40 14 36	19 35 0	6300, mark for Gremata cove.
Strada Bianca, extreme ...	40 8 45	19 37 30	⊤, but ⊥. ∠ M. Cicara, 5470, ∠⋋.
Port Palermo, fort	40 2 55	19 48 10	⊞. Var. 14° 30′ in 1818.
Santi Quaranta, dogana ...	39 53 46	20 0 14	δ_o by ⊤. ⚓′. ∠ to the town. ⚓.
Butrinto, guard-house	39 44 34	19 59 42	‖ of the Fishery and Lake. ⚓′. ⤙.
Gomenitsa, Prasudi rock...	39 30 13	20 9 7	⚓′ but ! the bank off ⊏ Kakuma. ⤙.
Gomenitsa, dogana	39 28 46	20 18 10	⊞. Magnetic Var. 14° 30′ in 1818.
Mourtso, Sybota rock	39 23 40	20 13 30	⊤ & ⊥. In the bay ⚓. ⚓. ⤙.
Parga, the citadel	39 16 29	20 23 29	⊡ and ⚓. Var. 13° 30′ in 1819.
Port Fanari, S. Giovanni	39 14 4	20 30 0	‖ of ancient Acheron and Cocytus.
Kastro-sikia, dogana	39 5 53	20 38 48	⚓,, ! the Ittisa reefs, β.
Prevesa, fort Pantakratera	38 56 17	20 45 14	‖ gulf of Arta. ⊤. Inside ⊞. $⚓_o$.
Vouvalos rock	38 58 25	20 55 0	Near centre of gulf. Var. 13° 10 (1820). ⤙.
LIVADIA.			
Vonitsa, the pier	38 54 26	20 53 14	⚓′ ⊡. ⚓. ∠⋋ inland hill, 1480.
Fort Giorgi	38 47 57	20 43 40	194. ⊞. ‖ of Santa Maura. ⤙.
Vurko bay, Mytika point...	38 40 10	20 56 44	⊞. ∠ M. Kandili 5000, & M. Bumisti 4950. ⋋∠.
Dragomestre, the skala ...	38 32 45	21 5 48	⊞. δ_o. M. Veloutzi, 2977. ⚓.
Port Plattea, inner point...	38 28 10	21 5 49	‖s ⊥. ⊞. ⚓. ⤙.
Port Skropha, the rock ...	38 18 55	21 9 0	⊡. ‖ Aspro-potamo, or Achelous.
Missolunghi, battery	38 21 50	21 26 30	Extensive lakes and marshes. ⤙.
Varasova point...............	38 20 15	21 39 0	⊥. ∠ to summit, 2830 feet.
M. Kako-skala, summit ...	38 21 20	21 42 40	3380. M. Koraka beyond, 6700.
Kastro Rûm-ili...............	38 19 28	21 47 0	‖ of Lepanto. ⊥. δ_o. $⚓_o$.
Lepanto, landing-place ...	38 23 15	21 50 10	Peaks of M. Rigani, ⋋∠, 4660 & 3950.
Galaxidi, building-yard ...	38 22 27	22 23 20	⊞. ∠ to 2500 ft. ⚓. M. Parnassus, 3970.
Dobrena, port Vathi	38 11 30	22 55 30	⚓. Peaks of Helicon, ⋋∠, 5200 & 5750.
IONIAN ISLANDS.			
Fano, west summit	39 50 20	19 19 50	1214 feet, ∠⋋ ⊥, except on the N.E.
Merlera, summit	39 53 28	19 31 57	⊤. ⚓. ⚓. but $⚓_o$. ⤙.
Samotraki, central hillock	39 45 44	19 28 5	! & ⊤ in approaching. δ. ⤙.
Diaplo isle, centre	39 45 37	19 32 40	⤚ and ! in the ‖s. δ. ⤙.
Tignosa rock, light	39 47 56	19 57 28	Middle of north Corfu ‖. ⌊.
Corfu, Santa Katerina ...	39 50 4	19 49 58	‖ of S. Spiridione. ⚓.
Corfu, M. Salvatore	39 43 30	19 49 20	S.W. peak 2590. Mark in the ‖.
VIDO I., fort Alexander ...	39 38 5	19 55 38	Var. 14° 33′; Dip 59° 10′; Int. 228, in 1818.

Place.	Latitude.	Longitude.	Notanda.
IONIAN ISLANDS.	° ′ ″	° ′ ″	
Corfu, Benitse villa.........	39 32 14	19 54 29	⚓, but open. ∠ Santa Dekka, 2000.
Corfu, citadel-flagstaff......	39 37 2	19 55 44	257 feet. ⊡. In the roads ⊞.
Corfu, Lefkimo point	39 27 20	20 4 27	⌐. Low, and δ at night.
Corfu, cape Bianco	39 20 50	20 6 50	Foot of the cliff. !. δ.
Corfu, Lagudia rock	39 24 19	19 54 50	⌐ with !. $\beta\beta_o$. ⌊. ⌐.
Corfu, port Ermones	39 35 30	19 45 32	⚓. ✻. ∠ M. S. Giorgio, 1326.
Corfu, Yliapades bay	39 40 0	19 41 10	Alipa point. ⊥. ⊞. ✻. ⌐.
Corfu, port Timona.........	39 42 20	19 36 30	⚓′. ∠ to Aphiona and M. Teodoro.
Paxo, Laka light	39 13 27	20 9 15	369. Off the point, δ. ⚘.
Paxo, port Gayo	39 11 40	20 12 19	Madonna light, 107. ⊞.
Anti-Paxo, point Novoro...	39 8 37	20 15 46	⊥, but !. Var. 13° 17′ (1820). ⌐.
LEUCADIA, Santa Maura...	38 50 19	20 42 58	⚓. In port Drepano, ⊞.
Leucadia, Sesola rock	38 41 50	20 32 30	δ_o in ‖. Opposite M. Nomali, 3700.
Leucadia, cape Dukato......	38 33 30	20 32 41	⊤, ⊥. ∠ to Sappho's Leap, 785.
Leucadia, Poro peak	38 38 14	20 43 0	1490 feet, ⌒. δ_o in ‖.
Leucadia, port Vliko	38 40 55	20 42 0	⊞. Var. 13° 40′ in 1820. ⚘.
Meganisi, Vathi mill	38 39 30	20 47 10	280. ⊞, but ! in approaching.
Arkudi, red cliff	38 33 16	20 42 30	⊤ & ⊥. δ_o around, but $⚓_o$. ⌐.
Atoko, summit	38 29 0	20 48 28	998 feet. ⊥. δ_o, but $⚓_o$. ⌐.
Kalamo, the port............	38 35 38	20 52 45	⊥. ⊡. ∠ to central summit, 2377.
Kastus, central height......	38 33 0	20 54 30	498. ⊥ all round. ⚘.
Dragonara, summit	38 29 0	21 1 40	⊥ in all the ‖s. ⌐.
Petala, summit	38 25 5	21 6 30	⊤. ⊥. ⊞. ⚘. Var. 13° 20′ (1820),
Vromona isle, summit	38 22 23	21 0 12	676 feet. ⊥ around, but $⚓_o$. δ_o. ⌐.
Oxia isle, summit............	38 18 54	21 7 10	1247. Off ‖ Achelous. Var. 13°32′ (1820).
Ithaca, point Marmaka ...	38 30 0	20 39 5	50. ✻. δ_o. M. Neritos, or Anoï, 2350.
ITHACA, port Vathi.........	38 22 5	20 42 47	Lazzaretto. Var. 13° 44′ (1820). ⊞.
Ithaca, point Joanni	38 19 28	20 46 20	⊤ & ⊥, ∠ M. Stefano, or Aito, 2170.
Cephalonia, port Viscardo	38 27 15	20 34 20	730. ⊥ in Daskalio ‖.
Cephalonia, Samos	38 14 30	20 38 0	⚓′. Ruins to the east. ⚘.
Cephalonia, point Atros ...	38 10 20	20 45 30	⊥, δ_o. ∠ to Napier's turret, 2656.
Cephalonia, cape Skala ...	38 2 55	20 46 38	δ. ∠ to M. Elato, or Nera, 5260. ⚘.
CEPHALONIA, port Argostoli	38 11 13	20 28 33	The station Hook-light, 35 feet. ⊞.
Cephalonia, castle S. Giorgio	38 8 20	20 33 52	998 feet. Var. 13° 24′, in 1820.
Cephalonia, Guardiana isle	38 8 13	20 25 30	Lighthouse 122 feet. ⊤. !.
Cephalonia, cape Aterra...	38 21 30	20 24 33	⊤ & ⊥. ⚓. δ_o. ∠ to 1655 feet.
Cephalonia, fort Asso	38 23 5	20 32 26	Height 410. ⊤ & ⊥.
Zante, cape Skinari.........	37 56 28	20 41 24	260. ⊥, δ_o. ✻. Var. 13° 21′ (1823).
Zante, mount Yeri	37 50 0	20 44 0	2274 feet. Mark in the ‖.
ZANTE, city mole-head ...	37 47 27	20 54 58	⊡. ⚓′. Var. 13° 12′; Dip 58° 50′ (1820).
Zante, mount Skopò.........	37 44 41	20 57 5	Convent 1489. Mark in the ‖.
Zante, Kieri bay	37 41 15	20 51 19	⚓. Station at the Pitch-wells.
Zante, port Vromi	37 49 0	20 39 8	⚓, but $⚓_o$. ✻. ∠ to M. Vrakiona, 2390.

Place.	Latitude.	Longitude.	Notanda.
IONIAN ISLANDS.	° ′ ″	° ′ ″	
Stamfani isle, convent......	37 15 12	21 1 27	Light 127. ⊥, δ_o with !. ⌒.
Prodano isle, summit	37 1 58	21 34 0	570. ⊥ in ‖, δ_o. Var. 13° 44′ (1820). ⌒.
Sphagia, summit	36 55 35	21 39 37	(*Sphacteria*). 480. ⊤. Ψ$_o$.
Sapienza, port Longona ...	36 43 42	21 41 30	⊥ & ⊥. ∠ to summit, 730 feet. ⌒.
Cabrera, Skhitsa cove	36 43 28	21 47 38	⊤. ⊥. δ_o in the ‖s, but ⚓$_o$. ⌒.
Venetiko, chapel	36 40 47	21 55 30	In ‖ to Cape Gallo. !. Ψ$_o$. ⌒.
Murmiki, or Ants	36 38 30	21 56 10	⊥. ⌊. Var. 12° 58′ (1820). ⌒.
Servi, point Franco	36 27 15	22 59 30	⊥. Summit, 950 feet. ⌒.
Cerigo, cape Spati...........	36 22 40	22 57 10	⊤ & ⊥. δ_o in Cervi ‖. Ψ$_o$.
Cerigo, port S. Nikolo......	36 13 14	23 5 9	The Castle. ⊡. ⚓′. M. S. Giorgio, 1000.
CERIGO, Kapsali dogana...	36 8 35	23 0 18	⊡. ⚓. Port of Tserigo, or Kythera.
Cerigo, cape Lindo	36 12 5	22 55 0	δ_o with !. ∠ to 1540.
Ovo isle, summit	36 5 5	23 0 10	⊤ & ⊥. ⌊. Height 550. ⚓$_o$. ⌒.
Koupho-nisi, north islet ...	36 7 17	23 6 12	⊤ & ⊥. δ_o. ⚓$_o$. Ψ$_o$. ⌒.
Porri islet, centre...........	35 58 10	23 15 0	⊥ & ⊥, but ⚓$_o$. δ_o with ! Ht. 410. Ψ. ⌒.
Nautilus rock	35 55 54	23 13 20	δ. Var. 12° 10′ in 1823. ⌊.
Cerigotto, Potamo fort ...	35 51 56	23 18 20	⊡. Var. 12° 20′; Dip 55° 24′ (1823).
Cerigotto, M. Turko-vouno	35 51 0	23 18 0	1100. Mrk. for ‖. M. Dometha, 980. Ψ$_o$.
Grabusa, Kastro	35 35 37	23 33 18	In Candia, but to complete the ‖s. ⌒.
MOREA.			
Athens, the Parthenon ...	37 58 10	23 43 50	In Greece, but to connect.
Corinth, dogana	37 55 46	22 53 52	⊥. τ. ⚓′. Site of Lechæum.
Corinth, citadel	37 53 20	22 52 58	Acro-Corinthus. 1850 feet.
Kamari, landing-place ...	38 5 45	22 34 54	⊥. τ. ∠ M. Koryphi, 2450. Ψ.
Vostitsa, beach fountain...	38 15 10	22 6 12	⊥. ⚓′. ∠ M. Pteri, 5900. ∠.
Kastro Morea, flagstaff ...	38 18 24	21 49 5	‖ of Lepanto. ⊥. δ_o.
Patras, mole-head	38 14 27	21 45 30	⚓′. δ_o by τ. Var. 13° 10′ in 1820.
Patras, castle flag...........	38 14 34	21 45 35	∠ M. Voïdeah, 6500. M. S. Nicolo (*Olonos*) 7100.
Cape Papa, ruined fort ...	38 12 40	21 25 5	‖ Lake Kalogria. Hill to S. 2980.
Cape Papa, sandy spit ...	38 13 3	21 24 10	τ. !. ∠ Mavro-vouna, 800. Ψ. ⌒.
Konoupoli, rock	38 5 29	21 22 0	⊥. ⚓. Ψ. M. Santa Meriotoko, 3420.
Klarentsa, old castle.........	37 56 24	21 9 35	τ. ⚓. Var. 13° 15′ (1820). ⌒.
Kastro Tornese...............	37 53 44	21 9 33	795. Commands plain of Elis.
Montague shoal	37 55 0	21 1 0	⊥ but δ. ‖s safe, but !.
Cape Katakolo...............	37 38 48	21 20 5	⊤. ⚓. Ψ. Var. 12° 35′ (1820).
Roufea river, skala	37 36 20	21 29 0	‖ Alpheius. Var. 12° 50′ (1820.) Ψ.
Arcadia, citadel	37 14 30	21 41 49	540. ∠ to 4000 feet. Ψ.
Navarin, ruins of Pylos ...	36 56 40	21 39 42	520. Boat ‖ to Sphaghia.
Navarin, Kulonisi rock ...	36 54 50	21 40 53	20 ft. Var. 13° 58′; Dip 57° 54′ (1820).
NAVARIN, castle flag	36 53 35	21 41 20	⊡. M. S. Nikolo, 1600, mark for ‖.
Modon, mole-tower	36 48 30	21 41 36	72. ⊡. ⚓. Var. 13° 27′ (1820).
Cape Gallo, the pitch	36 41 50	21 54 5	∠ 1390, Var. 12° 15′ (1823). ⌒.

Place.	Latitude.	Longitude.	Notanda.
MOREA.	° ′ ″	° ′ ″	
Koron, castle flagstaff......	36 46 35	21 59 12	220. ⚓′. Var. 11° 56′ (1820).
Mount Lykothimo, summit	36 54 0	21 53 0	2995 feet. Good sea-mark. Ψ.
Kalamata, dogana	37 0 25	22 8 36	‖ of the Nedon. ⊤. ⚓′.
Mount Makryno, S. Elias	36 58 0	22 22 0	Ancient Taygetus, nearly 8000 feet.
Cape Kephali, pitch.........	36 53 43	22 8 57	⊥. δ_o, but $⚓_o$. ∠ 1158, and 4250. ~.
Port Limeni, Vitylo skala	36 41 0	22 23 15	⚓. Ψ. Var. 11° 50′ (1823).
Port Djimova, Dyko point	36 38 57	22 22 53	⊤ & ⊥. ∠ 3450. M. Sanghia, 3990.
Cape Grosso, Kastro Orias	36 29 57	22 22 44	950 feet. ⊤ & ⊥. $⚓_o$. ~.
Cape Matapan, the pitch...	36 23 55	22 29 56	1020. ⊥. ∠ to Kaka-vouni, 4000. ~.
Cape Stavri, extreme	36 37 0	22 32 20	⊥. ⚓′ in Skutari bay. Ψ. ~.
Marathonisi, Crane islet ...	36 44 24	22 34 50	⚓′. Ancient port of Sparta. ∠ 510. ~.
Potamo Vasili, beach station	36 47 45	22 41 40	‖ Eurotas. Var. 12° 15′ in 1820. ~.
Kokino, beach tower.........	36 45 30	22 48 0	⚓′. ∠ to M. Kurkola, 3000.
Xyli bay, Rupina peak ...	36 40 35	22 49 27	⊤ & ⊥. ⚓. M. Kimatitsa, 1500.
Cape Malea, S. Angelo point	36 26 14	23 12 10	⊤ & ⊥. $Ψ_o$. $⚓_o$. ∠ to M. Krithyna, 2600. ~.
Monembasia, citadel	36 41 10	23 2 50	⊥. Magnetic var. 13° 10′ in 1820.
ARCHIPELAGO, BLACK SEA, AND LEVANT, TO ALEXANDRIA, *follow this Table.* (*See p.* 460.)			
EGYPT.			
Rosetta, fort Raschid	31 26 55	30 27 0	‖ of the Nile. ⊤ & !. †.
Nelson's isle, Burial bay ...	31 21 54	30 8 10	Low and δ. ⊤. ~. $Ψ_o$.
Al Bekur, castle tower ...	31 20 17	30 5 57	‖ to Nelson's isle, δ. Inside, ⚓.
Alexandria, Pharos castle	31 12 40	29 53 28	δ. β. In new port, ⚓,.
ALEXANDRIA, point Eunostos	31 11 31	29 51 58	New light, 180. ⊤. !. ⊞.
Alexandria, Pompey's pillar	31 10 45	29 53 47	99·5ft. Lat. by △. By an observation on the summit 31° 9′ 49″, but mercury tremulous.
Alexandria, Cleopatra's bath	31 9 55	29 52 5	Necropolis. Var. 11° 0′; Dip 57°45′ (1822).
Alexandria, Marabut isle	31 8 50	29 47 37	!, but inside ⚓′. Var. 11°15′(1822). $Ψ_o$. ~.
Abusir, Arab's tower	30 57 40	29 33 20	Approach δ. $\beta\beta_o$. †.
Al Amaïd, ruins	30 56 5	29 11 0	~. Broken grnd. Var. 10° 55′ (1822).
Iumeïmah point	31 2 7	28 47 0	⚓, but !. Var. 11° 20′ (1822).
Tanhoob, marabut	31 8 16	28 23 19	∠ to sandy spit. Var. 10° 52′ (1822).
Ras al Kanaïs..............	31 16 52	27 52 15	∠ 'Akabah-el-Soughaïr, 490. ~.
MARMARICA.			
Marsa Moháderah	31 12 7	27 39 30	⊞. Ψ. Var 11° 0′ in 1822. ~.
Ras al Harzeït, or Baratún	31 22 54	27 23 40	Baratún from Parætonium. †.
Marsa Labeït, Mhaddra rk.	31 23 47	27 16 33	⊞. Var. 11° 40′ (1822). Ψ. ~.
Ishailah rocks, east one......	31 31 18	26 39 44	58 feet. ⊤. ⌊. (*Scopuli Tyndarei*). ~.
Tifah rocks, centre	31 35 15	26 16 10	‖s ⊥, but !. Var. 12° 0′ (1822). ~.
Ras Haleimah, pitch	31 36 18	26 0 0	⊤ & ⊥. East point Gulfal Milhr.

Place.	Latitude. ° ′ ″	Longitude. ° ′ ″	Notanda.
MARMARICA.			
Port Sollum, cove...........	31 30 0	25 10 0	[⚓]. ⚓′. $\delta_{\circ}$. 'Akabah-el-Kibír, 840.
Ras al Milhr, or C. Lukkah	31 53 5	25 3 30	$\delta_{\circ}$ with !. (*Ardanaxes Prom.*) Ψ$_{\circ}$.
Tebruk, Saracenic gate ...	32 2 51	24 3 31	[⚓]. Var. 12° 40′; Dip 56° 58′; Int. 240 (1822).
Bombah, Seal isle...........	32 14 27	23 18 50	‖ Marsa Enharit Khuzitah. Ψ$_{\circ}$. ⌒.
BOMBAH, Bhurdah isle......	32 22 36	23 16 23	[⚓]. Var. 14° 55′; Dip 56° 24′ (1822). ⌒.
Bombah, Oum al Gharami	32 27 35	23 12 58	Ship rock. !. ⌒. Var. 14° 45′ (1822).
BARCAH.			
Ras et Tyn, beach...........	32 33 56	23 11 53	[⚓], but !. ❋. ⌒. Ψ$_{\circ}$.
Dernah, marabut...........	32 46 10	22 40 44	⚓′. Var. 13° 39′ in 1821. 570 feet.
Ras Halál, beach...........	32 55 29	22 10 2	⊥. ⚓. ❋, but near Ψ. ⌒.
Marsa Sousah, Cothon ...	32 54 51	21 56 27	⚓. Var. 14° 27′ (1821). Inland, Ψ.
Ras al Razat, or Ras Sem	32 56 56	21 38 0	⊤. ⊥. ∠ to Gureïnah, 1575. Ψ. ⚓$_{\circ}$.
Cyrene, near small theatre	32 49 38	21 49 5	2012 feet. Beechey's tent station.
Point Dolmeïtah	32 50 0	21 8 8	⊥. ∠ to lower Gureïnah range, 1050. ❋.
Dolmeïtah, the cothon......	32 43 7	20 54 52	⚓. In the offing ⚓. ❋. ⌒.
Taukrah, ruins...............	32 31 50	20 32 10	⊥. ∠ 950 feet. ❋ and Ψ.
Ben-Ghazi, castle..	32 6 51	20 2 40	[⚓]. Outside, ⚓. Var. 14° 50′ (1821). Ψ.
Ras Teyonas, sandy point	31 58 0	19 55 57	⊥. ⌐. Low, but $\delta_{\circ}$.
Marsa Kharkarah	38 28 30	19 58 25	⊥. ⚓. ∠ extensive sand-hills. ⌒.
Shahwan marabut	31 2 50	20 13 0	⚓, in off-shore winds. !.
Gharah isle	30 47 32	19 56 48	⊤. ⊥ with !. ⌊. Var. 15° 1′ (1822). ⌒.
Ishaifah rock	30 36 30	19 52 45	⊤. 50 feet. ⌊$_{\circ}$. ⌒.
Marsa Buraïgah, old fort	30 27 47	19 38 10	⚓ with !. ∠ high and white sand-hills.
Busheïfah islet	30 17 52	19 11 58	In summer, ⚓, with !. Ψ$_{\circ}$.
TRIPOLI.			
Muktahr, boundary pile ...	30 17 40	18 59 50	The frontier of Tripoli.
Ras al Omjah, or Licontah	30 55 58	17 58 0	Bluff rock to Ben-Jawad, ⌒. ⌊.
Abu-Saida, landing-place	31 0 15	17 39 0	⚓ with off-shore winds. ❋.
Marsa Zaphran, point ...	31 12 50	16 40 52	[⚓]. Approach !. Var. 16° 42′ (1821).
Jerid rocks	31 26 0	15 54 0	⌒. Outside ⚓ with land winds.
'Isá, Jebbah ruin...........	31 33 20	15 35 0	On a moderate ⌒. ⌒.
'Isá, beach station	31 35 25	15 37 56	Sand hummocks. Vr. 16° 50″ (1821). ⚓′.
Tawarkah village...........	32 1 30	15 13 50	Over the drowned lands.
Kharrah, or Aarār.........	32 9 58	15 25 5	Ψ. This single tree was here in 1770.
Sidi Buschaïfa, marabut...	32 21 26	15 16 45	⊥. ⌐. Var. 16° 40′ in 1816.
Misratah, mosque...........	32 22 30	15 9 0	Town is inside the point. Ψ.
Cape Misratah..............	32 25 15	15 10 24	⊤ & ⊥. $\delta_{\circ}$. Var. 16° 48′ (1816).
Marsa Zoraik, Youdi rock	32 26 50	14 48 25	[⚓]. ⌐. Village nearly 4′ east. ⌒.
Marsa Ziliten, marabut ...	32 30 5	14 32 58	⚓, ⊥, W. of Orir cliff. Vr. 16° 30′ (1817)
Marsa Ougrah, Tabia point	32 32 50	14 22 0	⚓. ‖ of the Khahan, or Kanafa. ∠ 350.
LEPTIS MAGNA, citadel ...	32 38 40	14 15 40	⚓′ in offing. Var. 16° 20′; Dip 55° 0′ (1817).

Place.	Latitude.	Longitude.	Notanda.	
TRIPOLI.	° ′ ″	° ′ ″		
Lebidah, mosque	32 39 30	14 11 40	Village, with olive groves. †.	
Marsa Ligatah..............	32 40 40	14 13 0	ɫ. In summer, ⚓′. ↼.	
Merkib tower.................	32 39 10	14 9 21	Commanding station.	Results of journeys into the interior.
Selineh, Roman ruin	32 37 56	14 10 0	Fortified eminence.	
Emsalatah, mosque	32 35 30	13 58 0	1250 feet.	
Medina Dugha, the gussar	32 32 0	13 40 0	Extensive ruins.	
Garatila hills, S.W. ex. ...	30 37 30	14 8 45	Peak in distant range.	
Ghirrza, high tomb	31 7 17	14 40 50	Var. 16° 10′ (1817).	
Wadi Zemzem, Roman well	31 35 0	14 38 0	85 feet deep.	
Benhoulat tower	31 28 10	14 18 15	Of the lower ages.	
Beniolid (*Beni Walid*) castle	31 45 38	14 12 10	Var. 16° 0′ (1817). †. 870 f.	
Wadi Denahr, Orfilli tents	31 52 10	14 3 50	Fertile spot.	
Mhaddra, spring...........	32 8 49	13 47 40	Var. 17° 5′ (1817). 990 f.	
Wadi Tinsiwah	32 15 0	13 43 0	Cultivated in patches.	
Weled-bu-Merian pass......	32 21 40	13 34 22	(Aúlád' ebn Maryun).	
Tarhouna, Melghra rocks	32 23 15	13 32 20	Summit, 920 to 1150 f.	
Rom. well, 2′ from Melghra	32 24 52	13 31 40	170 feet deep.	
Saiah grounds	32 28 37	13 16 40	Var. 16° 40′ (1817), †.	
Intzarrah (Nasárá?)	32 49 25	13 16 35	First wells.	
Wahryan Hills, castle......	32 7 50	13 2 10	Fine country. ∠ 3300.	
Ras Buswarah, pitch	32 44 40	14 1 30	‖ of stream Sidi Abdellata.	
Ras al Hamra, cove ruins	32 46 29	13 53 50	ɫ. Var. 16° 18′ (1817). ↼.	
Wad al Ramil, marabut...	32 47 30	13 35 0	‖ of the Ramil, or Sand river.	
Ras Tajourah, pitch.........	32 54 28	13 21 10	⊥. Town is within the cape. †.	
Tripoli, Consul's villa......	32 54 15	13 12 28	Var. 16° 35′; Dip 55° 14′; Int. 230 (1821). †.	
Tripoli, central rock.........	32 54 47	13 11 23	(Setif). Chronometer-sight station.	
TRIPOLI, Basha's castle ...	32 53 56	13 10 58	⊞, but enter !. In the road ⚓′.	
Tripoli vecchio, fort.........	32 49 50	12 26 26	⊡. ⊥. In off-shore winds ⚓.	
Zoarah, marabut	32 54 46	12 3 59	ɫ. Towards the S.E. †. ↼.	
Ras al Makhabez	33 7 20	11 42 35	‖ of ⊡. Outside ⚓′. Var. 16° 20′ (1822).	
Al Biban rock	33 15 57	11 22 20	‖ large lake ⊡. Outside ⚓′. ↼.	
Zera spit-rock	33 24 0	11 21 0	δ. $\beta\beta_o$. Inside ⚓′. ↼.	
Fort Zarsis, turret	33 29 50	11 10 10	Boundary of Tripoli in 1816.	
TUNIS.				
Ougla, Ras Mamorah	33 31 40	11 9 45	ϒ. ⊥. Off it, ⚓. ✱.	
Jerbah, Boukal castle	33 41 8	11 0 21	‖ of Al Kantarah; to fine ⊞ but for bars. ↼.	
JERBAH, Castle Zoug	33 52 54	10 52 58	Var. 15° 58′; Dip 55° 0′ (1822). ⚓′.	
Jerbah, fort Jelis	33 51 57	10 44 30	Outside the fishery flats, ⚓′. †.	
Kashr Natah, ruin	33 35 40	10 27 0	ϒ. ⚓. Hi. wat. $3^h\,10^m$. Rise & fall $5\frac{1}{2}$ft.	
Khabs, or Kabes, fort......	33 52 58	10 4 16	‖ Wad al Rif, or Khabs river. ⊥.	

Place.	Latitude.	Longitude.	Notanda.
TUNIS.	° ′ ″	° ′ ″	
Taflamah anchorage.........	34 4 45	9 57 0	⚓ of vessels for Khabs. Var. 16° 40′ (1822).
Sidi Midhil, landing-place	34 17 0	10 1 58	⊞ inside Zurkenis. ∠ to Jebel Thelj.
Sfakus, mole-head	34 43 56	10 39 50	⊞. Var. 17° 10′ in 1822. ↟. ☨.
Sidi Masour tower	34 48 21	10 47 0	60 ft. S. point of Karkenah ‖. !.
Ras Kadija, or Cape Vada	35 9 58	11 10 0	Tower 54 ft. N. pnt. of Karkenah ‖. !.
Karkenah Is. Dazak tower	34 48 10	11 15 30	40 feet. Low and δ. γ. ☨. ⌒.
Karkenah Is. Gherba tower	34 38 0	10 54 16	40 feet. γ. ☨. ⌒. Var. 17° 0′ (1822).
C. Africa, Mehdiyah castle	35 30 26	11 6 51	⊥. ⚓. Var. 16° 55′ in 1822.
Leptis Parva, ruins of......	35 39 43	10 51 40	∠ to town of Lamta. ☨.
Monastir, fort Akdir	35 45 23	10 48 53	⊞. Var. 16° 38′ in 1822.
Kuryah isles, outermost ...	35 47 20	11 3 30	Coniglieri. γ. ⌒. Var. 17° 10′ (1822).
Susah, the castle flag	35 50 0	10 35 56	⚓ off the moles, with !. ☨.
Herklah, minaret............	35 59 10	10 30 0	⊥. On an eminence. Var. 17° 0′ (1822).
Jebel Zawan, or Zaghwan	36 23 0	10 5 0	∠△ 3900 feet. A good sea-mark.
Hammamat, mosque	36 23 27	10 38 15	⚓′ by γ. Var 17° 10′ (1822).
Cape Mahmur, pitch	36 26 58	10 58 58	⊥. ∠ to Nabal and Mahmur.
Ras Mustafa, Kalybia fort	36 49 57	11 8 30	⊥. At point, β. Var. 16° 44′ (1822).
Cape Bon, summit tower...	37 4 50	11 3 36	1176 feet. ⊤ & ⊥, but $⚓_o$. ⌒.
Sidi Daoud, marabut	37 0 20	10 55 10	⊞ in Hamar cove. Var. 16° 50′ (1816).
Zembra, or Zowámir	37 6 37	10 48 29	Landing place. ∠ to 1560 ft. ⊥. δ_o.
Tail of Keith's reef	37 50 0	11 8 0	Adventure ⚓, 20 fms. Blown off at night (1822). Lat. by run from Maretimo; Lon. assumed from Captain Durban.
Skerki shoals..................	37 44 53	10 45 15	My ⚓age 1816, 3′ W.S.W. of S. Bank.
Adventure Bank	37 32 0	11 44 40	My ⚓age in 13 fms., 1816. Brokn. bttm.
Adventure Bank	37 11 30	12 7 0	My ⚓age in 8 fms., 1816.
Tunis bay, Cape Zafran ...	36 52 0	10 36 10	⊥. By natives, Ras al Durdas. 1075.
Mount Hammam Lynf ...	36 39 10	10 20 0	1217. ∠ to Jebel Irsas (Piombo), 1720 ft.
GOLETTA, Halk-al-wad fort	36 48 25	10 16 40	‖. ⊞. ⚓′. Var. 17° 40′, Dip 56° 48′, 1822.
Tunis, kobbeh, Kabeira ...	36 45 50	10 39 0	Marabut on ∠△ nr. lndng. place. ⌒.
Tunis, Cape Carthage	36 52 3	10 19 29	Light 406. ∠ to Sidi Buseïd (*Byrsa*). ⚓.
Port Farina, arsenal	37 10 10	10 8 10	‖ of the Majerdah, & Bushatta. ⚓.
Cape Farina, marabut......	37 10 43	10 14 25	δ_o in ‖ to isle, with ! $↟_o$.
Kamla, Piana, or Watiah I.	37 10 48	10 17 56	⊥. Var. 17° 20′ in 1822. ⌒.
Cape Zebib, the pitch	37 16 20	10 0 45	‖ to Kelb, δ_o. ∠ M. Shapta, 2000.
Kelb, Cane, or Dog rocks...	37 21 12	10 4 15	Brkn. grnd. but δ_o. $↟_o$. Var. 18° 10′ (1822). ⌒.
Bizertah, or Benzert castle	37 16 36	9 49 20	‖ two lakes. ⚓. Var. 18° 0′ (1822).
Bizertah, Jebel Ishkil	37 7 12	9 36 0	1750 feet. Mark for inner lake.
Ras Abiad, or Cape Bianco	37 19 32	9 47 2	Tower. ∠ to 950 feet. $↟_o$.
Akwat-kebir (Fratelli)......	37 18 14	9 22 24	279. ⊤ & ⊥, except N.E. $!_o$.
Galita, the Gallo rock	37 33 7	8 57 38	⊤ & ⊥. $!_o$. δ_o with !. ⌒.
Galita, Sugar-loaf peak ...	37 30 56	8 54 17	△ 1038. ⚓′. M. Guardia, 1173. Vr. 18° 9′ (1822).
Galitona, centre	37 29 45	8 52 54	484. Agúglia, 379. ⊤ & ⊥, $⚓_o$. δ_o in ‖s. ⌒.
Sorelle rocks, nearly awash	37 24 0	8 36 30	⊥, therefore very δ. !.

Place.	Latitude.	Longitude.	Notanda.
TUNIS.	° ′ ″	° ′ ″	
Ras al Munshihar	37 13 54	9 10 0	(Cape Serrat). ⊤&⊥. Var. 17° 58′ (1822).
Cape Negro, summit........	37 5 0	8 55 20	Coast ⊥. ✱. ⚓$_0$.
Tabarkah, castle	36 56 25	8 42 19	▣. ⚓′. Var. 17° 40′ (1822). 375 feet.
Alkaláh, or La Cala	36 51 57	8 24 43	▣. Boundary of Tunis & Algeria. ⚓$_0$.
ALGERIA.			
Ras al Bufahal, or Rosa...	36 55 15	8 13 0	⊤ & ⊥, except at base. ✱. ⤳.
BONAH, citadel..............	36 54 2	7 47 53	398, ∠ to 2600. ⚓′. Var. 18° 0′ (1813).
Ras al Hamrah (Mavera)	36 57 58	7 49 30	⊤ & ⊥. Guardia new lighths. 466. ⤳.
Tukush islet, centre	37 5 56	7 22 30	δ_0, with !. ⚓$_0$. ⤳.
Ras Hadid (Cape Ferro)...	37 5 10	7 10 57	The islet. ⊤ & ⊥. Inside ⚓, with !.
Cape Filfillah (Pepper) ...	36 54 0	7 6 0	Summit, 2500 feet. ✱. Base ⊥.
Storah, old pier...............	36 54 53	6 53 5	⚓′. Var. 17° 50′ in 1813.
Ramadi isle, summit........	36 58 45	6 43 15	⊤ & ⊥. Height 200 feet. ⤳.
Kolah, Hussein chapel......	37 0 59	6 34 55	500 feet. Var. 18° 0′ in 1813. ⚓.
Ras al Ferjan (Bujaroni)	37 6 58	6 28 6	Nrthmst. of the 7 Capes. ⊥. ⚓$_0$. ∠ to 2800. ⚓.
Al Imam rock	37 0 38	6 15 42	In bay ⚓ off ‖ wad al Kabir. ∠ to 3600. ⚓.
Jijeli, minaret	36 50 0	5 46 50	▣. Var. 18° 37′ in 1813. ⤳.
Ras Jemel (Cape Cavallo)	36 47 0	5 36 30	⊤ & ⊥. Summit 1500 feet, ⌒.
Mount Babora, summit ...	36 34 0	5 30 0	6300. Mark for Mansuryah cove. ⚓.
Bujeyah, castle..............	36 45 45	5 8 22	480 feet. Var. 18° 20′ (1813).
Cape Carbon................	36 46 43	5 8 30	630 ⌒. ∠ to 4000. ⊤ & ⊥. ⚓$_0$.
Pisan isle, the well	36 49 31	5 2 15	⊥. δ_0 in ‖. ⤳. Var. 18° 20′ (1823).
Cape Sigli, the pitch........	36 53 0	4 48 0	⊤ & ⊥. δ_0. Var. 18° 17′ in 1823.
Mount Jujerah..............	36 25 0	4 11 0	7000, summit of inland range.
Mars-el-Fahm, cove	36 53 15	4 25 10	⚓′ in the road. ⊤. ⚓.
Cape Tedlés, the pitch	36 54 12	4 11 15	⊥. ∠ to 3500 feet. ⤳.
Dellys, landing-place	36 55 10	3 56 25	1300. Summer, ⚓′. Var. 18° 25′ (1823).
Cape Bengut, summit	36 56 0	3 54 0	⊥. M. Bubarak 980 feet, ∠ to 2000.
Ras Temedfus (Matifuz) ...	36 48 58	3 13 0	Octagon fort. $\beta\beta_0$ at pitch.
ALGIERS, mole-light	36 47 31	3 4 18	▣. ⚓′. Var. 19° 10′ in 1813.
Algiers, Emperor's castle...	36 46 50	3 2 57	390 feet. Mark in taking ⚓age.
Ras Akkonada, C. Caxine	36 49 36	3 0 17	⊥. ⚓$_0$. ∠ to Mount Abu-Zariah.
Ras al Hamous, tarf Batal	36 37 50	2 23 54	Islet at base. ⊥. ∠ to about 3000.
Zerzahal, or Chershel	36 36 31	2 10 53	▣. In land winds ⚓′.
Ras Nakkus (Cape Tenez)	36 32 40	1 21 48	⊥. ∠ to 3500 ft. Var. 18° 37′ (1813).
Dnis, or Tenez, minaret ...	36 30 0	1 19 15	A mark for the watering-place.
Jezeïr al Hamman (Palomas)	36 25 58	0 55 36	80 feet. In ‖, δ_0. ⤳.
Ras Jebel Iddis (Cape Ivi)	36 5 45	0 12 40	⊥. ∠ to 1000 feet.
Mosta-ghanem, centre	35 57 0	0 6 20	Coast-line ⊤ & ⊥. δ_0.
Marsa Arzaw, fort	35 51 36	0 16 10w	⚓′. New lighthouse, 62 feet.
Ras Mishat (Cape Ferrat)	35 54 50	0 21 40w	⊥. ⚓$_0$. Summit, upwards of 2000.
Aguglia rock................	35 54 0	0 26 25w	180. ⊥. Pharaoh's finger, by Moors.

Place.	Latitude.	Longitude.	Notanda.
ALGERIA.	° ′ ″	° ′ ″	
Waharan, or Oran	35 40 49	0 39 18w	Hill fort. ⦣ to 1500.
Marsa Kebir, light	35 44 17	0 41 58w	118, ⦣ to 1500. ⊞.
Ras Harshfah (*C. Falcon*)	35 46 10	0 48 5w	⊥ but !. ⦣ to 1800. Station, E. pitch.
Habiba isles, largest	35 43 15	1 7 56w	280 feet. Var. 20° 30′ in 1813. ⤙.
Ras Ishgún (*Cape Fegalo*)	35 34 22	1 11 0w	⊤ & ⊥, but islet rocks at base.
Karakal islet	35 18 30	1 29 52w	190 feet. ‖ for small craft. ⤙.
Mount Noé	35 8 0	1 42 0w	Nrly. 3000. Mark for capes Noé & Hone.
Cape Malonia, pitch.........	35 7 50	2 8 40w	⊥. ⦣ to a flank of Atlas. ⦣. 3500.
River Mahala, or Mulwia	35 6 55	2 14 30w	‖. Boundary Algiers & Morocco. ⚓.
MOROCCO.			
Cape Agua, pitch............	35 9 10	2 24 25w	⊥. ⚓. ⦣ towards the Atlas range.
Zafrin isles, centre one ...	35 10 50	2 26 0w	130 feet. West isle, 440. ⚓. ⚓ₒ. ⤙.
Mount Partz, summit	35 2 30	2 36 0w	2600. Mark for Restinga, and Zafrin.
Melila, the baradero.........	35 20 55	2 54 58w	⊞. ⚓. Var. 20° 49′ (1813). ⚓ₒ.
Ras-ud-Deir (*Tres Forcas*)	35 28 10	2 57 16w	⊤ & ⊥, but ⚓ₒ. ⦣. ⤙.
Al Boran isle, summit......	35 57 48	3 0 58w	68 feet. Var 20° 30′ in 1813. ⤙.
Khozamah, or Al Buzema	35 16 45	3 47 36w	Fortified rock. ‖ of wad Nekkor.
Peñon de Velez, flagstaff...	35 12 20	4 15 39w	Fortified rock. ⊤ & ⊥. ⚓ₒ. ⤙.
Point Pescador, tower......	35 16 41	4 42 0w	⊥, ⊤. Town beyond. ⚓.
Tetuan, dogana tower	35 37 7	5 18 38w	‖ river Tetuan. In W. winds, ⚓′.
Ceuta, Acho flagstaff	35 53 58	5 17 35w	⊥, except N.W. point. ⚓.
Peregil isle, summit.........	35 54 48	5 25 23w	Under Ape's hill. ⊥ but ⚓,. ⌊ₒ. ⤙.
Ape's Hill, western summit	35 53 30	5 24 50w	Jebel Moussa, or Sierra Bullones, 2200.
Cape al Kazar, pitch	35 52 50	5 34 0w	⊥, except close under. ⦣.
TANGIER, citadel minaret	35 47 25	5 48 26w	⊞. Var. 21° 50′ in 1810.
Cape Spartel, the gap	35 47 39	5 55 30w	(Ras Shakkah). ⊤ & ⊥. 150 ft. ⦣. ⚓.
Djeremias bay, landing-pl.	35 43 0	5 56 18w	δ_o. ⚓′ in East winds. ⚓.
Arzila, minaret..............	35 28 57	6 0 45w	Near ‖ of Ayasha. Var. 22° 5′ (1810).
Jebel Habib	35 28 0	5 43 0w	2700. Mark for the coast. ⚓.
El Araïsh, citadel...........	35 12 45	6 8 32w	‖ wad al Khos (Lucos). ⚓.

Names of Places.	Latitudes North.	Longitudes East.
NORTHERN COAST OF CANDIA:	° ′ ″	° ′ ″
Spada, Cape, the summit	35 40 30	23 44 10
St. Theodore, island, North point	35 31 20	23 55 10
Canea, town, the Castle	35 28 40	24 00 30
Meleca, cape, North point	35 35 5	24 08 28
Drapano, cape, S.E. of Suda Bay	35 27 10	24 17 00
Retimo, town, centre	35 22 17	24 28 17
Retimo, cape	35 25 52	24 41 15
Santa Croce, cape, North point	35 25 54	24 58 36
Candia, principal minaret	35 21 00	25 08 05
Standia islet, North summit	35 27 20	25 14 20
Paximada, islet, the summit	35 26 40	25 19 02
Ovo rock, summit	35 37 50	25 35 00
Maglia point	35 19 15	25 35 50
San Giovanni cape, summit	35 19 10	25 46 50
Spinalonga port, fort	35 17 00	25 44 45
Sitia, cape	35 14 20	26 01 40
Yanis islands, summit of Cosua, northern islet	35 22 00	26 10 05
EAST END OF CANDIA:		
Sidera cape, summit	35 17 40	26 18 45
Lassa islet, S.E. point	35 15 25	26 21 40
Paleo-Castro, ruins	35 10 10	26 15 25
Salomone cape, East point	35 09 13	26 19 25
Yala cape	35 03 00	26 15 30
SOUTH COAST OF CANDIA:		
Christiana islets (Koupho-nisi), the southernmost	34 53 05	26 07 45
Calderoni islets (Gœduro-nisi), N.E. pt. of westernmst.	34 52 35	25 43 20
Matala, cape	34 55 05	24 45 10
Paximadi islets, summit of the largest	34 59 40	24 34 55
St. John, cape (Krio)	35 15 35	23 30 35
Western mountain of Candia	35 22 48	24 08 20
Mount Ida	35 13 19	24 47 01
Eastern mountain of Candia	35 06 46	25 30 38
Gozzo, great, of Candia, West point	34 52 00	24 02 05
Gozzo, small, of Candia, middle	34 56 15	23 59 30
St. John, cape, Candia	35 27 45	23 32 40
Sordi, middle of isle, Candia	35 34 20	23 27 08
Buso cape, Candia	35 36 38	23 35 35
Garabusa islet, Candia	35 35 00	23 33 40
ISLANDS:		
Caravi rock, summit	36 46 25	23 35 35
Falconera, summit of the island	36 50 40	23 53 05
Ananas rocks, the highest	36 32 45	24 09 15

Names of Places.	Latitudes North.	Longitudes East.
ISLANDS:	° ′ ″	° ′ ″
Milo, summit of Mount St. Elia	36 40 27	24 23 19
Paximado islet, S.W. of Milo	36 37 40	24 19 10
Anti-Milo, summit	36 47 42	24 14 38
Pettini rocks, few feet above water, S.E. of Milo ...	36 38 00	24 35 35
Argentiera	36 49 20	24 33 28
St. Istada, island, anchorage of Argentiera	36 46 16	24 36 00
Polino, highest point of the island	36 46 10	24 39 02
Siphanto, highest point of the island	36 58 04	24 42 40
Policandro, highest point of the island	36 37 03	24 55 10
Miconi, summit of the highest western mountain ...	37 29 15	25 21 27
Anti-Paro, highest point of the islet	36 59 39	25 03 32
Strongilo, highest point of the islet	36 56 40	24 58 20
Paros, summit of Mount St. Elia	37 02 46	25 11 23
Naxia, summit of Mount Jupiter	37 01 50	25 31 09
Raclia, summit of islet	36 49 28	25 28 03
Karo, summit of islet	36 53 29	25 39 56
Amorgo-Poulo islet, summit	36 36 54	25 42 39
Nios, highest summit	36 42 44	25 20 54
Sikyno, highest summit	36 39 51	25 06 53
Santorin, highest summit	36 20 52	25 28 26
Christiani, summit of the highest islet	36 14 40	25 12 50
Anaphi, summit of the island	36 22 21	25 47 14
Anaphi-Poulo, summit of the largest islet	36 16 00	25 51 00
Ponticusa, summit	36 31 48	26 17 08
Fidulce island, South point	36 31 25	26 09 45
Stamphalia, summit of Monte Veglia	36 32 12	26 19 40
Miconi, summit of Mount St. Elias	37 29 06	25 21 18
Tino, summit	37 35 01	25 14 21
Andros, summit	37 50 08	24 50 27
Syra, summit	37 28 56	24 55 33
Jura, summit—(better *Ghiour*)	37 36 36	24 43 18
Zea, summit of Mount St. Elias	37 37 18	24 21 45
Piperi, summit of the rock	37 18 15	24 31 53
Hydra, summit	37 19 58	23 28 44
Serpho-Poulo, summit	37 15 17	24 36 00
S. Giorgio d'Arbora, summit	37 28 14	23 55 47
Egina, summit	37 42 05	23 29 53
GREECE:		
Athens, monument of Philopappus	37 57 57	23 43 24
Pieræus, tomb of Themistocles	37 55 51	23 37 44
Corinth, castle	37 53 37	22 52 10
Colonna, cape—temple of Sunium	37 39 12	24 01 39

Names of Places.	Latitudes North.	Longitudes East.
GREECE:	° ′ ″	° ′ ″
Provençale, summit of islet	37 39 06	23 57 07
Raphti, port, summit of islet	37 52 51	24 02 31
Marathon, cape	38 10 47	24 05 09
Mandri, port, sugar-loaf	37 44 23	24 03 31
Makronisi, or Long island, northern summit	37 45 0	24 08 30
Negropont, St. Ella d'Oro, highest summit	38 03 36	24 28 22
Kaloyeri, centre of the rock	38 09 59	25 18 05
Skyro, San Giorgio, summit of Mount Cochilo	38 49 46	24 37 10
Negropont, Mount Delphi	38 37 43	23 51 23
Jura-nisi, or Devil's island	39 24 0	24 11 18
Skopelo, Mount Delphos	39 08 25	23 41 55
Trikeri, mount, gulf of Volo	39 06 58	23 10 24
Fetio, port, tower entering gulf of Volo	39 01 59	23 00 54
Trikeri, old, East side of gulf of Volo	39 09 42	23 06 19
Halata islet, East side of gulf of Volo	39 10 11	23 13 53
Pelion, mount	39 26 17	23 03 00
Ossa, mount	39 47 53	22 42 09
Olympus, mount	40 04 32	22 21 58
TURKEY IN EUROPE:		
Drepano, cape, or Trapano, summit	39 56 53	23 57 22
Mulliani, summit of the islet in the gulf of Mte. Santo	40 19 59	23 54 59
Limpiada, summit, in Contessa gulf	40 37 03	23 48 27
Strati or Strachi, St., summit of islet	39 31 00	25 01 36
Lemnos, summit of Mount Therma	39 53 42	25 08 37
Imbros, summit of the island	40 10 36	25 51 25
Samothrace, summit	40 26 57	25 35 59
Tarapia, French palace, N.E. terrace	41 08 31	29 02 48
Constantinople, palace of France at Pera	41 01 44	28 59 02
Constantinople, dome of St. Sophia	41 00 12	28 59 07
WESTERN SHORES OF THE BLACK SEA:		
Pharos of Europe	41 14 10	29 07 05
Kilios, castle	41 15 30	29 02 55
Kara-Bourou, cape	41 19 20	28 40 25
Kaliondjik	41 25 40	28 27 55
Malhatrah, cape	41 29 55	28 17 50
Taliangiéri	41 33 05	28 12 25
Mediah, town	41 36 45	28 06 20
TURKEY IN EUROPE, *continued*:		
Serves, cape	41 39 00	28 07 00
Sandal Limani, point	41 45 30	28 01 05
Ayo-Paoli, river	41 48 45	27 58 40
Tersanah, village	41 52 35	27 58 55

Names of Places.	Latitudes North.	Longitudes East.
TURKEY IN EUROPE, *continued*:	° ′ ″	° ′ ″
Kouri, cape, East of Inada anchorage	41 52 43	28 03 02
Resveh, cape	41 56 40	28 02 55
Babiah, mount	42 04 40	27 50 50
Ahteboli, town	42 04 30	27 59 20
Vassicos, village	42 07 40	27 51 50
Zaïtan, cape	42 17 55	27 47 40
Bagral-Altoun, cape	42 24 45	27 44 50
St. John's isle, entrance of Bourghaz gulf	42 25 54	27 41 27
Bourghaz town, minaret	42 29 20	27 28 05
Ahiouli island	42 32 10	27 38 45
Mesembria, town	42 39 15	27 44 25
Emenéh, cape	42 41 40	27 53 35
Djoski, village	42 49 55	27 53 20
Kara-Bouroun, cape	42 55 00	27 54 40
Ak-Bouroun, cape	42 58 20	27 54 25
Ilidjah-Varni, cape	43 05 20	27 55 50
Galata, cape	43 10 10	27 58 20
Varna, great Eastern tower	43 12 15	27 56 15
Soughanlik, cape, islet	43 13 25	28 02 05
Batouvah, cape	43 19 15	28 05 05
Baldjik, town and port	43 23 15	28 10 10
Kavarna, town and port	43 24 00	28 22 05
Calagriah cape, ruins	43 21 25	28 27 10
Chabler-Saghi, cape and old pharos	43 32 10	28 35 20
Khas-Elias, mouth of Danube	44 52 45	29 36 30
Soulinéh, mouth of Danube, light	45 10 15	29 40 55
RUSSIA:		
Isle of Serpents, summit	45 15 00	30 11 00
Dniester, N.W. mouth	46 10 00	30 33 35
Fontan, cape and light	46 22 20	30 43 40
Fountain point	46 26 50	30 44 30
Odessa, lazaret	46 28 54	30 43 27
Odessa, highest dome	46 29 10	30 41 45
Odessa, theatre	46 29 15	30 42 20
Odessa, Custom-house	46 29 50	30 41 25
Odessa, N.E. point of the roadstead	46 33 25	30 47 40
Bérézan islet, South bastion	46 35 34	31 22 47
Adji Hassan, cape	46 35 55	31 19 20
Bérézan, mouth of the river	46 37 40	31 23 30
Kinbourn, N.W. sandy point	46 35 00	31 26 55
Kinbourn, barracks	46 33 20	31 29 55
Balise, North point of Tendra island	46 21 40	31 29 25

Names of Places.	Latitudes North.	Longitudes East.
RUSSIA:	° ′ ″	° ′ ″
Fort on low point South of Otchakof	46 35 50	31 31 00
Otchakof, the dome	46 36 25	31 30 55
CRIMEA:		
Karamnoune, cape	45 25 35	32 31 05
Tarkhan, cape and light	45 21 35	32 31 20
Kazelof, low S.W. point, four miles distant	45 06 55	33 14 10
Kazelof, principal dome	45 09 05	33 19 45
Krasnoiars, village	45 00 45	33 37 25
Zamrouk, village	44 54 45	33 36 40
Alma, river	44 50 50	33 32 30
Loukoul, cape	44 50 45	33 32 15
Katcha, cape	44 46 15	33 29 40
Belbek, river	44 39 50	33 32 05
Outchquikal, point	44 37 55	33 29 25
Sevastopol, highest house of the Lazaret	44 35 58	33 29 11
Sevastopol, dome of hospital	44 34 55	33 31 20
Sevastopol, steeple of St. Nicholas	44 35 25	33 31 35
Chersonese lighthouse	44 34 25	33 20 50
Fiolente, cape	44 29 15	33 27 35
St. George, village	44 29 30	33 29 05
Balaklava, entrance of the port	44 28 55	33 34 40
Aïa, summit of the cape	44 24 40	33 39 10
Saritche, cape	44 22 00	33 44 20
Kerkines, cape	44 22 05	33 56 35
Aïtodor, cape	44 23 30	34 05 10
Nikita, point	44 29 25	34 13 45
Tchandirdag, mount, S.W. point of the table	44 44 40	34 18 20
Lioudag, cape, South point	44 32 10	34 19 50
Lioudag, summit	44 33 05	34 11 20
Alouchti, town	44 41 00	34 26 00
Limani, cape	44 48 05	34 56 25
Soudak, village	44 50 10	34 59 35
Alcessan, cape	44 49 45	35 00 10
Méganome, cape	44 46 40	35 06 40
Karadof, cape	44 53 10	35 15 10
Kiatlama cape, the rock	44 54 35	35 23 05
Caffa, East point of Lazaret	45 01 24	35 24 47
Caffa, town-house	45 01 37	35 23 33
Theodosia, cape	45 00 43	35 26 15
Tchaouda, cape	44 59 54	35 52 30
Jeltchankaléh, rock	45 01 31	36 16 24
Karak, cape	45 02 25	36 18 04

Names of Places.	Latitudes North.	Longitudes East.
CRIMEA:	° ′ ″	° ′ ″
Takli, cape	45 04 30	36 27 36
Ak-Bouroun, tumulus on the point	45 19 05	36 29 45
Kertch, town	45 21 29	36 28 54
Jénikaléh, town	45 21 12	36 36 16
KOUBAN:		
Taman, town	45 13 40	36 43 50
Taman, cape, islet off	45 09 10	36 37 35
Kiheli, cape	45 06 52	36 43 55
Kouban, river, low point	45 05 30	36 54 40
COAST OF THE ABASES:		
Anapa, West part of the town	44 54 21	37 16 04
Isussup, cape, peninsula 10 miles S.S.E. of Anapa	44 45 15	37 22 40
Sougoujak, S.W. entrance of the bay	44 39 00	37 46 40
Guelinjik, mid-entrance to the port	44 31 00	38 07 20
Pchiat, East point of entrance	44 22 20	38 19 35
Voulan, mid-entrance	44 20 25	38 31 00
Kodos, West point	44 16 55	38 42 20
Soubachi, river	44 09 25	38 59 45
Vardan, N.W. point of entrance	44 06 15	39 02 05
Peak of the Caucasus	43 56 30	39 51 35
Mamaï, river	43 53 25	39 18 45
Soutchali, N.W. point	43 42 35	39 33 00
Zengui, cape	43 30 40	39 44 40
Ardler, cape	43 22 55	39 56 20
Kentchili, river	43 20 35	40 10 20
Pitsiounta, low point, 2 miles S.W. of	43 08 20	40 19 40
Pitsiounta, end of the gulf	43 09 45	40 21 50
Soukoum-Kaleh, N.E. bastion	42 59 20	41 00 13
Soukoum village, or ruins of Dandar	42 58 10	41 02 35
MINGRELIA:		
Kodor, mouth of the river	42 50 34	41 04 20
Iskouria, cape	42 47 00	41 10 00
Jenichéri, village	42 43 50	41 29 30
Isiret, cape and river	42 27 00	41 30 24
Ilori, fort	42 24 20	41 32 20
Koulé, redoubt	42 14 12	41 38 35
COAST OF THE LAZES:		
Phase, new fort on the island	42 07 30	41 40 00
Tckehétil, village and redoubt	41 54 40	41 45 40
Tchourouk, town	41 49 15	41 46 20
Sikindsi, cape	41 46 10	41 43 40
Méandjour, tower	41 43 40	41 42 35

Names of Places.	Latitudes North.	Longitudes East.
ANATOLIA:	° ′ ″	° ′ ″
Batoum, town	41 38 40	41 39 00
Batoum, tower on the cape	41 40 00	41 35 30
Gouniéh, town	41 36 00	41 33 45
Gouniéh, cape	41 35 15	41 32 00
Makria, town	41 30 15	41 31 15
Khoppa, village	41 24 50	41 24 20
Arkava, town	41 23 00	41 16 50
Vitzé, village	41 17 25	41 06 45
Boulep, village	41 12 25	40 55 35
Laros, fort	41 10 30	40 48 50
Kemer, cape	41 09 20	40 45 20
Mapavreh, village	41 06 20	40 46 15
Rizéh, town	41 02 25	40 30 15
Foudji, cape	41 02 30	40 21 50
Mahané, village	40 56 10	40 11 50
Komourkiando, village	40 55 45	40 08 30
Héraclia, cape	40 58 05	40 00 45
Falcos, village	40 57 00	39 53 50
Trebizonde, French Consulate, East of the town	41 01 00	39 44 57
Platana, village	41 02 05	39 33 15
Akché-Kaléh, village	41 05 30	39 28 42
Ioroz, cape	41 06 55	39 23 45
Skiéfié, town	41 04 30	39 15 20
Koureléh, cape	41 05 45	39 09 35
Héléhou, village	41 03 30	39 04 15
Kara-Bouroun, cape	41 03 40	38 55 50
Tirboli, town	41 01 00	38 49 15
Espey, village	40 57 50	38 42 50
Zéphira, cape	40 59 30	38 36 45
Kessap, village	40 56 30	38 32 20
Arhentias, island	40 57 35	38 25 50
Kérésoun, town	40 57 10	38 23 45
Aio-Vassil, village and cape	40 58 35	38 19 15
Aio-Vassili, cape	41 00 40	38 07 50
Postipey, cape	41 01 40	37 52 30
Vona, cape	41 07 05	37 48 45
Yason, cape	41 08 15	37 39 40
Fatsah, town	41 02 45	37 29 05
Ouniéh, town	41 09 50	37 19 15
St. Nichola, point	41 10 30	37 18 45
Thermé, mouth of river	41 13 15	37 04 20
Thermé, cape	41 18 30	36 58 00

Names of Places.	Latitudes North.	Longitudes East.
ANATOLIA:	° ′ ″	° ′ ″
Kiatli-Bassi, cape	41 21 20	36 51 45
Tcherchembéh, cape	41 22 35	36 39 20
Samsoun, town	41 20 31	36 21 52
Samsoun, cape	41 12 30	36 22 05
Kizil-Irmack, point	41 45 20	35 57 48
Aladjam, village	41 38 40	35 39 20
Guerzéh, town	41 48 45	36 13 10
Sinope town, the castle	42 02 30	35 09 50
Boz-dépéh, cape	42 03 00	35 13 10
Pachi, cape	42 06 40	35 01 00
Indgéh, cape	42 07 57	34 56 30
Kérempéh, cape	42 02 01	33 19 10
Kidros, village	41 56 09	32 59 24
Sagra, mountain	41 48 01	32 50 20
Délikli-Chili, village	41 49 19	32 38 26
Amasserah, cape, 5½ miles N.E.	41 48 50	32 27 00
Amasserah, summit of peninsula	41 45 27	32 21 20
Bartin, village	41 33 52	32 14 04
Filiouz, village on peninsula	41 34 10	32 02 15
Guélimili, cape	41 32 27	31 53 36
Baba, cape	41 20 54	31 26 28
Heraclea, light	41 17 08	31 24 52
Kara river, the mouth	41 06 55	30 56 20
Sakaria river, mouth	41 09 24	30 39 10
Melin town	41 06 54	31 07 00
Kefken, centre of island	41 14 15	30 17 02
Kerpen, cape	41 13 36	30 16 10
Chili, tower	41 10 48	29 36 52
Cianée of Asia, northern	41 14 20	29 15 00
Pharos of Asia	41 13 00	29 09 20
EASTERN ARCHIPELAGO:		
Tenedos, summit of Mount St. Elias	39 50 14	26 03 50
Metelin, summit of Mount Ordymnus	39 15 00	25 57 42
Metelin, summit of Mount Olympus	39 04 17	26 22 13
Ipsera, summit of Mount St. Elias	38 35 38	25 36 04
Scio, summit of St. Elias, at North end	38 33 42	26 01 00
Hourlac islet, summit, gulf of Smyrna	38 26 32	26 47 01
Carabourno, mount, entrance of Smyrna gulf	38 31 33	26 31 38
Samos, summit of Mount Querki	37 43 46	26 38 26
Nicaria, highest summit	37 31 15	26 02 55
Nicaria, West summit	37 31 09	26 02 43

Names of Places.	Latitudes North.	Longitudes East.
EASTERN ARCHIPELAGO:	° ′ ″	° ′ ″
Nicaria, East summit	37 36 26	26 17 07
Miletus point, or Cape Tree	37 21 11	27 13 13
Patmos, summit of the island	37 17 02	26 35 19
Bove rock, middle	37 14 24	25 56 25
Lero islet, summit of Mount Clido	37 10 44	26 51 22
Zinaro, summit	36 58 42	26 17 38
Cos, or Stancho, summit of Monte Cristo	36 49 56	27 14 09
Crio, summit of the cape	36 44 05	27 34 50
Nicero, summit of the island	36 35 16	27 11 02
Madonna, summit of the island	36 30 31	26 57 28
Piscopi, summit of the island	36 26 22	27 20 53
Safrania, summit of the largest	36 25 11	26 38 24
Placa, summit of the island	36 04 11	26 25 14
San Giovanni, summit of the island	36 20 51	26 41 43
Plana island, summit	35 51 25	26 15 30
Adelphi or Fratelli islets, largest southernmost	35 49 40	26 29 00
Stazida island, middle	35 53 20	26 51 00
Caxo islet, South point	35 18 20	26 52 40
Scarpanto, North point	35 50 30	27 11 30
Scarpanto, South point	35 23 30	27 13 00
Scarpanto-poulo, North point	35 54 20	27 12 30
Yali, S.E. of Piscopi, summit	36 22 15	27 28 55
Crio, cape on the main, S.W. point	36 39 20	27 25 00
Carki, summit of island	36 13 20	27 35 05
Clalavalda, West point of Rhodes	36 07 35	27 41 20
Limonia island, summit	36 17 25	27 43 05
Simia island, West point	36 34 40	27 47 15
Diamond, summit of Simia	36 34 40	27 52 05
St. Catherine's island, South of Rhodes	35 52 00	27 45 35
St. George's Cape, N.W. point of Rhodes	36 22 50	27 56 40
Volno, cape	36 34 15	27 57 55
Adelphi, or Three Brothers, 4 ft. abv. water, S. of Rhodes	35 50 20	27 55 15
Chevalier, cape	36 34 10	28 02 20
Citadel of Rhodes town, Cape St. John	36 30 50	28 04 05
Barbanicolo, summit of island	36 36 15	28 07 20
Rhodes town, Mill point	36 27 35	28 12 05
Rhodes town, end of mole, North of light	36 26 53	28 13 33
Marmara cape, South point of entrance to the port	36 42 40	28 16 55
Ginacri cape, West point of entrance to Gulf of Macri	36 34 25	28 48 55
Macri gulf, S.E. point	36 32 10	28 58 25
Baba island, summit	36 38 40	28 38 35
Caraguachi island, entrance of Porto Fisquo	36 41 50	28 26 45

Names of Places.	Latitudes North.	Longitudes East.
CARAMANIA:	° ′ ″	° ′ ″
Seven Capes, South point, Caramania	36 20 00	29 11 30
Red Castle island, South point	36 06 35	29 35 00
Cacamo isle, East point	36 10 25	29 54 30
Khelidonia, islet off the cape, South point	36 10 30	30 26 15
Khelidonia, cape	36 12 45	30 25 55
Karabournou, point	36 40 00	31 37 50
Alaya nova	36 31 20	32 00 40
Célitibournou, cape	36 10 55	32 21 35
Anamouzi-vecchio, South point of Caramania	36 00 50	32 50 15
CYPRUS:		
Salizano, cape	35 06 20	32 16 35
Cormachiti, cape	35 23 50	32 57 10
Cerina, peak	35 19 30	33 09 05
Cerina, town	35 19 30	33 23 20
St. Andrew, cape	35 41 40	34 37 30
Famagosta, town	35 07 40	33 59 10
Grego, cape	34 57 05	34 06 30
Larnaca town, French Consul's garden	34 55 13	33 39 37
Larnaca, N.E. point of the town, Mr. Rey's house	34 54 31	33 40 20
Chiti cape, tower	34 49 55	33 38 20
Limasol, town	34 41 15	33 03 50
Gatto cape, S.E. point	34 32 50	33 01 40
Bianco, cape	34 39 20	32 40 20
Paphos, town	34 47 20	32 26 25
CARAMANIA, *completed*:		
Cavalier cape, South point of the peninsula	36 07 30	33 43 45
Provençale island, South point	36 10 30	33 47 20
Bagascia tongue, South point	36 12 45	33 57 40
Lamas, town at the mouth of the	36 31 35	34 18 50
Tarsús town, beach	36 46 30	34 46 50
Malo, cape, S.W. point	36 29 45	35 23 15
SYRIA:		
Canzir cape, Syria	36 16 00	35 49 35
Possidi, cape	35 52 10	35 51 00
Lataquié, town	35 30 30	35 48 00
Caria, or Gibili, town	35 19 45	35 55 55
La Marca, town	35 09 00	35 56 30
Tortosa, island and town	34 50 25	35 51 55
Tripoli town, French Consulate, North of castle	34 26 22	35 51 33
Madone, cape	34 19 30	35 42 30
Barut, cape	33 49 45	35 28 05
Seïde, town	33 34 05	35 23 45

Names of Places.	Latitudes North.	Longitudes East.
SYRIA:	° ′ ″	° ′ ″
Sour, or ancient Tyre	33 17 00	35 14 40
Bianco, cape	33 05 10	35 07 35
St. Jean d'Acre, town	32 54 35	35 06 25
Carmel, cape	32 51 10	34 59 40
Cesaréa, ruins	32 32 25	34 54 50
Jaffa, town	32 03 25	34 46 15
Ascalon, ruins	31 39 00	34 33 00
El-Arish, fort	31 05 30	33 48 30
EGYPT:		
Kacazoïm, cape, Egypt	31 10 40	33 03 30
Aboukir, tower	31 20 35	30 06 20
Alexandria, light-house	31 12 53	29 54 50

APPENDIX.

I.

THE OPENING OF A ROAD INTO CENTRAL AFRICA.

FROM the great political changes, and increased intercourse among nations in late years, it is difficult to recollect how much the coast of Barbary was dreaded by seamen and sojourners some forty years ago; insomuch that travellers on its shores were all but unknown. When the general peace of 1815 took place, my attention was strongly drawn to that quarter; and I had frequent conferences with Sir C. V. Penrose—my commander-in-chief—on the subject. One of my first consultations with that excellent admiral, was upon the feasibility of examining the ports, and numerous ancient relics of Tunis; for it was impossible to be at work in this sea without imbibing an antiquarian taste, and my recent operations in Sicily had almost brought me to the shores of Carthage. But being still in the borrowed Sicilian gun-boat, my movements could naturally be only on a very confined scale.

While in a state of suspense on these matters, Lord Exmouth made his memorable visit to the Barbary States, in the spring of 1816, for the abolition of Christian slavery; on hearing which, I immediately went to Valetta, and conferred with Sir Charles on the subject, hinting that such a vessel as mine—for with a light draught of water she mounted a 68-pounder carronade, and two Congreve rocket ladders—might be welcome to his lordship. My wishes were at once strengthened by the warm recommendations of the Admiral, insomuch that I started off that same evening, and quickly joined the squadron in Tunis Bay, being there most kindly received by Lord Exmouth. Here matters being amicably adjusted with the Bey, as they had just been with the Dey of Algiers, we sailed for Tripoli, where affairs were also satisfactorily settled; and this beautifully moral cruise for ever quashed the odious white slavery which had been so long and so shamefully submitted to.

On the terms being ratified, I accompanied Lord Exmouth when he made his take-leave visit to the Bashaw of Tripoli, and prevailed on him to make a formal request—which in this instance almost amounted to a condition—for me to be permitted to visit Lebida after the departure of the squadron, there to examine some ancient architectural relics, which the Bashaw, at the instance of our Consul-general, Colonel Warrington, had recently offered for the acceptance of our Prince Regent. This enabled me to enter upon a long-meditated field of inquiry; and the consequences were so successful as to enlarge our geographical knowledge, and to lead to the journeys of Ritchie and Lyon—Oudney, Denham, and Clapperton—and, lastly, of Richardson, Overweg, Barth, and Vogel.

It was therefore thought, that the circumstances which opened so important a highway might be added as an Appendix to this work; and last August (1853), having requested permission to consult the original letters in the archives of the Admiralty, my wishes were so considerately met, that the reader is presented with nearly all the correspondence relating to the steps which led to the above-mentioned expeditions. To these papers is added the copy of a letter to Lord Melville, which was written two or three years afterwards; which, though it did not relate to the question then most prominent, was meant to procure the examination of the Syrtis Major and the Cyrenaica. (*See page* 376.)

In these documents the orthography of the names might now be improved, for it was very difficult, with my imperfect knowledge of Arabic, to spell from the enunciation of the natives; but it is deemed best to give the copies as they were written on the spot. They thus show the impression at the time; and though they exhibit conclusions which mere hearsay led us to adopt, even when the responses to set questions were obtained with considerable difficulty, still they comprise the only information then obtainable in that quarter. Here follow the letters in chronological order:—

Satellite, *at Malta, May* 21, 1816.

SIR,—I request you to inform their Lordships that I had recommended to Captain Smyth (on the surveying duty) to proceed to Tunis and Tripoli, to take advantage of Lord Exmouth's countenance and presence; and, at the same time, having heard that the Bey of Tripoli had offered the antiquities of Lebida to H.R.H. the Prince Regent, and the communication had been made to H.M. Government on the subject, I directed his inquiries to them, and feel it my duty to forward a copy of the statement made to me by Captain Smyth, to their Lordships, in justice to the merit of that officer; and to state to their Lordships, in case application is made to remove any part of the said antiquities, that it must be done in summer, as there is anchorage only on the open coast, except for vessels of small burden.—I have the honour, &c.,

C. V. PENROSE, *Rear-Admiral.*

John Wilson Croker, Esq.
&c. &c. &c.

THE ENCLOSURE.

Marsamuscetto, May 14, 1816.

SIR,—I have to report my arrival in this harbour from Tripoli, which place I left on the morning of the 10th inst., having taken on board two Christian slaves—Bruno Spagnolo and Marco Polto, who were brought from Zoaro, and a deserter from H.M. ship *Myrmidon*—Robert Lee, who had been picked up in the country, since the departure of the squadron under Lord Exmouth.* By several observations which I

* The Bashaw declining to let him embrace Mahomedanism, on the ground that a bad Christian would never make a good Turk.

made, the whole coast of this Regency is laid down in the charts several miles too far to the eastward; for the martello tower in Tripoli is in latitude 32° 53′ 45″ north, and longitude 13° 6′ 25″, by a good chronometric run from Tunis, compared with some angular lunar distances.

I took the opportunity of my introduction to the Bashaw, by Lord Exmouth, to request permission to visit the ruins of Lebida, agreeably to your instructions, and found him extremely ready to grant every assistance I might require, as in addition to the camels, mules, and dragomen, he sent two of his own chiaus or couriers, without whom it would be very difficult to procure an article of subsistence from the Arabs.

After a journey of about seventy miles, partly over a desert and partly over a fine country, the approach to Lebida is indicated for several miles to the W.N.W. of it, by the many remains of ancient edifices.

The city formerly called Ptolemæa,* Leptis, and Lepida, with its immediate suburb, occupies a space of about 8000 yards, the principal part of which is covered with a fine light sand, which drifting with the wind along the beach, has been arrested in its progress by the ruins, and has doubtlessly been the means of preserving many fine specimens of art, if a judgment can be formed by the beautiful scattered capitals, cornices, and fragments of arabesque sculpture, which are lying in every direction. The materials are also the richest I have seen in such extensive quantity, for it appears to have been a profuse mass of porphyry, granitic porphyry, oriental granite, and gial-antique, and marbles of every description. Most of the walls, gates, and public buildings were composed of massy blocks of freestone and breccia, in layers without cement; and the temples have been executed in a style of the utmost grandeur, adorned with immense columns, all of a single piece in the shaft, and were generally of the Ionic or Corinthian orders; but I saw also several blocks of architrave ornamented with triglyphs, and two or three cyathiform capitals, which led me to suppose that a Doric temple of earlier date than the others had been erected here; and, on a triple plinth near them, I observed what I deemed to have been a species of socle used in those structures, as the base of a Doric column, part of the walls of a cella to the same may also be distinguished, of which the columns forming the peristyle stood outside.

The city was encompassed by strong walls of solid masonry, ornamented with magnificent gates and spacious porticoes. It abounded with splendid public edifices, the remains of which are so numerous, that without excavation there may be immediately removed upwards of thirty complete shafts of columns, in single blocks of variegated marble, from 18 to 20 inches diameter, four of 26 inches, and three of the immense circumference of 11½ feet. Of oriental granite may be obtained upwards of twenty shafts, from 14 to 18 inches diameter, and there are about eighteen of porphyry, from 22 to 30 inches, exclusive of large blocks of entablature, cornice, and architrave. It was divided from its principal suburb to the east, by a river, the mouth of which formed a spacious basin, and was the harbour. This was defended at the narrow entrance by two stout fortifications, which are in considerable preservation, and a small rivulet occupies the bed of the river, and falls into the sea between them.

At a little more than a mile and a half to the westward of these ruins, and about half a mile from the beach, are situated the two small villages now called Lebida, inhabited by a race of inoffensive and civil Moors, who attend to the cultivation of the adjacent country, and the rearing of cattle, sheep, and goats.† On the beach is a small

* I was here misled by a local antiquary.

† A great part of this plain is laid out in fields of corn, durrah, pulse, &c., interspersed with groves of olive, pomegranate, and date trees, among which are a few vineyards; but a great portion of the produce is annually destroyed by the gundy rat, and a species of jerboa (probably the μῦς δίπους of Cyrenian coins), which greatly infest all the grounds, yet no means are used to destroy them.

port, open to the S.E. winds, formed by several projecting rocks, where a couple of transports, and the lighters requisite to be employed in embarking the columns, might be secure. From this bay to the entrance of the river, the beach is shallow; but at the ruins of the westernmost fortress, I imagine those weighty masses might be removed, provided the necessary machinery and tackles were at hand. The whole coast would offer a summer anchorage.

But it is not so much the columns that I imagine the value of this present to consist in, as in the fine field offered for a gratifying and productive system of excavation; for temples constructed with such costly materials must also have been enriched with statues and bas-reliefs. I therefore, on my return to Tripoli, mentioned to the Bashaw the necessity there would be for a party remaining some time in the neighbourhood; when he not only gave his full permission to that effect, but promised also to send some of his own chiaus and janizaries to attend on, and be useful to such a party.

It would be an eligible thing to appoint a person from the consul's office to remain on the spot, as he would not only give information relative to the supplies of the country, but would prevent the natives from mutilating the columns, particularly those of granitic porphyry, which they frequently do to make millstones, &c., of; and not long ago a fine statue was discovered, which being too ponderous to remove, the head was struck off and taken away.

Colonel Warrington, the Consul-General, accompanied me, and manifests great zeal to exert himself in procuring men as labourers, or in any way you may think proper to direct, on this service.—I remain, &c.,

W. H. Smyth, *Commander, R.N.*

Rear-Admiral Sir C. V. Penrose, K.C.B.
&c. &c. &c.

An echo of this letter was written to Sir Thomas Maitland, the Governor of Malta, of which a copy was also forwarded to the Admiralty; but as the General had directed my attention to another object of inquiry, the communication contains the following addition:—

On the subject of the mountain of salt near Tripoli, mentioned by your Excellency to my attention, from a number of considerations and inquiries, I imagine it to be the only one of the several that have been long supposed to exist in these parts; and it would appear that even the ancients were fully acquainted with the saline properties of the soil, than which no country on the face of the earth was stated to produce so much nitric salt. Herodotus, Pliny, and Strabo, give minute details of several of these hills; which descriptions seem to have been confirmed beyond doubt in 1726, by Dr. Shaw, in his account of the Lesser Syrtis. But as I had not sufficient leisure to examine the place myself, I only call your Excellency's attention to this supposition (*of that being the site*), in order that, if an opportunity should offer, samples from each hill might be collected and compared, for the purpose of ascertaining by analysis, whether each of the masses is impregnated with a similar proportion of nitre.

On leaving Tripoli, my operations at Malta and its neighbourhood were resumed; and, as Colonel De Bosset, of De Rolle's regiment, was appointed by the home-government to embark the antiquities of Lebida for England, I gave him all the information in my power; together with copies of my notes and sketches. In the autumn, however, for some reason or other, the colonel declined the mission; and on Sir Charles Penrose sounding me, my services were at once volunteered, because it was so favourable an opportunity for gaining acquaintance with a part of the Mediterranean which had hitherto been, as it were, hermetically sealed against us. On the 29th of October a messenger brought me an official instruction from that officer, enclosing an extract of a despatch from Sir Thomas Maitland, of the 28th, expressing his surprise that Lieutenant-Colonel De Bosset

> Did not fulfil that mission to which he was originally recommended to me by Sir Henry Bunbury. . . . . It is impossible that we can now look to that officer any further upon this subject. . . . Looking, however, merely at the public interest, I am of opinion that the proposal you made to me is at once the most economical and the most fitting that could be adopted upon this interesting subject—viz., that Captain Smyth, of whose abilities on such a matter I have no doubt, should again proceed there to make a more detailed report than he has hitherto done, and make an economical, but at the same time, a fair experiment, of what may be the possible value of proceeding further in this business. I have given upon this head, my sanction to Major-General Layard, to advance him the money that may be requisite for this undertaking, and to supply him with tents and such articles as may be indispensably necessary; but I must suggest, that I think from 400 to 600 dollars is as much as is worth risking upon the occasion, and my subsequent proceeding must be regulated by his report, and the future instructions of Government.

On the 29th, Sir Charles also wrote to the Admiralty, mentioning that "Captain Smyth sails to-morrow for Lebida, with full powers, agreeably to their Lordships' Secretary's letter of the 27th of June last, being at my request furnished with such money and other means, by Sir Thomas Maitland, as were to have been given to Colonel De Bosset, had he not declined the mission."*

As the mission was of a novel nature, I was desirous of securing a coadjutor in case of accident, and in Captain (now Colonel) M. C. Dixon, of the Royal Engineers, an efficient companion was secured: but to my surprise and regret, after the eleventh hour obstacles were thrown in the way, and the Captain's leave of absence was recalled. Having hastily embarked a few tents, mattocks, and spades, I left Malta on the 30th; but on arrival at Tripoli found that the plague had preceded me, insomuch that Colonel Warrington considered we ought not to land. On this I returned to Malta, of which Sir Charles informed the Admiralty, by a letter of the 9th of November. In December, however, in consequence of favourable accounts arriving from the Consul, I again repaired to Tripoli, reinforced with the fine little schooner, the *Wellington* (gunboat No. 28); but we encountered a severe winter gale, which compelled me to bear up, though the schooner, having no ordnance to labour under, got into port. Of this Sir Charles informed the Admiralty:—

---

* This is giving the affair its warmest tint,—for the 600 dollars barely paid the working Arabs. My own expenses—as keep of servants and an assistant officer, instruments, horses, and camels—never were a shilling's cost to the country.

Albion, *at Malta, Jan.* 23, 1817.

SIR,—I have had the honour before to state, for their Lordships' information, that Captain Smyth, in one of the Sicilian flotilla and the gun-boat No. 28, sailed for Tripoli on the 15th of last month; and since that, Captain Smyth having put back for damage received in a severe gale, I had since sent that officer to Tripoli, in the *Express*, tender. That vessel returned yesterday, with information that after the gale before-mentioned, the gun-boat succeeded in getting into Tripoli, but very leaky, and obliged to be calked and otherwise repaired, before she could return here;* and that Captain Smyth had a very flattering reception from the Bashaw, and was about to proceed in the execution of his instructions.—I have the honour, &c.,

C. V. PENROSE, *Rear-Admiral.*

J. W. Croker, Esq.

Albion, *at Malta, Feb.* 10, 1817.

SIR,—Although nothing requiring their Lordships' attention has occurred respecting the squadron under my orders since my last letters by packet, yet I think it right not to miss a favourable opportunity which presents itself, of stating to their Lordships, Captain Smyth's progress on the coast of Africa, and to enclose a letter which I had already prepared for their Lordships on that subject.

My last information is dated Lebida, Feb. 1st, by which I found, that since it was suspected by the Arabs that the porphyry columns, &c., were likely to be removed, considerable mutilations had taken place for the purposes of millstones, &c., but that the captain had begun excavation, and found an entire statue (but his letter being written in a hurry, he omits to say of whom), and that a guard of 50 horse and a mameluke were to come to him about this time, to proceed to the other ruins stated to be of such value.

Captain Smyth does not apprehend that a traveller would find any difficulty but hard living, in proceeding with a caravan to Tombuctoo; but except that the brother of the Vice-Consul of Bengazy, mentioned in my former letter, is qualified for an interpreter of Italian and Arabic, there is no person on the spot fit for the business of interior research.

There are some travellers in the east parts of the Tripoline states, but not likely to interfere with the present or future plans of inquiry. I mentioned that Captain Smyth had been provided with 600 dollars by Sir Thomas Maitland, but the future means of supply are not ascertained.

I hope soon to forward an official detail from that officer.

In a gale which forced Captain Smyth to return, the gun-boat which sailed with him succeeded in getting to Tripoli, but with such damage as to require some repairs there.—I have the honour to be, &c.,

C. V. PENROSE, *Rear-Admiral.*

J. W. Croker, Esq.,
&c. &c. &c.

Lebida, *Feb.* 24, 1817.

SIR,—On my arrival in Tripoli on the 17th ultimo, I found the *Wellington* schooner lying there, but the gale she had encountered on her passage rendered an immediate repair necessary, and the foremast of the *Express* being badly sprung, to prevent loss of time I deemed it eligible to proceed to Lebida by land, to make a more detailed report on its antiquities and their rarity or value.

To this effect I waited on the Bashaw, in company with Colonel Warrington, and

* It proved that the *Wellington* was too seriously injured for repair; she was therefore sold to a Tripoline broker; her crew being sent back in the *Express*.

found him in the same disposition I had left him last May, which encouraged me to make an immediate request for permission to explore that part of his Regency near the Greater Syrtis; in which from its having once been populously inhabited by the Macæ, Hamanites, and Hasbitæ, and afterwards colonised by the Romans, I might, from its being less frequented, expect to find remains in better preservation than those of Leptis. His Highness instantly acceded to this request, and made a grant of all the works of taste, of whatever materials or value in his dominions, including the ruins of Zort, and the whole of that part of the Cyrenaica, forming the Pentapolis, to his Royal Highness the Prince Regent.

During the conference, the Mameluke Reis, of whom you have already heard, was introduced, and examined with respect to the journey he had recently made into the interior. It appears that this officer (Mukni), who is Bey of Fezzan, had departed from Mourzouk at the head of an army, with the professed intention of extending the Bashaw's dominions, and procuring slaves. He proceeded South Eastward, and, passing Bornou, a government in commercial relation with Tripoli, entered a country inhabited by a race of fine negroes, on whom he made war, defeated them in every encounter, and finally drove them into a large river, on the banks of which they had fought, where the greater part of them perished. This river he styles the Nile, and describes as running to the eastward, and forming at this part a wide expanse of waters, with numerous shallows, on some of which the negroes got. The central part he imagines to be deep, as many boats were passing and repassing, similar in size and construction to the Jerba boats, that is, long and narrow, and from five to fifteen or twenty tons. On his return from this expedition, he had passed through some ruins abounding in large edifices, and furnished with such a number of statues as to have all the appearance of an inhabited place. This account would at once stamp it for the celebrated Raz Sem, that has so long engaged the attention of the learned; but from its direction, it would rather appear to be in the country of the Troglodítæ, at Thama or Adaugmagdum. The relation thus officially delivered, besides the lively interest it inspired, prompted me to repair thither, and ascertain whether the specimens of art were really worthy of the attention of his Royal Highness the Prince Regent, or, should it be the scene of the petrifactions, to confirm or contradict the bold paradox; aware at the same time that such a journey would be in the spirit of my instructions, and would not only forward the views of modern geography, but also illustrate a portion of the Melpomene of Herodotus, and the works of Abulfeda, Ptolemy, Leo, and Edrisi.

To accomplish this I determined to repair to Lebida and employ a party of Arabs in excavating certain spots I conceived would be productive, and leaving charge with one of the officers of the *Wellington*, and the Bashaw's chiaus, to proceed in the mean time to the Interior; but the Bashaw intimated through Colonel Warrington, that he was just on the point of declaring war against his eldest son, the Bey of Bengazi, and that though his dominions were in general pacific, still some partisans might be lurking about who, by securing the Consul-General and myself, would demand terms with which he should feel bound to comply. He therefore wished me to remain two or three weeks longer, when he was going to send away an expedition to Bengazi, on the movement of which all the factious people would retire to the eastward, and he would then despatch a party of Janissaries to be entirely at my disposal, and conduct me to any part of his dominions.

Upon this I departed for Lebida, where I arrived on the 23rd; and, on riding over the ruins, was surprised to find that most of the valuable columns I had left last April, had either been removed or were lying broken on the spot. I discovered, on inquiry, that a report had been circulated by the chiaus on my former visit, of an intention to embark them during the summer, and, as it had been a quarry from whence they supplied themselves, they had, in the interval, been busily employed in breaking them into millstones, not only for the present but for a future supply. They had, of course, selected the most durable substances, the oriental granite and the red Egyptian granite, which, from its compact base, fine felspars, and small admixture of mica scales, I was induced to style granitic porphyry. This destruction was assisted by the peculiar construction of the Moorish oil-mills; they being built with a circular surface, having a gentle inclination towards the centre, round which a long stone, at least one-third of a shaft, traverses. However, there yet remain several very fine ones, principally of variegated Egyptian marble, which, being of greater frangibility than the others, have been spared. There are also several of grey granite, and of coarse red Egyptian granite, which may be removed immediately, should it be deemed desirable; but some of the granite ones have undergone the first process for the millstones, which is chipping off the astragal and torus, and many of the marbles show the corrosive effects of sea-spray.

To secure these from further damage, Colonel Warrington, who accompanied me hither, set off on his return to Tripoli the following morning, where he made the Bashaw acquainted with the loss; and I sent for the Sheiks of the Moorish towns and Arab villages in the neighbourhood, to admonish them, and on every side received assurances that the remains should be respected in future. Should it therefore be the intention of Government to remove them, I beg leave to suggest that an immediate order be sent to the Consul-General to prepare them for embarkation, which he, from having studied mechanics and engineer tactics, and possessing many local advantages with the natives, is so much better capable of effecting than a stranger. For the time of sending a vessel or vessels, I am of opinion, that after the equinoctial gales of September is much the most eligible time of the year, as the coast being unsheltered from the heavy gales of winter, which may be expected even in April, would render it unsafe in the spring. In the summer the heat is oppressive, and, I apprehend from an examination of the situation, and the result of inquiry, that the dangerous marsh fever, prevalent in the Mediterranean under the name of *malaria*, would be experienced here during the months of June, July, and August.

On the 25th I commenced an excavation with upwards of eighty Arabs, whom I increased the following day to a hundred; and as they quickly gained the use of the English spade and mattock, the work proceeded with spirit. But I had soon the mortification of perceiving, from every local evidence, that Leptis had been completely ravaged, and its public edifices demolished with diligent labour, owing, perhaps, to the violent contentions of the Carthaginian bishops, and the introduction of the Vandal Christians of Genseric, who zealously destroyed the Pagan monuments of places under their control. This opinion, many of the coins I have found bearing the Labarum, would countenance; or perhaps it was from the vengeance of the barbarians, for the memorable treachery of the Leptitani. From whatever cause it proceeded, the destruction is complete; the statues either broken to pieces or chipped to a shapeless mass, the arabesque works defaced, the acanthus leaves and volutes knocked off the fallen capitals, and even parts of the pavement and floorings torn up, the massy shafts alone remaining entire. Willing, however, to give it as fair a trial as was consistent with the economy recommended, I continued excavating till the 12th of February, when, having explored the principal basilica, a triumphal arch, a peristyleum, an arcade, and several minor places, with only a strengthened conviction of the distant existing chance of recovering any specimen of art worth the expense of enlarged operations, I determined to desist and prosecute, during the interval of the army's arriving, a geographical research of the parts formerly comprehended under the title of Cyniphi Regio.

In the course of the excavation I had an opportunity of observing, that the period of the principal grandeur of the city must have been posterior to the Augustan age, and when taste was on the decline; for, notwithstanding the valuable materials with which it was built, it appears to have been overloaded with bad ornaments, and three colossal statues I found (but without heads, arms, or feet) are in the very worst style of the Lower Empire. I send by this opportunity for your further information, several of the best pieces of sculpture I discovered, which will also show you the manner in which they have been mutilated, as these are selected from nearly two hundred fragments. I must also remark that there are evident appearances, in several places, of a former excavation, which most probably was carried on by the agents of Louis XIV.

There are many evidences of the city having been occupied after its first and violent destruction, from several of the walls and towers being built of various pieces of architecture, cornice, inscription, and other portions heaped together; but I have found nothing to indicate great antiquity, except an almost obliterated inscription on the wall of a mosque, at Ziphaar, in which the mixture of Roman and Greek characters would indicate an early age of the Republic. The pieces of Doric entablature I had before noticed, and the cyathiform capitals prove but poor and ill-proportioned imitations. I also opened three distinct burial-grounds, with a view of gaining further information, but with little success, as I found neither urns, vases, nor lachrymatories; only a coarse species of jar and a few coins. Even these were neither valuable nor rare, being wholly copper, greatly corroded, and principally those of Nerva, Commodus, Constantine, Alexander Severus, Balbinus, and Faustina.

On a fragment I observed near the ruins, I found the common Moslem prayer inscribed, which, as I understand, is to be found in the remotest places resorted to by caravans, and is also copied for amulets; I thought this would be curious, and its removal, far from being a matter of contention with the natives, appeared to give satisfaction, particularly to the Bashaw's son, Sidi Achmet Bey. This prince visited my tent and examined my instruments, the trenches I had made, and the marbles I had collected, with great attention. In fact, although for fear of interrupting the present friendly disposition of the Government, I have not pointedly introduced religious subjects, I find that Mahometan intolerance has greatly subsided in this Regency, for I was allowed freely to enter the mosques, with only the condition of pulling off my shoes; and the Moors have both ate and drank with me, even in country places, where they are more observant of their tenets, and less familiar with Christians than in the capital. I have shown the Arabic translation of the Bible to several; but though the characters are well understood, none even of their marabuts can read it, as the vernacular tongue of all these parts is the jargon called lingua-franca; and this, I am informed, is understood by some one in most of the interior caravans.

While on this subject, I must inform you of my having had several remarkable conversations relative to the existence of certain Christian tribes in the interior of Africa; and, it would appear, in the neighbourhood of Wangara and Gooba. They are described as a very muscular race of negroes, but I cannot discover that any sign of the Cross or other characteristic symbol has been observed, and their tenets are so slightly impressed, that on their arrival in the market they readily embrace Mahometanism. A French captain (Lautier) in the service of the Bashaw, who has resided in Tripoli twenty-five years, circumstantially related to me that several years ago some of them were brought from the interior, and that twenty-eight of the finest being selected to be sent to Algiers, he was appointed to transport them thither. As he was bringing his vessel to an anchor, an evening bell was heard on board one of the Christian ships, when, to his infinite surprise, those on deck manifested the utmost delight, and calling up their companions fervently embraced them, pointing at the same time towards the vessel the sound issued from, and repeating the word *campani*. As this appeared a corruption of the Italian, or more properly of the Latin itself, he made his interpreter inquire concerning their congratulations, and found that in their native town a large building occupied a central space, having a bell on it, which every morning and evening summoned them to prayers; and that in this building there were neither idol, mat, nor divan; but that their priest exhorted them. Another curious fact is, that the late Bey of Bengazi, who in his boyhood was brought a slave to Tripoli, recollected some ceremony similar to the celebration of mass, and the use of consecrated wine. I could not, in the course of my inquiry, find whether a manuscript, or portion of one, had ever been observed in any caravan; but the absence of circumcision, combined with the circumstance of the bell and the wine, sufficiently indicate that Mahometan doctrines are not the only ones prevalent. I therefore conceive that by procuring a man, and educating him for the purpose, important results may be anticipated, and a road opened to the full discovery of those regions in the vicinity of the Lunar Mountains.

Yesterday the Bashaw's army passed and encamped about a quarter of a mile from my tents; and as it is my intention to set off to-morrow into the interior, I deemed it requisite to make the arrangements for sending you the pieces of sculpture, and my opinion of the actual state of Leptis. These I shall leave in charge of a dragoman, to be embarked on board the *Wellington* the moment of her arrival. Should the place the Bey of Fezzan passed through be Raz Sem, or Ghirrza, in the neighbourhood of the Syrtis, I compute it will occupy about three weeks; should it be towards the S.W. longer, but as I feel in a great measure ignorant of the precise spot to which I am going, I am unable to state the probable time I shall be absent.—I am, &c.,

(Signed) W. H. SMYTH, *Commander, R.N.*

Rear-Admiral Sir C. V. Penrose,
&c. &c. &c.

Tripoli, March 27, 1817.

SIR,—After having written my letter of the 24th ultimo, I was prevented from following the army as I then proposed, by the report a courier brought me from the Consul-General, relative to the state of the *Wellington* schooner; in consequence of which, I deemed it absolutely necessary to repair to Tripoli, and inspect into the nature of her defects.

I availed myself of that opportunity of waiting on the Bashaw, to thank him for the attentions I had received at Lebida; and he then intimated that he wished me to proceed to a town on the mountains called Benuleat (*or Beniolid, from Beni Walid*), where a company of Janissaries had received orders to reinforce our party, as he would wish our safeconduct to be indubitable: however, from the situation of his affairs, and from what I have seen since of the country, I am persuaded his appointing so large an escort had political influence in view. His Highness also signified his desire that Sidi Amouri, his son-in-law, and Sidi Mahomet, his nephew, should accompany us, and furnished them with his teskerah, authorizing the whole of us to subsist gratuitously on the Arabs; but as I deemed such a paper detrimental to future undertakings, it was never used without a present, proportionate to the comparative value of the article, being made.

On the morning of the 28th I left Tripoli, accompanied by the Consul-General, the two Sidis, three dragomen, twenty-six Moorish horsemen, and several camels.* Proceeding by the fertile grounds of Sahaal, and afterwards over a hilly and almost uncultivated country, we arrived on the noon of the 3rd of March, at Benuleat, a place com-

* I ought here to have mentioned, that on the 2nd of March we passed Gusserkzab, an old tower in the plain of Frussa, where, about three years before, a number of gold and silver coins had been discovered. Of these I was unable to procure a single specimen, they having been all taken to Tripoli, where they were most probably melted, and their date and story lost for ever.

posed of several straggling villages, in a fertile ravine five or six miles in length, bounded by barren rocks of difficult access. The population, exclusive of the government guards, consists of about 2000 inhabitants, who subsist principally by agriculture, and a trifling manufacture of nitre; they are accounted industrious, hardy, and brave, and amongst them the present Bashaw, when in rebellion against his father, resided eight years.

A large and ill-proportioned building, called the Castle, near one of the pleasantest spots in the ravine, was prepared for our reception, and a plentiful supply of provisions and forage provided. Though I visited another species of blockhouse, this may be deemed the principal fortress; it is situated in a valley, and commanded at almost every point; it contains several rooms, good stabling, and a large courtyard, but the water is a musket-shot distant. The walls are badly perforated for musketry, and furnished with round bastions, too weak however to bear artillery. Its position by two meridian altitudes of the sun, in a quicksilver horizon, and bearings carefully corrected, is in latitude 31° 45′ 38″ North, and longitude 1° 01′ East of Tripoli.

Here having found several people who had recently arrived from the place I was bound to, and which I found was called Ghirrza, I repeated my inquiries respecting its remains, and again received assurances that I should find perfect figures of men, women, children, camels, horses, tigers, ostriches, and dogs; and the belief of their being petrefactions was so prevalent, that a doubt was expressed whether I should be able to remove one of those whom it had pleased Providence thus to punish for their sins. This revelation, while it wound my expectation to a high pitch, afforded me considerable gratification, in finding the report did not vary on approaching the scene, as I had apprehended.

On the 6th, having been joined by three mountain chiefs, Mahomed, Abdallah, and Hadji Alli, with twenty-five janissaries, and fifteen camels laden with water, barley, tents, &c., we proceeded to the south-eastward over a dreary mountainous country, nearly uninhabited;* and on the 8th arrived at a part of the Valley of Zemzem, which was within three or four miles of Ghirrza. It was then night, and such was my impatience to ascertain the cause of the extraordinary story so universally promulgated, that I anxiously watched for the approach of day. Early on the following morning, having left a party to guard the tents and baggage, I proceeded over the hills in company with Colonel Warrington, the Sidis, twenty janissaries, and a camel bearing my instruments.

I quickly perceived the mention of cold springs and shifting sands by some authors to be erroneous, as the situation is mountainous and barren, presenting only fatiguing masses of sandstone, quartz, and limestone; with occasionally a remarkable vitrified pyrite resembling porous lava. The scene is sometimes varied by ravines, which, though neglected, are evidently capable of great fertility, from the luxuriant talhr trees, lotus, and other shrubs, which spontaneously cover them. But, as might be expected of a government where despotism and bigotry have united to depress the exertions of private interest, destroy public spirit, and retard the progress of improvement, so as to keep its subjects at a vast distance from civilization, large tracts of country are almost left waste; and the little cultivation occasionally exhibited, is carried on in the most primitive mode, without deriving the slightest benefit from the many agricultural improvements which successive ages have effected.

---

* I might have mentioned, that on the 7th we encamped in an open space at a well of bad water, called Kanaphis, where we found a small kaffle from Fezzan. The district was remarkable, since we were exceedingly tormented by swarms of ticks, that teased alike both our horses and ourselves.

After a short ride, we came suddenly upon the ruins of Ghirrza, when, although I had not suffered my imagination to rise at all in proportion to the accounts I had received, I could not conceal the mortification I experienced in observing a few ill-constructed houses of comparatively modern date, on the break of a rocky hill; and across a ravine at a small distance several tombs. On approaching the latter I found them, in very bad taste, ornamented with ill-proportioned columns and clumsy capitals; and neglecting the divisions of architrave, frieze, and cornice, nearly the whole depth of the entablature was loaded with absurd representations of warriors, huntsmen, camels, horses, and other animals, in low relief—or rather scratched on the freestone of which they are constructed, and certainly forming the very worst attempt at sculpture I ever beheld. The pedestals are generally without a dye, and the space between the base and cornice bears a wretched attempt at arabesque ornament; while, after the manner of the Romans, a violation of decency is observable in several places.

Across a fine but neglected valley to the south-eastward, in which were great numbers of wild antelopes and ostriches, is a monumental obelisk of heavy proportion; and near it are five tombs, similar in style and ornament to the first. There are but three inscriptions, nor can other reference be had, as the whole have been opened, in search, I suppose, of treasure; but as no person resides near the spot, I was deprived of the benefit of local information. The reliefs are nearly perfect; and, as this ridiculous collection lies near the Fezzan road, people from the interior occasionally tarried to examine it, being the only specimen of sculpture they ever saw, and representing objects familiar to them, made them describe on their arrival in Tripoli, in glowing colours, what they had seen. This account, warmed perhaps by the story of Nardoun, increased to a *Petrified City;* and at length gained such celebrity, as not only to attract the attention of Europe, but in Africa to obtain universal belief; for it has been deemed a species of pilgrimage to resort thither as the caravan passes, and inscribe a blessing for the unfortunate petrified Moslems; and with these the pedestals are actually covered.*

Ghirrza is situated near some barren hills called Garatilia, and from its want of water, and sterile, comfortless appearance, could only have been a military station in communication with Thabunte. Its situation by two good altitudes of the sun, in an artificial horizon, is in latitude 31° 7′ 17″ North, and longitude, deduced from bearings carefully carried from Benuleat and tried back, 1° 29′ 52″ East of Tripoli. The ruins of the houses are neither indicative of greatness nor opulence; on the tombs the largest figures are about three feet and a half high, and the specimen I brought down with me is of the smallest; yet, notwithstanding the diminutive size and despicable execution, the Turks who accompanied me eyed them with admiration and respect. Never, in fact, has a palpable instance been brought before me so strongly proving the degraded state a Mahometan education, destitute at once of liberality and emulation, reduces its disciples to: nor could I but regret to see men, in many other respects estimable, so glaringly deficient in the necessary discernment acquired easily by the pursuit of general knowledge.

On the 11th, I wished to proceed by the road to Succa and Mesurata to Lebida; but as I had so many men and camels belonging to Benuleat, it became necessary to return thither. From thence I went to the north-eastward, in hopes of finding some

---

* A wandering Bedoween, who had been sometime in the Wadie, brought me a good large brass medal of the elder Faustina, which he had found in the immediate vicinity of Ghirrza. This is no criterion for date, and from the mixture of Egyptian and Roman taste in the architecture of the principal building, the uncertainty is still greater. We copied the rudely-cut inscriptions, but no name of a known family rewarded the trouble.

remains of Taliti, Tenadassa, Mespe, and Syddenis, which were in the chain of communication with Cydamus, and the stations of the Tritonis; but I met with nothing but a few dilapidated towers, and some indifferent ruins which, from their situation, were probably those of Mespe. On the 19th I passed along the banks of the Cyniphus, and from thence returned to Lebida. Having there made arrangements for preparing some of the variegated marble columns (the *cipollino*) and some of granite, for embarkation, I departed for Tripoli, and arrived on the 26th instant.

Although this journey has not wholly answered the sanguine expectations I had formed, it has been the means of showing the disposition of the Bashaw and his subjects with respect to Christians travelling in the Regency; and from the various information I have collected, it appears that so favourable an opportunity of prosecuting the investigation of those unknown regions—which yet remain the disgrace of geography and knowledge—has never before occurred. The Bey of Fezzan is preparing for another journey to the south-eastward, where he procured his slaves before, and would no doubt receive any person in charge from the Bashaw; by which an opening would be made for a direct route to the probable source of the Nile, as the Bashaw's influence extends to a great distance from his frontier. The Bey asserts, that an Englishman went on an expedition with him about seventeen years ago, from Fezzan; and that he died in consequence of a fever, and was buried near Aucalas.

The shores of the Syrtis, the Cyrenaica, and coast as far as Egypt, which may be almost styled unknown (and actually are in a late French chart), his Highness has already granted me permission to examine, and I hope to add them to the portion I have already explored. But I think attention ought to be directed to the important object of the Tombuctoo caravans from Mourzouk; by one of which the Prince of Tombuctoo, a few years ago, came to Tripoli, and settled a commercial treaty, which, from the exorbitant duties imposed, has almost expired. Still, however, caravans occasionally go from Fezzan, sometimes direct, and at others to Twat, and from thence to Tombuctoo. From every inquiry I really conceive it to be a practicable route; and with the protection, not only of the sovereign of Tripoli, but those of Tunis, Morocco, and Algiers, I believe a person would be perfectly secure, as none of the powers intervening between the Niger and their dominions would like to infringe on so formidable a guarantee.

I regret to be under the necessity of recommending that a person be sent to accompany the Bey of Fezzan, on so depraved a mission as that of dragging away the natives of a country for sale, which of course must be repugnant to every humane heart; but as it is the only method of acquiring a knowledge of their actual condition and distresses, with the mode of alleviating them, it becomes a necessary measure to be an eye-witness. This traffic has rapidly increased since the destruction of the slave-trade on the Western Coasts, and has also augmented since the suppression of the Barbary pirates. The slaves formerly embarked at the factories, are now driven over the country to these ports, and are from thence exported to different parts of Turkey and Syria; and I learn with surprise, that since I have been in the Regency, two vessels have sailed from the port of Tripoli, laden with a number unknown, even to the most hardened Liverpool traders of similar burthens.

Leaving these considerations to your better judgment, I remain,

Your obedient humble servant,

(Signed) W. H. SMYTH, *Commander, R.N.*

To Sir C. V. Penrose, K.C.B.
Commander-in-Chief.

Tripoli, April 5, 1817.

SIR,—Immediately after despatching my letter on the morning of the 27th ult., and having had another interview with the Bashaw, I hurried away again for the purpose of visiting a set of greatly-vaunted ruins, about fifty or sixty miles off, in a south-east direction; as I much wished to see them before sailing for Malta, having had much discourse about them with the Arab sheiks of the hills round Lebida. By the zealous care of Colonel Warrington, everything was soon ready; and, accompanied by a special red-burnoose chiaus, and the brother of Sidi Amouri, we started on the evening of the 29th, taking a route towards the Tarhounah mountains, which we had so recently passed on their north side. On the road we were well treated by the Duffa-surat Arabs, and crossed the first range of hills by the pass named from them; where we observed that the scarps on either side exhibited coarse sandstone above a bed of finer grain reposing on limestone. But some curious and remarkable specimens of fulgurites, or tubes vitrified by lightning in the sands, were shown us, as having been produced in the severe storm which had set us all afloat in the tents at Lebida, about three weeks before.

From this neighbourhood we passed, on the morning of the 31st, towards the well of Radwa-Weled-Busaid, and from thence travelled in a direction a little south of east; somewhat parallel to our accustomed Lebida route. Every here and there were scattered vestiges of former days, giving an indication of a greater and more important population than at present; and I was much struck with the remains of one tower, evidently Roman, which had been rendered defensible by some Arab insurgents, if such a designation can be applied to men endeavouring to escape a grinding and oppressive tyranny, the right of which they had never owned. At length, on crossing the upper part of the well-known Wady Ramel, we arrived at Milah, or Medina Dugha, the remains of which, I had been assured, would interest me more than those of Lebida had done, But—not to my surprise—they fell very short of report, and even the chiaus himself seemed to be disappointed. However, from the massy foundations, ashler blocks, and numerous architectural fragments among the brushwood and talhr trees, it is clear that an ancient city of considerable importance stood here; and that, however busy it might have been, and however its citizens might have plumed themselves, it is probable no record of it remains. It seems to have occupied more than three miles in length by about two in breadth; and there are indications of its having been strongly fortified, much in the style of Lebida, and of about the same date—i.e., if Spartian be right, about the time of the Emperor Severus, a native of that city, who ordered the fortifications to be built. While examining the ruins, the Arabs told us marvellous stories about the wonderful extent of the subterranean chambers and passages; but I had neither time nor inclination to explore them, and young Amouri was too much alarmed at the prospect of *dubbahs* and *dib-a-dibs* (hyænas and jackals) being there, to descend. The *badinage* upon this point occasioned the utmost good-humour; and I am more than ever convinced that men of the right cast can easily secure good fellowship in Africa.

Having satisfied myself that no result of sufficient value would follow my remaining here, we returned to the Consul's hospitable mansion, where I had the pleasure of receiving your packet. On this journey, as before, I rode on an English saddle, in an undress uniform; and found, despite of certain objections made by some consuls against showing the Christian dress, that our appearance excited as little surprise as it had done around Lebida. But it is a *questio vexata* here,* whether travellers should assume the

* This question has been pretty well settled in the last few years. When I was in the Levant, turbans, mantles, satins, silks, embroidery, fire-arms, yatagans, caftans, flowing robes, and loose Turkish petticoat-trousers were the universal wear; but now, *on a changé tout cela*, for even the Sultan in Constantinople, and all the officials, appear

Turkish attire or not; the opponents declaring that Moslems were never yet deceived by it, and that even Aly Bey was detected the moment he entered a bath, by the corns on his feet: yet when such men as Hornemann and Burckhardt found it necessary to adopt that costume, the topic merits serious consideration. Still, I think a *mezzo termine* is offered in some of the Frank dresses of Barbary.

Yesterday and this morning have been devoted to a continuation of my former inquiries on this head; and from all that can safely be trusted, without fully relying upon any one, I am becoming still more convinced that here—through this place, and by means of these people—is an open gate into the interior of Africa. By striking due south of Tripoli, a traveller will reach Bornú before he is out of Yusuf's influence; and wherever his power reaches, the protecting virtues of the British flag are well known. In fact, looking to the unavoidable causes of death along the malarious banks of the rivers on the western coast, I think this ought to be the chosen route, because practicable into the very heart of the most benighted quarter of the globe. Indeed, I feel more than half inclined to offer myself as a volunteer, but for the welcome news in your last, of a ship's being on her way from England for me. Were it not for this, I would assuredly beg you to allow of my exploring the flanks of the Atlas range at least, for the chance of there finding a tribe still speaking the language in which Hannibal wrote his despatches. Should you be successful in procuring an efficient and duly qualified person, he will be received here with open arms, both by Colonel Warrington and M. Carstensen, the Danish consul, a gentleman well-versed in the English language, and also acquainted with Arabic. With the friendship of these two, and the aid of the Bashaw, an explorer has all chance of success; but I would advise him, as his Highness, as well as all his chiefs, has much of that Numidian *fourberie* which Bacon termed 'sinister wisdom,' to permit a little forereaching upon him in small things—even with open eyes—to gain a desirable end.

Bashaw Yusuf is certainly a strange compound of virtue and vice; and he waded through a brother's blood to the throne, so far back as the year 1795. He is of social habits and intelligent, an affectionate father, and a warm friend. Nature seems to have intended him for a good man, and the awful instances to the contrary must be attributed to unbridled despotism and uncultivated mind. Owing to these, and other misfortunes of station, he has exhibited profusion and avarice, courage and timidity, temperance and excess, mercy and vengeance, credulity and scepticism, in singular antitheses. Still, a traveller need not trouble his head about this; for Colonel Warrington—by a manly firmness and judicious bearing—has acquired a complete influence over him. To myself his Highness has always been remarkably attentive, and has shown great patience in getting my repeated, and perhaps tiresome, inquiries answered: on these occasions I usually carried my principal queries written, and procured the replies deliberately, one at a time; and though we were often in confusion, a little information was gleaned. It may illustrate his character to mention, that when we were surveying the harbour, and measuring the outside of his own castle, a busy-body 'wondered' to him that he allowed such operations: 'Oh,' said his Highness, 'if the English wish to attack it, they need not come first and measure the wall.'*

In regard to Sir Thomas Maitland's desire, Colonel Warrington kindly got old

---

in frock-coats and close pants, with a Fez cap on their heads. They moreover find that they can walk about without lumbering themselves with a brace of pistols on each side of their belt. *Tempora mutantur!*

* On my finally taking leave of him, his Highness exhibited considerable feeling, and presented me with his own sword, a wavy-bladed scimetar, which had been blessed at Mecca. It is now in the Museum of the United Service Institution.

D'Ghies and Sidi Amouri to make inquiries for me; and others have assisted. From all that can be collected, both ancients and moderns have been greatly misled respecting *mountains* of salt, though there are most extensive saline tracts in this Regency. I myself have seen numerous camels loaded with solid indurated salt, in long blocks, from the shores of the Syrtis, and it is also found elsewhere. There is no doubt but that saltpetre—good and bad—is manufactured to a great extent; but the principal place I can hear of, as meeting the description which Sir Thomas had received, is an elevated plain about 400 miles south of Tripoli, the elevation being described merely 'as if a town had been smothered with salt, and the salt covered with sand to the depth of about a foot.' A poor mountain this! In like cases, from the defective information of the natives, and our own tendency to put leading traits which beg the question, we become mutually wrong, whence no strong reliance is to be placed on any of the information thus obtained; but at present we can gather no other. In my constant inquiries after the great river to the south, which I mentioned to you at Malta, the contradictory assertions are sorely perplexing. Without exact ideas of time or distance, they state the bearings of a place from the direction of a first day's journey; and any running stream is sometimes called a lake and at others a river. Now the river which should be the western Nile of Herodotus, gets farther and farther to the south after every consultation, and has even been carried several months' journey off. This would not be objected to, provided there was any dependance to be placed on the replies; for assuredly we have no warrant for keeping the historian's river on this side of the line. It is to be hoped that few years will pass ere a clear light will be thrown on this interesting problem.—I have the honour to remain, &c.

(Signed) W. H. Smyth, *Commander, R.N.*

To Sir C. H. Penrose, K.C.B.
&c. &c. &c.

His Majesty's Ship Aid, *off Lebida, Nov.* 9, 1817.

Sir,—I anchored off Tripoli on the 14th ultimo, with the *Weymouth* store-ship in company; and, in order to prevent loss of time, pressed an immediate audience with the Bashaw, who received me with a very marked attention, and readily entered into all the views I proposed. His Highness also directed that Sidi Amouri, his son-in-law, should be embarked on board the *Aid*, in order to render me every facility for expediting the departure of the *Weymouth*.

Having effected all the requisite arrangements, we made sail, and arrived at this anchorage on the 18th; and the same day commenced towing the spars on shore, and preparing the store ship's derrick and holds for the reception of the architectural remains which Colonel Warrington had during the summer brought down to the beach, under the ruins of the western fortress. As we had a continuance of fine weather, and the seamen were unanimous and cheerful in their exertions, I had the satisfaction to perceive these weighty masses embarked and stowed, at the rate of at least sixty tons a day; which, when you consider the open roadstead, the distance the ships necessarily were from the beach, and our limited crews, I trust will meet your approbation. It is but justice to add that Mr. Turner, the commander of the *Weymouth*, has been indefatigable in his exertions to complete; and by his judicious arrangements on board, no accident has occurred.

On the 22nd his Highness Sidi Achmet Bey, the presumptive successor to the Musnud, arrived with his army from Bengazi, whither he had been with intent to have brought his elder brother, an abandoned cruel character, prisoner to Tripoli; but he, finding very few partisans, had fled into Egypt, and quiet was restored with little bloodshed. His Highness, accompanied by his principal officers, rode down to the beach, when he was received with repeated cheers by the boats' crews; and I deemed it expedient to salute him from the ships. With this attention he was much gratified; and we experienced the good effects in the orders he gave, and the additional good-humour it inspired the Moors with. It is on many accounts satisfactory to state, that the service has been performed without the occurrence of a single quarrel, or dispute, between the seamen and the natives.

I was sorry to find that neither the raft-ports, nor hatchways, of the *Weymouth*, would admit the three large cipollino columns; and in embarking the others, I have been under the necessity of selecting those of various dimensions, in consequence of the destructive mutilation that has taken place since my first visit. I have, besides, sent pieces from which drums might be cut, to fit the damaged columns. With the same view I have put several fragments of marble slab and cornice on board, that fractures in the capitals, &c., might be repaired with stone of the same quality. But the specimens of sculpture are only embarked in order to show the style of execution, and the manner in which they have been defaced. The small stone with the horseman on it, is from one of the tombs of Ghìrrza; and the inscriptions are on specimens of the marmoric conglomerate with which the public edifices of Leptis Magna were constructed. Of those five columns which I styled granitic porphyry—and of which I sent you a fragment—not one remains above ground; and on examining the ruins, I found that notwithstanding the threats of the Bashaw, the promises of the Sheiks, and the whole business that took place last winter, a number of the finest columns I then left have been broken; and there is actually a Tunisian vessel now loading with the pieces.

I observed also that several of those, the astragal of which just appeared above the sand, had been cleared down a few feet, and struck off; consequently none worth removal (except the three large cipollino ones on the beach) remain visible; and though a quantity might, perhaps, be procured by excavation, yet as it would be in those spots where the sand is deepest, their removal would be very expensive. I therefore judged it expedient to discharge all the working party, until a determination on the subject would be made on the arrival of the store-ship in England, or a communication to the contrary should be sent by you; in either of which cases, the Consul-General can immediately procure the necessary assistance again.

The anchorage we are lying at has much better holding-ground than that of Tripoli Roads, being sand and rocky patches inshore, and mud in the offing. It is entirely clear of shelves and shoals; and I should think the winds, even in heavy gales, seldom blow home, from the existence of numerous phosphorescent medusæ and other mollusca, generally natives of smooth water. In standing inshore, a berth may be taken at pleasure in from twenty to ten fathoms, which last will be about a mile from the shore.

The bearings from the *Aid*, in latitude 32° 38' 50" North, and longitude 14° 15' 15" East, were Raz-al-Scian N.W.¼N.—Tabia Point S.S.E.—Ziliten Point S.E.—and the ruins of the ancient fortress of the harbour, S.W., about a mile and a quarter distant. Provisions might be procured in the greatest abundance, if the Bashaw's teskera is obtained for a supply, and the shores afford good fish; but water is scarce, and what there is is brackish.

(*Here follows a set of sailing directions for the coast.*)

I have the honour to remain, &c.,

W. H. SMYTH, *Commander*.

Rear-Admiral Sir C. V. Penrose, K.C.B.
&c. &c. &c.

H. M. Ship Aid, *Nov.* 26, 1817.

SIR,—After the departure of the *Weymouth* for Malta, I obtained an audience with the Bashaw, in order to make arrangements for my further proceedings; and also to take into consideration the practicability of researches being made in the interior of Africa, through his influence.

I found that, although his Highness was ready to grant my request for exploring the greater Syrtis, yet his sea-officers were utterly against the measure, at this advanced season of the year; deducing arguments by which I perceived that this celebrated gulf still retains its imaginary terrors. I say imaginary, because on close questioning them all, I could find no one who had been further in it than Bengazi, and consequently could form no opinion, but from traditional report. I therefore, on deliberate consideration, determined to persevere in my original resolution.

But his Highness entered on the subject of the interior of Africa with the most encouraging frankness; and, as an object so highly important ought to be circumstantially related, I subjoin the principal questions as proposed by me, and answered by the Bashaw, or officers present.

Q. His Royal Highness, the Prince Regent, by a magnanimous perseverance in the cause of humanity and justice, having bestowed peace to Europe, is now solicitous to extend his benevolent views to the natives of those regions lying to the southward of the dominions of your Highness, and the several kings, your allies. Will your Highness, therefore, assist so laudable an object, by affording your powerful protection?

A. I shall be happy to render every assistance to such an undertaking. I have already shown that to two Englishmen, who came here some years ago.

Q. Is your Highness certain they were Englishmen?

A. They said they were, and that they came from Egypt by way of Fezzan.

Q. Does your Highness, or any person in the Divan, recollect either of their names?

No answer was given to this question for some time; on which I asked if the name of one might not be Horneman, when Mourad Reis (Peter Lyell) said he now recollected it was.*

Q. How long is it since they were in Tripoli?

A. About fifteen or sixteen years.

Q. What became of them after they left Tripoli, and where were they bound to?

A. They returned to Fezzan with intent to penetrate southward to the Nile (Niger), and thence by the river to Tombuctoo; but one of them, who had been ill of a fever occasioned by drinking too much bad water, after fatigue, died at Aucalas.

Q. Was that the same person mentioned to me last winter by the Bey of Fezzan?

A. The same—the Bey had charge to conduct them to Bournou.

Q. Does your Highness know what became of the other?

A. He continued the journey, but fell ill at Houssa, in the dwelling of a Tripoline merchant established there, and resuming his travels before he was properly recovered, relapsed, and died at Tombuctoo.

Q. Does your Highness know whether either of them left any papers, books, or effects?

A. No; but I will direct an inquiry—Moors never destroy papers.

Q. Does your Highness imagine it difficult for a party to reach the Nile (Niger), through the dominions of your friend, the King of Burnu?

A. Not in the least,—the road to Burnu is as beaten as that to Bengazi.

Q. Will your Highness grant protection to a party wishing to proceed that way?

A. Any person wishing to go in that direction, I will send an embassy to Burnu to escort him thither; and from thence the king will protect him to the Nile. But I must first clothe him as a Turk.

Q. Will he be subject to much troublesome inquiry on that head?

A. No; but he must not say he is a Christian. People in the interior are very ignorant. I will clothe him myself in a particular way.

---

* This statement appeared in print, and enabled Horneman's heirs, through Baron de Zach's appealing to me for its authenticity, to succeed to considerable property.

Q. But will your Highness guarantee the perfect safety of such a person against all accidents, except sickness or unavoidable casualties?

A. I do guarantee.

Q. Will your Highness undertake to produce, in the event of disaster, the papers and effects of the deceased, with a particular note, written by himself, commencing on the day he might be taken ill, stating his opinion, &c. of the cause, and continued daily until he shall be rendered incapable of writing? This question is not to be considered by your Highness as a doubt of safeconduct, but it is absolutely necessary for the consolation of the friends of the defunct.

A. I do undertake to produce all such papers; but there ought not to be less than four persons in case of misfortunes by sickness.

Q. Will your Highness give directions that a party shall not be obliged to proceed at the will of the escort, nor to travel in the heat of the sun, nor in the summer, unless they like?

A. The strangers shall be masters. From September to May is the time I recommend for an Englishman; but travellers have a fault of generally hurrying a caravan.

Q. Will you answer for the assistance and guarantee of the King of Burnu?

A. Most certainly.

Q. Would not a small present be acceptable to that sovereign?

A. Yes, he would take it as a great compliment.

Q. What does your Highness think would be most gratifying to him?

A. Broad-cloth (but need not be the finest), showy muskets, pistols, daggers, swords, and cutlery.

Q. To what amount should your Highness think it necessary to send of such articles?

A. Twelve or fourteen hundred dollars.

Q. Can your Highness afford protection to a party going to the south-westward?

A. Nearly the same as through Burnu.

Q. Are there many boats passing and repassing that part of the Nile (Niger), south of Burnu, and what is their object?

A. They are numerous, and carry effects and passengers to the several towns on the banks of the river.

Q. What are the names of the towns in that direction your Highness has the greatest commerce with?

A. In Wangurra, Cuthorra, Cashna, Zangarra, Gooba, Bombarra, Houssa, and Tombuctoo, there are always some resident Tripoline merchants.

Q. Next to Burnu, what place has your Highness most direct communication with?

A. Souat, which is the principal station for caravans that proceed to Tombuctoo, by way of Gadam.

Q. What is the form of government at Souat?

A. Republican, with a sort of head chief, or prince, the same as Houssa and Tombuctoo.

Q. In what manner do the subjects of your Highness obtain leave to pass those countries, at a great distance from your frontier.

A. The travelling merchants insure themselves, by giving presents—trifling ones—to the head of the country they arrive at, who affords them safeconduct to the next.

Q. How is the usual trade between Tripoli and Tombuctoo conducted?

A. It is mostly carried on by Fezzan and Gadam merchants.

Q. What number of camels does the Tombuctoo caravan usually consist of?

A. Not so many as formerly—not above a hundred and fifty:—the caravan to Morocco is the largest, as they have not so far to go; it is generally composed of three or four thousand camels.

Q. When does the Fezzan caravan proceed for Tombuctoo?

A. The direct road is rather by Gadam, as the nearer one. They set out commonly in March, travel greatly by night, and return towards November, when there is a very extensive fair held at Gadam, resorted to by immense numbers.

Q. What are the principal articles of traffic?

A. Slaves, gold, gum, hides, dates, barracans, nitre, natron, salt, cotton, cloth, and great quantities of a fruit resembling coffee.

Q. What is the greatest length of time the caravan is without the means of replenishing their water?

A. Eight days.

Such is the substance of the principal questions I asked of the Bashaw; whose patience and good nature during the long conference, were eminently conspicuous, particularly as the discussion of several of them required both time and reference. I trust such conduct will be duly appreciated, when it is considered that this prince, by the communication thus made, and the free access to his several towns already given to me, has fully proved himself above the mean intolerance that actuates the generality of Turks; and more especially as he is acting thus in defiance of the memorable prophecy which states that all these countries are to be restored to the Christians. This prediction is so universally believed, that the gates of the several towns and fortresses are closed every Friday from 11 A.M. till 1 P.M., the day and hour predicted for the event: to this, in a great measure, may be ascribed the jealous anxiety with which the Turks watch our desire of exploring those countries.

I remain, your obedient humble servant,

W. H. Smyth, *Commander, R.N.*

To Rear-Admiral Sir C. V. Penrose, K.C.B.
&c. &c. &c.

I open this letter again to add, that the Bashaw, pursuant to his promise, directed an inquiry to be made relative to the effects of the late Mr. Horneman: and it appears that his books, papers, instruments, sealed letters, and clothes, were brought to Tripoli by the Bey of Fezzan's orders, and were to be all delivered to a Mr. McDonnagh, formerly surgeon to the Consulate, by an intriguing man at the Bashaw's court, a Signior Naudi, but his notoriously infamous character leads me to suspect fraud, and the Consul-General is now actively and zealously employed in investigating the whole transaction.

Albion, *at Malta, Nov.* 21, 1817.

Sir,—I have the honour to acknowledge the receipt of your letter of the 9th inst. communicating the result of your labours at Lebida, as well as other highly interesting particulars, which I shall not fail to transmit to the Lords Commissioners of the Admiralty by the first opportunity, with the high sense I entertain of your ability and indefatigable exertions.

I remain, Sir, your most obedient humble servant,

C. V. Penrose,
Rear-Admiral and Commander-in-Chief.

W. H. Smyth, Esq.
Commander of H.M. Ship Aid.

P.S.—I enclose an acting order for the second master of the *Weymouth* to act as master of the *Aid.*

This order was considerately sent in consequence of my warm approbation of the spirit and skill of Mr. Thomas Elson, of whom mention is so often made in this work. Soon afterwards I received another letter from Sir Charles, dated the 23rd of December, 1817, of which the following extract may be said to conclude the Lebida correspondence:—

I have to acknowledge the receipt of your letter of the 26th ultimo, and to assure you that it gave me sincere satisfaction to observe the manner in which you have continued your valuable researches, and also to express my entire approbation of your proceedings with the Bashaw of Tripoli.

I have felt equal pleasure in forwarding a copy of your letter to the Admiralty, as I am assured that it will be a further proof to their Lordships of the zeal as well as ability, with which you execute the service entrusted to you.

I forwarded that copy by packet on the 14th instant, and, as the packet now in port sails on the 29th, I request you would, previous to that day, furnish me with any further information which you may have obtained, and which you deem useful to the furtherance of the great object of understanding and exploring the interior of Africa.

Shortly after my return to England, the First Lord of the Admiralty consulted me upon several of the above points, when I dwelt upon the advantage of a land journey round the Syrtis (*See page* 376), contemporaneously with my examination by sea. Having been desired to reduce my views to writing, the following letter is the result:—

35, *Soho-square, Dec.* 31, 1820.

My Lord,—In obedience to your Lordship's desire I venture to place before you my idea on that part of North Africa lying between Tripoli and Egypt; and which, notwithstanding it constituted one of the most interesting sites of antiquity, is unaccountably a perfect blot in the geography of the present day.

In consequence of a strict attention to the subject, I had reason to think that on my visit to Tripoli in 1816, no other knowledge existed of those countries extending along the coast from the city of Tripoli to the Arab's tower in Egypt, than what was gleaned from the Melpomene of Herodotus—excepting indeed the part now called the Gulf of Sidra, which is evidently deduced from the old map in Ptolemy.

From my numerous inquiries in various quarters, touching the present state and resources of those parts, and from the aggregate of a variety of conflicting statements, I have reason to imagine that material benefit is likely to accrue from a proper investigation thereof; for it appears that there are certainly several harbours almost unknown to us, of which the principal are those of Bomba, Toubrouk, and Tabraka; and my representation of them appeared in so favourable a light to that excellent officer, Sir Thomas Fremantle, that he directed my utmost attention to them, and to the facilities of procuring timber from certain forests reported to exist in that neighbourhood.

But as the protection of his Highness the Bashaw of Tripoli does not extend beyond Derna; and indeed is only precarious at any distance from Mesurata, a thorough investigation of the shores of the Syrtis, and the whole of the Cyrenaica, becomes an object of serious difficulty, and is, perhaps, impracticable to a Christian, though the attainment of it certainly promises the gratification of much geographic and historic inquiry.

I could myself soon fix all the important points on scientific data for the commencement of a coast survey; and a person properly qualified, would not only forward the hydrography, but, from thence, could continue those journeys and researches that would be most conducive to add to our general knowledge; and from my long acquaintance with him, I make no hesitation in recommending Lieutenant Lyon as singularly eligible for such a mission, from his natural ardour, his attainments, his professional habits, and, above all, his very complete assumption of the Moorish character. After the naval and military objects are considered, a research could be made for the two great Roman roads that led to Cydamis, the present Gadam; a town, I am led to believe, of the utmost importance to travellers in the interior, as being the resort of numerous trading caravans.

The site of the celebrated altars of the Philæni would form a satisfactory point; for though they appear no longer to have existed in the time of Strabo, their situation might, perhaps, be placed by approximation.

Inquiries might be made respecting the Silphium, a famous shrub which must have existed in abundance, as sugar was made from it; though others report that it bore benzoin and asafœtida;—that marked on the ancient coins bears a strong resemblance to the large apocynum which grows on most parts of this coast.

We have no proof respecting the fossil called sal ammoniac, said by Pliny to have been found in great quantity below the sand, in a district of Cyrenaica.

Rare coins and medallions of the Pentapolis may, perhaps, be procured, of which the most valuable are these erroneously named Ophellas, especially when large; the usual types are the head of Ammon, with the Silphium as a reverse, and the legend ΚΥΡΑ or ΒΑΡΚ; but those of the state, and not belonging to any individual city, had the word ΚΟΙΝΟΝ; there is also a silver coin with the Punic characters  of tolerable execution.

Inquiries could also be directed towards the celebrated scarlet dye possessed by those countries so many ages, and of which the Cynomosium coccineum is supposed to form the principal ingredient.

Attention could be paid to the petrified palms and fossils in the vicinity of Augila, and in fact to the whole detail of the deserts of Lybia. Of these the vicinity of Cyrene was reported as fertile, well watered, and possessed of forests and pasturages. It is plain to me that the remains of the city of Cyrene (now called Grenna) are extensive, and that its famous fountain still affords a constant supply of the purest water; views, plans, and copies of inscriptions therefore, in this important place, appear to promise a gratifying illustration of the invaluable writings of Herodotus.

The situation of the Garden of the Hesperides, reported to have been near Berenice, would also be a desirable object; as would the complete exploration of Taukra, the ancient Teuchira, and of Tolometa that formed the Port Barca, which I believe possesses fine remains of the magnificence of the Ptolemies.

After the examination of Cyrenaica and the Deserts of Barca and Augila, the grand question of the junction of the Nile and the Niger could be considered; and, if confidence, ability, and perseverance are applied, I see no chance of a failure. In fact, I must here state my regret that the late expedition for the interior was so hastily formed.

With a view of further illustrating this matter, I beg leave to subjoin the substance of some inquiries I made from the officers of the Bashaw's army, who went on an expedition to chastise the Bey of Bengazi, a rebellious son of his Highness, and with whom I was on the point of proceeding, but that the operations at Leptis required my personal attendance, I have many reasons for placing considerable confidence in their replies.

Q. What towns are there between Ziliten and Mesurata, and what are their names?

A. Between Ziliten and Mesurata there are no towns, but frequent remains of large buildings.

Q. What description of buildings?

A. The original forms cannot be observed, the Moors have preserved only some wells of good water.

Q. Have you observed any ruins near Ziliten?

A. Part of an aqueduct near Wadie Khahan, and a sort of arch a little inland.

Q. What is the probable population of Mesurata?

A. About 900 or 1000; though the Aga who governs can put 1000 cavalry and 2000 infantry of the province into a state of service.

Q. Where are the salterns of Mesurata?

A. The principal are between Zafran and Nahim, though there are others along the Gulf.

Q. Is the salt mineral or marine?

A. The salt is not mineral, but produced by evaporation in summer; in winter it melts again by more water flowing in.

Q. But that which I have seen was in long bars.

A. Yes, they cut it in bars for trading, for it is very hard and solid.

Q. What great towns are there between Mesurata and Bengazi?

A. There is no town or place worthy the name between Mesurata and Bengazi, nor from thence to Derna.

Q. How are the shores of the Gulf of Sidra?

A. Generally hard, sandy beach, with a low country adjacent, in some parts very rocky.

Q. Does the gulf marked on this chart, and called Suca, exist?

A. There is no gulf of that name; the army passed close to the sea, where it is marked, and the beach is continuous.

Q. Are there any ruins on the shores of the Syrtis?

A. Near the above-mentioned salterns there are frequent ruins; the most remarkable are to the S.E. of Zafran called Elbenia, and those of Medina Sultan.

Q. What is their appearance?

A. The former consists of two pilasters, with bases of gritstone, and Greek inscriptions much injured. The latter offers vestiges of a large city. There are other ruins at Jhimines and Quabia, two days' journey from Bengazi.

Q. Does the gulf at the bottom of the Syrtis, called Tinch, exist?

A. It does not; we still continued along the beach; there is, however, a large maremma or marsh inside where our route led, but it is very hilly beyond it.

Q. Do you know of any quicksands in that neighbourhood?

A. There is a considerable tract of fine impalpable sand that moves with tempests.

Q. What is the situation of the moving sands, and are there marshes there?

A. The moving sands extend from Ain-Agan to Areys, occupying a greater or less width along the coast from the sea towards the interior; but at Albasce there is a long streak, stretching many miles inland, very fine, and of the colour of brick, whereas the other is white as snow; there are some very extensive salt-marshes at Ain-Agan and Bagomara, two hours S.E. of Manhool.

Q. What is the nature of the coast in the direction of the moving sands?

A. Only the surface of the coast is covered with sand; below, it consists of a hard grit-stone.

Q. Which is the site of the Garden of the Hesperides?

A. They lie about two hours from Bengazi, and have no trees, only a few shrubs grow there.

Q. But what is there remarkable to point the place out?

A. Many deep grottoes, some wells of excellent water, and vestiges of canals to carry water all over the gardens.

Q. Is there not wood in the vicinity?

A. No timber fit for building, nothing but a grove of stunted cypress.

Q. But I have heard from the Bey of Derna, Murad Reis, and others, that a large forest existed somewhere in that part.

A. I believe there is further towards Bomba, but we did not go so far.

Q. Have you heard of this forest ?

A. Very frequently; and that the wood is fit for large ships.

Q. What kind of a town is Bengazi ?

A. Not so flourishing as formerly; it has a tolerable castle and small port, mud houses, and about 1000 inhabitants.

Q. Are there any vestiges of the ancient Berenice ?

A. A few slight ones ;—cameos and intaglios are frequently found, and a hill near the sea is supposed to contain riches, as gold is often picked up after heavy storms.

Q. Can refreshments be procured there ?

A. Sheep, cattle, and corn, but no fruit.

Q. Why have they not oranges, as they grow so well at Tripoli ?

A. They never had any, so do not feel the want of them.

Q. What kind of places are Tolometa and Taukra ?

A. Taukra is a walled town, with many inscriptions; but has few things of architectural beauty, except some vine branches entwined in low relief on the pieces of a pediment of grit, or stone of the country. It is built on the sea-shore, on a plain, bounded on the south by stony mountains bearing the low cypress-tree. Tolometa is at the foot of the chain of mountains that extends from Bengazi to Bomba; it offers few vestiges, except some columns of gritstone belonging to a Corinthian portico, and the tombs of the Kings in the Elysian Fields.

Q. What is there at Barca, and are there any inhabitants ?

A. Barca is now only a mountain of stones and ruins, at the head of a fine valley, with a great many wells of good water, for which reason it is much frequented by the Arabs.

Q. Are the Arabs as trusty as those of Mesurata ?

A. No; they are exceedingly treacherous, and capable of committing murder for a mere gilt button.

Q. Would they respect the usual laws of hospitality ?

A. Most probably they would, even against their desire.

Q. Have you seen the harbour called Marza Suza ?

A. I have seen Suza; the sea has intersected almost all the town; there are many ruins, but of moveable things there are now only to be seen a few columns of marble, granite, and gritstone belonging to its temples.

Q. Is it easy to reach Cyrene on the side of Bengazi ?

A. From Bengazi to Cyrene is six summer days' journey, and the road leads through cypress woods and fine mountain-valleys; it is not difficult.

Q. Is Cyrene far from Derna ?

A. Cyrene is a long day and a half from Derna, over some stony mountains of extremely difficult ascent, through woods of cypresses, and places inhabited by wandering Arabs.

Q. What aspect has the land about here from the sea ?

A. The sea is almost everywhere bounded by steep mountains of rock, in the fissures of which grow cypresses and some other trees.

Q. What state is Cyrene in. I have heard the town is entire ?

A. The town is nearly destroyed, but the ruins and isolated tombs, or mausoleæ, are extensive; the finest part is the Camp of Mars, on account of the numerous streets of tombs cut in the rocky mountains. The various ruins make it extremely easy to determine the limits of the city.

Q. Do you recollect any temples there ?

A. The ruins of a temple near the fountain are partly buried, and all there is remaining in sight are some columns and several statues, the latter so mutilated, that they look like amorphous blocks of marble. Excavation in this part would, probably, be very productive.

Q. Does the fountain still afford good water, and are there any inhabitants in Cyrene?

A. The fountain always gives abundance of the purest water, for which reason there are always upwards of four or five hundred Arab tents in the town.

Q. What is the population of Derna?

A. Emigration and the plague have reduced it to about 360 souls.

Q. Are there still any troglodytes, or inhabitants of caves, and are they numerous?

A. The district between Marza Suza and Cyrene is full of caverns in the very heart of the mountains, into which whole families get by means of ropes; and many are born, live, and die, in these dens, without ever going out of them; their Bedouin relations in the neighbourhood provide them with food, and there preserve their property from the rapine of inimical tribes; the friendly Arabs collect in these holes a sufficiency of water for all their wants.

Q. What is the disposition of these people?

A. They are savage, untractable, and dangerous, the government of the country itself never having been able to reduce them.

Q. Do you consider a landing at Bomba as safe?

A. Being situated on the limits of Tripoli and Cairo, it is inhabited by tribes that have been driven away by their respective governments, so that they continually molest pacific tribes, and the caravans destined for Mecca.

Such, my Lord, is the sum of the most direct and credible information I have been able to collect; besides which, I have made many other inquiries, and have also constructed a map of the march of the said army, by inference; but I hope I have shown your Lordship that this interesting portion of geography (seated so near to civilized Europe), need no longer remain a blank; and also that its examination may lead to satisfactory ulterior results as to the confluence of the Nile and the Niger, and the actual state of the level of the countries south of Bournou, compared with Abyssinia and the west coast of Africa. And this, if I may be allowed to express my opinion, is the only practicable road to Europeans—for I have ever considered the difficulties and diseases incident to the swampy banks of rivers in a tropical climate (at all times replete with decomposing vegetable substances), so insurmountable, that I have never been surprised at their failure.

I have the honour to subscribe myself, my Lord, your Lordship's most obedient humble servant,

W. H. SMYTH, *Commander, R.N.*

Right Hon. Lord Viscount Melville,
&c. &c. &c.

II.

ON GRAHAM ISLAND.

THIS is the Paper mentioned on page 112; and is extracted from the CXXIInd volume of the *Philosophical Transactions.* H.R.H. the Duke of Sussex, in submitting it to the Meeting of the Royal Society, stated that it was written as part of a report to the Council upon Dr. Davy's paper on the same subject, which had been read a few weeks before; but that the Council viewed it as containing so much original and important matter, that they had determined upon its being treated as a separate communication.—(*See the Literary Gazette, No.* 786, *page* 90.)

Some Remarks on an Error respecting the Site and Origin of Graham Island. By Captain W. H. SMYTH, R.N., F.R.S., F.S.A. Read February 9, 1832.

In consequence of accounts recently published concerning the rise and progress of this island, which I conceive to have been stated materially in error, and in order that physical inquiry may receive as exact data as can be afforded, I beg leave to offer the following remarks to the Royal Society.

It was stated, in the first letters which arrived from Malta, that an officer on the Mediterranean station was in possession of an old chart, whereon was 'a shoal with only four fathoms on it, and called Larmour's Breakers'—and this being asserted to be 'within a mile of the latitude and longitude' of the new island, was consequently announced as its nucleus. On reading some of these letters I saw at once that the chart was mistaken for a valuable document; but being aware that its particulars were well known to navigators, I should not have deemed it to require notice, had not the erroneous inference been repeated both in the *Journal of the Geographical Society* and in the *Quarterly Review.*

The danger alluded to as existing upon the 'old chart,' was never ascertained or verified; it was only thought to have been seen by Captain Larmour, when in command of the *Wassanaer*, a troop-ship, on the Egyptian expedition. But the same impression did not strike all the officers and passengers; and, on the commander-in-chief despatching two or three vessels to examine it for a more detailed report, no shoal-water could be found. The present Captain Richard Spencer, C.B., then a lieutenant on board the *Wassanaer*, was one of the officers sent to assist in the search; and from him I had these particulars. Yet the minute which had been forwarded to me from the Admiralty, being written in these decided terms—

'H.M. Ship *Wassanaer*, 11th of December, 1800, P.M. The island of Pantellaria S.W. by W. nine or ten leagues, saw a reef of rocks S.S.E. distant three or four miles, extending N.N.W. and S.S.E., about one mile in length. Hauled up S. by W., to clear them. Saw something on the reef like a ship's mast. Bearings by compass.

I examined the spot with a rigorous strictness; and from the various traverses which I made in every direction, with the lead going by night and by day, I feel prepared to assert, that no reef of the nature described by Captain Larmour in 1800, and no shoal of four fathoms water, could have existed in 1814. How the said 'four fathoms' crept into our charts is best known to the ship-chandlers, who too long purveyed to the scientific wants of seamen; but from the absence of positive testimony, from the careful search made by order of Lord Keith, from my own several cruizes, and from the material fact of its being in the high road which is annually beaten by hundreds of ships, it is not presuming greatly to say, that neither the one nor the other had any existence.

Nor is the assigned place 'within a mile' of the position of the volcanic islet, though it may accidentally have been so marked upon the 'sea-cards;' for it should be remembered, that the true site even of the principal headlands around, was not then decided. According to the minute just quoted, corrected for magnetic variation, Larmour's supposed reef is no less than sixteen miles W. by N. from it, on a part of the subaqueous plateau (which I named Adventure Bank), uniting Sicily to Africa by a succession of ridges,—about a spot where I found from forty to fifty fathoms of water. Graham's Isle, however, is not upon this bank; it arose between it and a knoll some miles to the eastward, which, from a shell brought up by the arming, I called Nerita; and, if the observations which determine the latitude and longitude of the stranger as in 37° 8′ 25″ N. and 12° 43′ 50″ E. be correct, it must have been elevated through more than a hundred fathoms of water.

In thus doubting the actual existence of the Larmour Shoal, it is not my intention to dispute the appearance and disappearance of natural phenomena, nor that stupendous alterations may occur by the subsidence and uplifting of strata,—because an obstinate scepticism would be absurd, especially in a part of the globe where, to use a well-expressed Italian metaphor, the whole ground is 'tremblingly alive.' But it is reasonable and proper to question such rumours as have been made without due examination. In the instance before us no endeavour was made to establish the truth by either shortening sail, lowering a boat, or even getting a cast of the lead; moreover, they were three or four miles from the supposed object, and opinions on board the *Wassanaer* were not at all unanimous. By similar indecision a teasing knot of perils has gained random insertion upon our charts, to the disquietude of sea commanders; but it is a fault which is fast disappearing, and it may be trusted that there are few officers who would not think themselves liable to the imputation of culpable carelessness, did they not seek to verify such 'dangers' as they might accidentally encounter.

I do not think subaqueous volcanic explosions are of such rare occurrence as is generally supposed; and extremely sudden intumescence may arise from the expansion of an inferior lava bed. It is not at all improbable that gaseous fluids and ejectamenta may have been seen, before the accumulation of solid matter, protruded from the vent, was sufficient to form a crater of eruption. A volcanic apex may become visible, and again be quickly destroyed by trituration, the solution of mineral substances, and the repressive force of the column of water over the vent. Now, as there was a chance that something of the kind had occurred in the neighbourhood assigned to Larmour's reef,—breakers having been reported near the same spot by the *Greyhound* frigate, and shoals having been immemorially marked there under the names of La Ajuga and B. Scoglio,—I laboriously explored the whole vicinity. In examining the chart which resulted from this undertaking, it will be found that a knoll, with only seven fathoms upon it, was discovered not far from the site of all these reports, and that the Adventure Bank extends from Sicily nearly to Pantellaria, where the water deepens at once from 76 fathoms to no bottom with 375 fathoms of line. A further inspection will show that the Phlegræan islands of Pantellaria and Linosa, have been protruded from the greatest depths, where, perhaps, the fires found the least resistance.

All these considerations led me to suppose that, though the reports were exceedingly vague, volcanic agency might still have given grounds for them. I therefore made particular inquiries, both in Sicily and Pantellaria, as to local earthquakes, and whether any volumes of smoke, ferilli, or jets of flame, comminuted ashes, or other fragmentary ejectments, had been noticed in that direction ; but I could hear of none. Yet we are told, as a 'fact' of weight, that a tradition is current, which says :—'A volcano existed in the same spot about the commencement of the last century.' It would be difficult to say how this tradition was preserved amongst a people little given to letters ; and I never, in my long residence and systematic researches at the above place, and in Malta, heard the slightest hint of it.

I am, therefore, led to the conclusion,—firstly, that no shoal or danger has lately existed in that channel, excepting only an occasional overfall in very heavy weather on the seven fathom knoll where I anchored H.M. ship *Adventure*, and which is sufficiently near for bearings taken at random, and without suspicion of the existence of a local attraction, to be placed in identity with the reports above-mentioned. Secondly, that even if what Captain Larmour became persuaded he saw was actually a temporary volcanic effect, it had no possible relation to breakers with 'four fathoms' upon them. And it follows, that the assertion of Graham Island having been formed by the mere 'lifting up' of such shoal, must be utterly destitute of foundation.

INDEX

THE END.